Springer Tracts in Electrical and Electronics Engineering

Springer Tracts in Electrical and Electronics Engineering (STEEE) publishes the latest developments in Electrical and Electronics Engineering - quickly, informally and with high quality. The intent is to cover all the main branches of electrical and electronics engineering, both theoretical and applied, including:

- Signal, Speech and Image Processing
- Speech and Audio Processing
- Image Processing
- Human-Machine Interfaces
- Digital and Analog Signal Processing
- Microwaves, RF Engineering and Optical Communications
- Electronics and Microelectronics, Instrumentation
- Electronic Circuits and Systems
- Embedded Systems
- Electronics Design and Verification
- Cyber-Physical Systems
- Electrical Power Engineering
- Power Electronics
- Photovoltaics
- Energy Grids and Networks
- Electrical Machines
- Control, Robotics, Automation
- Robotic Engineering
- Mechatronics
- Control and Systems Theory
- Automation
- Communications Engineering, Networks
- Wireless and Mobile Communication
- Internet of Things
- Computer Networks

Within the scope of the series are monographs, professional books or graduate text-books, edited volumes as well as outstanding PhD theses and books purposely devoted to support education in electrical and electronics engineering at graduate and post-graduate levels.

Review Process

The proposal for each volume is reviewed by the main editor and/or the advisory board. The books of this series are reviewed in a single blind peer review process.

Ethics Statement for this series can be found in the Springer standard guidelines here https://www.springer.com/us/authors-editors/journal-author/journal-author-helpdesk/before-you-start/before-you-start/1330#c14214

Manish Kumar · Abhishek Raj
Editors

Perovskite Solar Cells

Reshaping the Future Energy Landscape

Springer

Editors
Manish Kumar
Department of Physics
Atma Ram Sanatan Dharma College
University of Delhi
New Delhi, Delhi, India

Abhishek Raj
National Environmental Engineering
Research Institute
CSIR-NEERI
Nagpur, Maharashtra, India

ISSN 2731-4200 ISSN 2731-4219 (electronic)
Springer Tracts in Electrical and Electronics Engineering
ISBN 978-981-96-9462-4 ISBN 978-981-96-9463-1 (eBook)
https://doi.org/10.1007/978-981-96-9463-1

This Springer imprint is published by the registered company Springer Nature Singapore Pte Ltd.
The registered company address is: 152 Beach Road, #21-01/04 Gateway East, Singapore 189721, Singapore

Preface

We are delighted to present this book, entitled "Perovskite Solar Cells: Reshaping the Future Energy Landscape", sum up the exceptional research and innovative advancements on white light emitting materials. This book aims to provide a comprehensive overview of perovskite solar cells, covering both fundamental principles and the latest advancements in the field. Beginning with an introduction to perovskite materials and their unique properties, it delves into various synthesis techniques and device structures that define these cutting-edge solar cells. It addresses critical challenges in stability and interfacial engineering while presenting strategies for bandgap modification and the integration of 2D materials in perovskite solar cells (PSCs). A significant focus is placed on the environmental impact, particularly with lead-free alternatives, and the potential for semitransparent and tandem solar cells. Additionally, the book covers practical applications, from indoor use to outdoor environments, and examines the commercial scalability and life cycle assessment. Concluding with insights into the future outlook, this book serves as an essential resource for researchers, students, and industry professionals interested in the rapidly evolving field of perovskite solar cells.

As per the requirement of this perovskite solar cell technology for the day-to-day life, this book is divided into the seventeen dedicated chapters:

Chapter 1, "Introduction: An overview of perovskite solar cells", highlights the advancements in interfacial engineering, and approaches to improve stability and scalability of perovskite solar cells. The major focus is given to environmental issues, for example lead toxicity and durability in outdoor conditions. Here, by analysing the current advancements and prospects, this overview seeks to offer meaningful insights into the transformative capabilities of PSCs and their contribution to a sustainable energy future.

Chapter 2, "Perovskite materials and synthesis techniques", present the details about the perovskite materials explored for applications in solar cells, their properties, classifications, different synthesis methods/techniques for its possible applications in the devices.

Chapter 3, "Device Structure of Perovskite Solar cell", provide the advancements in the structural development of PSCs, including the implementation of advanced interfacial layers and their impact on performance. This chapter discussed the various types of perovskite structure used to enhance the efficiency of PSCs.

Chapter 4, "Perovskite Solar Cells without Hole and Electron Transport Layers", discussed about the various structural configurations of PSCs without a hole transport layer (HTL) or electron transport materials. The configurations introduce a unique approach that simplifies the cell architecture, reduces manufacturing cost, and facilitates the fabrication process.

Chapter 5, "Study of the Electrode Materials for Perovskite Solar Cells", provides the comparison of various electrodes materials including metal electrodes, transparent conductive oxides (TMO) based electrodes, conductive polymers, and carbon-based electrodes based on their cost, efficiency, and stability.

Chapter 6, "Scaling up perovskite photovoltaic from lab to commercial market: Progress and challenges" highlights the recent advancement in large area perovskite solar cells and various several challenges, including lowering production costs, producing large area panels, maintaining high production rates, maintaining environmental concern, and achieving high efficiency in converting sunlight to energy.

Chapter 7, "Stability of Perovskite Solar Cell: A Real Challenge", addresses the primary challenge lies in their instability and degradation, particularly under moisture and environmental exposure, limiting their operational lifespan to well below the 10-year requirement for commercial viability of PSCs. This chapter includes the key factors influencing PSC stability, highlights recent advancements, and explores future directions.

Chapter 8, "Lead free Multiferroic materials for Perovskite based Solar Cells: Recent Advancements and Future insights", revied the properties of perovskite multiferroics, advancement in the field of multiferroics based lead free perovskite solar cells, their challenges and future insights.

Chapter 9, "Latest Updates on Materials, Device Designs, and Performance Optimization in Low Band Gap Perovskite Solar Cells", explores the latest developments in low band gap perovskite absorber-based PSCs, detailing the different materials involved, the device architecture and the factors determining performance enhancement.

Chapter 10, "Role of 2D Materials in Perovskite Solar Cells", provide in-depth knowledge of various factors affecting solar cell efficiency with possible remedies such as by using two-dimensional (2D) materials as electrodes, additives, and charge transport layers to overcome them.

Chapter 11, "Future of Perovskite Solar Cells and Challenges in Large Scale Energy Production", discussed the advancement and technological challenges deterring the large-scale development of perovskite solar cells including intermittency, grid integration, storage solutions, and material sustainability have been discussed.

Chapter 12, "Evaluation of Toxicity of Lead in Perovskite Solar Cells: Current Status and Future Challenges", summarizes the available acquaintance on lead (Pb) toxicity in PSCs, concentrating on physical and chemical properties of Pb in those materials, its environmental hazards, and future perspectives for both solar cell industry and the global ecosystem.

Chapter 13, "Perovskite-perovskite tandem solar cells: Basics to latest updates with challenges" reviews recent accomplishments, stability studies and scalable manufacturing methods for tandem devices and the advancements made in power conversion efficiencies, surpassing 30% recently.

Chapter 14, "Perovskite-silicon tandem solar cells: Basics to latest updates with challenges", aims to analyze the trends and developments in field of perovskite-silicon tandem solar cells (PSTSC), covering last ten years and presenting in detail the state of the art, persisting issues, and manufacturing processes of tandem solar cells.

Chapter 15, "Solar derived Halide Perovskites Photocatalysts for Water Splitting and CO_2 Reduction", provide the basic understanding for the utilization of halide-based perovskite (HLP) for water splitting and CO_2 reduction to generate valuable chemical energy using HLP-based solar cell devices.

Chapter 16, "Deposition Methods of Perovskite Absorber", offers a detailed look at the different methods used to create perovskite layers, from traditional solution-based techniques to newer methods like vapor deposition and hybrid approaches.

Chapter 17, "Perovskite Solar Cells: Summary and Future Directions", evaluates the development of PSC research, highlighting significant breakthroughs and pointing out outstanding issues.

In conclusion, this book offers a comprehensive viewpoint on the prospects and avenues for upcoming innovations of PSCs, including investigating environmentally friendly and lead-free substitutes, incorporating perovskites with tandem architectures, and developing recycling and life-cycle assessment techniques.

We believe that this book will be greatly useful for the undergraduate, postgraduate students and research scholars, working in the field of materials science, physics, chemistry and various interdisciplinary engineering branches along with the industry and academia.

New Delhi, India — Manish Kumar
Maharashtra, India — Abhishek Raj

Acknowledgments

It gives us great pleasure in expressing our gratitude to all who have supported us and made their contributions in making this book possible. First and foremost, we must acknowledge and thank to "GOD" for their blessings, giving us good health, and protection throughout this period. We could never have accomplished this without the faith; we have, in "GOD".

The editors extend the deepest gratitude to Prof. Gyantosh Kumar Jha, Principal, ARSD College, University of Delhi, New Delhi, India.

Editors would also like to express the appreciation to whole Editorial team of Springer Nature, for their resolute initiate and hard work during the process of the publication of this book. Moreover, in continuation, it's our pleasure to acknowledge the support and assistance provided by the production team and their staff through their devotion, proficiency, and commitment to excellence.

We would like to express our deepest gratitude to our family members, friends, and colleagues who have provided us with unwavering support, inspiration, and understanding throughout this journey.

Dr. Manish Kumar is also thankful to his grandparents and parents (Late Shri K. L. Vats (Grandfather), Late Smt. Shima Devi (Grandmother), Shri Pravind Kumar (Father) and Smt. Anita Sharma (Mother)) for the endless support and understanding of my goals and aspirations. Their constant words of inspiration and confidence in me has been a source of encouragement and enthusiasm. I want to specially mention the names of my lovely wife Tanu and cute daughter Vanya for the all-time support and understanding towards my work without any kind of complain. I am really very thankful to both of my life lines. I also want to owe my deepest gratitude towards my uncle and aunty, Shri Mithelesh Sharma, Smt. Suman Sharma, Shri Sanjay Sharma and Smt. Sarika Sharma cute and caring sisters, Shikha and Reshu, cousin brothers Ankit, Priyanshu and Upanshu, for support and encouragement in any situation.

Dr. Abhishek Raj would like to express his sincere gratitude and appreciation to his Ph.D. supervisor, Dr. Avneesh Anshul, Principal Scientist at CSIR-NEERI, for his invaluable guidance, and insightful suggestions during his Ph.D. journey. He wants to thank his grandparents and parents (Late Dr. Rajendra Prasad (Grandfather), Late Smt. Sabuja devi (Grandmother), Shri Nand Kishor Prasad (Father), Smt.

Krishnamani Devi (Mother) and his younger brother (Aniket Raj) for their unwavering love, endless patience, and constant support. In the face of every obstacle, regardless of its difficulty, they remained by my side, maintaining their belief in me when I struggled to believe in myself. Their confidence has been my most significant source of strength, and he genuinely cannot imagine completing this journey without their love brightening my path. He dedicated his achievement to his parents and his wonderful younger brother; this achievement belongs to them as much as it does to him. He extends his heartfelt gratitude to all who have generously shared their time, insights, and support.

We also want to acknowledge to one and all involved, whether directly or indirectly, in the foundation and publication of this book. Your assistance, support, and commitment have made this effort a worthwhile and satisfying experience.

New Delhi, India Manish Kumar
Maharashtra, India Abhishek Raj

Contents

About the Editors

Dr. Manish Kumar obtained his M.Tech. degree in Ceramic Engineering from IIT(BHU), Varanasi and Ph.D. degree in Physics from Institute of Science, Department of Physics, Banaras Hindu University, Varanasi, India. At present, he is serving as Assistant Professor at Department of Physics, ARSD College, University of Delhi, New Delhi, India since last more than 12 years. His research area focuses on environmental friendly multifunctional materials for energy and memory devices, materials for green energy solutions, multiferroics, magnetoelectrics, composites, perovskite materials for photovoltaic applications, supercapacitors, white light emitting materials, solar simulation via different software and DFT calculation. Recently, he was listed among top 2% scientists worldwide by Stanford University, USA, Ranking-2021 & 2025 and published more than 131 research publications in international journals of high repute, one authored book on White Light Emitting Materials, five edited books (in press) in Springer Nature, 35 international book chapters and one Indian Patent with h-index of 30, i10-index of 71 including more than 2750 citations as per Google-Scholar. He has delivered the talks in a number of national and international conferences. He is also a member of various research committees of different universities. In addition, he has completed two national research projects and received the best research award from his institute. He is guiding/ guided number of Ph.D., M.Tech./M.Sc. and B.Sc./B.Tech. students as supervisor and/ co-supervisor. Dr. Kumar is serving and/ served as an Editor in number of highly reputed journals such as, Scientific Reports (Nature), Materials Today Communications (Elsevier), ECS Sensors Plus (IOP, ECS), High Performance Polymers (SAGE), Materials Today: Proceedings (Elsevier) and Springer Nature Books. He has organized one national conference (RAFM-2020) as co-convener and three international conferences (RAFM-2022, RAFM-2024 and APGC-2025) as convener.

Dr. Abhishek Raj received his Bachelor of Technology degree in electrical and electronics engineering from Cochin University of Science and Technology, Kochi, and completed his M.Tech. and Ph.D. (M.Tech.-Ph.D. dual degrees) from Academy of Scientific and Innovative Research (AcSIR), at CSIR-NEERI, Nagpur, India. He has published 28 international research papers and 5 book chapter in an international research book of high repute. His work has received wide recognition, reflected in an h-index of 17, an i10-index of 23, and over 1120 citations as per Google Scholar. He also received Late Shri P. S. Dutt Memorial Award for "Best Young Researcher" for his significant contribution to R & D in Environmental Science and Engineering presented during the 65th CSIR-NEERI Foundation Day celebration on 8th April 2023. His research focuses on materials for energy and environmental applications, with a particular emphasis on the design, synthesis, and development of lead-free perovskite solar cells through an integrated experimental and computational approach. He combines materials engineering, device fabrication, and theoretical modeling to investigate environmentally benign alternatives to conventional lead-based perovskites, aiming to achieve high efficiency, enhanced stability, and scalability in next-generation photovoltaic technologies.

Chapter 1
Introduction: An Overview of Perovskite Solar Cells

Abhishek Raj, Manish Kumar, and Avneesh Anshul

Abstract Perovskite solar cells (PSCs) have emerged as one of the advanced technologies within the photovoltaics sector, characterized by their exceptional efficiency, customisable properties, and straightforward manufacturing processes. The distinctive structural and optoelectronic characteristics of perovskite materials, including high absorption coefficients, extended carrier diffusion lengths, and adjustable bandgaps, render them highly suitable for next-generation solar energy applications. Over the past decade, PSCs have achieved remarkable progress in power conversion efficiencies rising from just 3.8% in 2009 to over 26% in recent years. This book chapter investigates the progress and innovations in PSCs, covering aspects such as perovskite material, device design, and performance enhancement. Here discussion undergoes significant advancements in interfacial engineering, and approaches to improve stability and scalability in production. Particular attention is given to environmental issues, such as lead toxicity and durability in outdoor conditions. Here, by analysing the current advancements and prospects, this overview seeks to offer meaningful insights into the transformative capabilities of PSCs and their contribution to a sustainable energy future.

Keywords Perovskite absorber · Electron transport layer · Hole transport layer · Indoor photovoltaic · Stability

A. Raj · A. Anshul (✉)
CSIR-National Environmental Engineering Research Institute (CSIR-NEERI), Nagpur, India
e-mail: anshul@neeri.res.in

Academy of Scientific and Innovative Research (AcSIR), Ghaziabad, India

M. Kumar (✉)
Experimental Research Laboratory, Department of Physics, ARSD College, University of Delhi, New Delhi, India
e-mail: mkumar2@arsd.du.ac.in

M. Kumar and A. Raj (eds.), *Perovskite Solar Cells*, Springer Tracts in Electrical and Electronics Engineering, https://doi.org/10.1007/978-981-96-9463-1_1

Introduction

The traditional energy sources such as fossil fuels are steadily diminishing, with the rise in global energy demand. The burning of fossil fuels affects the environmental air quality and releases toxic gases into the atmosphere (Traverse et al. 2017; Yun et al. 2019; Arango-Aramburo et al. 2020). Therefore, a considerable portion of the energy must be produced from environmentally friendly renewable sources. Accordingly, it is essential to investigate new energy alternatives for future renewable and sustainable energy technologies (Samiul Islam et al. 2021; Jalalian et al. 2019; Hao et al. 2014). Various energy sources, including tidal energy, wind energy, biomass energy, and solar energy, are considered sustainable as they can be harnessed repeatedly (Rahman et al. 2022; Salhi et al. 2018). Solar energy, in particular, is accessible globally, making it a superior and viable alternative to fossil fuels. Solar cells are one of the most versatile renewable energy resources in recent times. Solar cells can be broadly categorized into different types based on their materials and structures (Dambhare et al. 2021). First-generation solar cells include monocrystalline and polycrystalline silicon cells, known for their high efficiency and durability, and amorphous silicon cells, which are flexible but less efficient. Second-generation solar cells, such as thin-film technologies, include cadmium telluride (CdTe), copper indium gallium selenide (CIGS), and organic photovoltaics (OPVs), which are lightweight and cost-effective but face challenges like toxicity and lower efficiency (Novoselov et al. 2004; Tafazoli et al. 2018; Jena et al. 2019; Ghosh and Singh 2019; Dong et al. 2015). Third-generation solar cells feature innovative designs like perovskite cells, known for rapid efficiency improvements, dye-sensitized solar cells (DSSCs), and quantum dot solar cells, which offer tuneable bandgaps and flexibility (Frost et al. 2014; Kramer et al. 2015). Additionally, multi-junction solar cells achieve very high efficiencies by combining materials to capture a broader spectrum of sunlight, while emerging technologies like bifacial, flexible, and transparent solar cells are being developed for specialized applications (Zeng et al. 2019; Li et al. 2014; Xiao et al. 2020). Among various solar cells technologies, perovskite material-based photovoltaic technology become more popular due to superior device efficiency. PSCs are named after the unique crystal structure of the mineral perovskite. These materials are mostly based on the formula ABX_3, where 'A' and 'B' represent cations and 'X' represents an anion (Li et al. 2008; Tong et al. 2018). These materials are very effective in converting light into energy because of their perovskite structure, which enables remarkable light absorption and charge transport capabilities. The emergence of organometal halide perovskites, including methylammonium lead iodide ($CH_3NH_3PbI_3$), which showed great promise for application in solar cells, marked the start of the journey of perovskite materials in photovoltaics. Since their first application in dye-sensitized solar cells in 2009, the efficiency of PSCs has seen a meteoric rise, from an initial 3.8% to over 26% in laboratory settings to date (Akhiro et al. 2009). This unprecedentedly quick increase in efficiency highlights the special potential of perovskite materials and is a first in the history of solar cell manufacturing. The capacity to generate high-quality films by low-temperature solution processing, a high absorption coefficient, and long carrier diffusion lengths are the essential characteristics of perovskite materials that contribute to their high efficiency (Zhang et al. 2023; Raj et al. 2023; Bati et al. 2023). These characteristics allow PSCs to be fabricated using very straightforward and affordable methods, resulting in excellent power conversion efficiency (PCE). Furthermore, the tunability of the perovskite's bandgap through compositional engineering allows for the optimization of light absorption and the development of tandem solar cells, where perovskites can be paired with other photovoltaic materials to surpass the efficiency limits of single-junction cells (Akhil et al. 2021; Futscher and Ehrler 2016; Cheng and Ding 2021).

The mineral perovskite was initially discovered in the nineteenth century, which is when the foundational study on perovskites began. On the other hand, perovskites were first used in solar cells in the late 2000s. Under the direction of Tsutomu Miyasaka, University of Tokyo researchers released the first study on a perovskite solar cell in 2009. Lead halide perovskite materials were employed in the dye-sensitized solar cell (DSSC) architecture of these early devices. Akhiro et al. (2009) creatively developed $CH_3NH_3PbBr_3/TiO_2$ and $CH_3NH_3PbI_3/TiO_2$-based dye-sensitized solar cells (DSSCs) with liquid electrolyte sensitizer and achieved PCE and V_{OC} as 3.1% and 0.96 V for $CH_3NH_3PbBr_3/TiO_2$ and 3.81% and 0.61 V for CH_3NH_3PbI3/TiO_2. However, the following research did not take much attention from the research community due to liquid electrolytes in hole transport material (HTM). Recently perovskite-based solar cells achieved device efficiency of more than 26%. In 2012, Kim et al. (2012) designed a PSC device with 9.7% efficiency, V_{OC} of 0.888 V, and an FF of 62% for the first time by using spiro-OMeTAD as HTL and mp-TiO_2 was incorporated as ETL. Further, Lee et al. (2012) reported a low-cost, solution-processable solar cell, with PCE 10.9% and V_{OC} more than 1.1 V in a single-junction device and showed that mixed-halide perovskite on a mesoporous Al_2O_3 photo anode acts together as an electron conductor and light absorber. Further, Noh et al. (2013) developed a PSC with the incorporation of Spiro-OMeTAD/co-complex as HTL and delivered a device efficiency of 10.4%. Docampo et al. (2013) developed a low-temperature single thin-film solar cell in which a perovskite absorber was deposited on glass and flexible substrate using a solution-processable technique and exhibited the device efficiency of 10% and 6% respectively. Bolink and their colleague demonstrated a PSC with device heterostructure of PEDOT: PSS/PolyTPD/ $CH_3NH_3PbI_3$/PCBM in which organic ETL and HTL are used for the transportation of charge carriers which achieved a device efficiency of 12%. The perovskite layers were made via sublimation in a high-vacuum room to guarantee a high level of homogeneity and purity. Numerous conducting substrates, such as plastic and textiles, can be used with this straightforward planar device structure and room-temperature deposition techniques (Malinkiewicz et al. 2014). In 2016, Li et al. (2016) fabricated a PSC using simple vacuum-flash solution processing method, and obtained a maximum device efficiency of 20.5% (certified PCE of 19.6) with an aperture area greater than 1 cm^2 having virtually no hysteresis. Zettl et al. documented a PSC which, having graded band gap, managed to maintain a stable device efficiency of 18.4% and a peak device efficiency of 21.7% without the use of reflective coatings. These PSCs are made up of two layers of perovskite material, namely $MASnI_3$ and $MAPBI_{3-X}Br_x$

which include GaN, a monolayer of hexagonal boron nitride, and a graphite aerogel, setting a record for the highest device efficiency for perovskite material-based solar cells (Ergen et al. 2016). In 2017, Palma et al. (2017) reported a PSC module based on laser processed technique with a larger area of 14.5 cm^2 with an aperture ratio of 95% and delivered a conversion efficiency of 9.3%. Niu et al. demonstrated a solar cell by Cs doping to enhance the thermal stability of the device. They designed $Cs_{0.09}MA_{0.91}PbI_3$ as a light-absorbing layer, which reached device efficiency, V_{OC}, fill factor, and current density as 18.1%, 1.06 V, 0.76, and 22.57 mA/cm^2 (Niu et al. 2017). Zhang et al. (2017) fabricated rubidium (Rb)-incorporated PSCs by applying a gas quenching method to achieve 20% power conversion efficiency on a 65 mm^2 device.

Arora et al. (2017) created a solar panel using copper (I) thiocyanate (CuSCN) as the layer for extracting holes, achieving a PCE of over 20%. Yang et al. developed flat-type PSCs by using an EDTA-complex tin oxide (SnO_2) as the extraction layer, setting a recording device of 21.60% (certified at 21.52% by Newport). When the device was exposed to an ambient environment for 2820 h, it only degraded by 8%, and only 14% under 100 mW cm^{-2} of irradiation (Yang et al. 2018). Recently perovskite-based solar cells achieved device efficiency of more than 26%. Liu et al. reported an interfacial modification technique for the development of high-quality PSCs. They incorporated popular self-assembled molecule [4-(3,6-dimethyl-9H-carbazol-9-yl)butyl]phosphonic acid (Me-4PACz) with the multiple aromatic carboxylic acids 4,4′,4″-nitrilotribenzoic acid (NA) to enhance the interfacial properties of the heterostructure. The developed inverted perovskite exhibited a superior efficiency of 26.54% and 22.74% for large-area PSCs with an aperture area of 11.1 cm^2 (Liu et al. 2024).

This chapter provides an overall overview of PSCs. Here, initially, the types of PSCs, the properties of perovskite absorber materials, and device configurations of PSCs are discussed. Further, the issues associated with the stability and toxicity of PSCs are examined. Finally, the summary and future outlook of PSCs are presented.

Solar Cell Technologies

Solar cell technologies have evolved significantly since their inception, driven by the need for more efficient, cost-effective, and sustainable energy solutions. The primary types of solar cell technologies include silicon-based, thin-film, and emerging third-generation solar cells, each with unique characteristics and progress trajectories. Various types of solar cell technologies are shown in Fig. 1.1. Silicon-based solar cells are the photovoltaic technology that is most utilized and widely recognized. They fall into three main categories: amorphous silicon cells, polycrystalline silicon cells, and monocrystalline silicon cells. Monocrystalline silicon solar cells are those that are produced with a single, continuous crystal structure, according to the definition (Shah et al. 1995; Sun et al. 2021; Rathore et al. 2021). They are known for their high efficiency, often exceeding 20%, due to the purity of the silicon used. Monocrystalline cells also have a longer lifespan and better performance in low-light conditions. However, they are relatively expensive to produce due to the complex manufacturing process. Polycrystalline silicon solar cells are made from silicon crystals that are melted together; polycrystalline cells are less efficient than their monocrystalline counterparts, typically achieving 15–17% efficiency. However, they are cheaper to produce and have seen widespread adoption due to their balance of performance and cost. Amorphous silicon solar cells use non-crystalline silicon and are often employed in thin-film technologies. While they have lower efficiencies, usually around 6–10%, they are flexible and can be applied to a variety of surfaces. Advances in deposition techniques have improved their efficiency and stability over time. Thin-film solar panels are created through the process of applying single or multiple thin layers of photovoltaic material onto a base. Common types of thin-film technologies consist of amorphous silicon, copper indium gallium selenide (CIGS), and cadmium telluride (CdTe) (Lee and Ebong 2017; Lee et al. 2015; Ramanujam et al. 2020). In the case of thin-film solar cells, one or more thin layers of photovoltaic materials are layered on a substrate. Key types of thin-film technologies include cadmium telluride (CdTe), copper indium gallium selenide (CIGS), and amorphous silicon (Isabela et al. 2021). CdTe cells are among the most commercially successful thin-film technologies, offering a good balance of cost and efficiency, with efficiencies around 18%. They are easier and cheaper to produce than silicon-based cells but face challenges related to the toxicity of cadmium. CIGS cells offer higher efficiencies, often around 20%, and can be applied to flexible substrates. They promise high performance and versatility, but their complex manufacturing process and material scarcity have hindered widespread adoption. As mentioned earlier, amorphous silicon is used in flexible thin-film applications. Continuous improvements in deposition and manufacturing techniques have enhanced their performance and commercial viability (Reinhard et al. 2012).

Third-generation solar cells have drawn a lot of attention lately from the photovoltaic groups. Third-generation solar cells use technology designed to outperform first- and second-generation cells in terms of efficiency and cost. Perovskite solar cells, organic photovoltaics, and multi-junction cells are a few notable examples. Because of their constantly increasing efficiencies, which have reached over 26% in lab settings, perovskite cells have attracted a lot of attention. They are relatively inexpensive to produce and offer potential for high efficiency and flexibility (Hartono et al. 2023). However, stability and long-term durability remain challenges that researchers are actively addressing. Organic compounds are used by OPVs to capture light and produce energy. They have benefits including low production costs, flexibility, and lightweight design. Even though their efficiencies are often lower, between 10 and 15%, continuous research is being done to increase their performance and stability. Several layers of various photovoltaic materials are stacked in multi-junction solar cells, each of which is tuned to absorb a particular range of the solar spectrum (Fan 2023). Multi-junction cells can achieve efficiencies exceeding 40%, making them ideal for space applications. However, their complexity and high cost limit their use in terrestrial applications.

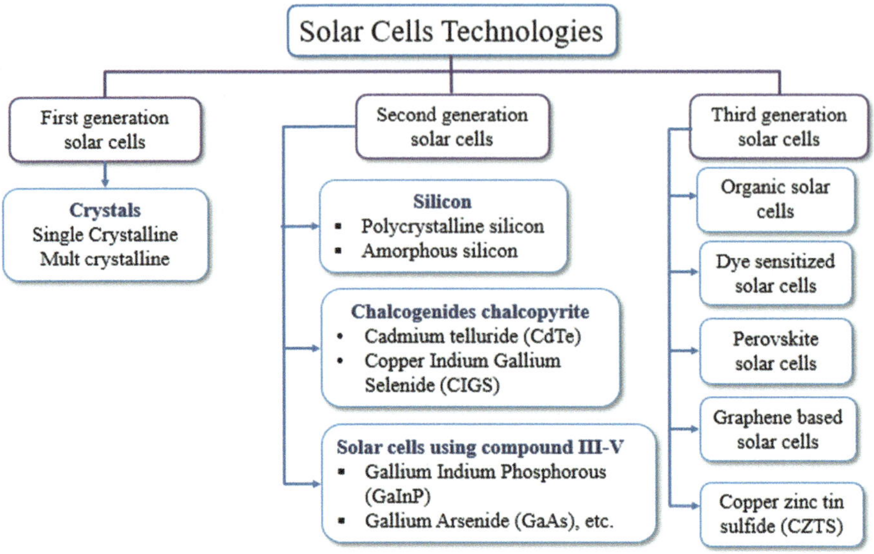

Fig. 1.1 Schematic diagram of various types of solar cell technologies

Perovskite Material and Device Configurations of PSCs

The hybrid halide perovskite materials, represented by the general formula ABX_3, possess a crystal structure that is equivalent to the structure of calcium titanate ($CaTiO_3$) (Steele et al. 2020; Raj et al. 2024). This shared structural arrangement reflects the perovskite framework, characterized by a three-dimensional network in which the A-site cation occupies the central position, the B-site cation is surrounded by six anions forming an octahedron, and the X anions link these octahedra to create a repeating lattice. The unit cell and crystal structure of perovskite absorber are shown in Fig. 1.2a, b (Clabel et al. 2022). The octahedral factor (μ) and the Goldschmidt tolerance factor (t), two crucial empirical parameters that determine the stability of the perovskite crystalline structure, are represented as follows:

$$\mu = \frac{r_A}{r_B} \tag{1.1}$$

$$t = \frac{r_A + r_X}{\sqrt{2}(r_B + r_X)} \tag{1.2}$$

The ionic radii of the A, B, and X ions are represented by r_A, r_B, and r_X, respectively. The degree of distortion of the octahedron is quantified by the octahedral factor, while the tolerance factor evaluates the compatibility between the A cation and the BX_6 octahedron. Furthermore, values of $t = 1$ and $\mu = 1$ indicate the ideal perovskite structure; any deviation from these values signifies a lack of stability and

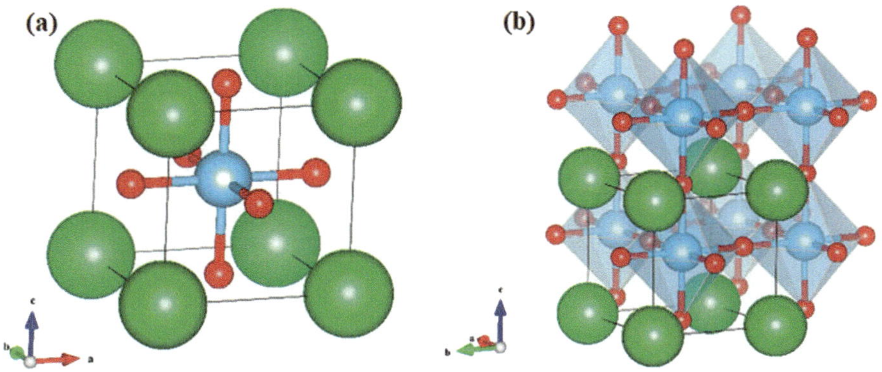

Fig. 1.2 **a** Unit cell of perovskite absorber materials. **b** Crystal structure of perovskite material in which A-site, B-site, and X-site are represented by blue, green, and red colours respectively. Reproduce from ref. Clabel et al. (2022) from open access licence

distortion within the structure. Typically, in perovskite structures if the values 't' lie between 0.8 and 1 and the value of 'μ' lies between 0.44 and 0.94, then the perovskite solar cell is considered as a stable structure (Travis et al. 2016; Kour et al. 2019).

In the ideal cubic perovskite structure, the tolerance factor (t) is equal to 1. However, in most of the perovskites, the symmetry deviates from this ideal cubic configuration and takes on lower symmetry forms, such as tetragonal, orthorhombic, rhombohedral, and monoclinic structures. These deviations from cubic symmetry appear due to structural distortions, which play a significant role in the physical properties of materials. Such distortions are frequently associated with the emergence of ferroelectric or antiferroelectric phases (Raj et al. 2022a; Pitaro et al. 2022; Dawa and Sajjadi 2024). Ferroelectric phases exhibit spontaneous electric polarization, a property where the material generates an electric dipole moment without an external electric field. In contrast, antiferroelectric phases are characterized by a periodic alternation of electric polarization, where adjacent dipoles align in opposite directions, cancelling each other out on a macroscopic scale. These characteristics significantly influence the electronic and functional properties of perovskite materials.

Perovskite films grown through solution-based fabrication techniques are primarily composed of earth-abundant materials. However, the most efficient devices currently available are developed using gold electrode and organic hole transport materials (HTMs), such as 2,2′,7,7′-tetrakis(*N*,*N*-di-p-methoxy-phenylamine)-9,9′-spirobifluorene, commonly referred to as spiro-OMeTAD. PSC heterostructures consist of light harvesting material which as inserted between hole transport layer (HTL) and electron transport layer (ETLs). In perovskite-based solar cells, p-type semiconducting materials act as HTL and n-type semiconducting materials act as an electron transport layer (ETL). The various device configurations of perovskite solar cells are shown in Fig. 1.3a–d. In mesoscopic perovskite solar cells (PSCs), which have evolved from dye-sensitized solar cells,

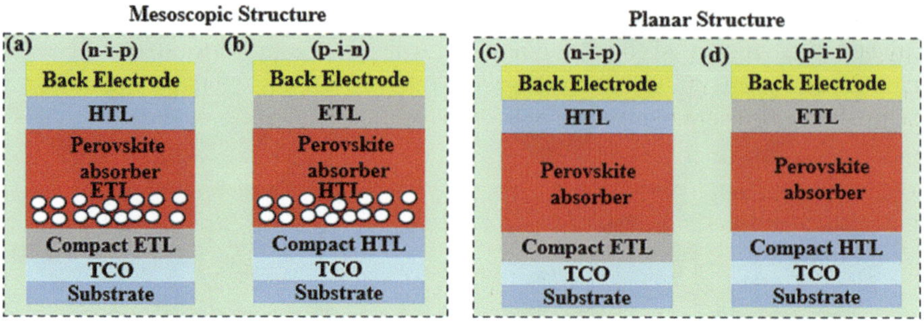

Fig. 1.3 Various device configurations of PSCs. **a** Mesoscopic formal n-i-p structure. **b** Mesoscopic inverted p-i-n structure. **c** Planar formal n-i-p structure. **d** Planar inverted p-i-n structure

the perovskite absorber activates a mesoporous metal oxide layer, like mesoporous titanium dioxide (meso-TiO_2), that acts as a supporting framework. In planar PSCs, the perovskite layer is inserted between a planar ETL such as compact TiO_2 (c-TiO_2), SnO_2, and C_{60} and a HTL, which includes materials like spiro-OMeTAD, poly(triarylamine) (PTAA), or poly(3,4-ethylenedioxythiophene)-poly(styrenesulfonate) (PEDOT:PSS). The device architecture is categorized as either conventional (standard) or inverted, depending on the order in which the ETL and HTL are deposited (Rong et al. 2018). Notably, designing inverted devices entails more than merely reversing the standard structure, as each configuration requires detailed development of suitable ETLs and HTLs, particularly considering the limitations posed by processing solvents. In triple mesoscopic PSCs, the perovskite layer is applied onto a three-layer scaffold made up of a screen-printed mesoporous TiO_2 layer, a zirconium dioxide (ZrO_2) spacer layer, and a carbon electrode. These devices substitute noble metal back contacts with carbon electrodes, thereby removing the necessity for a hole transport layer. Recently, multiple solar cells with varying band gaps combine to form perovskite tandem solar cells (TSCs), which absorb and transform a larger portion of the spectrum of sun into electrical energy (Raj et al. 2021).

Stability of PSCs

Stability of PSCs is one of the biggest obstacles to the widespread commercialization of this technology. Perovskite materials degrade quickly under operating conditions because they are extremely sensitive to environmental elements like heat, light, oxygen, and moisture. The perovskite structure breaks down into lead iodide and other byproducts when exposed to humidity, which effectively reduces the performance of solar cells. Similarly, the low decomposition temperature of organic components in the perovskite, especially in hybrid formulations with methylammonium ions, causes thermal instability. Additionally, extended exposure to UV light or high-energy photons can cause phase segregation and accelerate ion migration in mixed-halide perovskites, which further reduces the lifespan and efficiency of the device (Liu et al. 2021; Rong et al. 2015; Urbina 2020).

To tackle these challenges, researchers have investigated a wide range of strategies focused on improving both the intrinsic and extrinsic stability of perovskite solar cells (PSCs). Techniques for passivation, including the application of two-dimensional perovskites or self-assembled monolayers (SAMs) at interfaces and grain boundaries, have demonstrated effectiveness in diminishing trap states and reducing ion migration. Additive engineering, which involves the incorporation of small molecules or ions such as guanidinium, cesium, and potassium into the perovskite precursor solution, has shown potential in enhancing film quality and thermal stability. Furthermore, encapsulation technologies utilizing durable polymeric or glass coatings offer external protection against moisture and oxygen infiltration, thereby facilitating extended operation in real-world environments. Additionally, advancements in materials, such as substituting volatile methylammonium with formamidinium or cesium-based alternatives, have played a significant role in enhancing the thermal and structural stability of perovskites (Wang et al. 2023; Parida et al. 2020; Raj et al. 2022b).

Recent developments have concentrated on enhancing operational stability through the optimization of device architectures and interfaces. For example, inverted (p-i-n) structures are becoming increasingly favoured due to their excellent resistance to hysteresis and improved stability when exposed to continuous light. The creation of UV-resistant electron transport layer (ETL) materials, such as SnO_2, alongside thermally stable HTL materials, like inorganic NiO_X, has been crucial in extending the lifespan of these devices. Additionally, research on encapsulated PSC modules has yielded encouraging outcomes, with certain devices retaining over 80% of their original efficiency after 1000 h of operation under standard testing conditions. Nevertheless, the pursuit of stability benchmarks that match those of silicon solar cells continues to be a significant challenge, necessitating interdisciplinary strategies that integrate materials science, chemistry, and engineering advancements (Zhou and Pang 2020; Sanglee et al. 2022; Hidalgo et al. 2023). The various stability issues associated with PSCs are discussed below.

Moisture and Oxygen

PSCs are highly affected by moisture and oxygen content when they are exposed in ambient atmosphere during fabrication and running process. Perovskite material-based light-absorbing layers such as methylammonium lead iodide are extremely sensitive towards water and oxygen leads to be potential degradation in perovskite photovoltaic devices and affecting the stability. The exceeding limit of oxygen in the devices leads to oxidation of organic compounds, which are present in the PSC. Some research demonstrated that the stability of perovskite devices is not affected by oxygen when it is stored in a dry and dark atmosphere. The degradation mechanism pursues the following series of reactions:

$$CH_3NH_3PbI_3(s) \leftrightarrow CH_3NH_3I\ (aq.) + PbI_2(s) \quad (1.3)$$

$$CH_3NH_3I\ (aq.) \leftrightarrow CH_3NH_2(aq.) + HI\ (aq.) \quad (1.4)$$

$$4HI(aq.) + O_2(g) \leftrightarrow 2I_2(s) + 2H_2O\ (l)\ (\text{Presence of oxygen}) \quad (1.5)$$

$$2HI\ (aq.) \leftrightarrow H_2(g) + I_2(s)\ (\text{UV light}) \quad (1.6)$$

In the above-mentioned reaction mechanism, methylammonium lead iodide is degraded as CH_3NH_3I and PbI_2 in the presence of water, as demonstrated in Eq. (1.3). It can be observed that oxygen, ultraviolet light, and moisture are essential for the degradation of PSC. Moreover, Eq. (1.4) illustrates the co-existence of CH_3NH_3I, HI, and CH_3NH_2 at equilibrium.

In the next step, HI can be degraded into I_2 and H_2O or I_2 and H_2 due to existence of oxygen and ultraviolet light. To resolve the degradation problem, most of the perovskite light-absorbing layers are prepared in a glove box under an inert atmosphere (Salhi et al. 2018; Rong et al. 2015; Shahbazi and Wang 2016; Niu et al. 2014). The schematic diagram of the degradation mechanism of halide-based perovskite material in the presence of moisture is illustrated in Fig. 1.4a, b (Frost et al. 2014; Fang and Jena 2016).

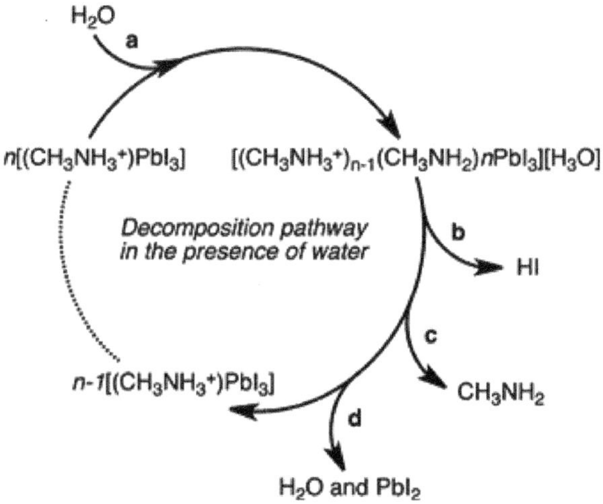

Fig. 1.4 Schematic diagram of degradation mechanism of halide perovskite due to moisture. Reproduced from Ref. Frost et al. (2014) open access license

UV Stability

Generally, TiO_2 is used as ETL in PSC devices because of its low cost, non-toxic property, and resistance to corrosion. Although TiO_2-based PSC dominated in photovoltaic technology but because of band gap of 3.2 eV and redox property, it acts as photocatalytic, which oxidized water molecule and produce OH radical. The degradation of PSC by light radiation can be prevented by using surface blocking layer. Ito et al. reported that the $CH_3NH_3PbI_3$ solar cell exhibited zero efficiency after storage for 12 h during light exposure test without encapsulation, and the colour of the fabricated perovskite thin film also completely changes from black to yellow because of $CH_3NH_3PbI_3$ turning into PbI_2 (Ito et al. 2014). The degradation process caused by UV light follows the following series of reactions:

$$2I^- \leftrightarrow I_2 + 2e^- \tag{1.7}$$

[At the interface between TiO_2 and $CH_3NH_3PbI_3$]

$$3CH_3NH_3^+ \leftrightarrow 3CH_3NH_2 \uparrow + 3H^+ \tag{1.8}$$

$$I^- + I_2 + 3H^+ + 2e^- \leftrightarrow 3HI \uparrow \tag{1.9}$$

First, electrons were extracted from I^- by TiO_2 followed by breaking the perovskite structure, leading to the production of I_2 as demonstrated in Eq. (1.7). After that, Eq. (1.8) explains that $CH_3NH_3^+$ decomposes in the form of CH_3NH_2 and H^+ and in Eq. (1.9) I^- reacts with I_2 and H^+, forming HI. The extracted electrons between TiO_2 and $CH_3NH_3PbI_3$ could return to TiO_2, and Eq. (1.9) occurs, while emitting HI and CH_3NH_2. In the end after losing the H^+ ion, the equilibrium of Eq. (1.8) is shifted to the right side and CH_3NH_2 is emitted because of their low boiling point (Zhang et al. 2018a).

Temperature Effect

Thermal stability of PSC material is one of the major concerns after environmental exposure. Normally, the low-temperature annealing process is used in the fabrication of PSC because high temperature causes degradation in perovskite light-absorbing layer. Several studies found that the temperature creates a high impact on the phase transition and crystal structure of the prepared material (Wang et al. 2016). Supasai et al. found that the performance of the perovskite thin film begins to deteriorate once the annealing temperature exceeds 140 °C. Their findings, based on surface photo-voltage spectroscopy (SPV), indicate that the band gap of $MAPbI_3$ shifts from 1.5 eV to 2.36 eV as a result of the breakdown of PbI_2 at temperatures above 140 °C (Supasai et al. 2013).

Toxicity

The downside of the PSC material contains heavy metals, which are responsible for several health issues and various other environmental problems at the time of installation at the commercial scale. Therefore, the photovoltaic community is very much conscious about the toxicity of newly developed PSC material. Recently developed lead (Pb)-based PSC exhibited excellent performance owing to its superior optoelectrical properties and can be employed as the perfect replacement for silicon (Si)-based solar cells. However, higher toxicity and poor stability are the major concern about the Pb-based perovskite light-absorbing layer. Even though very less amount of metal content (few 100 mg per square meter) is used in the fabrication process, the potential exposure occurs when combined with the large-scale implementation of this technology leading to severe hazards (Zhang et al. 2018b; Babayigit et al. 2016; Ke and Kanatzidis 2019).

The external aspects such as elevated temperature, moisture, humidity, and their combinations tend to degrade the perovskite light-absorbing layer, which leaches out into the environment through air and water causing structural failure of the subsequent photovoltaic devices. An important aspect based on the above concerns is to develop lead-free PSCs, which are always preferred in the photovoltaic market without compromising its performances. This is only possible when Pb-free candidates having properties such as long charge-carrier lifetime, narrow direct band gaps, low excitation binding energies, high optical absorption coefficients, and low toxicity are employed. Among the available materials, Sn-based PSCs has attracted much more attention of photovoltaic researchers to develop lead-free perovskite light-absorbing material due to their analogous properties and most promising performance achieved by their fabricated module. Based on the above discussion on the toxicity of perovskite material, Sn can play a significant role in the fabrication of lead-free PSCs. But Sn-based perovskite light-absorbing materials are still lagging due to the stability problem. In future, further research is required to address the toxicity issue and preparation of environmentally friendly PSCs (Su et al. 2020; Yang et al. 2023; Liu et al. 2019).

The emission of toxic lead elements into our surrounding environment causes major environmental concerns and pollutes the surrounding ecosystem. However, the recycling of lead from used PSCs can be a good solution, and it can be incorporated in the new perovskite-based devices. Therefore, nowadays sustainable recycling methods have become popular and unavoidable for the future development of PSCs. Figure 1.5 shows the various recycling techniques of PSCs using different methods. In the first case, PSCs are directly dissolved in the solvent and the lead is collected from the solvent. In the second case of in situ lead recycling technologies, the used perovskite layer, which contains lead, is directly recovered from the device. For instance, the perovskite layer that has decomposed into lead iodide (PbI_2) can be restored through treatment with methylammonium iodide (MAI). This process enables the reuse of the underlying charge transport layer and FTO substrates, thereby contributing to a reduction in production costs (Kim et al. 2023).

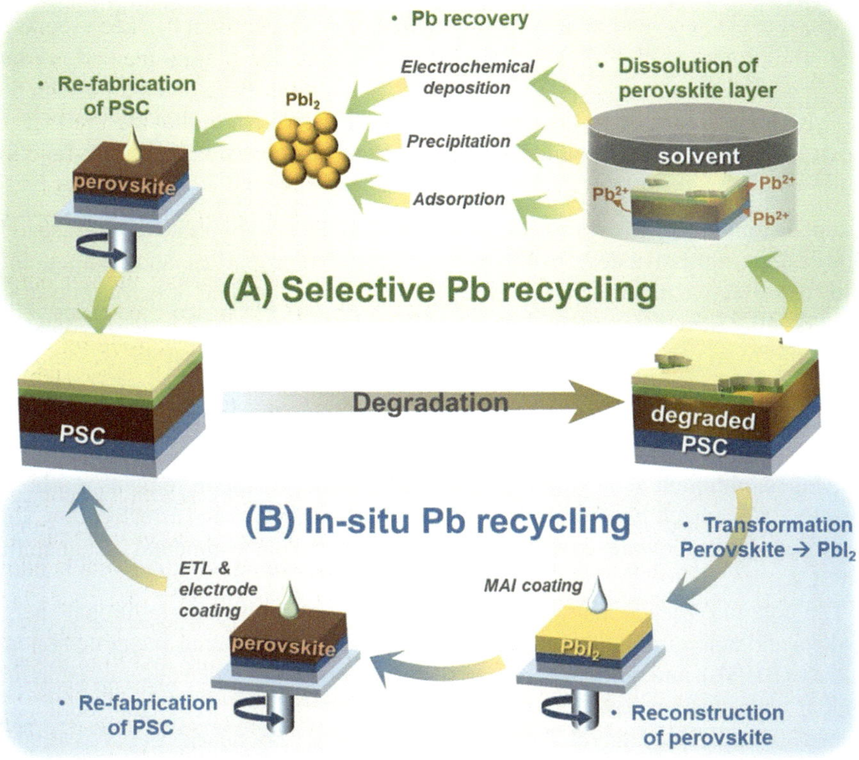

Fig. 1.5 Schematic diagram representing the various Pb-recycling techniques of PSCs. Kim et al. (2023 Reproduced from Ref.) open access license

Current Market Status of PSCs

The current landscape of PSCs indicates a shift from experimental research to commercial possibility. With remarkable PCE surpassing 26% and the potential for cost-effective, scalable manufacturing, PSCs have not yet reached the level of widespread commercial use seen with silicon-based photovoltaics. Start-ups and research organizations are at the forefront of pilot-scale production, targeting specialized markets such as building-integrated photovoltaics (BIPVs), flexible solar solutions, and energy harvesting for Internet of Things (IoT) applications (Skandalos et al. 2023). Nonetheless, issues concerning long-term stability, lead toxicity, and scalability continue to hinder broader adoption. There is growing support from governments and private investors for the advancement of PSC technology through funding and collaborative efforts, indicating a positive outlook for their future in the renewable energy sector. Companies such as Oxford PV and Saule Technologies are making significant strides in bringing PSC products to market, with early modules and tandem cells showing potential for mass production in the coming years (Yan et al. 2022).

Conclusion and Future Outlook

In conclusion, PSCs have significantly transformed the field of photovoltaic research, showcasing impressive device efficiencies, adjustable properties, and compatibility with cost-effective manufacturing techniques. The development of perovskite structures, ranging from basic configurations to mixed cation and double perovskites, has led to improved performance and stability. Various device configurations, including planar and mesoscopic designs, as well as tandem integrations, have further enhanced the adaptability and competitiveness of PSCs. Advances in stability through material engineering, passivation methods, and encapsulation strategies have tackled major challenges, although achieving longevity comparable to silicon remains a primary objective. Concurrently, efforts to create lead-free alternatives and effective recycling processes aim to alleviate toxicity issues, enhancing the environmental sustainability of PSCs.

The potential for PSCs is extensive with their applicability in traditional energy markets and emerging fields such as flexible electronics, building-integrated photovoltaics (BIPVs), and indoor energy harvesting. Innovative device designs, including tandem cells that incorporate silicon or other materials, present opportunities to exceed 30% efficiency, thereby redefining the efficiency boundaries of photovoltaics. Ongoing research into scalable manufacturing methods, such as roll-to-roll printing and inkjet coating, will support the transition from laboratory prototypes to commercial applications. Addressing stability and environmental challenges will necessitate interdisciplinary collaborations that merge materials science, chemistry, and environmental engineering. As PSC technologies continue to rise, they possess the potential to substantially impact the energy landscape.

Acknowledgements The author Abhishek Raj acknowledges the Council of Scientific and Industrial Research (CSIR), CSIR-HRDG, New Delhi, India, and its constituent laboratory CSIR-National Environmental Engineering Research Institute (NEERI), Nagpur, India, for supporting the research.

Declaration of Competing Interest The authors declare that they have no known competing financial interests or personal relationships that could have appeared to influence the work reported in this review article.

CRediT Authorship Contribution Statement **Abhishek Raj**: Writing—original draft, Formal analysis; **Manish Kumar**: Writing—review and editing, Formal analysis; **Avneesh Anshul**: Conceptualization, Writing—review and editing.

References

Akhil S, Akash S, Pasha A, Kulkarni B, Jalalah M, Alsaiari M et al (2021) Review on perovskite silicon tandem solar cells: status and prospects 2T, 3T and 4T for real world conditions. Mater des 211:110138. https://doi.org/10.1016/j.matdes.2021.110138

Akhiro K, Kenjiro T, Yasuo S, Tsutomu M (2009) Organometal halide perovskites as visible-light sensitizers for photovoltaic cells. J Am Chem Soc 131:6050–6051

Arango-Aramburo S, Ríos-Ocampo JP, Larsen ER (2020) Examining the decreasing share of renewable energy amid growing thermal capacity: the case of South America. Renew Sustain Energy Rev 119:109648. https://doi.org/10.1016/j.rser.2019.109648

Arora N, Dar MI, Hinderhofer A, Pellet N, Schreiber F, Zakeeruddin SM et al (2017) Perovskite solar cells with CuSCN hole extraction layers yield stabilized efficiencies greater than 20%. Science 358:768–771. https://doi.org/10.1126/science.aam5655

Babayigit A, Ethirajan A, Muller M, Conings B (2016) Toxicity of organometal halide perovskite solar cells. Nat Mater 15:247–251. https://doi.org/10.1038/nmat4572

Bati ASR, Zhong YL, Burn PL, Nazeeruddin MK, Shaw PE, Batmunkh M (2023) Next-generation applications for integrated perovskite solar cells. Commun Mater 4:2. https://doi.org/10.1038/s43246-022-00325-4

Cheng Y, Ding L (2021) Perovskite/Si tandem solar cells: fundamentals, advances, challenges, and novel applications. SusMat 1:324–344. https://doi.org/10.1002/sus2.25

Clabel HJL, Chacaliaza-Ricaldi J, Marega E Jr (2022) Potential application of perovskite structure for water treatment: effects of band gap, band edges, and lifetime of charge carrier for photocatalysis. Front Nanotechnol 4:827925. https://doi.org/10.3389/fnano.2022.827925

Dambhare MV, Butey B, Moharil SV (2021) Solar photovoltaic technology: a review of different types of solar cells and its future trends. J Phys Conf Ser 1913:012053. https://doi.org/10.1088/1742-6596/1913/1/012053

Dawa T, Sajjadi B (2024) Exploring the potential of perovskite structures for chemical looping technology: a state-of-the-art review. Fuel Process Technol 253:108022. https://doi.org/10.1016/j.fuproc.2023.108022

Docampo P, Ball JM, Darwich M, Eperon GE, Snaith HJ (2013) Efficient organometal trihalide perovskite planar-heterojunction solar cells on flexible polymer substrates. Nat Commun 4:1–6. https://doi.org/10.1038/ncomms3761

Dong X, Fang X, Lv M, Lin B, Zhang S, Ding J et al (2015) Improvement of the humidity stability of organic–inorganic perovskite solar cells using ultrathin Al_2O_3 layers prepared by atomic layer deposition. J Mater Chem A 3:5360–5367. https://doi.org/10.1039/c4ta06128d

Ergen O, Gilbert SM, Pham T, Turner SJ, Tan MTZ, Worsley MA, et al (2016) High current, high efficiency graded band gap perovskite solar cells. Arxiv Preprint 1–19

Fan X (2023) Advanced progress in metal halide perovskite solar cells: a review. Mater Today Sustain 24:100603. https://doi.org/10.1016/j.mtsust.2023.100603

Fang H, Jena P (2016) Super-ion inspired colorful hybrid perovskite solar cells. J Mater Chem A 4:4728–4737. https://doi.org/10.1039/C5TA09646D

Frost JM, Butler KT, Brivio F, Hendon CH, van Schilfgaarde M, Walsh A (2014) Atomistic origins of high-performance in hybrid halide perovskite solar cells. Nano Lett 14:2584–2590. https://doi.org/10.1021/nl500390f

Futscher MH, Ehrler B (2016) Efficiency limit of perovskite/Si tandem solar cells. ACS Energy Lett 1:863–868. https://doi.org/10.1021/acsenergylett.6b00405

Ghosh S, Singh T (2019) Role of ionic liquids in organic-inorganic metal halide perovskite solar cells efficiency and stability. Nano Energy 63:103828. https://doi.org/10.1016/j.nanoen.2019.06.024

Hao F, Stoumpos CC, Cao DH, Chang RPH, Kanatzidis MG (2014) Lead-free solid-state organic-inorganic halide perovskite solar cells. Nat Photonics 8:489–494. https://doi.org/10.1038/nphoton.2014.82

Hartono NTP, Köbler H, Graniero P, Khenkin M, Schlatmann R, Ulbrich C et al (2023) Stability follows efficiency based on the analysis of a large perovskite solar cells ageing dataset. Nat Commun 14:4869. https://doi.org/10.1038/s41467-023-40585-3

Hidalgo J, Kaiser W, An Y, Li R, Oh Z, Castro-Méndez A-F et al (2023) Synergistic role of water and oxygen leads to degradation in formamidinium-based halide perovskites. J Am Chem Soc. https://doi.org/10.1021/jacs.3c05657

Isabela CB, Lameirinhas RA, Torres JP, Fernandes CA (2021) Comparative study of the copper indium gallium selenide (CIGS) solar cell with other solar technologies. Sustain Energy Fuels 5:2273–2283. https://doi.org/10.1039/D0SE01717E

Ito S, Tanaka S, Manabe K, Nishino H (2014) Effects of surface blocking layer of Sb_2S_3 on nanocrystalline TiO_2 for $CH_3NH_3PbI_3$ perovskite solar cells. J Phys Chem C 118:16995–17000. https://doi.org/10.1021/jp500449z

Jalalian D, Ghadimi A, Kiani A (2019) Modeling of a high performance bandgap graded Pb-free HTM-free perovskite solar cell. Eur Phys J Appl Phys 87:10101. https://doi.org/10.1051/epjap/2019190095

Jena AK, Kulkarni A, Miyasaka T (2019) Halide perovskite photovoltaics: background, status, and future prospects. Chem Rev 119:3036–3103. https://doi.org/10.1021/acs.chemrev.8b00539

Ke W, Kanatzidis MG (2019) Prospects for low-toxicity lead-free perovskite solar cells. Nat Commun 10:1–4. https://doi.org/10.1038/s41467-019-08918-3

Kim HS, Lee CR, Im JH, Lee KB, Moehl T, Marchioro A et al (2012) Lead iodide perovskite sensitized all-solid-state submicron thin film mesoscopic solar cell with efficiency exceeding 9%. Sci Rep 2:1–7. https://doi.org/10.1038/srep00591

Kim G, Kim K, Kim HJ, Jung HS, Jeon I, Lee J (2023) Sustainable and environmentally viable perovskite solar cells. EcoMat. https://doi.org/10.1002/eom2.12319

Kour R, Arya S, Verma S, Gupta J, Bandhoria P, Bharti V et al (2019) Potential substitutes for replacement of lead in perovskite solar cells: a review. Glob Challenges 3. https://doi.org/10.1002/gch2.201900050

Kramer IJ, Minor JC, Moreno-Bautista G, Rollny L, Kanjanaboos P, Kopilovic D et al (2015) Efficient spray-coated colloidal quantum dot solar cells. Adv Mater 27:116–121. https://doi.org/10.1002/adma.201403281

Lee TD, Ebong AU (2017) A review of thin film solar cell technologies and challenges. Renew Sustain Energy Rev 70:1286–1297. https://doi.org/10.1016/j.rser.2016.12.028

Lee MM, Teuscher J, Miyasaka T, Murakami TN, Snaith HJ (2012) Efficient hybrid solar cells based on meso-superstructured organometal halide perovskites. Science 338:643–647. https://doi.org/10.1126/science.1228604

Lee TD, Ebong A (2015) Thin film solar technologies: a review. In: 2015 12th International Conference on High-capacity Optical Networks Enabling/Emerging Technology. IEEE, pp 1–10. https://doi.org/10.1109/HONET.2015.7395441

Li C, Lu X, Ding W, Feng L, Gao Y, Guo Z (2008) Formability of ABX_3 ($X = F, Cl, Br, I$) halide perovskites. Acta Crystallogr Sect B Struct Sci 64:702–707. https://doi.org/10.1107/S0108768108032734

Li Z, Kulkarni SA, Boix PP, Shi E, Cao A, Fu K et al (2014) Laminated carbon nanotube networks for metal electrode-free. ACS Nano 8:6797–6804

Li X, Bi D, Yi C, Décoppet JD, Luo J, Zakeeruddin SM et al (2016) A vacuum flash-assisted solution process for high-efficiency large-area perovskite solar cells. Science 353:58–62. https://doi.org/10.1126/science.aaf8060

Liu W, Wu T, Liu M, Niu W, Chueh Y (2019) Recent challenges in perovskite solar cells toward enhanced stability, less toxicity, and large-area mass production. Adv Mater Interfaces. https://doi.org/10.1002/admi.201801758

Liu H, Yu M, Lee C, Yu X, Li Y, Zhu Z et al (2021) Technical challenges and perspectives for the commercialization of solution-processable solar cells. Adv Mater Technol. https://doi.org/10.1002/admt.202000960

Liu S, Li J, Xiao W, Chen R, Sun Z, Zhang Y et al (2024) Buried interface molecular hybrid for inverted perovskite solar cells. Nature 632:536–542. https://doi.org/10.1038/s41586-024-07723-3

Malinkiewicz O, Yella A, Lee YH, Espallargas GM, Graetzel M, Nazeeruddin MK et al (2014) Perovskite solar cells employing organic charge-transport layers. Nat Photonics 8:128–132. https://doi.org/10.1038/nphoton.2013.341

Niu G, Li W, Meng F, Wang L, Dong H, Qiu Y (2014) Study on the stability of $CH_3NH_3PbI_3$ films and the effect of post-modification by aluminum oxide in all-solid-state hybrid solar cells. J Mater Chem A 2:705–710. https://doi.org/10.1039/c3ta13606j

Niu G, Li W, Li J, Liang X, Wang L (2017) Enhancement of thermal stability for perovskite solar cells through cesium doping. RSC Adv 7:17473–17479. https://doi.org/10.1039/c6ra28501e

Noh JH, Jeon NJ, Choi YC, Nazeeruddin MK, Grätzel M, Seok S (2013) Nanostructured $TiO_2/CH_3NH_3PbI_3$ heterojunction solar cells employing spiro-OMeTAD/co-complex as hole-transporting material. J Mater Chem A 1:11842–11847. https://doi.org/10.1039/c3ta12681a

Novoselov KS, Geim AK, Morozov SV, Jiang D, Zhang Y, Dubonos SV et al (2004) Electric field effect in atomically thin carbon films. Science 306:666–669. https://doi.org/10.1126/science.1102896

Palma AL, Matteocci F, Agresti A, Pescetelli S, Calabrò E, Vesce L et al (2017) Laser-patterning engineering for perovskite solar modules with 95% aperture ratio. IEEE J Photovolt 7:1674–1680. https://doi.org/10.1109/JPHOTOV.2017.2732223

Parida B, Yoon S, Jeong SM, Cho JS, Kim J-K, Kang D-W (2020) Recent progress on cesium lead/tin halide-based inorganic perovskites for stable and efficient solar cells: a review. Sol Energy Mater Sol Cells 204:110212. https://doi.org/10.1016/j.solmat.2019.110212

Pitaro M, Tekelenburg EK, Shao S, Loi MA (2022) Tin halide perovskites: from fundamental properties to solar cells. Adv Mater 34. https://doi.org/10.1002/adma.202105844

Rahman A, Farrok O, Haque MM (2022) Environmental impact of renewable energy source based electrical power plants: solar, wind, hydroelectric, biomass, geothermal, tidal, ocean, and osmotic. Renew Sustain Energy Rev 161:112279. https://doi.org/10.1016/j.rser.2022.112279

Raj A, Kumar M, Anshul A (2021) Recent advancement in inorganic-organic electron transport layers in perovskite solar cell: current status and future outlook. Mater Today Chem 22:100595. https://doi.org/10.1016/j.mtchem.2021.100595

Raj A, Kumar M, Kumar A, Singh K, Sharma S, Singh RC et al (2022a) Comparative analysis of 'La' modified $BiFeO_3$-based perovskite solar cell devices for high conversion efficiency. Ceram Int. https://doi.org/10.1016/j.ceramint.2022.09.112

Raj A, Kumar M, Anshul A (2022b) Recent progress in cesium-based lead-free halide double perovskite materials for photovoltaic applications. Phys Status Solidi 219:2200425. https://doi.org/10.1002/pssa.202200425

Raj A, Kumar M, Anshul A (2023) Topical advances in fabrication technologies of perovskite solar cell heterostructures: performance and future perspective. Mater Lett 340:134171. https://doi.org/10.1016/j.matlet.2023.134171

Raj A, Kumar M, Singh DV, Anshul A (2024) A critical review on transition metal dichalcogenides (TMDs): an efficiency booster for perovskite solar cells. FlatChem 44:100629. https://doi.org/10.1016/j.flatc.2024.100629

Ramanujam J, Bishop DM, Todorov TK, Gunawan O, Rath J, Nekovei R et al (2020) Flexible CIGS, CdTe and a-Si:H based thin film solar cells: a review. Prog Mater Sci 110:100619. https://doi.org/10.1016/j.pmatsci.2019.100619

Rathore N, Panwar NL, Yettou F, Gama A (2021) A comprehensive review of different types of solar photovoltaic cells and their applications. Int J Ambient Energy 42:1200–1217. https://doi.org/10.1080/01430750.2019.1592774

Reinhard P, Chirilă A, Blösch P, Pianezzi F, Nishiwaki S, Buechelers S et al (2012) Review of progress toward 20% efficiency flexible CIGS solar cells and manufacturing issues of solar modules. In: 2012 IEEE 38th Photovoltaic Specialists Conference PART 2, IEEE, pp 1–9. https://doi.org/10.1109/PVSC-Vol2.2012.6656789

Rong Y, Liu L, Mei A, Li X, Han H (2015) Beyond efficiency: the challenge of stability in mesoscopic perovskite solar cells. Adv Energy Mater 5:1–16. https://doi.org/10.1002/aenm.201501066

Rong Y, Hu Y, Mei A, Tan H, Saidaminov MI, Seok SI et al (2018) Challenges for commercializing perovskite solar cells. Science. https://doi.org/10.1126/science.aat8235

Salhi B, Wudil YS, Hossain MK, Al-Ahmed A, Al-Sulaiman FA (2018) Review of recent developments and persistent challenges in stability of perovskite solar cells. Renew Sustain Energy Rev 90:210–222. https://doi.org/10.1016/j.rser.2018.03.058

Samiul Islam M, Sobayel K, Al-Kahtani A, Islam MA, Muhammad G, Amin N et al (2021) Defect study and modelling of SnX_3-based perovskite solar cells with SCAPS-1D. Nanomaterials 11:1218. https://doi.org/10.3390/nano11051218

Sanglee K, Nukunudompanich M, Part F, Zafiu C, Bello G, Ehmoser E-K et al (2022) The current state of the art in internal additive materials and quantum dots for improving efficiency and stability against humidity in perovskite solar cells. Heliyon 8:e11878. https://doi.org/10.1016/j.heliyon.2022.e11878

Shah AV, Platz R, Keppner H (1995) Thin-film silicon solar cells: a review and selected trends. Sol Energy Mater Sol Cells 38:501–520. https://doi.org/10.1016/0927-0248(94)00241-X

Shahbazi M, Wang H (2016) Progress in research on the stability of organometal perovskite solar cells. Sol Energy 123:74–87. https://doi.org/10.1016/j.solener.2015.11.008

Skandalos N, Kapsalis V, Ma T, Karamanis D (2023) Towards 30% efficiency by 2030 of eco-designed building integrated photovoltaics. Solar 3:434–457. https://doi.org/10.3390/solar3030024

Steele JA, Lai M, Zhang Y, Lin Z, Hofkens J, Roeffaers MBJ et al (2020) Phase transitions and anion exchange in all-inorganic halide perovskites. Accounts Mater Res 1:3–15. https://doi.org/10.1021/accountsmr.0c00009

Su P, Liu Y, Zhang J, Chen C, Yang B, Zhang C et al (2020) Pb-based perovskite solar cells and the underlying pollution behind clean energy: dynamic leaching of toxic substances from discarded perovskite solar cells. J Phys Chem Lett 11:2812–2817. https://doi.org/10.1021/acs.jpclett.0c00503

Sun J, Zuo Y, Sun R, Zhou L (2021) Research on the conversion efficiency and preparation technology of monocrystalline silicon cells based on statistical distribution. Sustain Energy Technol Assessments 47:101482. https://doi.org/10.1016/j.seta.2021.101482

Supasai T, Rujisamphan N, Ullrich K, Chemseddine A, Dittrich T (2013) Formation of a passivating $CH_3NH_3PbI_3/PbI_2$ interface during moderate heating of $CH_3NH_3PbI_3$ layers. Appl Phys Lett 103:1–4. https://doi.org/10.1063/1.4826116

Tafazoli S, Timasi N, Nouri E, Mohammadi MR (2018) The role of a vapor-assisted solution process on tailoring the chemical composition and morphology of mixed-halide perovskite solar cells. CrystEngComm 20:4428–4435. https://doi.org/10.1039/c8ce00628h

Tong G, Li H, Li G, Zhang T, Li C, Yu L et al (2018) Mixed cation perovskite solar cells by stack-sequence chemical vapor deposition with self-passivation and gradient absorption layer. Nano Energy 48:536–542. https://doi.org/10.1016/j.nanoen.2018.04.012

Traverse CJ, Pandey R, Barr MC, Lunt RR (2017) Emergence of highly transparent photovoltaics for distributed applications. Nat Energy 2:849–860. https://doi.org/10.1038/s41560-017-0016-9

Travis W, Glover ENK, Bronstein H, Scanlon DO, Palgrave RG (2016) On the application of the tolerance factor to inorganic and hybrid halide perovskites: a revised system. Chem Sci 7:4548–4556. https://doi.org/10.1039/c5sc04845a

Urbina A (2020) The balance between efficiency, stability and environmental impacts in perovskite solar cells: a review. J Phys Energy 2:022001. https://doi.org/10.1088/2515-7655/ab5eee

Wang D, Wright M, Elumalai NK, Uddin A (2016) Stability of perovskite solar cells. Sol Energy Mater Sol Cells 147:255–275. https://doi.org/10.1016/j.solmat.2015.12.025

Wang K, Tong Y, Cao L, Yue S, Li Y, Li C et al (2023) Progress of inverted inorganic cesium lead halide perovskite solar cells. Cell Reports Phys Sci 4:101726. https://doi.org/10.1016/j.xcrp.2023.101726

Xiao YY, Meng Y, Gao H, Chen Y, Meng Q, Bai Y et al (2020) Flexible perovskite solar cells fabricated by a gradient heat treatment process. Sustain Energy Fuels 4:824–8311. https://doi.org/10.1039/C9SE00933G

Yan J, Savenije TJ, Mazzarella L, Isabella O (2022) Progress and challenges on scaling up of perovskite solar cell technology. Sustain Energy Fuels 6:243–266. https://doi.org/10.1039/D1SE01045J

Yang M, Tian T, Fang Y, Li W-G, Liu G, Feng W et al (2023) Reducing lead toxicity of perovskite solar cells with a built-in supramolecular complex. Nat Sustain 6:1455–1464. https://doi.org/10.1038/s41893-023-01181-x

Yang D, Yang R, Wang K, Wu C, Zhu X, Feng J et al (2018) High efficiency planar-type perovskite solar cells with negligible hysteresis using EDTA-complexed SnO_2. Nat Commun 9. https://doi.org/10.1038/s41467-018-05760-x

Yun S, Zhang Y, Xu Q, Liu J, Qin Y (2019) Recent advance in new-generation integrated devices for energy harvesting and storage. Nano Energy 60:600–619. https://doi.org/10.1016/j.nanoen.2019.03.074

Zeng L, Chen Z, Qiu S, Hu J, Li C, Liu X et al (2019) 2D–3D heterostructure enables scalable coating of efficient low-bandgap Sn–Pb mixed perovskite solar cells. Nano Energy 66:104099. https://doi.org/10.1016/j.nanoen.2019.104099

Zhang M, Yun JS, Ma Q, Zheng J, Lau CFJ, Deng X et al (2017) High-efficiency rubidium-incorporated perovskite solar cells by gas quenching. ACS Energy Lett 2:438–444. https://doi.org/10.1021/acsenergylett.6b00697

Zhang Y, Zhang H, Zhang X, Wei L, Zhang B, Sun Y et al (2018a) Major impediment to highly efficient, stable and low-cost perovskite solar cells. Metals (Basel) 8:1–33. https://doi.org/10.3390/met8110964

Zhang Q, Hao F, Li J, Zhou Y, Wei Y, Lin H (2018b) Perovskite solar cells: must lead be replaced—and can it be done? Sci Technol Adv Mater 19:425–442. https://doi.org/10.1080/14686996.2018.1460176

Zhang Z, Liu Y, Sun Q, Ban H, Liu Z, Yu H et al (2023) The importance of elemental lead to perovskites photovoltaics. Chem Inorg Mater 1:100017. https://doi.org/10.1016/j.cinorg.2023.100017

Zhou Z, Pang S (2020) Highly efficient inverted hole-transport-layer-free perovskite solar cells. J Mater Chem A 8:503–512. https://doi.org/10.1039/C9TA10694D

Chapter 2
Perovskite Materials and Synthesis Techniques

Arvind Kumar, Swati Jharwal, Brijmohan Prajapati, and Komal Verma

Abstract Perovskite materials are of interest in device applications due to their multifunctional properties. Perovskite materials crystallize having the chemical formula ABX_3, where *A* and *B* are two cations and *X* is an anion that bonds to both and can be C, N, O, or halogen. Perovskite materials display a wide variety of properties and applications. Physical properties of perovskite materials strongly depend on the synthesis methods. In this chapter, we will discuss in detail the perovskite materials explored for applications in solar cells, their properties, classifications, and different synthesis methods/techniques for their possible applications in the devices. The impact of synthesis methods on physical properties of perovskites and hence the performance parameters with respect to solar cells have been discussed.

Keywords Perovskites materials · Solar cells · Synthesis methods · Large-scale manufacturing

Introduction

The invention of silicon and silicon-based technology in the last century revolutionized the world of devices. Till then, many advancements and developments have occurred in the field of electronics and optics to fulfill the energy requirement across the world. In the present time as the need of energy demand is increasing across the world, we need sustainable and renewal energy sources. It was predicted that global energy demand will increase to 16 terawatts (Tws) by 2030 (Asghar 2012; Nema

A. Kumar (✉) · K. Verma
School of Physical Sciences, Jawaharlal Nehru University, New Delhi, India
e-mail: arvind@jnu.ac.in

S. Jharwal
Materials Science Research Lab (Theory and Experimental), Department of Physics, ARSD College, University of Delhi, New Delhi, India

B. Prajapati
Department of Physics, Deshbandhu College, University of Delhi, Delhi, India

M. Kumar and A. Raj (eds.), *Perovskite Solar Cells*, Springer Tracts in Electrical and Electronics Engineering, https://doi.org/10.1007/978-981-96-9463-1_2

et al. 2009; Lund and Kempton 2008; Panwar et al. 2011; Østergaard 2012). It needs more concern because the increased demand cannot be fulfilled by available non-renewal energy sources such as fossil fuels-based energy sources. Many efforts were made in the last few decades for power generation systems and devices, especially on solar cells based on photovoltaic effect (Lewis 2007). The most promising fact about the solar cell is that it can fulfill our annual power demands under one hour's continuous illumination of solar energy with 100% PCE value (Prasanthkumar and Giribabu 2016). This makes and draws special attention toward the photovoltaic solar cells for power generation having the efficient solar cell devices (large power conversion efficiency). In the last few decades, various materials were utilized for this purpose and continuous efforts are going on to serve this cause. In the last few decades various types of solar cell, i.e., based on silicon and perovskite materials were studied for its commercial applications. Solar cells based on amorphous and crystalline silicon, dye-sensitized solar cell, thin-film solar cell, perovskite solar cell, etc. are the best examples (Rand et al. 2007; T.I.R.E.A. (Irena) 2013; Hibberd et al. 2010; O'regan and Grätzel 1991). In the present era (from 2009 onwards), much focus is to develop Perovskite Solar Cell (PSC) due to its high potential (Kojima et al. 2009a) and high-power conversion efficiency (PCE) (> 25% to date).

In this chapter, a detailed discussion has been made in parts 2–4 on a variety of perovskite materials, their classifications, synthesis methods, and fabrication techniques of perovskite-based solar cells.

Perovskite Materials Used for Solar Cell Fabrication

The name perovskite was initially given to the naturally occurring mineral $CaTiO_3$ in honor of Russian mineralogist Aleksevich Lev Perovskite (Hazen 1988). Presently, perovskites are the materials that adopt the general formula ABX_3, where *A* is a larger cation usually occupied by an alkali earth metal, lanthanide, or an organic group, *B* is a smaller cation usually occupied by a transition metal, and *X* is a non-metallic anion typically oxygen or a halide ion. Perovskite materials have been at the forefront for diverse applications owing to the structural and compositional diversity they offer. Perovskites with long carrier diffusion lengths, tunable band gap, and good light absorption coefficient are suitable candidates to be used as light-harvesting active layer in fabrication of solar cell (Shi and Jayatissa 2018). Perovskite-based solar cells have exhibited extraordinary photovoltaic performance comparable to silicon solar cells. Compared to silicon-based devices, the fabrication of perovskite solar cells (PSC) is convenient and cost-effective. PSCs have demonstrated enormous potential with an increase in efficiency from 3.8% in 2009 to 25.7% in 2021 (Kojima et al. 2009b). In a significant development, recently researchers at KAUST have reported the efficiency of 33.7% for tandem perovskite/silicon solar cells (Ugur et al. 2024). The different perovskites that have been used in PSC can be broadly classified as shown in Fig. 2.1.

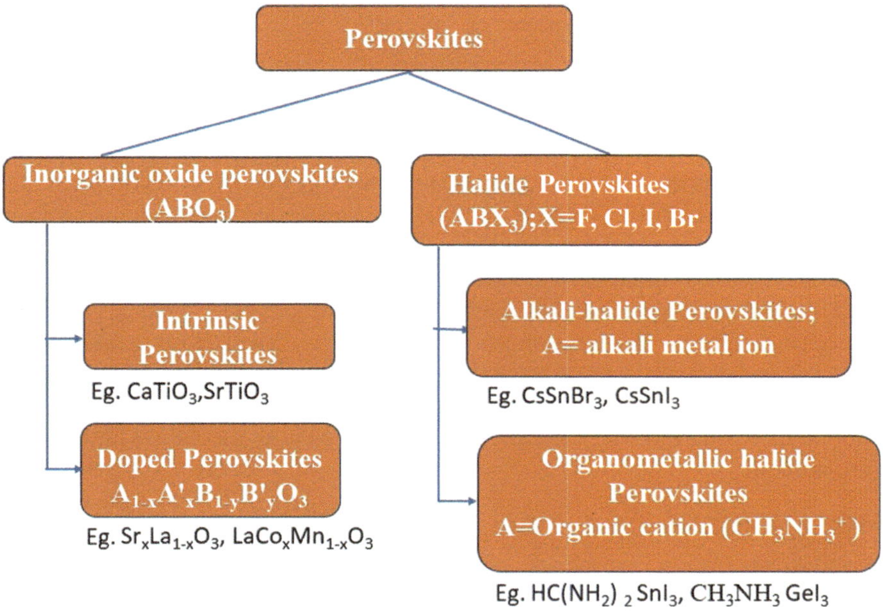

Fig. 2.1 Classification of perovskites

Halide Perovskite System

In the perovskite ABX_3 if the *X* site is occupied by a halide ion, i.e., F^-, Cl^-, Br^-, or I^- then we get halide perovskite system. Depending on the cation that occupies the A site, halide perovskite can be further classified in the following categories..

Alkali Halide Perovskite System

In alkali halide perovskite system, site *A* is occupied by an alkali metal ion like Cs^+, Rb^+, K^+, Li^+, etc. The *B* site is typically occupied by Sn^{2+} or Pb^{2+}. All inorganic halide perovskites such as $CsSnI_3$, $CsPbX_3$, and $CsGeX_3$ are widely explored as the organometallic halide perovskites as hybrid (organic/inorganic) perovskites have stability issues. Chen et al. have reported all inorganic lead-free perovskite $CsSn_{0.5}Ge_{0.5}I_3$ for light absorption and reported a PCE of 7.11% with very high stability and less than 10% degradation after 500 h of functioning (Chen et al. 2019). In the recent past, various theoretical calculations based on first principles methods have projected $CsGeX_3$ as a potential material for optoelectronic applications. Recently, lead-free and purely inorganic perovskite-based optimum structure of ITO/PCBM/$CsSnI_3$/CFTS/Se was shown to be capable of having PCE of 24% using SCAPS 1D numerical simulation methods (Hossain et al. 2023).

Organo-Metal Perovskite System

Also called hybrid (organic/inorganic) metal halide, perovskites are the most explored materials for the solar cell applications. They have an organic ion at the A site like methyl ammonium ($CH_3NH_3^+$, MA^+) or formamidinium ($(NH_2)_2CH^+$, FA^+). Cation at B site is usually an inorganic ion like Sn^{2+} or Pb^{2+}. The hybrid perovskite $CH_3NH_3PbI_3$ ($MAPbI_3$) is one of the widely investigated materials owing to its desirable optoelectronic properties, i.e., exciton B. E~16 meV, E_g ~ 1.51–1.55 eV, electron and hole diffusion lengths (0.1–1 μm) having high mobilities (Savill et al. 2021). They have high absorption coefficient ($\alpha = 81{,}448$ cm $^{-1}$) making it a good choice for the photoactive absorbing layer (Miyata et al. 2015). However, $MAPbI_3$ is not stable and undergoes degradation under ambient conditions. Hence, composition engineering of the perovskites has resulted in the following class of perovskites in an effort to increase the device efficiency and overcome the shortcomings of the halide perovskites.

Mixed A-Site Cation Perovskites

By introducing two cations at the A site, the crystal lattice is stabilized and an increase in PCE has been observed for various compositions. Pellet et al. (2014) have reported an increased PCE of 14.9% for $MA_{0.6}FA_{0.4}PbI_3$ greater than PCE of 12.5% and 11% for $MAPbI_3$ and $FAPbI_3$ respectively. Apart from the mixed FA/MA cation combinations at A site, the combination of an inorganic and organic ion has also been synthesized. Park et al. (2015) have shown that $FA_{0.9}Cs_{0.1}PbI_3$ has superior stability w.r.t. light and moisture. Rubidium (Rb) is another cation that has been shown to increase the stability of α-$FAPbI_3$ perovskite. In mixed cation systems by introduction Rb/Cs, the tolerance factor is modified in a way so as to stabilize the photoactive phase (Byranvand et al. 2022). Triple A site cation compositions Cs/FA/MA have also been prepared and are reported to be thermally stable, having high phase purity and less prone to the processing conditions.

Mixed B-Site Cation Perovskites

The toxicity of lead in these perovskites limits their application on a large scale due to the potential health hazards. A natural alternative to lead is tin, which belongs to the same group. The band gap of tin-based perovskites is found to be lesser as compared to lead-based ones (Zhou et al. 2018). A mixture of Sn and Pb at the *B* site modifies the band gap so that the absorption shifts to near-IR region (Byranvand et al. 2022). Ogomi et al. have studied the compound $MASn_{1-x}Pb_xI_3$ by varying the stoichiometric ratio of Pb to Sn and reported that the energy band gap could be tuned to the range 1.17–1.55 eV and absorption wavelength was increased to 1060 nm (Ogomi et al. 2014) However, in these mixed *B*-cation perovskites the stability is an issue as Sn^{2+} is easily oxidized to Sn^{4+} resulting in lower PCE of solar cells (Zhou et al. 2018).

Mixed Halide Perovskites

In this system mixed halogen element perovskites like $MAPbI_{3-x}Cl_x$ and $MAPbI_{3-x}Br_x$ are shown to exhibit various advantages like enhanced device stability, improved carrier transport, and reduced carrier recombination (Hamed and Mola 2020). Mixed halide perovskite has shown a longer charge diffusion length and enhanced stability in air compared to pure halides (Green et al. 2014). Snaith et al. have reported that the charge diffusion length in $MAPbI_{3-x}Cl_x$ increased by approximately ten times as compared to $MAPbI_3$ (~ 100 nm), which resulted in good carrier collection (Snaith et al. 2014).

Perovskite Single Crystal

Apart from the polycrystalline perovskite-based solar cells, monocrystalline perovskites are also at the focus of many researchers due to better optoelectronic properties. Owing to the low trap density and absence of grain boundaries in the case of single crystals, they exhibit higher efficiency and stable operating condition of the device (Tonui et al. 2018). Peng et al. (2016) reported for the first time the growth of monocrystalline films on substrates by cavitation-triggered asymmetrical crystallization (CTAC) strategy. The device exhibited superior optoelectronic properties for $MAPbBr_3$ as compared to its polycrystalline counterpart.

Inorganic Perovskite System

Perovskites having oxygen ion, i.e., compounds of type ABO_3 are referred to as inorganic perovskite systems.

Intrinsic Perovskite System

Perovskites of the type ABO_3 such as $BaSnO_3$, $(Ba/Sr/Ca/Pb)TiO_3$ have been explored for the electron transport layer (ETL) in the design of solar cells (Chander and Tripathi 2022).

Doped Perovskite

Composition engineering of the intrinsic perovskites at the A and B sites is included in this class of compounds. Grinberg et al. have reported that the oxide composition $[KNbO_3]_{1-x}[BaNi_{1/2}Nb_{1/2}O_{3-\delta}]_x$ (KBNNO) shows a large variation of band gap ranging 1.1–3.8 eV (Grinberg et al. 2013). For $x = 0.1$ the compound exhibited a band gap of 1.39 eV and a photocurrent density 50 times greater than the ferroelectric perovskite $(Pb,La)(Zr,Ti)O_3$.

Synthesis Techniques for Perovskite Materials

Solvothermal Method

Solvothermal method can be adopted for the synthesis of halide perovskites. Zhai et al. have reported the synthesis of $CsPbX_3(X = Cl,Br)$ nanowires of high quality by using the solvothermal method (Zhai et al. 2018). In this method, precursor PbX_2 and Cs-oleate are taken in teflon lined autoclave and maintained at 100 °C for 72 h in a hot air oven. The obtained product was centrifuged, and the substrate was washed with ethyl acetate.

Solid-State Reaction

Halide perovskites can be synthesized by the high-temperature solid-state synthesis procedure. In this method, typically a mixture of AX and BX_2 is taken and grinded either manually or ball-milled. The mixture obtained as a result of manual grinding usually has some unreacted precursors, whereas ball milling results in crystalline products indicated by X-ray diffraction pattern (Prochowicz et al. 2015). The purity of the crude products thus prepared can be further improved by annealing in a sealed evacuated ampule in the temperature range from 200 to 600 °C for a few hours for all inorganic halide perovskites (Rosales et al. 2019). For the synthesis of mixed halide and mixed cation halide perovskites, thermal annealing of the mixture is in the range 150–200 °C (Rosales et al. 2016).

Sol–Gel Method

A sol–gel-based technique is one of the very famous and simple methods to synthesize the high-quality nanostructures of perovskite materials. It is a low cost, and flexible and tunable synthesis route to synthesize the final product (perovskite materials)

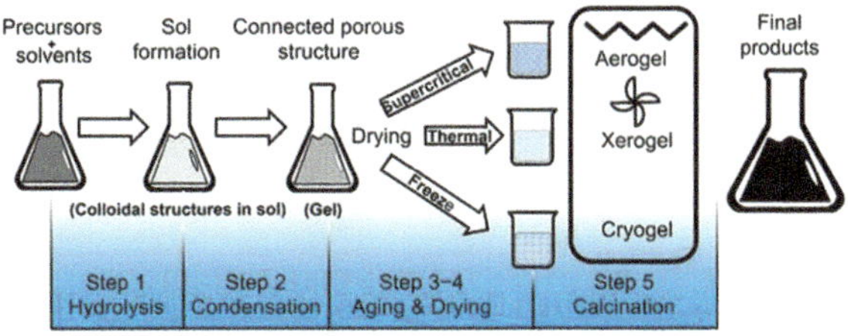

Fig. 2.2 Schematic representation of step-by-step Sol–Gel method (Navas et al. 2021). Reproduced with permission from reference (under open access). Copyright (2022) MDPI Publishing

in different morphologies having different characteristics. In this technique, one can easily modulate the shape, size, and physical properties by controlling different physicochemical parameters (pH, temperature, precursors, time, and pressure) at the time of synthesis (Athayde et al. 2016; Navas et al. 2021). The sol–gel method is a good technique to synthesize homogeneous and dense nanomaterials with uniform morphology and texture (Salavati-Niasari et al. 2009; Noori et al. 2014).

This method of synthesis involves different steps, such as hydrolysis, condensation, aging, drying, and crystallization. In the last few years, various oxide perovskite nanostructures were prepared using this technique and studied for its use in PSCs. In a recent work, $CsCaF_3$ perovskite was prepared using the well-established sol–gel route of synthesis (Pellegrino et al. 2022). Silva et al. also used the sol–gel mediated soft chemistry route to synthesize the photoluminescent $CsPbX_3$ (X = Br, Cl/Br) perovskite nanocrystals (Silva et al. 2023). Figure 2.2 represents the required steps during the sol–gel synthesis.

Hydrothermal Route

This method to synthesize has advantages in comparison to other available methods to synthesize perovskite materials with specific crystal structures, shape, and size (Zhang and Park 2024). This approach offers to synthesize compounds with low-temperature crystallization, having different oxidation states, etc., Krishnaiah et al. (2022) gave an extensive report on an optimized hydrothermal synthesis method to synthesize stable double-layer lead-free Cs_2SnI_6 (CSI) (Krishnaiah et al. 2022). A different research group optimized and worked on the hydrothermal method to synthesize efficient perovskite materials for its possible applications in PSCs and predicted that it is an effective method for the synthesis of perovskites (Zhang and Park 2024; Ban et al. 2019; Favacho et al. 2024).

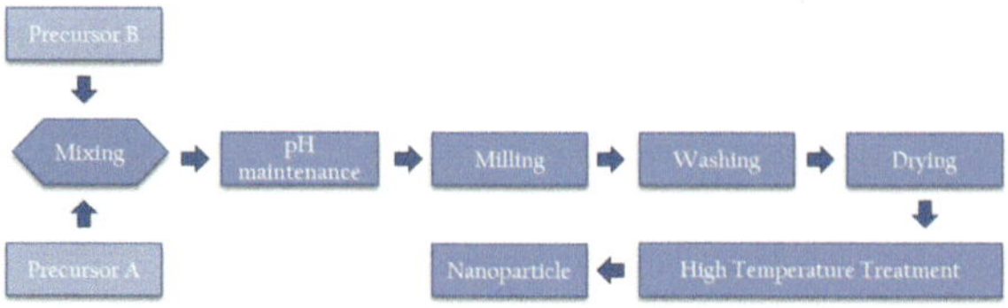

Fig. 2.3 Different steps during mechanochemical process

Mechanochemical Synthesis

This method is a combination of two different types of synthesis methods, mechanical and chemical methods to prepare the nanomaterials (Dutková et al. 2018; Galaburda et al. 2019) (Dutková et al. 2018; Galaburda et al. 2019). The method of synthesis can be considered as a *bottom-up approach* to synthesize the nanostructured compounds. This method includes different steps in the preparation of materials as shown in Fig. 2.3 (flowchart). Mechanochemical synthesis can be performed under different conditions, e.g., using a reactive atmosphere (reactive ball milling [BM]), under cryogenic conditions (cryomilling), or in a solvent (Huot et al. 2013; Suryanarayana 2001). In addition, other experimental parameters can be controlled that influence the characteristics of the final material (Huot et al. 2013; Suryanarayana 2001). García-Espejo et al. prepare the Cs-based double perovskites such as $Cs_2AgBi/SbBr_6$ and $(CH_3NH_3)_2TlBiBr_6$ using this technique which are good candidates for photovoltaic devices (García-Espejo et al. 2019).

Ball Milling Method

High-energy ball milling is a type of mechanical grinding of materials to obtain the required product. Ball milling method can be processed in dry, i.e., without solvents or in a wet environment. In the ball milling process, the milling process occurred due to mechanical friction between different sizes or same size balls for different time scales to ground the materials in the required scale (up to nm). Balls in this process are generally made up of high hardness materials such as zirconia, corundum, and stainless steel. During this milling process, rotary motion of the bowl provides the mechanical energy to the system. This ball milling process is a very popular method since the 1970s for producing metal oxides (such as Al_2O_3 and Y_2O_3) and other materials. Protesescu et al. synthesized lead halide perovskite nanocrystals, $APbX_3$ type [A = Cs, formamidinium (FA), methylammonium; X = Br, I] by Wet Ball Milling, and studied them for optoelectronic applications (Protesescu et al. 2018).

Co-precipitation Synthesis

This method is the most common and easy method for the synthesis of the nanomaterials. In this method of synthesis, generally, metal salts are used as a starting material to initiate the reaction with required liquid medium. This route of synthesis is very popular to synthesize simple oxides (Ecija et al. 2012). Precipitation forms after adding a chemical reagent during the reaction which decreases the solubility limit, and co-precipitation happens when different cations in solution precipitate simultaneously. In this method, we can control the different physicochemical parameters such as temperature, concentration, and pH to obtain a homogeneous mixture. The obtained mixture further is treated thermally to obtain the final desired powder (Athar 2015). Zhang et al. (2015) and Ha et al. (2017) prepared metal halide perovskite and studied it for its optoelectronic properties.

Preparation Techniques for Perovskite Solar Cells (PSCs)

The effectiveness of a perovskite-based solar cell device depends on the type of materials, such as the absorber, charge transporting materials acting as ETL and HTL, the electrode materials, and the process used in its fabrication with excellent characteristics and feasibility. Modern age perovskite-based solar (PSCs) cells have drawn great attention due to their low-cost production, high quality, scalability, and availability at commercial scale. The subsequent sections will provide a detailed explanation of the various fabrication procedures used to assemble the high-performance PSCs, each with unique benefits.

Wet Chemistry Processing Methods

It has been commonly observed that applying the conventional solid-state approach to synthesize the perovskite material has numerous drawbacks because of the repetitive heating and crushing of the used salt before calcination (Varma et al. 2016). As a result, the photoconversion efficiency of perovskite-based solar cells is typically not as improved by perovskite materials produced using these techniques as we would want at commercial scales. Consequently, the wet chemical processing approach, which employs liquid solutions to produce the perovskite layer and other species required to build perovskite-based solar cells, can be used to overcome these issues. This approach offers the cost-effectiveness over the vacuum-based technique, scalability for large area production, and customization in the processing and composition of PSCs for better performance. Nowadays, single-step and two-step deposition techniques are primarily used to produce perovskite layers and hence the PSCs via wet chemical method. Each approach has benefits, uses, and drawbacks, so one can choose the one that best suits the needed solar cell's desired characteristics. In the following section, we will discuss these two methods briefly.

Single-Step Deposition

In this method all the precursors are used in a single solution. A suitable spin coating method is applied for coating. The coated substrate finally goes for annealing. Figure 2.4a depicts the schematic representation of PSC formation via the one-step deposition technique. Initially, organic precursors (CH_3NH_3I) and PbI_2 were dissolved in a polar aprotic solvent like N, N-dimethylformamide (DMF), or N, N-dimethylacetamide (DMA) (acting as coating solutions). Prepared solution then is coated on a mesoporous TiO_2 layer. One must follow the drying and subsequent annealing after the spin coating process to get the high-quality perovskite layer. The basic advantage of this process is its straightforwardness and reduced processing time in quick steps. One can get high-quality perovskite films if the steps are well executed with all care. However, depending on the solvent, concentration, and annealing conditions used, the sensitivity and the growth of crystallization continue to present a significant challenge to researchers in recent decades.

Two-Step Deposition

This method yields the perovskite layers by employing two different solution precursors in different phases. A thin film of the first solution precursor, typically the salt that contains the ingredients of which the perovskite substance is based, has been coated on a substrate, usually by the method like spin coating. For example, for lead-based perovskites, researchers used lead iodide. When the first solution's thin film is annealed, the second precursor solution, which typically contains organic cations, reacts to produce the perovskite layer. Figure 2.4b represents the different steps involved in the fabrication of PSCs via the two-step deposition technique. Here, first, the PbI_2 solution in an appropriate amount is coated on the mesoporous TiO_2 layer. After this, an able solution of the CH_3NH_3I in 2-propanol is spun on the prepared PbI_2 film. The amount of CH_3NH_3I and the spin-coating condition need to be carefully controlled to limit the amount of PbI_2 that is deposited and hence the high-quality perovskite layer. The basic challenges to this method are its complexity, which takes longer time due to more steps involved during the process, and the risk of unwanted defects/incomplete reactions whenever the reaction conditions cannot be controlled.

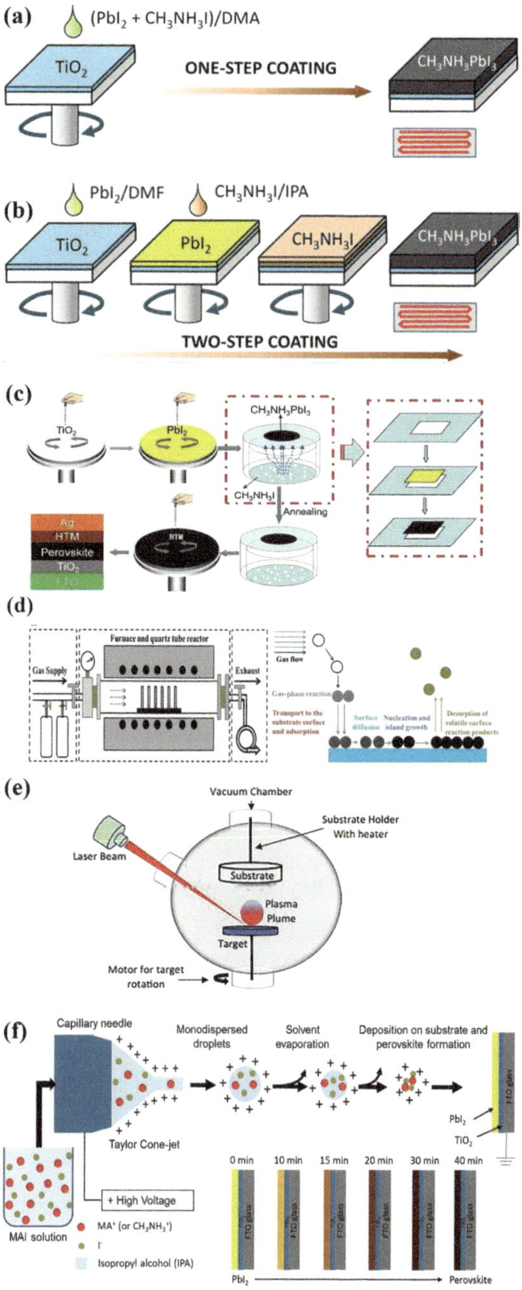

Fig. 2.4 **a** and **b** One-step and two-step wet chemistry-based coating processes for PSC fabrication (Im et al. 2014). Reproduced with permission from reference (under open access). Copyright (2014) AIP Publishing. **c** Schematic of perovskite solar cell preparation via modified Vapor-Assisted Solution Process (Liu et al. 2015). Copyright 2014. Reproduced with permission from ACS Publishing. **d** CVD system with the basic physicochemical reaction mechanisms (Luo et al. 2017). Copyright 2017. Reproduced with permission from Wiley Publishing (Open access). **e** A schematic illustration of PLD's basic working principle (Haider et al. 2022). Copyright 2022. Reprinted with permission from Springer Publishing. **f** A schematic illustration of electrospray-assisted deposition (Kavadiya et al. 2017). Copyright 2017. Reprinted with permission from Wiley Publishing

Vapor-Assisted Processing Methods

In this method formation of perovskite layer is made through vapor. The key aspects of this method involve vapor deposition (via thermal evaporation, chemical vapor deposition, etc.), solvent vapor-assisted technique, precursor conversion, improved film quality, scalability, and the hybrid approaches.

Vapor-Assisted Solution Process

Vapor-Assisted Solution Process (VASP) is a two-step sequential deposition technique. The vapor-assisted solution process (VASP), developed in 2013 by Yang and colleagues, is a substantially innovative technology that combines the advantages of gas-phase deposition and the solution process (Chen et al. 2014). The idea behind this process is the vapor phase interaction of various precursors with a substrate that has been coated in the perovskite precursor solution. The step-wise -modified vapor-assisted solution process of PSC fabrication has been shown in Fig. 2.4c. The preparation of the substrate, solution coating, vapor exposure, and annealing are hence the primary processes in this procedure (Liu et al. 2015). One employs properly cleaned conducting glass with DI water, acetone, and alcohol by turn, such as FTO or ITO, as a substrate. The substrate then goes to the N_2-bath for drying and is then exposed to oxygen plasma for up to 3 min. After this, titanium precursors along with a suitable solvent like alcohol or 2-propanaol in an appropriate proportion are used for the deposition of a compact layer of TiO_2 on the conducting glasses. The substrate is sintered for 30 min at 500 °C after drying at 125 °C. After this, the solutions utilized in proper stoichiometry for coating on the substrate through spin coating or drop-casting are essentially organic cations, like methylammonium iodide, and metal halides, like lead iodide. However, after the coating and before the annealing, a vapor phase of a complimentary or matching precursor material is exposed to the substrate for enhancing the crystallinity and uniformity of the perovskite layer. In the end, the charge transporting layer, like HTL/or ETL, is spin-coated by selecting suitable compounds that can further enhance the photoconversion efficiency of the device. These materials are basically the transition metal oxides, 2D materials like graphene, transition metal dichalcogenides (TMDs), black phosphorous, metal–organic frameworks (MOFs), carbon nitrides, etc.

Chemical Vapor Deposition

The Chemical Vapor Deposition (CVD) technique is one of the most used methods for large-scale production and fabrication of perovskite solar cells because of its great reproducibility and uniformity in producing high-quality films. After the evolution of VASP, different independent groups have developed the tubular CVD technology (Ha et al. 2014; Leyden et al. 2014). Till date, a variety of CVD techniques are subsequently developed, including modified chemical vapor transport (mCVT), in situ tubular CVD (ITCVD), one-step tubular CVD, aerosol-assisted CVD (AACVD), hybrid physical–chemical vapor deposition (HPCVD), and so on (Luo et al. 2017). The fundamental idea behind CVD is that solid materials deposit onto substrates because of the chemical reaction between gaseous precursors. Unlike other vapor-assisted methods, CVD is known for producing high-quality films with excellent uniformity and reproducibility. Figure 2.4d depicts the fundamental concept for the CVD method of perovskite layer deposition utilizing a round tube. The furnace, quartz tube reactor, gas supply, and exhaust system are the fundamental parts of CVD. The basic physicochemical reaction in the chemical vapor deposition technique involves several key steps. (i) Production and delivery of the vapor phase reactants from the main flow gas to the quartz chamber; (ii) In the chamber, gas-phase reactions produce reactive intermediates and byproducts; (iii) film precursor transportation to the substrate surface; (iv) film precursor's adsorption on the substrate surface; (v) surface diffusion, nucleation, and island expansion results in film formation; (vi) transporting the leftover pieces out of the chamber and desorbing the volatile surface reaction products.

In this way, it would be possible to deposit very pure, dense, and homogeneous coatings that have good adhesion. During the process, one can tune the possible reaction parameters affecting the crystal growth and film morphology. More significantly, the slow gas-phase reaction nature of perovskites can lower the excessively high intercalating reaction rate, resulting in smooth and uniform perovskite films. To date, perovskites have been fabricated using a variety of tubular CVD procedures, which may be broadly categorized into two groups: one- and two-step CVD. The former involves a single gas-phase chemical process inside the quartz tube to generate $MAPbI_3$, whereas the later involves two distinct steps: coating PbI_2 precursor films and then deposition of perovskites. Accordingly, continuous improvements in CVD processes could lead to even higher stability and efficiency levels for perovskite solar cells, which would increase their commercial feasibility.

Thermal Vapor Deposition

Thermal vapor deposition (TVD) is also known as a physical vapor deposition method in which a high-quality thin film is formed on a substrate by condensing heated vapor created from solid precursors. Since it can achieve high purity and controlled morphology, therefore one can use this method to produce perovskite layers in PSCs. In a typical procedure adopted by Yabin et al., $PbCl_2$ powder was used to deposit the $PbCl_2$ layer by thermal evaporation in a vacuum deposition system ($\sim 2.0 \times 10^{-6}$ Torr) at roughly 0.5 Å/s. Additionally, a quartz-crystal microbalance was used to determine the thickness, and a profiler. Methyl ammonium iodide (MAI) is loaded into the first zone of a multi-zone tube furnace while substrates pre-deposited with a $PbCl_2$ layer are loaded into the middle zone. After sealing the furnace, the nitrogen gas is continuously forced down to a close pressure. After 30 min of ramping up to the temperatures of 130 °C and 185 °C, respectively, the substrates and MAI are allowed to cool to ambient temperature. For post-annealing, one goes to a temperature of about 120 °C. In this series, a hole transport material for HTL was spin-coated to finish the manufacture of the solar cell device. Finally, a shadow mask outlining the solar cell active zones between 0.07 and 0.1 cm^{-2} was used to deposit the Au top electrodes via heat evaporation (< 1.0×10^{-6} Torr, at 0.1 Å/s) (Leyden et al. 2014).

Other Techniques

In addition to the methods mentioned above, there are several different ways to fabricate perovskite solar cells, each with unique benefits and drawbacks. There are a few typical techniques discussed below.

Pulse Laser Deposition (PLD)

Film deposition technique based on pulsed laser deposition is basically a kind of physical vapor deposition. In this method a target material is made incident by an energetic laser beam to produce plasma plume. This results in the deposition of theh desired uniform thin layer on the substrate. This technique consists of the following fundamental steps: film growth, plume dynamics, target preparation, laser ablation, and post-deposition treatment. This method is known to produce highly pure, thickness-controlled materials that stay away from complex chemical reactions. The components of a pulsed laser deposition system include a vacuum chamber with at least one window that lets in the incident laser wavelength. In the vacuum chamber, target material and the substrate on which the film is to be deposited have been placed maintaining the required space. This plasma produced by penetration of the target by an energetic laser beam is equivalent to energetic materials that are deposited as a thin film on the substrate. Spot size, or the area of the target that the laser ablates, is correlated with the number of species that are eliminated in every pulse. Spots typically have an area of a few square millimeters and are either square or rectangular in shape. A schematic illustration of PLD's basic working principle has been shown in Fig. 2.4e.

The exceptional photoelectric and photovoltaic qualities of halide perovskite materials garnered a lot of interest over the past decades. In a typical method, the target was placed in a vacuum chamber at ~ 10^{-6} Torr, and pulses from an excimer laser (λ ~ 248 nm; energy density: ~ 0.3 J/cm^2; pulse rate: ~ 5 Hz) were incident on it. In this process, the PbI_2 to MAI ratio should be 1:4 to get the perfect stoichiometry of pure phase perovskite.

Electrospray-Assisted Deposition

This method has been known to produce fine and monodispersed droplets which in turn form a uniform thin film compared to other available atomization techniques. A very high voltage is required to break the precursor solution, commonly made up of lead oxide and MAI, to very small droplets for the above-said procedure. These tiny droplets are made to settle down to form a uniform thin layer of perovskite on transparent conductive oxides such as ITO and FTO. It has been seen that electrospray is a more suitable process to get homogenous perovskite layers having minimal defect. Additionally, this method provides the control on the size of the droplets and the deposition rate. For example, the formation steps in the electrospray deposition method are shown in Fig. 2.4f.

Briefly, the Taylor cone-jet mode of the methyl ammonium iodide (MAI) precursor on the FTO/dense-TiO_2/mp-TiO_2/PbI_2 glass demonstrates perovskite production, solvent evaporation, and droplet generation. A uniform thin layer is produced by electrospray because it produces fine, monodisperse droplets in contrast to other atomization methods. As the deposition duration increases, the film's color changes from the yellow of PbI_2 to the dark brown of perovskite.

Rapid Deposition Crystallization

As the name suggests, it offers quick deposition of precursor solution, made up of lead halides and methyl ammonium halides in appropriate proportion, on the substrate. During deposition, the fast and anisotropic growth of crystal layer also takes place. One may opt for a different existing coating technique for the deposition of solution precursor on the substrate. Care must be taken for the fast deposition and hence a named rapid deposition technique. Before deposition, one must go through proper and quick heat treatment of the precursor solution. This rapid variation in the temperature offers a fast crystallization of precursor solution into the crystal grains. In this way one may get a compact and high-quality perovskite layer with minimal defect. Such layers are essential for high-performance perovskite solar cells. Therefore, with improved crystallinity and processing efficiency, rapid deposition crystallization (RDC) provides a specific route to high-efficiency devices, making it an attractive technology in the overall perovskite solar cell manufacturing scenario.

Large-Scale Manufacturing Techniques

The easy manufacturing techniques and commercialization of realized efficient perovskite-based PSCs are required globally to fulfill the energy demand. In the labs, spin coating is the best way to realize the required structures from an investigation point of view. But to fabricate the PSCs at a large-scale, various techniques are available and are discussed below.

Inkjet Printing

This is a very versatile, non-contact, additive patterning technique for fabrication and production of solar cells and other optical devices at a large scale. This technology is directly taken from newspaper printing and graphics technology (Huang et al. 2019). This technique is better as compared to other available techniques. Inkjet printing has the advantage as it offers high-resolution printing, low cost, better use of materials, scalability and freedom of design, etc. (Peng et al. 2017). Figure 2.5a and b shows the basic schematic of PSC fabrication and triple cation-based PSC using this method. Wei et al. in 2014 first predicted about this technique and fabricated the first planar PSC (Wei et al. 2014). In this process, there is selective ejection of the ink through the nozzle on the substrate in the chamber (Hwang et al. 2015; Singh et al. 2010). The main drawback of this method is blocking of the nozzle. This blocking of the nozzle is due to poor solubility of material solvents (Jiang et al. 2016). Wei et al. in (2014) also examined the performance and stability of inkjet-printed PSC, respectively (Wei et al. 2014; Yang et al. 2017a). They found good results with better stability.

Drop Casting

Figure 2.5c shows the simple schematic for the drop casting technique. It is a very basic low-cost method in the fabrication of PSCs and very similar to the spin coating method. But the difference is that no substrate spinning is required and thickness of the film and properties basically depend on the volume of the dispersion and concentration. In this technique, volatile solvents are preferred. Control over the thickness of the film thickness and its uniformity over large size wafers is the major challenge and one of the drawbacks of this technique (Chang et al. 2016). Han et al. in 2014 first used this method to fabricate the PSCs (Roy et al. 2020). They directly deposited the perovskite layer on m-ZrO_2 and m-TiO_2 double scaffold layer. Later Niu et al. used a solution-based hot casting method to fabricate the perovskite layer (Nie et al. 2015). Anye Mei et al. used this technique to fabricate a perovskite panel of cells. It was observed that the fabricated panel showed stable performance (up to 1000 h) with PCE of 10.4% which is less in comparison to PCE attained through conventional spin coating method (Mei et al. 2014; Yang et al. 2017b).

Doctor Blade Coating

Figure 2.5d shows the schematic for the doctor blade method to fabricate the perovskite films. This technique, among the various available techniques to fabricate perovskite at a large scale, is one of the simple, cost-effective, environment-friendly, and vacuum-free techniques (Burschka et al. 2013). In this method, blade coater

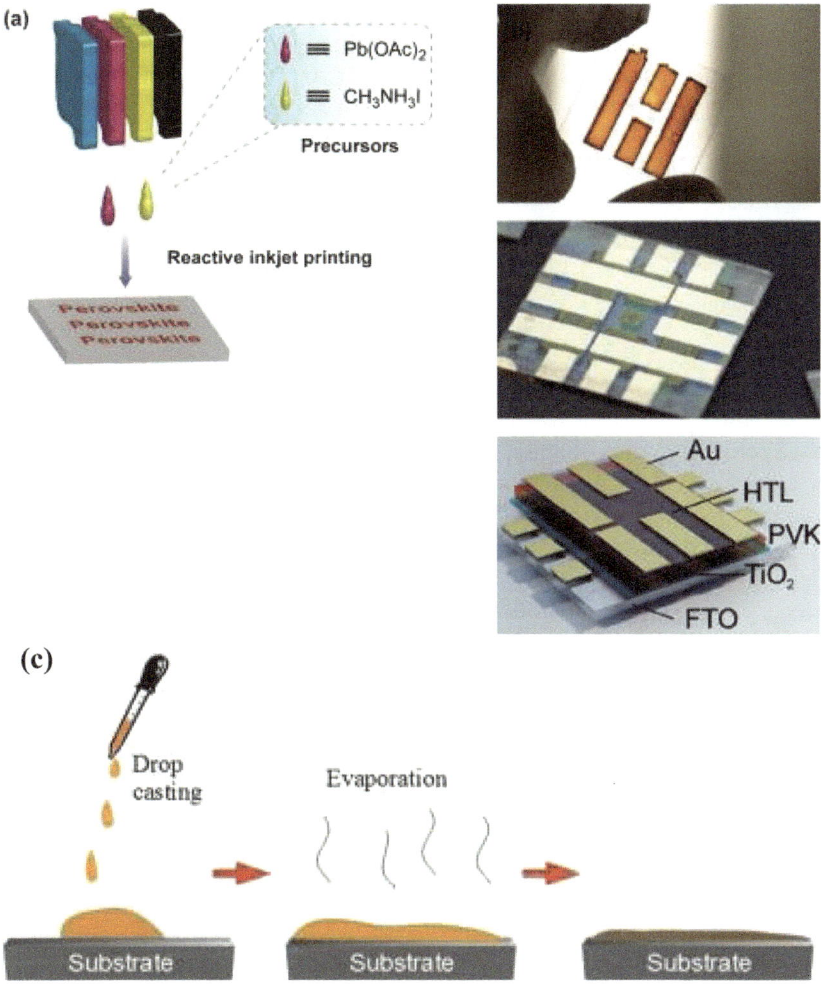

Fig. 2.5 Schematic for the fabrication of perovskite material using reactive ink-jet printing technique (Jiang et al. 2016). **a** Inkjet-printed perovskite (520 nm) layer on FTO/TiO_2-coated glass substrate Copyright 2016. Reproduced with permission from Hal (Open access). Science Publishing. **b** Photograph of inkjet-printed perovskite solar cell and Schematic diagram of the solar cell stack (Mathies et al. 2018). Copyright 2018. Reproduced with permission from ACS Publishing. **c** Perovskite deposition using drop-casting technique (Kajal et al. 2018). Copyright 2018. Reproduced with permission from Springer Publishing. **d** Perovskite solar cell fabrication by blade-coating technique at laboratory scale (Deng et al. 2015). Copyright 2017. Reproduced with permission from IOP Science Publishing. **e** Perovskite layer deposition using slot die coating technique (Cai et al. 2017). Copyright 2017. Reproduced with permission Chinese Institute of Electronics (Open access)

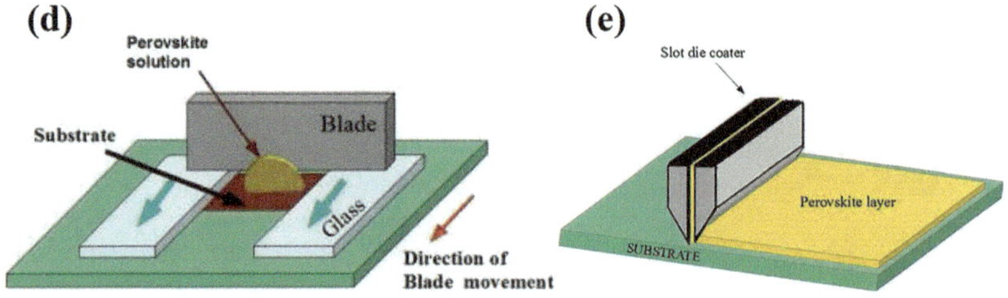

Fig. 2.5 (continued)

applicator consists of a screw arrangement used to adjust the height of the blade w.r.t. substrate surface during the deposition (Razza et al. 2016). In order to fabricate the perovskite films, required amount of perovskite precursor solution is poured over the substrate, followed by swiping the glass blade. Temperature of substrate should be kept higher to obtain the uniform, pinhole-free, and smooth film. The high temperature of the substrate enhances the solvent evaporation rate that promotes the nucleation and crystal growth. Maintaining the airflow over the substrates during the deposition helps in evaporation of solvent easily and rapidly (Wang et al. 2019). Doctor blade coating method is preferable as compared to other methods. It has more control over film morphology as compared to other methods to fabricate the perovskite-based PSCs.

Slot Die Coating

In the blade coating method, electron-transporting layer (ETL) and hole transporting-material (HTM) layer can be deposited along with perovskite layer. This makes it an available technique to be used in the fabrication of PSCs at a large scale. But, proper utilization and control of ink used are the main drawback of this technique. To remove this problem, use of slot die coating is an alternative method which has a similar operating principle as the blade-coating technique. In this technique, there is precise control over the amount of material used to prepare a well-defined thickness, uniform films of materials. This technique also known as the premetered coating process for this cause (Hwang et al. 2015) and has high potential to be used at the industrial scale for PSC manufacturing (Qin et al. 2017). This technique is advantageous over other techniques and achieved large popularity and success in the production of PVs. The basic schematic of the slot die coater is shown in Fig. 2.5e.

Conclusions

In this chapter, detailed discussion about the perovskite materials, its applications in solar cells, their properties, classifications, different synthesis methods/techniques for its possible applications in the devices has been included. The impact of synthesis methods on different physical properties of perovskites and hence the performance parameters with respect to solar cells has been discussed. This chapter thoroughly includes all the aspects of synthesis of perovskite materials and solar cell devices based on this. Different methods and technique were also explored in this chapter to fabricate the perovskite-based PSCs at large scale (industrial level).

Declaration of Competing Interests The authors do not have any conflict of interest to declare.

References

Asghar MI (2012) Stability issues of dye solar cells

Athar T (2015) Smart precursors for smart nanoparticles. In: Emerging nanotechnologies for manufacturing. William Andrew Publishing, pp 444–538. https://www.sciencedirect.com/topics/engineering/liquid-phase-synthesis

Athayde D, Souza DF, Silva A, Vasconcelos D, da Nunes EH, Costa JCD, Vasconcelos W (2016) Review of perovskite ceramic synthesis and membrane preparation methods. Ceram Int 42:6555–6571

Ban T, Kaiden T, Ohya Y (2019) Hydrothermal synthesis of layered perovskite-structured metal oxides and cesium tungstate nanosheets. Cryst Growth des 19(12):6903–6910

Burschka J et al (2013) Sequential deposition as a route to high-performance perovskite-sensitized solar cells. Nature 499(7458):316–319

Byranvand MM et al (2022) Recent progress in mixed a-site cation halide perovskite thin- films and nanocrystals for solar cells and light-emitting diodes. https://doi.org/10.1002/adom.202200423

Cai L, Liang L, Wu J, Ding B, Gao L, Fan B (2017) Large area perovskite solar cell module. J Semicond 38(1):014006

Chander S, Tripathi SK (2022) Recent advancement in efficient metal oxide-based flexible perovskite solar cells: a short review. https://doi.org/10.1039/d2ma00700b

Chang C, Huang Y, Tsao C, Su W (2016) Formation mechanism and control of perovskite films from solution to crystalline phase studied by in situ synchrotron scattering

Chen Q, Zhou H, Hong Z, Luo S, Duan H-S, Wang H-H, Liu Y, Li G, Yang Y (2014) Planar heterojunction perovskite solar cells via vapor-assisted solution process. J Am Chem Soc 136:622

Chen M et al (2019) Highly stable and efficient all-inorganic lead-free perovskite solar cells with native-oxide passivation. Nat Commun 10(1). https://doi.org/10.1038/s41467-018-07951-y

da Silva TC, Fernández-Saiz C, Sánchez RS, Gualdrón-Reyes AF, Mora-Seró I, Julián-López B (2023) A soft-chemistry route to prepare halide perovskite nanocrystals with tunable emission and high optical performance. J Sol-Gel Sci Technol 1–8

Deng Y, Peng E, Shao Y, Xiao Z, Dong Q, Huang J (2015) Scalable fabrication of efficient organolead trihalide perovskite solar cells with doctor-bladed active layers. Energy Environ Sci 8(5):1544–1550

Dutková E, Bujňáková Z, Kováč J, Škorvánek I, Sayagués MJ, Zorkovská A, Kováč J Jr, Baláž P (2018) Mechanochemical synthesis, structural, magnetic, optical and electrooptical properties of $CuFeS_2$ nanoparticles. Adv Powder Technol 29(8):1820–1826

Ecija A, Vidal K, Larrañaga A, Ortega-San-Martín L, Arriortua MI (2012) Advances in crystallization processes. InTech, pp 486–506 (chapter 19)

Favacho VS, Melo DM, Costa JE, Silva YK, Braga RM, Medeiros RL (2024) Perovskites synthesized by soft template-assisted hydrothermal method: a bibliometric analysis and new insights. Int J Hydrogen Energy 78:1391–1428

Galaburda M, Kovalska E, Hogan BT, Baldycheva A, Nikolenko A, Dovbeshko GI, Oranska OI, Bogatyrov VM (2019) Mechanochemical synthesis of carbon-stabilized Cu/C, Co/C and Ni/C nanocomposites with prolonged resistance to oxidation. Sci Rep 9(1):17435

García-Espejo G, Rodríguez-Padrón D, Luque R, Camacho L, de Miguel G (2019) Mechanochemical synthesis of three double perovskites: Cs 2 AgBiBr 6,(CH 3 NH 3) 2 TlBiBr 6 and Cs 2 AgSbBr 6. Nanoscale 11(35):16650–16657

Green MA, Ho-Baillie A, Snaith HJ (2014) The emergence of perovskite solar cells. https://doi.org/10.1038/nphoton.2014.134

Grinberg I et al (2013) Perovskite oxides for visible-light-absorbing ferroelectric and photovoltaic materials. Nature 503(7477). https://doi.org/10.1038/nature12622

Ha ST, Liu X, Zhang Q, Giovanni D, Sum TC, Xiong Q (2014) Synthesis of organic–inorganic lead Halide perovskite nanoplatelets: towards high-performance perovskite solar cells and optoelectronic devices. Adv Opt Mater 2:838

Ha ST, Su R, Xing J, Zhang Q, Xiong Q (2017) Metal halide perovskite nanomaterials: synthesis and applications. Chem Sci 8(4):2522–2536

Haider AJ, Alawsi T, Haider MJ, Taha BA, Marhoon HA (2022) A comprehensive review on pulsed laser deposition technique to effective nanostructure production: trends and challenges. Opt Quantum Electron 54

Hamed MSG, Mola GT (2020) Mixed halide perovskite solar cells: progress and challenges. https://doi.org/10.1080/10408436.2018.1549976

Hazen RM (1988) Perovskites. Sci Am 258(6):74–81

Hibberd CJ, Chassaing E, Liu W, Mitzi DB, Lincot D, Tiwari AN (2010) Non-vacuum methods for formation of Cu (In, Ga)(Se, S) 2 thin film photovoltaic absorbers. Prog Photovoltaics Res Appl 18(6):434–452

Hossain MK et al (2023) Achieving above 24% efficiency with non-toxic CsSnI3 perovskite solar cells by harnessing the potential of the absorber and charge transport layers. RSC Adv 13(34). https://doi.org/10.1039/d3ra02910g

Huang F, Li M, Siffalovic P, Cao G, Tian J (2019) From scalable solution fabrication of perovskite films towards commercialization of solar cells. Energy Environ Sci 12(2):518–549

Huot J, Ravnsbæk DB, Zhang J, Cuevas F, Latroche M, Jensen TR (2013) Mechanochemical synthesis of hydrogen storage materials. Prog Mater Sci 58(1):30–75

Hwang K et al (2015) Toward large scale roll-to-roll production of fully printed perovskite solar cells. Adv Mater 27(7):1241–1247

Im J-H, Kim H-S, Park N-G (2014) Morphology-photovoltaic property correlation in perovskite solar cells: one-step versus two-step deposition of $CH_3NH_3PbI_3$. APL Mater 2

Jiang Z, Bag M, Renna LA, Jeong SP, Rotello VM, Venkataraman D (2016) Aqueous-processed perovskite solar cells based on reactive inkjet printing

Kajal P, Ghosh K, Powar S (2018) Manufacturing techniques of perovskite solar cells. Appl Sol Energy 341–364

Kavadiya S, Niedzwiedzki DM, Huang S, Biswas P (2017) Electrospray-assisted fabrication of moisture-resistant and highly stable perovskite solar cells at ambient conditions. Adv Energy Mater 7

Kojima A, Teshima K, Shirai Y, Miyasaka T (2009a) Organometal halide perovskites as visible-light sensitizers for photovoltaic cells. J Am Chem Soc 131(17):6050–6051

Kojima A, Teshima K, Shirai Y, Miyasaka T (2009) Organometal halide perovskites as visible-light sensitizers for photovoltaic cells. J Am Chem Soc 131(17). https://doi.org/10.1021/ja809598r
Krishnaiah M, Kim S, Kumar A, Mishra D, Seo SG, Jin SH (2022) Physically detachable and operationally stable Cs_2SnI_6 photodetector arrays integrated with μ-LEDs for broadband flexible optical systems. Adv Mater 34(17):2109673
Lee JW, Kim DH, Kim HS, Seo SW, Cho SM, Park NG (2015) Formamidinium and cesium hybridization for photo- and moisture-stable perovskite solar cell. Adv Energy Mater 5(20). https://doi.org/10.1002/aenm.201501310
Lewis NS (2007) Toward cost-effective solar energy use. Science 315(5813):798–801
Leyden MR, Ono LK, Raga SR, Kato Y, Wang S, Qi Y (2014) High performance perovskite solar cells by hybrid chemical vapor deposition. J Mater Chem A Mater Energy Sustain 2:18742
Liu C, Fan J, Zhang X, Shen Y, Yang L, Mai Y (2015) Hysteretic behavior upon light soaking in perovskite solar cells prepared via modified vapor-assisted solution process. ACS Appl Mater Interfaces 7:9066
Lund H, Kempton W (2008) Integration of renewable energy into the transport and electricity sectors through V2G. Energy Policy 36(9):3578–3587
Luo P, Zhou S, Xia W, Cheng J, Xu C, Lu Y (2017) Chemical vapor deposition of perovskites for photovoltaic application. Adv Mater Interfaces 4
Mathies F, Eggers H, Richards BS, Hernandez-Sosa G, Lemmer U, Paetzold UW (2018) Inkjet-printed triple cation perovskite solar cells. ACS Appl Energy Mater 1(5):1834–1839
Mei A, Li X, Liu L, Ku Z, Liu T, Rong Y, Xu M, Hu M, Chen J, Yang Y, Gratzel M, Han H (2014) A hole-conductor-free, fully printable mesoscopic perovskite solar cell with high stability. Science 345(6194):295–298. https://doi.org/10.1126/science:1254763
Miyata A et al (2015) Direct measurement of the exciton binding energy and effective masses for charge carriers in organic-inorganic tri-halide perovskites. Nat Phys 11(7). https://doi.org/10.1038/nphys3357
Navas D, Fuentes S, Castro-Alvarez A, Chavez-Angel E (2021) Review on sol-gel synthesis of perovskite and oxide nanomaterials. Gels 7(4):275
Nema P, Nema RK, Rangnekar S (2009) A current and future state of art development of hybrid energy system using wind and PV-solar: a review. Renew Sustain Energy Rev 13(8):2096–2103
Nie W, Tsai H, Asadpour R, Neukirch AJ, Gupta G, Crochet JJ, Chhowalla M, Tretiak S, Alam MA, Wang H (2015) High-efficiency solution-processed perovskite solar cells with millimeter-scale grains. Science (80-)347:522–526
Noori E, Mir N, Salavati-Niasari M, Gholami T, Masjedi-Arani M (2014) J Sol-Gel Sci Technol 3:544
Ogomi Y et al (2014) $CH_3NH_3Sn_xPb_{(1-x)}I_3$ perovskite solar cells covering up to 1060 nm. J Phys Chem Lett 5(6). https://doi.org/10.1021/jz5002117
O'regan B, Grätzel M (1991) A low-cost, high-efficiency solar cell based on dye-sensitized colloidal TiO_2 films. Nature 353(6346):737–740
Østergaard PA (2012) Comparing electricity, heat and biogas storages' impacts on renewable energy integration. Energy 37(1):255–262
Panwar NL, Kaushik SC, Kothari S (2011) Role of renewable energy sources in environmental protection: a review. Renew Sustain Energy Rev 15(3):1513–1524
Pellegrino AL, Lo Presti F, Malandrino G (2022) A molecular route to fluoro-perov-skite materials: synthesis of $CsCaF_3$ films through a sol–gel/spin-coating process. Discov Mater 2(1):4
Pellet N et al (2014) Mixed-organic-cation perovskite photovoltaics for enhanced solar-light harvesting. Angew Chem-Int Ed 53(12):3151–3157. https://doi.org/10.1002/anie.201309361
Peng X, Yuan J, Shen S, Gao M, Chesman AS, Yin H, Cheng J, Zhang Q, Angmo D (2017) Perovskite and organic solar cells fabricated by inkjet printing: progress and prospects. Adv Func Mater 27(41):1703704
Peng W et al (2016) Solution-grown monocrystalline hybrid perovskite films for hole-transporter-free solar cells. Adv Mater 28(17). https://doi.org/10.1002/adma.201506292

Prasanthkumar S, Giribabu L (2016) Recent advances in perovskite-based solar cells. Current Sci (00113891):111(7)

Prochowicz D, Franckevičius M, Cieślak AM, Zakeeruddin SM, Grätzel M, Lewiński J (2015) Mechanosynthesis of the hybrid perovskite $CH_3NH_3PbI_3$: characterization and the corresponding solar cell efficiency. J Mater Chem A Mater 3(41). https://doi.org/10.1039/c5ta04904k

Protesescu L, Yakunin S, Nazarenko O, Dirin DN, Kovalenko MV (2018) Low-cost synthesis of highly luminescent colloidal lead halide perovskite nanocrystals by wet ball milling. ACS Appl Nano Mater 1(3):1300–1308

Qin T et al (2017) Amorphous hole-transporting layer in slot-die coated perovskite solar cells. Nano Energy 31:210–217

Rand BP, Genoe J, Heremans P, Poortmans J (2007) Solar cells utilizing small molecular weight organic semiconductors. Prog Photovoltaics Res Appl 15(8):659–676

Razza S, Castro-Hermosa S, Di Carlo A, Brown TM (2016) Research update: large-area deposition, coating, printing, and processing techniques for the upscaling of perovskite solar cell technology. APL Mater 4(9)

Rosales BA, Men L, Cady SD, Hanrahan MP, Rossini AJ, Vela J (2016) Persistent dopants and phase segregation in organolead mixed-halide perovskites. Chem Mater 28(19). https://doi.org/10.1021/acs.chemmater.6b01874

Rosales BA, Wei L, Vela J (2019) Synthesis and mixing of complex halide perovskites by solvent-free solid-state methods. J Solid State Chem 271. https://doi.org/10.1016/j.jssc.2018.12.054

Roy P, Sinha NK, Tiwari S, Khare A (2020) A review on perovskite solar cells: evolution of architecture, fabrication techniques, commercialization issues and status. Solar Energy 198:665–688

Salavati-Niasari M, Davar F, Farhadi M (2009) J Sol-Gel Sci Technol 1:48

Savill KJ, Ulatowski AM, Herz LM (2021) Optoelectronic properties of tin-lead halide perovskites. https://doi.org/10.1021/acsenergylett.1c00776

Shi Z, Jayatissa AH (2018) Perovskites-based solar cells: a review of recent progress, materials and processing methods. May 04, 2018, MDPI AG. https://doi.org/10.3390/ma11050729

Singh M, Haverinen HM, Dhagat P, Jabbour GE (2010) Inkjet printing-process and its applications. Adv Mater 22(6):673–685

Snaith HJ et al (2014) Anomalous hysteresis in perovskite solar cells. J Phys Chem Lett 5(9). https://doi.org/10.1021/jz500113x

Suryanarayana C (2001) Mechanical alloying and milling. Prog Mater Sci 46(1–2):1–184

T.I.R.E.A. (Irena) (2013) Solar photovoltaics technology brief. In: The international renewable energy agency, pp 1–28

Tonui P, Oseni SO, Sharma G, Yan Q, Tessema Mola G (2018) Perovskites photovoltaic solar cells: an overview of current status. https://doi.org/10.1016/j.rser.2018.04.069

Ugur et al (2024) Enhanced cation interaction in perovskites for efficient tandem solar cells with silicon. [Online]. Available: https://www.science.org

Varma A, Mukasyan AS, Rogachev AS, Manukyan KV (2016) Solution combustion synthesis of nanoscale materials. Chem Rev 116:14493

Wang D, Zheng J, Wang X, Gao J, Kong W, Cheng C et al (2019) Improvement on the performance of perovskite solar cells by doctor-blade coating under ambient condition with hole-transporting material optimization. J Energy Chem 207–213. https://doi.org/10.1016/j.jechem.2019.03.023

Wei Z, Chen H, Yan K, Yang S (2014) Inkjet printing and instant chemical transformation of a $CH_3NH_3PbI_3$/nanocarbon electrode and interface for planar perovskite solar cells. Angew Chem Int Ed 53:13239–13243. https://doi.org/10.1002/anie.201408638

Yang S, Fu W, Zhang Z, Chen H, Li C-Z (2017a) Recent advances in perovskite solar cells: efficiency, stability and lead-free perovskite. J Mater Chem A 5:11462–11482. https://doi.org/10.1039/C7TA00366H

Yang Z, Zhang S, Li L, Chen W (2017b) Research progress on large-area perovskite thin films and solar modules. J Mater 3(4):231–244

Zhai W et al (2018) Solvothermal synthesis of cesium lead halide perovskite nanowires with ultra-high aspect ratios for high-performance photodetectors. Nanoscale 10(45). https://doi.org/10.1039/c8nr05683h

Zhang C, Park NG (2024) Materials and methods for cost-effective fabrication of perovskite photovoltaic devices. Commun Mater 5(1):194

Zhang F, Zhong H, Chen C, Wu XG, Hu X, Huang H, Han J, Zou B, Dong Y (2015) Brightly luminescent and color-tunable colloidal $CH_3NH_3PbX_3$ (X= Br, I, Cl) quantum dots: potential alternatives for display technology. ACS Nano 9(4):4533–4542

Zhou D, Zhou T, Tian Y, Zhu X, Tu Y (2018) Perovskite-based solar cells: materials, methods, and future perspectives. https://doi.org/10.1155/2018/8148072.

Chapter 3
Device Structure of Perovskite Solar Cell

Prashant Kumar, Manish Kumar, and Bhupendra Singh

Abstract Perovskite solar cells (PSCs) have gained popularity in the energy storage area owing to their ease of manufacturing, extended diffusion lengths, cost-effectiveness, and excellent efficiency. Notably, the power conversion efficiency of PSCs has reached an impressive 25.7% using $FAPbI_3$ as the active layer. Current research is focused on enhancing the open-circuit voltage and fill factor through the utilization of alternative interfaces, charge-selective contacts, and innovative perovskite material shapes. Grain size, homogeneity, and covering area are important factors affecting PSC performance, and they all have a major effect on the perovskite solar cell's power conversion efficiency (PCE). Moreover, the stability and degradation mechanisms of these devices are closely associated with their structural design, highlighting the importance of tailored device architectures. This book chapter highlights the advancements in the structural development of PSCs, including the implementation of advanced interfacial layers and their impact on performance, while also addressing the challenges that continue to impede the efficiency of PSCs.

Keywords Perovskite · Electron transport layer · Hole transport layer · Electrode

Introduction

Perovskite solar cells (PSCs) have garnered attention over the last two decades owing the significantly increased power conversion efficiency (PCE) from hackneyed 3.8% to certifiable performance level 25.7% (Paul et al. 2006; Abhishek et al. 2022; Mahapatra et al. 2022; Kosar et al. 2021; Kumar et al. 2024). PSCs have several specular features such as lightweight, low-cost fabrication process, and tunable

P. Kumar (✉)
Department of Chemistry, GMDC, Maa Shakumbhari University, Saharanpur, India
e-mail: iitprashant412@gmail.com

M. Kumar · B. Singh
Experimental Research Laboratory, Department of Physics, ARSD College, University of Delhi, New Delhi, India

M. Kumar and A. Raj (eds.), *Perovskite Solar Cells*, Springer Tracts in Electrical and Electronics Engineering, https://doi.org/10.1007/978-981-96-9463-1_3

band gap. Hence PSCs have gained increased demand in recent years. Additionally, perovskite materials exhibit broad absorption spectra, enabling them to capture the entire solar spectrum (Feldmann etal. 2021a; Elangovan et al. 2024). According to various studies reported in the literature, PSCs have achieved a power conversion efficiency (PCE) of up to 25.7% to date. The PCE of solar cell is governed by various factors such as charge carrier recombination, interfacial trap state, and crystalline and morphology state of perovskite film. Moreover, the device geometry and related layers directly influence charge carrier transport to the contacts and the assembly of these charge carriers at the particular electrodes. In order to enhance the PCE of perovskite solar cell, the recombination losses originated through defect-mediated can be eliminated by changing the device architecture. The overall performance of PSCs can be improved by modification in interface properties (Oliver et al. 2020). The overall performance of the device can be improved by employing passivation methods such as treating FTO {fluorine-doped tin oxide} with TMAH{Tetramethylammonium hydroxide} (Huang et al. 2019a). Furthermore, the device performance can also be enhanced by substituting the current electrode with one that possesses superior electronic properties and a work function closely aligned with adjacent layers. The thickness as well as stacking pattern of PSC layers can also play a crucial role in determining device efficiency. Furthermore, PSCs face stability issue owing to the environmental factors such as moisture in air, temperature, and the direct connection of charge transport layer to the active layer (Quanrong et al. 2016). To address these degradation issues, controlling morphological changes and minimizing defects in the active layer can improve the overall performance and durability of PSCs. Figure 3.1 illustrates the schematic diagram of the perovskite solar cell's device architecture. The device structure primarily comprises an absorbing layer which is sandwiched between ETL {Electron transport layer} and HTL (Hole transport layer). Generally, the ETL has *n*-type semiconductor, while HTL has *p*-type semiconductor. Moreover, the device structure also comprises fluorine-doped tin oxide (FTO) and metal electrode which serve as a substrate. Furthermore, the PSC can be categorized into two more common configurations, namely mesoporous and planar structures.

This chapter gives an overview of device structure, and basic principle of perovskite solar cell subsequently by cutting edge of such devices presented here.

Progression in Device Architecture

The device configuration significantly influences the overall performance of perovskite solar cells (PSCs) (Yang et al. 2018a; Yang et al. 2018). On the basis of direction of photoirradiation relative to the electron or hole transport layer, PSCs can be categorized into regular (n–i–p) and inverted (p–i–n) structures. The fundamental structure of a PSC comprises a transparent conductive oxide (TCO) layer, such as fluorine-doped tin oxide (FTO) or indium tin oxide (ITO), followed by an electron transport layer (ETL), light-absorbing perovskite layer, hole transport layer

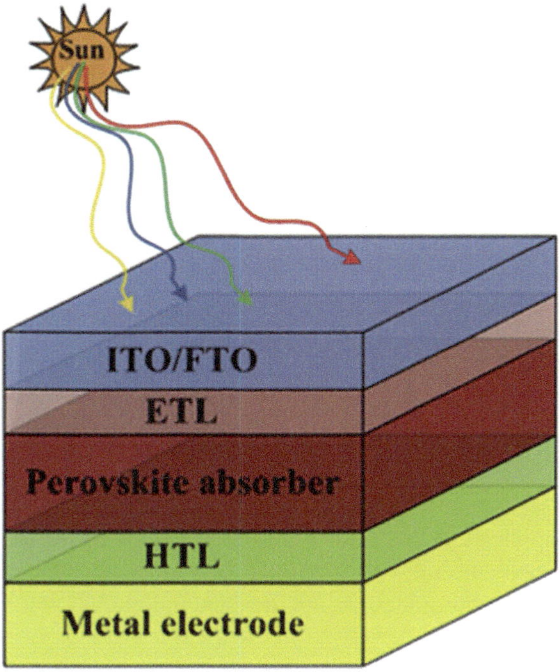

Fig. 3.1 Schematic representation of the device structure of a perovskite solar cell (PSC), illustrating the key layers including the substrate, electron transport layer, perovskite active layer, hole transport layer, and electrode. Reprinted with permission from Raj et al. (2021)

(HTL), and a final electrode. The following sections will delve into these specific configurations (Alipour and Alipour 2024; Ruxin et al. 2022).

Mesoporous (n–i–p) Configuration

The remarkable performance and efficiency of the mesoporous n–i–p configuration have made it a dominant structure in perovskite solar cells. A mesoporous layer, usually composed of titanium dioxide (TiO_2), is incorporated into this design to provide a high porosity and surface area. This mesoporous layer is covered with the perovskite material, enabling close contact and improved light absorption. The mesoporous layer's high pore size (2–50 nm) makes it easier for the perovskite material to penetrate, increasing its exposure to incident light and lowering charge carrier recombination. Because electrons can be moved across the interconnected network of pores before recombining with holes, this arrangement also facilitates effective electron collecting. Mesoporous solar cells' structural characteristics are described using X-ray diffraction. Maximum light transmission into the device requires a transparent conductive oxide (TCO) layer, such as fluorine-doped tin oxide (FTO). In order to facilitate electron transport and inhibit hole diffusion, the ETL (TiO_2) layer is essential in creating an integrated absorbent layer with the mesoporous oxide layer. Mesoporous n–i–p solar cells are a potential technology for upcoming renewable energy

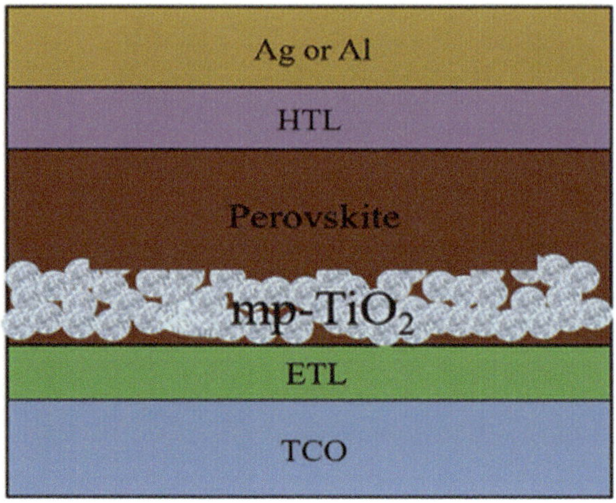

Fig. 3.2 Illustration of the mesoporous n–i–p structure in a perovskite solar cell, highlighting the arrangement of the *n*-type layer, intrinsic perovskite layer, and *p*-type layer. Reprinted with permission from (Deepika et al. 2023)

applications because of this combination of features, which significantly increases photocurrent generation and total efficiency. The first kinds of devices studied as perovskite solar cells are depicted in Fig. 3.2.

Planar n–i–p Structure

Planar perovskite solar cells don't have a porous metal oxide structure like mesoporous ones do. As shown in Fig. 3.2, (Zhou et al. 2018) the fundamental components of a planar solar cell are a metal electrode, a transparent conductive oxide (TCO) layer, an electron transport layer (ETL), a perovskite absorber layer, and a hole transport layer (HTL). The ETL and HTL then gather the charge carriers that are effectively separated by the perovskite layer's inherent electric field when it is lit. Power conversion efficiencies (PCEs) of up to 9.7% have been attained by researchers by substituting a compact TiO_2 layer for the mesoporous one (Liu et al. 2013). By using yttrium-doped TiO_2 (Y-doped TiO_2) as the ETL, additional advancements have been made, resulting in a PCE of 19.3% because of improved electron mobility (Liu and Kelly 2014). With recent developments, verified PCEs for planar perovskite solar cells with lower areas have reached 25.2% (Wu et al. 2021).

Mesoscopic p–i–n Structure

An alternative to the conventional n–i–p structure for perovskite solar cells is the mesoscopic p–i–n configuration. The electron transport layer (ETL), the mesoscopic material, the hole transport layer (HTL), the perovskite absorber layer, and the top electrode are deposited in this order. Although there have been reports of NiO_x-based p–i–n mesostructured devices, their efficiencies have often lagged behind their n–i–p counterparts. But using an Al_2O_3 mesoporous layer has produced significant gains, with devices achieving up to 16.9% efficiency (Danladi et al. 2021).

The physical confinement of the perovskite material within the mesoscaffold affects its electrical characteristics in a mesoscopic p–i–n configuration. The performance of the device can be greatly impacted by the mesoscopic material selection. Using FTO/NiO/meso-Al_2O_3/$CH_3NH_3PbI_3$/PC61BM/BCP/Ag, Chen et al. (2015) studied a p–i–n device structure and obtained a power conversion efficiency (PCE) of about 13.6%. This study shows that the p–i–n arrangement may achieve high efficiencies and is the greatest PCE ever recorded for NiO-based perovskite solar cells.

The HTL at the bottom of the device is responsible for the p–i–n structure's decreased hysteresis, which is one of its main benefits. This layer lessens the reliance of device performance on scan rate during measurements by preventing the buildup of charge carriers at the perovskite/electrode interface. Furthermore, when compared to n–i–p devices, the p–i–n arrangement may demonstrate better long-term stability, especially with regard to resistance to moisture and deterioration.

Finally, the mesoscopic p–i–n configuration offers a promising alternative to the traditional n–i–p structure for perovskite solar cells. While further research is needed to fully optimize this approach, the results achieved thus far demonstrate its potential for achieving high efficiencies and improved device performance. By carefully selecting the mesoscopic material and optimizing the device architecture, it may be possible to unlock the full potential of p–i–n perovskite solar cells for future renewable energy applications.

ETL-Free Configuration

ETL-free-based PSCs have presented significant challenges in achieving high efficiencies. Generally, ETL-free devices often suffer from hindered charge carrier transport toward the respective electrodes, leading to enhance recombination. As a result of this, the overall performance has deteriorated as compared to conventional ETL-based devices. For instance, Huang et al. (2019b) reported an ETL-free PSC with a power conversion efficiency (PCE) of 20.1%, utilizing a FTO layer treated with tetramethylammonium hydroxide (TMAH). Interestingly, this device exhibited promising parameters, including an open-circuit voltage (V_{OC}) of 1.17 V, a fill factor (FF) of 0.739, and a short-circuit current density (J_{SC}) of 23.33 mA/cm^2. Notably, in the absence of TMAH treatment, the PCE was dropped to 14.4%, highlighting the importance of surface modification for improving the device performance. Furthermore, the TMAH treatment effectively enhanced the interface between the FTO and the perovskite layer, refined the surface morphology of the active layer, and suppressed the charge recombination, ultimately contributing to the development of a highly efficient ETL-free perovskite solar cell.

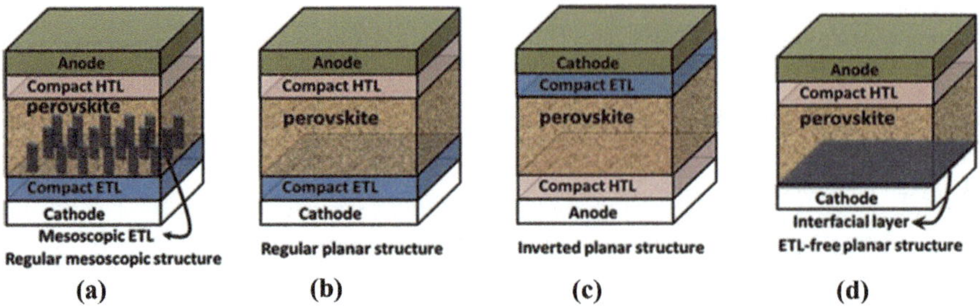

Fig. 3.3 Structures of perovskite solar cells (PSCs): **a** regular mesoscopic, **b** regular planar, **c** inverted planar, and **d** planar without an electron transport layer (ETL). Reprint with permission from (Lian et al. 2018)

Figure 3.3 represents the different structures of perovskite solar cell, viz. (a) regular mesoscopic, (b) regular planar, (c) inverted planar, and (d) planar without an electron transport layer (ETL).

HTL-Free Solar Cells

By gathering holes from the perovskite layer and moving them to the electrode, the hole transport layer (HTL) is an essential component of perovskite solar cells. However, because perovskite materials are ambipolar, the lack of an HTL may result in decreased carrier efficiency and higher recombination. Notwithstanding these difficulties, HTL-free perovskite solar cells provide a number of benefits, such as reduced production costs and the possibility of improved performance.

In order to minimize the recombination centers and improve the device performance without the utilization for an HTL, the energy band alignment, surface modification of the transparent conductive oxide (TCO), and doping within the perovskite material should be carefully optimized.

Another promising absorber material for HTL-free PSCs is $CsPbI_3$ (cesium lead iodide). Owing to its excellent chemical stability, optoelectronic properties, and suitable bandgap, utilizing a $CsPbI_3/CsSnI_3$ mixed halide perovskite absorber layer has shown improved performance and a wider absorption range in comparison to single-layer $CsPbI_3$ devices. Power conversion efficiency (PCE), for cells with a back-front Au electrode, has increased significantly as a result of this combination, going from 15.60 to 19.99% and from 13.87% to 19.59% for cells with a back-front C electrode (Fig. 3.4).

One important step in reducing the complexity and expense of these devices is the development of HTL-free perovskite solar cells. Researchers have shown that high efficiencies can be achieved without employing conventional HTL materials

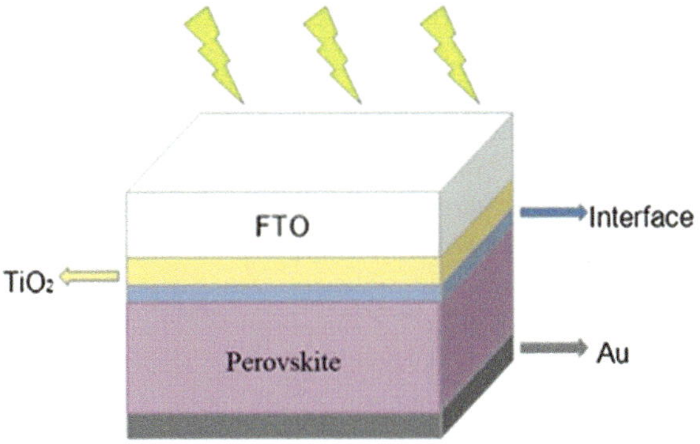

Fig. 3.4 Diagram depicting a configuration without a hole transport layer (HTL), showcasing the essential components of the system. Reprinted with permission from (Deepika et al. 2023)

by resolving issues related to charge transport and recombination. The commercialization of large-scale perovskite solar cells may be possible with additional developments in this area.

Replacing ITO with Gold

As indium-based electrodes are expensive and scarce, researchers are actively looking for alternate possibilities to replaced indium (Deepika et al. 2023). Transparent electrodes composed of conducting polymers, graphene, carbon nanotubes, thin metals like gold, or mixtures of these materials are promising advancements (Feldmann et al. 2021b). Gold has remarkable electrical conductivity and does not impede photon absorption, despite its high cost and scarcity. By serving as a seeding layer, chromium makes sure that gold nanoparticles are distributed evenly and keeps them from clumping together. The high surface energy of chromium makes it easier for thin gold films to grow continuously. Researchers were able to enhance the device efficiency to 19.8% by using gold as the top layer instead of indium tin oxide (ITO) (Yang et al. 2021). In this continuation,

Wang and his associates created a semi-transparent device structure by using indium tin oxide (ITO), zinc oxide (ZnO), molybdenum oxide (MoO_3), polymer PTB7-Th:IHIC, a thin film of gold (1 nm), and a larger layer of silver (15 nm). The power conversion efficiency (PCE) of this creative design was 9.77% (Wang et al. 2017).

Basic Component of Perovskite Solar Cell

The Perovskite Layer

The most important part of PSC is perovskite layer which will act as an active layer. The main function of this layer is to absorb the photons from the source. Usually perovskite contains organic–inorganic mixed halide, having the general formula of perovskite ABX_3 similar to calcium titanate ($CaTiO_3$), where *A*, *B* are monovalent and divalent cations respectively and *X* represents the halide ion (Mitchell et al. 2018; Raj et al. 2021; Kojima et al. 2009) (Fig. 3.5).

Materials having low exciton energy are suitable candidates for PSCs. Based on the literature reports, methylammonium lead iodide ($CH_3NH_3PbI_3$) was the most widely used material for perovskite layer as this material has low exciton energy and extended carrier diffusion length (Kumar et al. 2020). Herein, the low exciton energy refers to the dissociation of the charge carriers at less energy even at room temperature. Extended carrier diffusion length, on the other hand, describes the movement of charge carriers that prevent recombination. Perovskite material having these properties has shown high open-circuit voltage and short-circuit current density (J_{SC}) which is an essential parameter for higher performance of PSCs in terms of PCE. When the active layer is in contact with sunlight, the exciton formation takes place. The generated excitons get separated in the form of charge carrier under the influence of electric field. The excitons refers to the combination of electron and holes. The electrons are collected by electron transport layer (ETL), while holes were collected by hole transport layer (HTL). The carrier diffusion length of methylammonium lead iodide ($CH_3NH_3PbI_3$) {$MAPbI_3$} was at least 100 nm while for $MAPbI_3XCl_3$ it is more than 1 μm x, which is the most appropriate material for perovskite solar cell. According to the literature reports, formamidinium lead iodide ($FAPbI_3$) material was extensively utilized in PSCs (Banerjee and Chattopadhyay 2018; Singh and Nagarjuna 2014). The stability of FA cation in $FAPbI_3$ is quite large than MA cation in $MAPbI_3$. Owing to this, $FAPbI_3$ has higher thermal stability than $MAPbI_3$. However, at higher

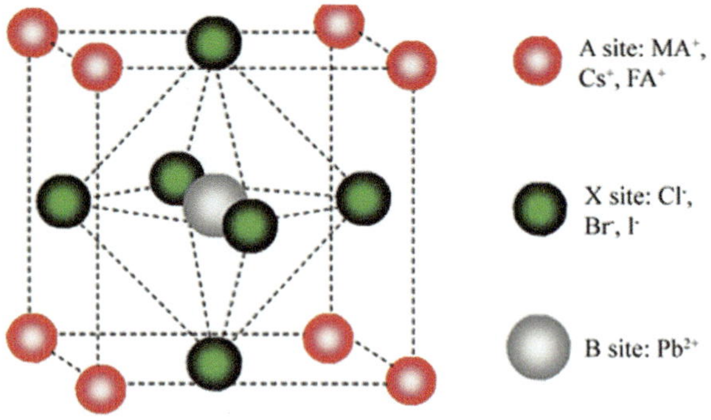

Fig. 3.5 Three-dimensional representation of the perovskite crystal lattice structure. Reprinted with permission from (Kojima et al. 2009

moisture situations, $FAPbI_3$ was transformed to yellow non-perovskite materials (δ-$FAPbI_3$) (Song et al. 2016). In order to prevent this transformation, small insertion of MA^+, Cs^+, and Rb^+ in FA-based mixed A-site perovskite was required (Turren-Cruz et al. 2018; Lee et al. 2019). After incorporation of these ions, the phase transition and thermal stability were enhanced.

Generally, in perovskite solar cell, the most common electron transport layer (ETL) materials were titanium dioxide (TiO_2), zinc oxide (ZnO), fullerene derivatives like phenyl-C61-butyric acid methyl ester (PCBM), and tin oxide (SnO_2). These materials transport electrons more efficiently from the perovskite layer to the cathode, contributing to enhanced device performance. In addition to this, the hole transport layer (HTL) materials, such as spiro-OMeTAD, 2,2′,7,7′-tetrakis (N,N-di-p-methoxyphenylamine)-9,9’-spirobifluorene (spiro-MeOTAD), and poly(triaryl amine) (PTAA), enable the extraction of holes from the perovskite layer to the anode. These materials play an important role to achieve high power conversion efficiencies in perovskite solar cells. The detailed discussion of these materials is given below.

Electron Transport Materials (ETMs)

The primary function of ETMs is to generate the electrons from the layer and allow the transport of electrons toward cathode and prevent the holes. Moreover, ETMs play a crucial role in enhancing the PCE of the device. Thus, choosing an ETM has received a lot of attention recently (Raj et al. 2022). Generally, the PCE of perovskite solar cell is directly related to the short-circuit current density (J_{sc}), open-circuit voltage (V_{oc}), and fill factor (FF), (Edri et al. 2014) where open-circuit voltage (V_{oc}) was originated as a result of splitting of electron and holes (Fig. 3.6). Furthermore, the term “short circuit current density” describes the device’s carrier recombination and reflection of the light harvester’s spectral response, whereas fill factor is directly dependent on the charge abstraction and transference. To chose the best ETMs, the following conditions should be satisfied:

(i) Since ETMs have a little lower energy level of the lowest unoccupied molecular orbital than perovskite materials, they have a strong electron-accepting characteristic that enables electron selection. The perovskite material has ambipolar transport property; owing to this it has wider bandgap; and the lowest HOMO than the active perovskite materials are essential to accomplishing the exciton confinement and hole-blocking function (Lee et al. 2012).

(ii) The morphology of film also affects the PCE of solar cell. As we know that perovskite materials have ambipolar nature, owing to which pinhole-free morphology of ETL is crucial for enhancing the PCE of perovskite solar cell. By eliminating pinholes and minimizing defects, the ETL can effectively stop current leakage and charge recombination at the electrode interface, resulting in a high shunt resistance.

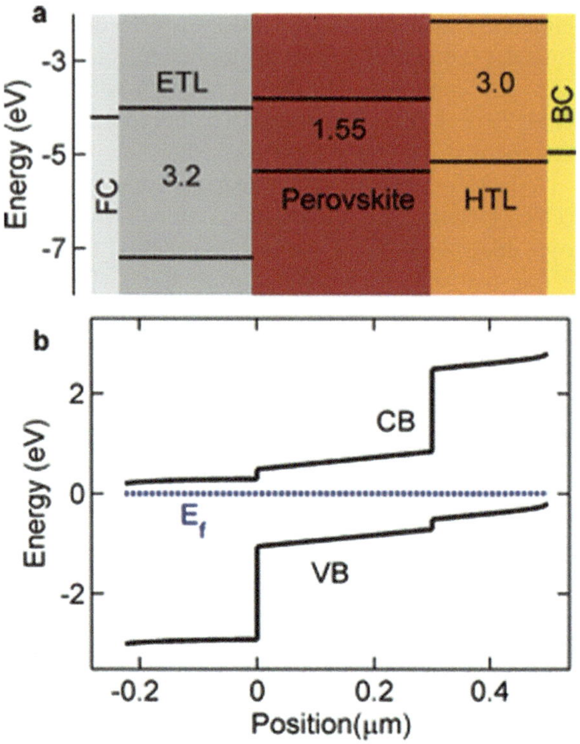

Fig. 3.6 **a** Schematic of a PSC device showing the energy levels of each material. **b** Energy band alignment of the device in equilibrium conditions. Reprint with permission from Agarwal et al. (2014)

(iii) The chemical and thermal stability of the ETLs are very good. Furthermore, hydrophobicity is a significant factor in improving PSC stability. Hydrophobicity prevents moisture from coming in contact with perovskite material.

Inorganic Electron Transport Material

TiO_2

Titanium dioxide (TiO_2) has emerged as a core electron transport material (ETM) in perovskite solar cells (PSCs) due to its excellent properties. The conduction band minimum (CBM) of TiO_2 was 4.1 eV, slightly lower than that of perovskite absorber ($MAPbI_3$) (Yang et al. 2015). Hence, it ensures the efficient electron injection from the perovskite to the TiO_2 layer. Furthermore, the valence band maximum (VBM) was located at a very low energy level having outstanding hole-blocking capabilities. Owing to this, it helps out to avoid the recombination of photo-generated holes with electrons in the TiO_2 layer. As a result, the overall efficiency of device was enhanced. e Ongoing research on TiO_2 has provided a deep understanding of its synthesis and morphology control (Agarwal et al. 2014). This adaptability allows for the fabrication of TiO_2 films with various structures, such as mesoporous, compact, and core–shell configurations. These distinct variations allow the fine-tuning of various ETM's properties such as surface area, porosity, and electron transport pathways. The ability to modify the TiO_2 morphology was a crucial factor for optimizing the performance of PSCs.

There were several restrictions attached to TiO_2, despite its many benefits. It has moderately low electron mobility which can delay the charge carrier transport, leading to potential losses in device efficiency. Moreover, the formation of trap state below the conduction band minimum (CBM) can behave as an recombination centers, dropping the overall device performance and stability, specifically under ultraviolet (UV) illumination (Gubbala et al. 2008). Furthermore, the TiO_2 fabrication requires high processing temperature which can impose limitations on the device manufacturing process. In order to overcome these challenges, various researchers have explored several strategies for modifying and improving TiO_2-based ETMs. These include the following:

- **Doping**: Doping was a successful tactic to improve the device's performance. Dopants like nitrogen (Zhang et al. 2017) or fluorine can be incorporated into the TiO_2 layer using the doping process, increasing electron mobility and decreasing the trap state density in TiO_2.
- **Hybrid structures**: Hybrid structures also help to improve the device performance. In order to obtain hybrid structures, TiO_2 combines with other materials, such as graphene or carbon nanotubes. The obtained hybrid structures have synergistic properties, such as enhanced electron transport and conductivity.
- **Surface modifications**: The surface functionalization of TiO_2 with organic ligands or inorganic nanoparticles can alter its electronic properties and interfacial interactions with the perovskite layer.
- **Alternative fabrication methods**: Alternative synthesis routes, such as sol–gel or hydrothermal methods, can enable the fabrication of TiO_2 films with tailored properties and reduce the processing temperatures.

The device performance of PSCs can be further enhanced with the help of doping technique as we have discussed above. By doping foreign atoms into the TiO_2 lattice, it is possible to alter its electronic properties, charge carrier transport, and interfacial interactions with the perovskite absorber.

The key benefits of doping TiO_2 are the potential to enhance its electrical conductivity. This enhanced conductivity can facilitate electron transport within the ETM layer, reducing the charge carrier recombination and improving the overall device efficiency. Additionally, doping can also help mitigate current hysteresis, a phenomenon that can lead to irregularities in device performance (Snaith et al. 2014). The presence of the trap states within the ETM and perovskite layers were responsible for the hysteresis, which can capture and release charge carriers, affecting the device' s response to varying measurement conditions. In order to reduce trap state density and minimize hysteresis, doping is an effective technique as the doping technique alters the electronic structure of TiO_2.

Furthermore, doping can also improve the device stability. Doping can help extend the operational lifetime of PSCs by suppressing the degradation mechanisms. For example, doping with certain elements can reduce the formation of oxygen defects, which can lead to the degradation of the perovskite absorber.

s Several studies in the literature have reported the effectiveness of TiO_2 doping in enhancing PSC performance. Li-doped TiO_2 scaffolds have shown exceptionally enhanced electron transport properties, leading to a significant increase in power conversion efficiency (PCE) from 17% to over 19% as compared to undoped devices. Furthermore, Li-doped PSCs have shown minimized hysteresis behavior, indicating a reduction in trap state density (Giordano et al. 2016).

In addition to this, doping TiO_2 with Zr^{4+} has been shown to reduce the density of deep trap states, which were supposed to be one of the primary causes of current hysteresis. By changing the CBM upward, Zr^{4+} doping can also help to suppress hysteresis and improve the overall device performance (Nagaoka et al. 2015). Additionally, doping with aluminum has been found to enhance the long-term stability of PSCs by reducing the formation of oxygen defects (Pathak et al. 2014).

Hence, doping strategies offer a promising avenue for improving the performance and stability of TiO_2-based PSCs. Furthermore, it was possible to enhance the electrical conductivity, reduce hysteresis, and prolong device lifetimes by careful selection of dopants and optimizing doping conditions. Further research and development efforts are necessary to fully explore the potential of doping TiO_2.

SnO_2

Owing to its excellent properties, SnO_2 has become a viable substitute for TiO_2, which has been the predominant electron transport material (ETM) in PSCs. Compared to TiO_2-based devices, SnO_2-based PSCs are more stable, especially when exposed to constant full sunshine. The high transparency of SnO_2 in the ultraviolet (UV) light spectrum, which reduces photodegradation of the perovskite absorber, was responsible for this enhanced stability. Furthermore, SnO_2 has a significantly greater bulk electron mobility than TiO_2, which provides better pathway to transport the charge carrier in the ETM layer more effectively (Lee and Kim 2022).

Numerous studies have been documented in the literature where the perovskite absorber's electron injection efficiency into SnO_2 is higher than that into TiO_2. This improved electron injection could lead to reduced charge recombination and improved device performance (Quan et al. 2023). r, Despite these developments, there were still certain challenges with mesoporous SnO_2. Its low electron selectivity, which can result in charge recombination and decreased device performance, was one of the drawbacks. The two major potential reasons for this low selectivity have been identified: the migration of fluorine ions and the cracking of the SnO_2 layer during the heating process. These elements have the ability to modify SnO_2's electrical characteristics and create new channels for charge recombination.

The maximum PCE for mesoporous SnO_2-based PSCs has been found less than 17% (Liu et al. 2016). This suggests that in order to attain performance comparable to

TiO_2-based devices, additional modification of the SnO_2 shape and interfacial characteristics was needed. Compact SnO_2 films, on the other hand, have demonstrated encouraging outcomes in planar-structure PSCs, indicating that even more potential could be unlocked by a superior mesoscopic SnO_2 structure.

As an ETM in PSCs, SnO_2 has benefits over TiO_2, including improved stability, increased electron mobility, and more effective electron injection. To fully realize the potential of SnO_2-based devices, however, issues with charge recombination and electron selectivity must be resolved. To maximize SnO_2's morphology and interfacial characteristics and make it a competitive material, more research and development are required.

Zinc Stannate

Because of its valuable photoelectric characteristics, zinc stannate (Zn_2SnO_4, ZSO) has become an appropriate electron transport material (ETM) for perovskite solar cells (PSCs). ZSO was ideally suited for effective charge carrier transport and light harvesting in PSCs owing to its tiny electron effective mass, high electron hall mobility, wide optical bandgap, and comparatively low refractive index. Furthermore, it ensures efficient electron injection since its CBM location is similar to that of hybrid perovskites. ZSO's remarkable chemical stability, which enables its compatibility with a range of processing methods, including solution-based processes, is one of its most alluring qualities.

It has been demonstrated that applying mesoporous ZSO to perovskite films greatly accelerates perovskite layer's crystallization, encouraging the development of bilayer perovskite structures with uniformly crystallized overlayers and bigger perovskite grains. Increased stability and performance of the device may result from this better perovskite morphology.

One-step synthesis of reduced graphene oxide (rGO) grafted onto extremely porous ZSO nanofiber scaffolds has produced a synergistic combination of the two materials. The addition of rGO increases the total device efficiency by facilitating effective electron injection from the perovskite to ZSO. Power conversion efficiencies (PCEs) of up to 17.89% have been attained by PSCs made with this ZSO–rGO composite ETM, indicating ZSO's promise for high-performance PSCs. To sum up, Zn_2SnO_4 presents a viable substitute for conventional materials (Mali et al. 2016).

In order to achieve the desired properties such as high electron mobility, good energy level alignment, hydrophobicity, and effective metal penetration blocking, extensive research has been focused on by various research groups to develop new ETL materials. The comprehensive overview of typical performance of inverted planar PSCs with several ETL materials highlighting the significant advancements made in this field, which was described in text.

Organic Electron Transport Materials

Fullerene-Based Compounds

In recent years, fullerenes and their derivatives, particularly PCBM, have become the most widely used electron transport layers (ETLs) in inverted planar perovskite solar cells (PSCs) (Chengbo et al. 2021; Tao et al. 2015; Kegelmann et al. 2017). As compared to TiO_2, fullerenes offer several advantages, such as low-temperature solution processing and faster charge transfer with the perovskite. When fullerenes are added at the perovskite–TiO_2 interface, the rate of photoluminescence decay increases, indicating a faster charge transfer. Furthermore, due to their faster charge transfer from perovskite to fullerene and their trap-passivation and ion-migration-blocking characteristics at the perovskite surface and grain boundaries, fullerene-based PSCs show very little current hysteresis. These advantages have resulted in notable increases in power conversion efficiency (PCE), with $MAPbI_3$-based PSCs surpassing 20% thanks to meticulous fullerene film structure and interface engineering.

Hexaazatrinaphthylene (HATNA)

Hexaazatrinaphthylene (HATNA) derivatives have emerged as promising candidate for electron transport materials (ETMs) in perovskite solar cells (PSCs). By substituting alkylsulfanyl chains for fluorine groups, Zhao et al. (2016) suggested a unique design approach to improve the solubility and trap-passivation characteristics of HATNA-based ETLs. The lowest unoccupied molecular orbital (LUMO) of the HATNA derivative may be successfully aligned with the conduction band minimum (CBM) of $MAPbI_3$ by changing the sulfur species from sulfide to sulfoxide and sulfone. Inverted PSCs with minimum hysteresis and an exceptional power conversion efficiency (PCE) of 17.6% were the outcome of this modification.

Overview of Hole Transport Materials (HTMs)

Hole Transport Layer (HTL) plays a crucial role in perovskite solar cells. It mainly acts as a hole-transporting material, which efficiently extracts and transports holes from the active layer to the electrode. Furthermore, the HTL provides an energy barrier to prevent electrons from stretching out the anode and confirm their proper flow. Furthermore, it physically splits up the perovskite layer from the anode. Moreover, it also assists the device protecting it from moisture and air, reducing degradation, and enhancing its overall stability. Furthermore, when the HOMO energy level of HTL's aligns with the perovskite layer, it can assist in enhancing the open-circuit voltage, leading to improving the overall efficiency of the solar cell (Schloemer et al. 2019; Urieta-Mora et al. 2018).

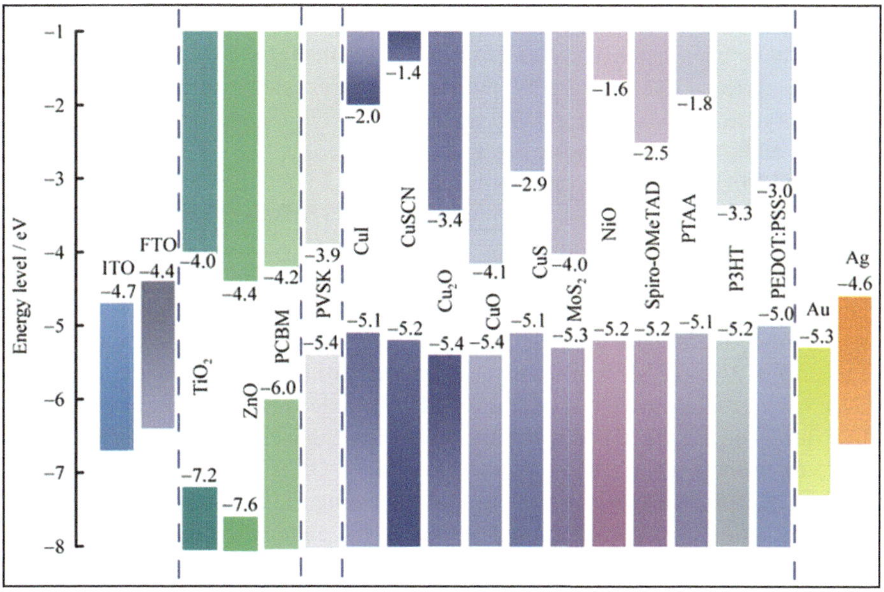

Fig. 3.7 Energy level diagrams of materials frequently utilized in perovskite solar cells. Reprint with permission from (Li et al. 2021)

For perovskite solar cell (PSC), the ideal hole transport material should consist of these properties: (i) The HTMs should have a good hole mobility to provide smoothly efficient charge carrier transport from the perovskite layer to the anode. Furthermore, the energy level of HTM should match with the perovskite layer to reduce energy barriers and facilitate charge transfer. In order to confirm the long-standing device performance, the HTM should exhibit good stability in air, along with photochemical and thermal stability. Furthermore, solution-processability is critical for the fabrication of HTLs, specifically in conventional (n–i–p) PSCs. For scalable manufacturing and commercialization, the HTM should be cost-effective and easy to develop. Figure 3.7 illustrates the energy levels of HTMs and other common PSC materials.

Spiro-OMeTAD

Based on the literature reports, many researchers had utilized Spiro-OMeTAD (full name) as chosen HTM for high-efficiency PSCs, but its performance demands the incorporation of additives like 4-tert-butylpyridine (tBP) and bis(trifluoromethane)sulfonimide lithium salt (Li-TFSI) (Hawash et al. 2018). In comparison, dopant-free HTMs such as PEDOT:PSS [poly(3,4-ethylenedioxythiophene):polystyrene sulfonate] and P3HT [poly(3-hexylthiophene-2,5-diyl)] have developed as promising substitutes (Liu et al. 2020; Li et al. 2021;

Rakstys et al. 2019). Although organic molecules-based hole transport materials (HTMs) have been extensively utilized in perovskite solar cells (PSCs), however, they have suffered from instability and complex multi-step synthesis. On the other hand, inorganic HTMs offer several advantages such as good stability, high hole mobility, and cost-effectiveness. For example, materials like Cu_2O, CuO, CuSCN, NiOx, and MoS_2 have been broadly researched and have demonstrated promising performance in PSCs. Additionally, other inorganic HTMs like CuS, $CuCrO_2$, MoO_x, and WOx have also been explored.

The hole movement of the pristine spiro-OMeTAD, a frequently used organic hole transport material (HTM) in perovskite solar cells (PSCs), is relatively less. However, its performance can be significantly improved after insertion of additives like Li-TFSI and tBP. Li-TFSI mainly increases the disorder and extends the density of states, while tBP improves the solubility of Li-TFSI, leading to a more homogeneous film. Moreover, electrical conductivity was further enhanced by the incorporation of cobalt- or antimony-based salts. Furthermore, spiro-OMeTAD has become the most extensively used HTM in PSCs as properly adjusting the combination of additives, light-absorbing materials, contribute to remarkable enhancements in overall power conversion efficiency (PCE).

As we stated above, spiro-OMeTAD is commonly doped with Li-TFSI and tBP; research has investigated alternative methods to improve device performance and stability. Studies have shown that the positioning of methoxy substituents on the spiro-OMeTAD molecule itself can significantly improve the power conversion efficiency of perovskite solar cell. Moreover, after incorporating a modest amount of residual PbI_2 in the perovskite layer, it has been demonstrated to significantly enhance the stability and power conversion efficiency (PCE) in planar PSCs. Various researchers have also utilizing carbon electrodes (graphene anode and carbon nanotube cathode) with spiro-OMeTAD as the HTM, achieving a PCE of 11.9% (Luo et al. 2018a). Furthermore, low-temperature processed all-inorganic $CsPbI_2Br$ perovskite solar cells were constructed using spiro-OMeTAD as the HTL, with the addition of a MoO_3 interface layer to enhance the efficiency (PCE up to 14.05%) (Zhou et al. 2019).

Finding better dopants has led to the investigation of Zn-$TFSI_2$ as a replacement for Li-TFSI, achieving a PCE of 22.0% and improved operational stability and moisture resistance (Seo et al. 2018). More recently, studies utilizing Mg-$TFSI_2$ and Ca-$TFSI_2$, which are more hydrophobic than Li-TFSI, have shown promise. These dopants, along with tBP and FK209, resulted in PSCs with a PCE exceeding 20% and good stability due to their ability to limit additive aggregation and hydration, eventually enhancing hole mobility and moisture resistance (Pham et al. 2020).

Poly[bis(4-phenyl)(2,4,6-trimethylphenyl)amine] (PTAA)

PTAA has another emerged hole transport material (HTM) in the field of perovskite solar cells (PSCs). Its broad use has been facilitated by its effectiveness in allowing charge transfer and its compatibility with different perovskite compositions.

PTAA utilized as HTM in PSCs started in 2013 when Heo et al. (2013) and Noh et al. (2013) established its potential, achieving efficiency of 12%. In order to achieve effective output, PTAA also combine with additives like LiTFSI and tBP in surpassing the performance of other HTMs such as spiro-OMeTAD (Heo et al. 2013). Moreover, PTAA has formed smooth films and made strong interaction with the perovskite layers as key aspects contributing to its better performance. PTAA exhibited notable characteristics, i.e., it was amorphous in nature, examined by differential scanning calorimetry (DSC) analysis up to 300 °C (Ko et al. 2018). This amorphous state of PTAA was responsible for the formation of smooth films, leading to enhancing the efficiency of charge carrier transport (Ko et al. 2018). Further, in order to improve the PTAA performance, researchers have utilized the use of additives and dopants. For instance, Luo et al. (2018) introduced a fluorine-containing hydrophobic Lewis acid dopant (LAD), which effectively replaced tBP and Li-TFSI. This groundbreaking approach led to a significant enhancement in power conversion efficiency (PCE), reaching up to 19.01%.

Furthermore, PTAA versatility extends beyond conventional (n–i–p) PSC architectures. It has also been utilized in inverted (p–i–n) configuration-based PSCs, where it was paired with (Feldmann et al. 2021a. -phenyl-C61-butyric acid methyl ester (PCBM) as the electron transport material (ETM). Generally inverted PSCs containing PTAA often show longer operating lifetimes and reduced hysteresis, making them attractive for practical applications. However, the PCE of inverted PTAA-based PSCs may be slightly lower than that of conventional devices. The ongoing research efforts focused on enhancing the interfacial layers and perovskite compositions to bridge this gap.

Recently, researcher have explored various strategies to further enhance the performance of PTAA-based PSCs. For example, Liu et al. (2018) have successfully fabricated planar $CH_3NH_3PbI_3$ PSCs by using PTAA and PCBM(-phenyl-C61-butyric acid methyl ester), where PTAA, PCBM act as an HTM, ETM respectively. Moreover, they achieved a remarkable V_{OC} of 1.26 V and a PCE exceeding 20% by precisely optimizing the perovskite precursors and the interfacial layer between PTAA and PCBM. These results showed that PTAA was the promising candidate for HTM to achieve high-efficiency PSCs (Liu et al. 2018). In order to significantly improve the device performance, the optimization of interfacial layer between PTAA and PCBM played an important role. After the optimization of these layers the charge carrier transport was significantly increased by minimizing recombination losses, leading to improved device performance. These findings open the way for further research and development of PSCs with higher efficiencies and broader applications. In another attempt, Zheng et al. have investigated the effect of surface-anchoring alkylamine ligands (AALs) to the perovskite precursor solution. These AALs effectively suppressed the non-radiative carrier recombination, leading to a remarkable improvement in PCE and long-term stability (Zheng et al. 2020).

Finally, PTAA has a versatile and high-performance HTM for PSCs. Furthermore, it was adaptable with various other perovskite compositions, ability to form smooth films, and strong interaction with the perovskite interface leading to achieve high performance of PSCs.

Another fascinating hole transport material was PEDOT:PSS. It is widely utilized in inverted (p–i–n) planar perovskite solar cells (PSCs). The straightforward fabrication process and minimal hysteresis of PEDOT:PSS make it an attractive choice for PSC growth. However, the open-circuit voltage (V_{OC}), short-circuit current (J_{SC}), and stability of PEDOT:PSS-based PSCs were comparatively incomplete owing to several factors. The substantial energy barrier between PEDOT:PSS and perovskite layers was significantly challenging, which hampers efficient charge transfer. The formation of energy barrier was due to charge recombination at the interface, which leads to decreased overall efficiency of the device. Furthermore, the high acidity of the PEDOT:PSS solution was responsible for degradation of the perovskite layer, negotiating the long-term stability of the PSCs. To address these issues, various strategies have been explored by the researchers, for instance, the modifying form of PEDOT:PSS by adding functional groups or introducing interfacial layers to enhance charge transfer e and decrease the energy barrier (Liu et al. 2017)., In order to enhance the stability of the perovskite layer and improve the overall performance of PEDOT:PSS-based PSCs, a less acidic formulation is under investigation.

P3HT {Poly(3-hexylthiophene)}

In perovskite solar cells, another promising organic hole transport material (HTM) is Poly(3-hexylthiophene) (P3HT). It offers several advantages such as suitable bandgap aligned with the solar spectrum, high charge carrier mobility, and cost-effectiveness. However, the P3HT-based devices have relatively low conductivity which limits their overall conversion efficiency. In order to overcome this challenge, various strategies have been employed for enhancing P3HT conductivity.

Optimizing the molecular structure and morphology of P3HT is one of the approach to improve its conductivity. For instance, the packing efficiency and intermolecular interactions can be influenced by incorporating alkyl side chains with variable lengths, which affects charge transport (Poelking et al. 2014).

Furthermore, the controlled crystallization process and annealing conditions can also help to achieve a more ordered crystalline P3HT film, leading to enhanced carrier mobility. Additionally, in order to facilitate charge transport, P3HT can combine with other organic materials, such as fullerene derivatives or conjugated polymers, forming interfacial charge-transfer complexes which enable charge transport.

Another strategy to enhance the P3HT conductivity was through chemical modification. Herein, the polymer backbone can modify by electron-withdrawing groups which alter the electronic properties of P3HT leading to improve its conductivity (Maher Al-Ibrahim et al. 2005). Moreover, electrical conductivity can also be enhanced by doping. Doping P3HT with small molecules or ions generates additional charge carriers, which are responsible for increasing the conductivity. However, the doping method has certain drawbacks such as increased trap states or decreased device stability.

Previously we have discussed molecular engineering and chemical modification to enhance the P3HT conductivity. Furthermore, P3HT conductivity can also be enhanced by interfacial engineering. The interface between P3HT and the perovskite layer was responsible for efficient charge transfer and collection (Zhang et al. 2016).

The interfacial properties can be modified by introducing electron-transport layers, or using self-assembled monolayers can enhance the charge-transfer kinetics and minimize interfacial recombination.

Finally, the optimization of device architecture can also contribute to higher P3HT conductivity and overall device performance. For example, using inverted device structures with a hole-blocking layer can minimize charge recombination and improve the fill factor (Gu et al. 2022). Moreover, the thickness of various layers and the annealing conditions can help ensure efficient charge collection and transport within the device.

Furthermore, P3HT possesses several desirable properties for PSCs, however, its relatively low conductivity remains a significant challenge. Researchers have made significant progress in enhancing P3HT conductivity and achieving higher conversion efficiencies in PSCs by employing strategies such as molecular engineering, chemical modification, interfacial engineering, and device architecture optimization.

Figure 3.8 represents the chemical structures of widely used organic hole transport materials (HTMs).

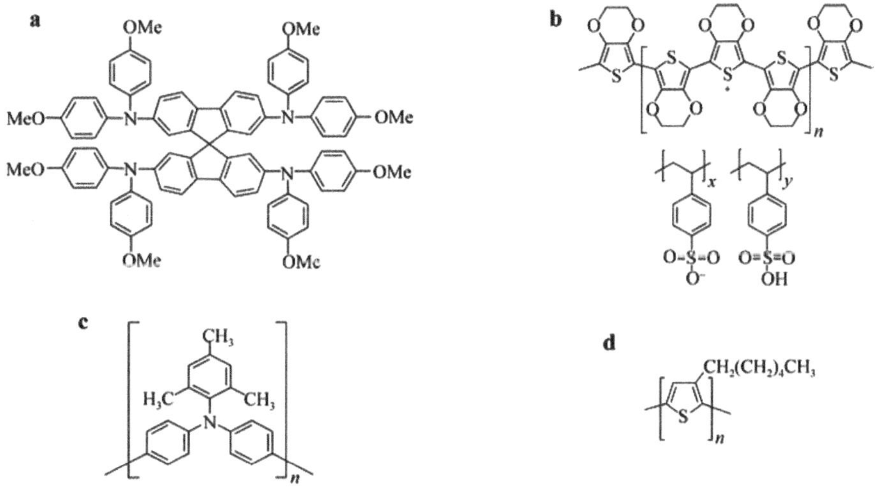

Fig. 3.8 Chemical structures of widely used organic hole transport materials (HTMs): **a** spiro-OMeTAD, **b** PEDOT, **c** PTAA, and **d** P3HT. Reprint with permission (Li et al. 2021)

Other Organic Hole Transport Materials

In order to enhance the performance of PSCs, various researcher have developed novel organic HTMs. These novel HTMs often exhibit remarkable power conversion efficiencies (PCEs) having enhanced stability or cost-effective manufacturing. In this continuation, Saliba and coworkers (Saliba et al. 2016) have developed bis(bis(4methoxyphenyl)amino)spiro[cyclopenta[2,1-b:3,4b0]dithiophene4,90-fluorene] (FDT), a novel organic HTMs. A remarkable PCE of up to 20.2% was observed, when it was incorporated into PSCs with dopants like Li-TFSI, FK209, and tBP. In another attempt, Zhu et al. (2018) introduced SCZF-5, a novel carbazole-based single-spiro HTM. PSCs fabricated with SCZF-5, along with spiro-OMeTAD and SAF-5 (another single-spiro material), were compared. All the HTMs were doped with tBP and Li-TFSI. The PCE of SCZF-5-based devices was dramatically enhanced reaching up to 20.10% compared to other devices incorporated with spiro-OMeTAD's 19.11% and SAF-5's 13.93%. The rigid molecular backbone of SCZF-5 enhanced hole-transporting capability and lowered the HOMO energy level. Owing to this, it showed better PCE compared to others.

Inorganic Hole Transport Materials

Generally, organic hole transport materials (HTMs) have impressive power conversion efficiencies (PCEs) achieved by PSC; however, stability remains a significant hurdle. In order to overcome this issue, researchers have developed alternative approaches. One of the promising strategies to resolving this issue was the utilization of dopant-free organic HTMs, which can enhance stability. Usually, inorganic HTMs often exhibit several advantages such as easy preparation, excellent chemical stability, high hole mobility, and cost effectiveness (Palilis et al. 2020; Yu and Sun 2018). Such outstanding properties of inorganic based HTMs make them attractive candidates for replacing organic HTMs in PSCs, leading to the development of more stable and efficient devices. Some of the inorganicbased HTMs are given below.

Copper Oxides

The copper oxides {Cuprous oxide (Cu_2O) and cupric oxide (CuO)} have considerable significant interest as inorganic HTMs due to their attractive properties. The main advantage of Cu_2O and CuO is that they have high-absorption coefficients. Hence, they allow them to effectively capture a significant portion of the incoming sunlight, contributing to the overall efficiency of the PSC. r, Excessive absorption can also be detrimental due to unwanted light losses within the HTM layer itself.

Despite this challenge, Cu_2O-based HTMs are a promising candidate for inverted PSCs, which possess a different device architecture compared to conventional PSCs. Inverted PSCs have several advantages such as enhanced stability and compatibility with flexible substrates. In 2016, Yu et al. (2016) demonstrated the effectiveness of ultrathin Cu_2O films, which was prepared by a simple thermal oxidation process.

They achieved a PCE of 11.0% by optimizing the film thickness through controlled sputtering time. This success can be attributed to several factors:

(i) The energy levels of Cu_2O closely matched with the commonly used perovskite material $CH_3NH_3PbI_3$, which facilitate the efficient hole transfer at the interface.
(ii) Furthermore, Cu_2O exhibits excellent hole transporting properties, confirming efficient movement of these carriers within the HTL.
(iii) The ultrathin films also minimized the recombination losses, allowing photo-generated holes to reach the electrode before recombining with electrons.

Cu_2O-based HTMs can further be improved by surface modification strategies. For instance, Liu et al. (2019) in 2019 have proposed a method for modification of Cu_2O quantum dots using a silane coupling agent. This modification not only improved the compatibility of the Cu_2O layer with the perovskite film but also led to a significant improvement in PCE, reaching up to 18.9% compared to 11.9% for unmodified devices.

PSC performance can also be enhanced by optimizing device architecture. For example, in 2020, Tseng et al. (2020) have successfully reduced the Cu_2O HTL and SiO_2 electron transport layer (ETL) thickness via a novel hetero contact synthesis engineering technique. As a result, significant improvement in the device's open-circuit voltage (V_{OC}) was observed. Moreover, to further reducing parasitic absorption within the device, they have employed a gallium-doped zinc oxide layer as the transparent conducting oxide (TCO) film. By utilizing these combined strategies, they achieved a remarkable PCE of 18.4%.

Overall, these studies demonstrate the immense potential of Cu_2O as a highly efficient HTM for PSCs. With continued research efforts, Cu_2O holds significant promise for achieving even higher efficiencies and contributing to the development of next-generation photovoltaic technologies.

Copper Iodide

Another most promising HTM for PSCs is CuI. It has several characteristic features such as cost-effectiveness, wide bandgap, brilliant chemical stability, and high hole mobility. In 2014, Christians et al. (2014) have utilized CuI in mesoporous PSCs, achieving a prominent power conversion efficiency (PCE) of 6.0%. However, the open-circuit voltage (V_{OC}) of CuI-based devices was slightly lower than that of spiro-OMeTAD-based devices. Moreover, CuI-based HTMs exhibited superior fill factor (FF) which was attributed to significant higher electrical conductivity. In another attempt, Li et al. (2017) in 2017, have constructed the mesoporous PSCs by employing the modified TiO_2 as the electron transport layer (ETL) and a CuI film as the HTL. A significant enhancement in electron conductivity and mobility was observed when the TiO_2 layer was being modified by Na species. Additionally, for CuI layer preparation, they have employed a spray-deposition method to achieve an impressive maximum PCE of 17.6%, along with excellent stability and reduced hysteresis. These findings reveal the potential of CuI as a feasible and promising HTM for high-performance PSCs, offering a convincing alternative to organic HTMs.

Copper Thiocyanate

Another promising candidate of HTMs is copper thiocyanate. Since CuSCN has transparent nature within the visible region of electromagnetic spectrum, this unique feature has made it a potential candidate for hole transport layers (HTLs) in perovskite solar cells (PSCs).

However, the utilization of diethyl sulfide (DES) as a solvent in CuSCN solution processing has been recognized as a detrimental factor for the perovskite layer, hampering its broader application in n–i–p-based PSCs. In order to overcome this limitation, researchers have explored alternate strategies, such as the p–i–n configuration to incorporate CuSCN into PSCs. For instance, Ye et al. (2015) have successfully fabricated CuSCN-based PSCs by employing electrodeposition method to create the CuSCN layer, and subsequently the CuSCN layer was deposited on methyl ammonium lead iodide ($CH_3NH_3PbI_3$) perovskite film via a rapid deposition–crystallization method. The resulting perovskite films demonstrated negligible interface contact resistance and a smooth surface morphology, contributing to the exceptional performance of the CuSCN-based PSCs. Notably, the developed device achieved an average power conversion efficiency (PCE) of 15.6%, with a peak PCE of 16.6%. The remarkable efficiency can be attributed to the high hole mobility exhibited by CuSCN, further confirming its potential as a promising HTL material for PSCs.

In 2019, Yang et al. (2019) concentrated on improving the open-circuit voltage (V_{OC}) of CuSCN-based PSCs, building on previous developments in this field. They sought to passivate flaws and enhance the interfacial contact between $MAPbI_3$ and CuSCN by carefully inserting functional molecules onto the $MAPbI_3$ perovskite layer's surface. Phenylene-1,4-diisothiocyanate (Ph-DITC) and 3-pyridyl isothiocyanate (Pr-ITC) worked especially well together to accomplish this. When compared to clean devices, the V_{OC} increased significantly after this chemical mixture was added, surpassing 40 mV. This improvement resulted in a peak power conversion efficiency (PCE) of 19.17% and an average PCE of 18.57%. Additionally, the interfacial alteration decreased the J-V hysteresis and increased the device's long-term stability in addition to improving performance, proving that this strategy is helpful in resolving important issues with CuSCN-based PSCs.

Transition Metal Chalcogenides (TMCs)

Two-dimensional transition metal chalcogenides (2D TMDs) have another prominent class of hole transport materials (HTMs) for perovskite solar cells (PSCs) owing to their exceptional properties. Among transition metal chalcogenides, molybdenum disulfide (MoS_2) has paved specific attention owing to its high hole mobility and compatibility with perovskite layers. Recently, MoS_2-based PSCs have witnessed outstanding advancements, achieving exceptional power conversion efficiencies (PCEs) up to 20%.

Early research showed that 2D TMDs might be used as HTMs in PSCs. As an example, Kim et al. (2016) effectively used graphene oxide (GO), WS_2, and atomically thin MoS_2 as HTMs in inverted PSCs. These materials were used to manufacture devices that performed similarly to conventional spiro-OMeTAD-based devices, with PCEs of 8.02%, 9.53%, and 9.62% for WS_2, MoS_2, and GO, respectively. The possibility of 2D TMDs as workable substitutes for organic HTMs was demonstrated by these findings.

The characteristics and functionality of 2D TMDs in PSCs were further investigated in later studies.

Electrodes

The electrode selection has a major effect on the overall performance of perovskite solar cells (PSCs) by affecting charge carrier transport. Power conversion efficiency (PCE), fill factor (FF), and series resistance (R_s), all these parameters are impacted by electrode resistivity (Nath et al. 2022). Taking into account either organic or inorganic materials, the cathode electrode's work function should correspond to the LUMO level of the electron transport layer (ETL), while the anode electrode's work function should match the HOMO level of the hole transport material (HTM). In PSCs, back electrodes made of Al, Au, and Ag have been employed extensively. Because Au's work function aligns with HTMs such as CuSCN, spiro-OMeTAD, and NiO_x, it performs well in n–i–p structures. Without encapsulation, copper electrodes provide long-term stability, corrosion resistance, and affordability (Xue 2022). Nickel electrodes can be utilized in place of gold since they have a well-matched work function. Separate HTLs are not necessary because of carbon's hydrophobic nature and hole self-extraction capabilities (Jong et al. 2022). However, carbon electrodes have problems including complicated fabrication procedures, high cost, and instability at high temperatures, while copper electrodes can be oxidized by direct contact with perovskite.

Conclusion

Perovskite solar cells (PSCs) are emerging as a leading alternative to traditional silicon-based solar cells, showcasing a remarkable increase in power conversion efficiency (PCE) from 3.8% in 2009 to over 25.7% in 2022. This chapter highlights the significance of various dopants used in the electron transport layer (ETL), hole transport layer (HTL), and the engineering of perovskite compositions. A critical concern is the impending scarcity of indium, prompting the search for alternative transparent conducting electrodes to replace indium tin oxide (ITO). The chapter discusses several HTMs, such as spiro-OMeTAD, PEDOT, and P3HT, noting that while inorganic HTMs generally offer better stability and cost-effectiveness, organic

HTMs tend to achieve higher efficiencies. However, the performance of organic HTMs often depends on the use of dopants to enhance their conductivity and hole mobility, which can compromise device stability. Despite advancements in dopant-free organic HTMs, spiro-OMeTAD remains a standout choice in high-performance PSCs, demonstrating efficiencies that rival those of silicon solar cells.

References

Abhishek R, Kumar M, Anshul A (2022) Recent progress in cesium-based lead-free halide double perovskite materials for photovoltaic applications. 219:20222270047. https://doi.org/10.1002/pssa.202270047

Agarwal S, Seetharaman M, Kumawat NK, Subbiah AS et al (2014) On the uniqueness of ideality factor and voltage exponent of perovskite-based solar cells. J Phys Chem Lett 5:4115

Alipour A, Alipour H (2024) Device modeling of high performance and eco-friendly $FAMASnI_3$ based perovskite solar cell. Sci Rep 14:15427. https://doi.org/10.1038/s41598-024-66485-0

Banerjee D, Chattopadhyay KK (2018) Perovskite photovoltaics. In: Thomas S, Thankappan A (eds) Academic press books. Elsevier, Cambridge, p 123

Chen W, Wu Y, Liu J, Qin C, Yang X, Islam A, Cheng Y-B, Han L (2015) Hybrid interfacial layer leads to solid performance improvement of inverted perovskite solar cells. Energy Environ Sci 8:629

Chengbo T, German B, Solis ZN, Kaikai L, Kebin L et al (2021) Efficient and stable inverted perovskite solar cells enabled by inhibition of self-aggregation of fullerene electron-transporting compounds. Sci Bull 66:339–346

Christians JA, Fung RC, Kamat PV (2014) An inorganic hole con ductor for organo-lead halide perovskite solar cells. Improved hole conductivity with copper iodide. J Am Chem Soc 136:758

Danladi E, Dogo DS, Michael SU, Uloko FO, Salawu AAO (2021) Recent advances in modeling of perovskite solar cells using SCAPS-1D: effect of absorber and ETM thickness. East Eur J Phys 4:5

Deepika, Singh A, Verma UK, Tonk A (2023) Device structures of perovskite solar cells: a critical review. Physica Status Solidi a 220(9):2200736

Edri E, Kirmayer S, Henning A, Mukhopadhyay S et al (2014) Why lead methylammonium tri-iodide perovskite-based solar cells require a mesoporous electron transporting scaffold (but not necessarily a hole conductor). Nano Lett 14:1000

Elangovan NK, Kannadasan R, Beenarani BB, Alsharif MH, Kim MK, Inamul ZH (2024) Recent developments in perovskite materials, fabrication techniques, band gap engineering, and the stability of perovskite solar cells. Energy Rep 11:1171–1190

Feldmann S, Neumann T, Ciesielski R, Friend RH, Hartschuh A, Deschler F (2021a) Tailored local bandgap modulation as a strategy to maximize luminescence yields in mixed-halide perovskites. Adv Opt Mater 9:2100635

Feldmann S, Neumann T, Ciesielski R, Friend RH, Hartschuh A, Deschler F (2021b) Tailored local bandgap modulation as a strategy to maximize luminescence yields in mixed-halide perovskites. Adv Opt Mater 9:2100635

Giordano F, Abate A, Correa Baena J et al (2016) Enhanced electronic properties in mesoporous TiO_2 via lithium doping for high-efficiency perovskite solar cells. Nat Commun 7:10379. https://doi.org/10.1038/ncomms10379

Gu WM, Jiang KJ, Li F, Yu GH, Xu Y, Fan XH, Gao CY, Yang LM, Song Y (2022) A multifunctional interlayer for highly stable and efficient perovskite solar cells based on pristine poly(3-hexylthiophene). Chem Eng J 444:136644

Gubbala S, Chakrapani V, Kumar V, Sunkara MK (2008) Band-edge engineered hybrid structures for dye-sensitized solar cells based on SnO_2 nanowires. Adv Funct Mater 18:2411

Hawash Z, Ono LK, Qi YB (2018) Recent advances in spiro-MeOTAD hole transport material and its applications in organic-inorganic halide perovskite solar cells. Adv Mater Interfaces 5(1):1700623

Heo JH, Im SH, Noh JH, Mandal TN et al (2013) Efficient inorganic–organic hybrid heterojunction solar cells containing perovskite compound and polymeric hole conductors. Nat Photonics 7:486–491

Huang C, Lin P, Nianqing F, Chang L, Baomin X, Kaiwen S, Danyang W, Xierong Z, Shanming K (2019a) Facile fabrication of highly efficient ETL-free perovskite solar cells with 20% efficiency by defect passivation and interface engineering. Chem Commun 55:2777–2780

Huang C, Lin P, Fu N, Liu C, Xu B, Sun K, Wang D, Zeng X, Ke S (2019b) Facile fabrication of highly efficient ETL-free perovskite solar cells with 20% efficiency by defect passivation and interface engineering. Chem Commun 55:2777

Jong Y-H, Kim P, Ryu K-I, Sonu K-S et al (2022) Effect of ionic liquid methylammonium acetate on the performance of hole-transport material-free mesoporous perovskite solar cells with carbon electrode. J Electron Mater 51:7085

Kegelmann L, Wolff CM, Awino C, Lang F et al (2017) It takes two to tango-double-layer selective contacts in perovskite solar cells for improved device performance and reduced hysteresis. ACS Appl Mater Interfaces 9:17245

Kim YG, Kwon KC, Le QV, Hong K et al (2016) Atomically thin two-dimensional materials as hole extraction layers in organolead halide perovskite photovoltaic cells. J Power Sources 319:1

Ko Y, Kim Y, Lee C, Kim Y, Jun Y (2018) Investigation of hole-transporting poly(triarylamine) on aggregation and charge transport for hysteresisless scalable planar per ovskite solar cells. ACS Appl Mater Interfaces 10:11633–11641

Kojima A, Teshima K, Shirai Y, Miyasaka T (2009) Organometal halide perovskitesas visible-light sensitizers for photovoltaic cells. J Am Chem Soc 131:6050

Kosar S, Winchester AJ, Doherty TAS, Macpherson S, Petoukhoff CE, Frohna K, Anaya M, Chan NS, Madéo J, Man MKL, Stranks SD, Dani KM (2021) Unraveling the varied nature and roles of defects in hybrid halide perovskites with time-resolved photoemission electron microscopy. Energy Environ Sci 14:6320

Kumar M, Raj A, Kumar A, Anshul A (2020) An optimized lead-free formamidinium Sn-based perovskite solar cell design for high power conversion efficiency by SCAPS simulation. Opt Mater 108:110213. https://doi.org/10.1016/j.optmat.2020.110213

Kumar M, Pundir SK, Singh DV, Kumar M (2024) Effect on green energy conversion and stability with 'Er' modification in multiferroic based perovskite solar cell devices. Mater Today Commun 38:107841. https://doi.org/10.1016/j.mtcomm.2023.107841

Lee J, Kim J (2022) Compact SnO_2/Mesoporous TiO_2 Bilayer electron transport layer for perovskite solar cells fabricated at low process temperature. Nanomaterials 12:718

Lee MM, Teuscher J, Miyasaka T, Murakami TN, Snaith HJ (2012) Efficient hybrid solar cells based on meso-superstructured organometal halide perovskites. Science 338:643

Lee J-W, Kim S-G, Yang J-M, Yang Y, Park N-G (2019) Verification and mitigation of ion migration in perovskite solar cells. APL Mater 7:041111

Li X, Yang JY, Jiang QH, Chu WJ et al (2017) Synergistic effect to high-performance perovskite solar cells with reduced hysteresis and improved stability by the introduction of Na-treated TiO_2 and spraying-deposited CuI as transport layers. ACS Appl Mater Interfaces 9:41354

Li S, Cao Y-L, Li W-H, Bo Z-S (2021) A brief review of hole transporting materials commonly used in perovskite solar cells. Rare Met 40:2712–2729

Lian J, Lu B, Niu F, Zeng P, Zhan X (2018) Electron-transport materials in perovskite solar cells. Small Methods 1800082

Liu D, Kelly T (2014) Perovskite solar cells with a planar heterojunction structure prepared using room-temperature solution processing techniques. Nature Photon 8:133–138. https://doi.org/10.1038/nphoton.2013.342

Liu M, Johnston M, Snaith H (2013) Efficient planar heterojunction perovskite solar cells by vapour deposition. Nature 501:395–398. https://doi.org/10.1038/nature12509

Liu Q, Qin MC, Ke WJ et al (2016) Enhanced stability of perovskite solar cells with low-temperature hydrothermally grown SnO_2 electron transport layers. Adv Funct Mater 26:6069
Liu DY, Li Y, Yuan JY, Hong QM et al (2017) Improved performance of inverted planar perovskite solar cells with F4-TCNQ doped PEDOT:PSS hole transport layers. J Mater Chem A 5(12):5701
Liu C, Zhou XY, Chen SM, Zhao XZ et al (2019) Hydrophobic Cu_2O quantum dots enabled by surfactant modification as top hole-transport materials for efficient perovskite solar cells. Adv Sci 6:1801169
Liu Z, Krückemeier L, Krogmeier B, Klingebiel B, Márquez JA, Levcenko S, Öz S, Mathur S, Rau U, Unold T, Kirchartz T (2018) Open-circuit voltages exceeding 1.26 V in planar methylam monium lead iodide perovskite solar cells. ACS Energy Lett 4(1):110
Liu C, Zhang LZ, Li Y, Zhou XY, She SY, Wang XZ et al (2020) Highly stable and efficient perovskite solar cells with 22.0% efficiency based on inorganic-organic dopant-free double hole transporting layers. Adv Funct Mater 30(28):1908462
Luo Q, Ma H, Hou QZ, Li YX et al (2018a) All-carbon-electrode-based endurable f lexible perovskite solar cells. Adv Funct Mater 28(11):1706777
Luo JS, Xia JX, Yang H, Chen LL et al (2018) Toward high-efficiency, hysteresis-less, stable perovskite solar cells: unusual doping of a hole trans porting material using a fluorine-containing hydrophobic Lewis acid. Energy Environ Sci 11(8):2035
Mahapatra A, Kumar S, Kumar P, Pradhan B (2022) Recent progress in perovskite solar cells: challenges from efficiency to stability. Mater Today Chem 23:100686. https://doi.org/10.1016/j.mtchem.2021.100686
Maher Al-Ibrahim H, Klaus Roth M, Schroedner AK et al (2005) The influence of the optoelectronic properties of poly(3-alkylthiophenes) on the device parameters in flexible polymer solar cells. Org Electron 6:65–77
Mali SS, Shim CS, Kim H, Hong CK (2016) Reduced graphene oxide (rGO) grafted zinc stannate (Zn_2SnO_4) nanofiber scaffolds for highly efficient mixed-halide perovskite solar cells. J Mater Chem A 4:12158
Mitchell RH, Welch MD, Chakhmouradian AR (2018) Nomenclature of the perovskite supergroup: a hierarchical system of classification based on crystal structure and composition. Mineral Mag 81:411
Nagaoka H, Ma F, DeQuilettes DW, Vorpahl SM, Glaz MS et al (2015) Zr incorporation into TiO_2 electrodes reduces hysteresis and improves performance in hybrid perovskite solar cells while increasing carrier lifetimes. J Phys Chem Lett 6:669
Nath B, Ramamurthy PC, Hegde G et al (2022) Role of electrodes on perovskite solar cells performance: a review. ISSS J Micro Smart Syst 11:61
Noh JH, Im SH, Heo JH, Mandal TN et al (2013) Chemical management for colorful, efficient, and stable inorganic-organic hybrid nanostructured solar cells. Nano Lett 13(17):64–1769
Oliver RDJ, Lin Y-H, Horn AJ, Xia CQ, Warby JH, Johnston MB, Ramadan AJ, Snaith HJ (2020) Thermally stable passivation toward high efficiency inverted perovskite solar cells. ACS Energy Lett 5:3336
Palilis LC, Vasilopoulou M, Verykios A, Soultati A, Polydorou E et al (2020) Inorganic and hybrid interfacial materials for organic and perovskite solar cells. Adv Energy Mater 10:2000910
Pathak SK, Abate A, Ruckdeschel P, Roose B et al (2014) Performance and stability enhancement of dye-sensitized and perovskite solar cells by al doping of TiO_2. Adv Funct Mater 24:6046
Paul A, Sarkar B, Paul S, Moyez SA, Jung HS, Roy S (2006) From 3.8% to over 23.8% power conversion efficiency: commercial perovskite solar cells, significant manufacturing techniques, and future prospects. Academic Press, pp 418–433. https://doi.org/10.1016/B978-0-12-819728-8.00044-9
Pham ND, Shang J, Yang Y, Hoang MT et al (2020) Alkaline-earth bis(trifluoromethanesulfonimide) additives for efficient and stable perovskite solar cells. Nano Energy 69:104412
Poelking C, Daoulas K, Troisi A, Andrienko D (2014) Morphology and charge transport in P3HT: a theorist's perspective. In: Ludwigs S (eds) P3HT revisited—from molecular scale to solar cell

devices. Advances in polymer science, vol 265. Morphology and charge transport in P3HT: a theorist's perspective. Springer, Berlin, Heidelberg. https://doi.org/10.1007/12_2014_277
Quan L, Fei F, Yibo X, Leilei G, Xin D et al (2023) SnO_2:TiO_2 hybrid nanocrystals as electron transport layer for high-efficiency and stable planar perovskite solar cells. Org Electron 120:106815
Quanrong D, Yiqi L, Lian C, Shenggao W, Geming W, Yonglong S, Guosheng S (2016) The effects of electron and hole transport layer with the electrode work function on perovskite solar cells. Mod Phys Lett B 30:1650341
Raj A, Kumar M, Anshul A (2021) Recent advancement in inorganic-organic electron transport layers in perovskite solar cell: current status and future outlook. Mater Today Chem 22:100595. https://doi.org/10.1016/j.mtchem.2021.100595
Raj A, Kumar M, Kumar A, Laref A, Singh K, Sharma S, Anshul A (2022) Effect of doping engineering in TiO_2 electron transport layer on photovoltaic performance of perovskite solar cells. Mater Lett 313:131692. https://doi.org/10.1016/j.matlet.2022.131692
Rakstys K, Igci C, Nazeeruddin MK (2019) Efficiency vs. stability: dopant-free hole transporting materials towards stabilized perovskite solar cells. Chem Sci 10(28):6748
Ruxin G, Zhao Y, Quanrong D, Yongshang Z, Zhiheng W, Yanyan D, Wei Z, Yonglong S, Guosheng S (2022) Energy level matching between transparent conducting electrodes and the electronic transport layer to enhance performance of all-inorganic $CsPbBr_3$ solar cells. Vacuum 200:111028
Saliba M, Orlandi S, Matsui T, Aghazada S, Cavazzini M, Correa-Baena JP et al (2016) A molecularly engineered hole-transporting material for efficient perovskite solar cells. Nat Energy 1:15017
Schloemer TH, Christians JA, Luther JM, Sellinger A (2019) Doping strategies for small molecule organic hole-transport materials: impacts on perovskite solar cell performance and stability. Chem Sci 10(7):1904
Seo JY, Kim HS, Akin S, Stojanovic M et al (2018) Novel p-dopant toward highly efficient and stable perovskite solar cells. Energy Environ Sci 11(10):2985
Singh SP, Nagarjuna P (2014) Organometal halide perovskites as useful materials in sensitized solar cells. Dalton Trans 43:5247
Snaith HJ, Abate A, Ball JM et al (2014) Anomalous hysteresis in perovskite solar cells. J Phys Chem Lett 5:1511
Song Z, Abate A, Watthage SC, Liyanage GK et al (2016) Perovskite solar cell stability in humid air: partially reversible phase transitions in the PbI_2-CH_3NH_3I-H_2O system. Adv Energy Mater 6:1600846
Tao C, Neutzner S, Neutzner L, Colella S et al (2015) 17.6% stabilized efficiency in low-temperature processed planar perovskite solar cells. Energy Environ Sci 8:2365
Tseng CC, Chen LC, Chang LB, Wu GM, Feng WS et al (2020) Cu_2O-HTM/SiO_2-ETM assisted for syn thesis engineering improving efficiency and stability with heterojunction planar perovskite thin-film solar cells. Sol Energy 204:270
Turren-Cruz S-H, Hagfeldt A, Saliba M (2018) Methylammonium-free, high-performance, and stable perovskite solar cells on a planar architecture. Science 362:449
Urieta-Mora J, Garcia-Benito I, Molina-Ontoria A, Martin N (2018) Hole transporting materials for perovskite solar cells: a chemical approach. Chem Soc Rev 47(23):8541
Wang W, Yan C, Lau T-K et al (2017) Fused hexacyclic nonfullerene acceptor with strong near-infrared absorption for semitransparent organic solar cells with 9.77% efficiency. Adv Mater 29:1701308
Wu T, Qin Z, Wang Y et al (2021) The main progress of perovskite solar cells in 2020–2021. Nano-Micro Lett 13:152. https://doi.org/10.1007/s40820-021-00672-w
Xue J (2022) Nat Energy 7:476
Yang IS, Lee S, Choi J, Jung MT et al (2019) Enhance ment of open circuit voltage for CuSCN-based perovskite solar cells by controlling the perovskite/CuSCN interface with functional molecules. J Mater Chem A 7:6028

Yang WS, Noh JH, Jeon et al (2015) High-performance photovoltaic perovskite layers fabricated through intramolecular exchange. Science 348:1234

Yang F, Hirotani D, Kapil G, Kamarudin MA, Ng CH, Zhang Y, Shen Q, Hayase S (2018) All-inorganic $CsPb_{1-x}Ge_xI_2Br$ perovskite with enhanced phase stability and photovoltaic performance. Angew Chem Int Ed 57:12745

Yang F, Zhang P, Kamarudin MA, Kapil G, Ma T, Hayase S (2018a) Addition effect of pyreneammonium iodide to methylammonium lead halide perovskite-2D/3D heterostructured perovskite with enhanced stability. Adv Funct Mater 28:1804856

Yang D, Zhang X, Hou Y, Wang K, Ye T, Yoon J, Wu C, Sanghadasa M, Liu S, Priya S (2021) 28.3%-efficiency perovskite/silicon tandem solar cell by optimal transparent electrode for high efficient semitransparent top cell. Nano Energy 84:105934

Ye SY, Sun WH, Li YL, Yan WB et al (2015) CuSCN-based inverted planar perovskite solar cell with an average PCE of 15.6%. Nano Lett 15:3723

Yu Z, Sun LC (2018) Inorganic hole-transporting materials for perovskite solar cells. Small Methods 2:1700280

Yu WL, Li F, Wang H, Alarousu E, Chen Y et al (2016) Ultrathin Cu_2O as an efficient inorganic hole transporting material for perovskite solar cells. Nanoscale 8:6173

Zhang ZL, Li JF, Wang XL et al (2017) Enhancement of perovskite solar cells efficiency using N-doped TiO_2 Nanorod arrays as electron transfer layer nanoscale. Research Letters 12:43. https://doi.org/10.1186/s11671-016-1811-0

Zhang Y, Elawad M, Yu Z, Jiang X, Lai J, Sun L (2016) Enhanced performance of perovskite solar cells with P3HT hole-transporting materials via molecular p-type doping. RSC Adv 6(110):108888–108895. https://doi.org/10.1039/C6RA2177C

Zhao D, Zhu Z, Kuo MY, Chueh CC, Jen AKY (2016) Hexaazatrinaphthylene derivatives: efficient electron-transporting materials with tunable energy levels for inverted perovskite solar cells. Angew Chem Int Ed 55:8999

Zheng XP, Hou Y, Bao CX, Yin J, Yuan FL et al (2020) Managing grains and interfaces via ligand anchoring enables 22.3%-efficiency inverted perovskite solar cells. Nat Energy 5(2):131

Zhou D, Zhou T, Tian Y, Zhu X, Tu Y (2018) Perovskite-based solar cells: materials, methods, and future perspectives. J Nanomater 2018:8148072

Zhou L, Guo X, Lin ZH, Ma J et al (2019) Interface engineering of low-temperature processed all-inorganic CsPbI2Br perovskite solar cells toward PCE exceeding 14%. Nano Energy 60:583

Zhu XD, Ma XJ, Wang YK, Li Y, Gao CH et al (2018) Hole-transporting materials incorporating car bazole into spiro-core for highly efficient perovskite solar cells. Adv Funct Mater 29:1807094

Chapter 4
Perovskite Solar Cells Without Hole and Electron Transport Layers

Parvesh K. Deendyal, Shweta Dhakla, Renu Singla, Ankur Taya, and Manish K. Kashyap

Abstract Perovskite solar cells (PSCs) have emerged as a promising photovoltaic technology due to their high photovoltaic performance and low production costs. Charge transport layers are often essential to generate high-performance PSCs, yet the high cost and risk of instability of these layers make it difficult to produce efficient and stable PSCs on a large scale economically. Therefore, among various structural configurations, PSCs without a hole transport layer (HTL) introduce a unique approach that simplifies the cell architecture, reduces manufacturing cost, and facilitates the fabrication process. A lot of research is currently going on to design and fabricate HTL-free PSCs and about 20% power conversion efficiency (PCE) has been achieved to date for the same. After 1000 h of continuous illumination, it successfully maintained 90% of the initial PCE, confirming its enormous practical application soon. Similarly, a PSC design without an electron transport layer (ETL) has also been proposed, and 23% PCE has been reported to date. This chapter delves into the basic principles, recent advancements, and cost analysis of HTL- and ETL-free PSCs. In addition to the difficulties associated with HTL- and ETL-free PSCs, the directions for further research and development in this field of PSCs have also been outlined.

Keywords Photovoltaics · Perovskite · Solar Cells · Power conversion efficiency · Stability

P. K. Deendyal · S. Dhakla · M. K. Kashyap (✉)
Renewable Energy Laboratory, School of Physical Sciences, Jawaharlal Nehru University, New Delhi, India
e-mail: mkkashyap@jnu.ac.in

P. K. Deendyal
Government Polytechnic for Women, Faridabad, Haryana, India

R. Singla
Department of Physics, Daulat Ram College, University of Delhi, Delhi, India

A. Taya
Department of Physics, Mukand Lal National College, Yamunanagar, Haryana, India

M. Kumar and A. Raj (eds.), *Perovskite Solar Cells*, Springer Tracts in Electrical and Electronics Engineering, https://doi.org/10.1007/978-981-96-9463-1_4

Introduction

Perovskite solar cells (PSCs), by achieving significantly high efficiencies in the past few years, have attained a milestone in solar energy research and emerged as one of the contemporary revolutionary technologies in the photovoltaic landscape. Despite their modest start in the dawn of the twenty-first century, their present potential has secured them as one of the leading contenders for future energy solutions. The term perovskite originated from calcium titanate ($CaTiO_3$), whose crystal structure was later found to be applied to a wide range of materials having the same ABX_3 configuration, where X is an anion, A and B are cations (Jung et al. 2015). These materials attracted a lot of attention for their applications in ferroelectricity, superconductivity, photovoltaics, and magnetism due to their extensive and flexible electronic, structural, magnetic, and optical properties (Suresh Kumar et al. 2021). In the 1990s, methylammonium lead halide ($CH_3NH_3PbX_3$), one of the materials from the organic–inorganic hybrid class of perovskites, was noticed for its application in photodetectors and light-emitting diodes (LEDs). But from 2009 onwards, after their use as light absorbers in a dye-sensitized solar cell (DSSC) by Kojima et al. (Kojima et al. 2009) with reported photon conversion efficiency (PCE) of ~ 3.8%, the door was opened for further investigations, and the surge of papers seen afterward in this domain was unprecedented. The watershed moment came with the introduction of solid-state PSCs by the Miyasaka group with elevated stability and efficiencies of ~ 10% (Song et al. 2017). This breakthrough brought a change in the direction of perovskite research altogether. The period from 2013 to 2017 witnessed remarkable improvements in these solar cell efficiencies due to refinement in various aspects of material stability, crystal growth, and fabrication techniques. The upcoming innovation in interfacial layer modifications, solvent engineering, and compositional tuning also contributed to boost PCEs from about 10% to over 20% in a span of a few years only. With PCE surpassing the 25% mark, the next push is naturally toward their commercialization with primary emphasis to enhance stability, upgrade scaling and streamline the production by addressing the hurdles on the way (Lan et al. 2024).

The primary underlying architecture of PSCs typically consists of five primary layers of electrodes, a perovskite absorber layer, and two (electron/hole) transport layers, which work together as a unit to generate electrical energy from sunlight (Roy et al. 2020). Transparent Conductive Electrode (TCE) constitutes the first layer, which is usually Indium Tin Oxide (ITO) or Fluorine-doped Tin Oxide (FTO). This is a transparent conductive window for solar cells which not only permits the photons to pass to underlying absorber layer but also acts as a front electrode for electrons/holes collection. The materials like Zinc oxide (ZnO), Titanium dioxide (TiO_2), and various other doped/undoped metal oxides are used as the electron transport layer (ETL), whose primary responsibility lies in the extraction and transportation of generated electrons from the absorber layer to the TCE and blocking holes to prevent charge carrier recombination. The epicenter of PSC lies in its absorber layer, which is made by choosing various compositions of perovskites, generally formamidinium lead iodide ($FAPbI_3$) and methylammonium lead iodide ($CH_3NH_3PbI_3$) that converts sunlight into excitons. Meanwhile, the hole transport layer (HTL), consisting of Poly(triarylamine) (PTAA) or Spiro-OMeTAD, blocks electrons and takes holes from the absorber layer and transports them back to TCE, resulting in reduced recombination losses. The architecture is completed with the use of Silver (Ag)/Gold (Au)/Carbon (C)/Copper (Cu) electrodes as back contact, which forms the rear contact for PSC. Furthermore, two specific architectures for PSCs are mesoporous and planar, with a difference that in mesoporous architecture, a scaffold of mesoporous TiO_2 is employed to infiltrate the perovskite, making room for better interaction between ETL and perovskite, and improving the charge separation and efficiency in PSCs.

As explained above, ETL is crucial in streamlining electron extraction and transportation in the PSC. However, a thrust to reduce the solar cell cost and ease the fabrication processes has led the researchers to consider eliminating ETL (Huang et al. 2019; Huang et al. 2019). Surprisingly, this approach produced a simplified and cost-effective alternative to PSCs, still maintaining acceptable PCE levels. It is a known fact that high-purity metal oxides used in the ETL require high-temperature annealing and costly deposition techniques, and bypassing ETL drastically brings down the PSC manufacturing cost. Second, the issues of energy loss arise due to imperfect charge extraction due to misalignment in energy levels at the perovskite/ETL interface, and UV instability in traditionally used ETLs like TiO_2 deters the functions of PSCs. From work perspectives, in such PSCs, TCE makes a direct interface with the absorber perovskite layer, which mandates careful engineering of the perovskite/electrode interface to suppress recombination losses while facilitating efficient electron collection from the absorber perovskite layer. Also, fine-tuning of perovskite/TCE is needed to permit the HTL and back electrodes to function as a selective contact, thereby allowing only one kind of charge carrier, i.e., holes, to pass through.

Another variation to the standard PSC structure involves simplifying device architecture and reducing its potential cost by excluding the HTL layer from PSCs. Although, by extracting holes from the absorber layer and transporting them back to the electrodes, the importance of HTL is undebatable yet it has been established that under certain conditions, PSCs can work efficiently even in the absence of HTL (Miyasaka et al. 2020). Like the case of ETL-free PSCs, eliminating HTL in PSC also quenches the production cost and simplifies the complexities involved in the fabrication processes. Organic materials like PTAA and Spiro-OMeTAD are not only costly but also susceptible to degradation on exposure to high temperature, moisture, and oxygen. They also involve additional processes involving doping and coating. Bypassing HTL can also help to reduce the recombination at several interfaces by simplifying the device architecture and reducing the device stack. In this case, TCE is in direct contact with the absorber perovskite layer. It depends on the intrinsic properties of the perovskite employed for hole transport and acts as a hole collector layer. It is imperative to select a suitable electrode material to ensure the suitability of band alignment so that holes can be collected efficiently while preventing them from recombining with electrons (Huang et al. 2020).

Challenges of ETL-Free and HTL-Free PSCs

Notably, while ETL-free and HTL-free PSCs simplify the structure, the overall PCE gets compromised due to inefficient charge collection. Also, one of the biggest hurdles to surmount in such cases pertains to minimizing the charge recombination at the perovskite/electrode interface, as there is no ETL or HTL in these PSCs to aid that (Liao et al. 2020). Continuous efforts to ensure effective charge collection, modification of interface with self-assembled monolayers (SAMs), and use of fullerene (C_{60}) derivatives are some of the successful surface passivation techniques that can be used to prevent charge recombination at perovskite/electrode interface as far as ETL-free PSCs are concerned. Various strategies like selecting proper back electrodes, C-based materials, and metals such as Au with properly aligned energy levels are being used to provide a way for efficient collection of holes in HTL-free PSCs. Besides that, using a thin interfacial layer that can selectively block electrons or holes is also a good option. Typically, CuSCN is used between perovskite and electrode in HTL-free PSCs to help hole collection without much adding to architecture complexity (Woong Jung et al. 2015). These approaches, when employed in ETL/HTL-free PSCs, add architectural simplicity to the device and open up avenues to scale up the industrial production processes of such PSCs.

Thus, undoubtedly employing ETL- or HTL-free PSCs has the significant potential for improving the stability and reducing the cost of PSCs. While the challenges pertain to recombination losses and efficiency reduction persist, the ongoing research efforts for the development of efficient and stable electrode materials alongside consistent probing of interface engineering are addressing these issues. These simplified structures and efficient structures can play an important role, especially for low-cost and large-scale photovoltaic applications, in the future commercialization of PSCs. In this chapter, we aim to examine various key aspects of charge transport layer-free perovskite solar cells that include ETL-free, HTL-free, and both ETL- and HTL-free PSCs, besides evaluating and highlighting their current status and exploring their potential in shaping the future of commercialization of perovskite photovoltaic technology.

Charge Transport Layer-Free PSCs

PSCs without charge transport layers (CTLs) are a kind of PSCs that do not have the standard CTLs commonly employed in traditional architectures. In solar cells, these layers—also referred to ETLs/HTLs—are frequently employed to enable effective charge carrier (electrons/holes) extraction and transport to their respective electrodes. In high-performance PSC devices, an active layer of perovskite is sandwiched between CTLs, viz. ETL and HTL (Jiang et al. 2019; Ravishankar et al. 2018). Despite their numerous benefits, these CTLs can pose significant hurdles. It is noteworthy that the introduction of additional layers raises the manufacturing cost and complicates the fabrication processes of the solar cells (Wu et al. 2018). Furthermore, some metal oxide ETL and HTL, viz. TiO_2 and NiO_x, necessitate high-temperature sintering, which is incompatible with flexible substrates and requires more energy (Chowdhury et al. 2023). Additionally, these layers may also have hydrophilic or photocatalytic capabilities, which can accelerate the degradation of PSC devices (Chowdhury et al. 2023). To address the aforementioned challenges, many researchers have done a lot of efforts in developing innovative charge transport materials (Shin et al. 1979). In contrast, CTL-free design paves a new path for simplified device architecture along with the lower production cost and better stability by simply relying on the inherent characteristic of the perovskite layer to allow charge transfer and extraction. The perovskite layer, which is principally responsible for light harvesting, is commonly regarded as one of the most important components, influencing the solar cell performance. Despite being an excellent light absorber, perovskite has been discovered to efficiently allow the flow of charge carriers in solar cells, making it attractive for high-performance optoelectronic devices (Liao et al. 2020). Despite ongoing research efforts, the recently reported CTL-free PSCs remain significantly behind standard PSC devices (Liao et al. 2020). There are some materials or techniques that can be used to enhance the performance of CTL-free PSCs, such as interface engineering with self-assembled monolayers (SAMs) or using passivation layers to reduce surface defects, recombination, etc., which are discussed in the subsequent sections.

Device Architecture

Conventional PSCs generally consist of five main layers, viz. perovskite absorber layer for photon absorption, CTLs (ETL/HTL) for charge carrier transportation as well as transparent conductive oxide (TCO), and counter electrode (CE) to collect and extract charges that are assembled using layer-by-layer stacking architecture. Usually, the absorber layer, i.e., perovskite is placed in between the two CTLs. CTL-free PSCs have a simpler device architecture that eliminates the requirement for typical charge transfer layers, possibly leading to cheaper production costs and increased stability (Chowdhury et al. 2023). As shown schematically in Fig. 4.1, CTL-free PSCs are categorized into three types: (i) ETL-free, (ii) HTL-free, and (iii) both ETL- and HTL-free.

For instance, in HTL-free PSCs, the typical configuration begins with a substrate, often glass or a flexible material, coated with a TCO layer (ITO/FTO). This layer serves to facilitate both electron transport and light transmission. An ETL is sometimes applied directly above this, followed by the perovskite absorber layer, which is crucial for absorbing light and generating electron–hole pairs. Instead of employing separate HTL, some designs opt for HTL-free approaches, relying on the inherent properties of the perovskite material or a simple interface layer to efficiently extract and transport charge carriers. The structure is finalized with a metal electrode, such as gold or silver, which collects the generated holes. This streamlined design not

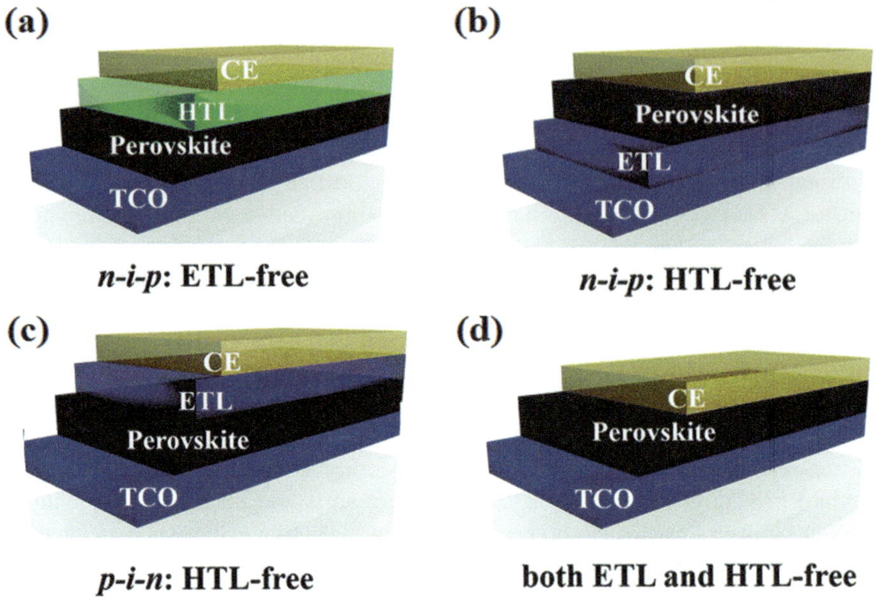

Fig. 4.1 Device architecture of CTL-free PSCs: **a** n–i–p ETL-free, **b** n–i–p HTL-free, **c** p–i–n HTL-free, and **d** both ETL- and HTL-free (adapted from Ref. Liao et al. (2020) with permission)

only simplifies the manufacturing process but also aims to enhance the stability and longevity of PSCs while maintaining their competitive PCEs.

Working Principle of CTL-Free PSCs

The regular device architecture of PSCs contains layers of ETL, Perovskite, and HTL on a transparent conductive oxide (TCO)-coated glass substrate. In normal PSCs, light is absorbed by the perovskite layer, generating electron–hole pairs. These electrons and holes are extracted by ETLs and HTLs respectively, which are further transferred to their respective electrodes, resulting in the photocurrent. The working mechanism of these PSCs primarily depends on the energy level alignment of all these layers. The advancement of CTL-free PSCs presents a novel way to reduce production costs while maintaining high efficiency. CTL-free PSCs are categorized into three types: ETL-free, HTL-free, and those without both ETL and HTL. The operational mechanisms of ETL-free, HTL-free, and both ETL- and HTL-free PSCs depend on the device structure, interface properties, and heterojunctions (Liao et al. 2020).

As elucidated in the previous section, we studied the various architectures of CTL-free PSCs. The sole difference between the CTL-free and normal PSC operation is that the charge carriers are directly transferred to the electrode. The vital factor for transfer is the energy level alignment and the meticulous properties of perovskite itself, e.g., long diffusion length, ambipolar charge transport, and fast exciton dissociation (Liao et al. 2020; Deendyal et al. 2024). In ETL-free PSCs with an n–i–p configuration, perovskite is directly deposited on the TCO-coated glass substrate, following the deposition of HTL and then finally electrode. The energy level alignment at the TCO/Perovskite layer interface is adjusted to allow for the efficient extraction of electrons while inhibiting hole movement by the built-in potential at the interface (Huang et al. 2020; Liao et al. 2020; Zhou et al. 2020). Further, these holes can be extracted by HTL. The whole process is systematically shown in Fig. 4.2 (Liao et al. 2020). Likewise, for HTL-free PSCs with p–i–n structure, holes are transferred to the TCO electrode and electrons are extracted by ETLs. In addition, owing to the favorable band bending at the interfaces, such a type of passage and inhibition of charge carrier at the interface can occur efficiently. In HTL-free PSCs with n–i–p configuration, the ETL/perovskite heterojunction generates a depletion region at the interface, providing the driving force for charge separation and transfer. The built-in potential from this region enhances charge separation and extraction while reducing charge recombination. For both ETL- and HTL-free PSCs, the electric field gets generated at the interface between perovskite and electrode of different work functions, which assists in the transportation of electrons and holes toward FTO/ITO (front electrode) and metal (back electrode), respectively (Bhattarai et al. 2021).

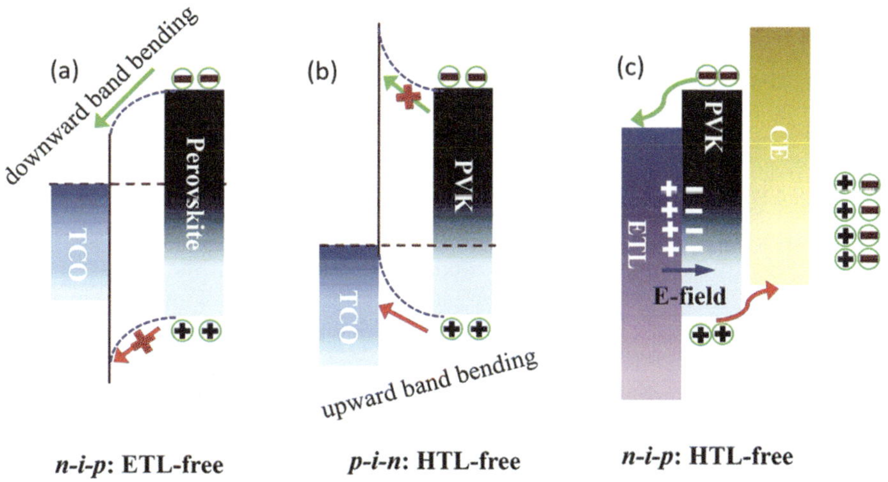

Fig. 4.2 The working mechanism of **a** ETL-free, and **b** HTL-Free devices showing the downward and upward band bending, respectively, at the TCO/Perovskite interface for extracting one type of carrier and inhibiting the other; **c** the establishment of the built-in electric field at ETL/Perovskite heterojunction for efficient charge division. (Reprinted with permission from reference (Liao et al. 2020))

Current Status of ETL-Free PSCs

The electron transport layer (ETL), as the name indicates, is particularly effective for extracting electrons and blocking holes from the active layer of the PSCs. N-type semiconductors with high electron mobility and the favorable band alignment with the perovskite form the basis for ETL selection in PSCs. The most prevalent ETLs used in PSCs are n-type metal oxides (TiO_2, SnO_2, ZnO) and fullerene derivatives (PCBM, C_{60}). Metal oxide film deposition may need high-temperature sintering or a sophisticated synthesis procedure, which, in turn, unfavorably increases the fabrication cost and the energy consumption (Chen et al. 2019; Huang et al. 2016; Zhu et al. 2016). Furthermore, TiO_2 and ZnO can accelerate perovskite degradation, whereas fullerene and its derivatives are susceptible to the ambient environment (Yu et al. 2017), resulting in an adverse effect on PSC stability. If ETLs are not required, PSCs can be manufactured using simpler procedure with increased stability. Thus, the simplified structural ETL-free PSCs with equivalent performance to their counterparts are considered to be potential candidates for commercialization in a photovoltaic market.

In comparison to ETL-containing PSCs, ETL-free PSCs showed much lower PCEs in the early stage. Numerous attempts have been made in recent years to improve the performance of ETL-free PSCs, but their stability has yet to be thoroughly examined. Table 4.1 summarizes the photovoltaic performance of ETL-free PSCs reported over the years. In 2014, Liu et al. (2014) proposed the first ETL-free PSCs employing Spiro-OMeTAD as HTL and $CH_3NH_3PbI_3$ ($MAPbI_3$) as absorber material, as shown in Fig. 4.3a. The device achieved a PCE of 13.5%, which is comparable to that of a ZnO-based PSC, indicating decisively that the ETL is not required to achieve good device efficiencies. Although the achieved PCE was low, this was a delightful encounter with simplified architecture and potentially low-cost ETL-free PSCs.

After following their pioneering effort, numerous studies on ETL-free PSCs were reported thereafter. Several milestones are noteworthy in the ETL-free PSC roadmap: In 2015, Xu et al. (2015) fabricated an ETL-free PSC with a device architecture ITO/$FAPbI_{3-x}Cl_x$/Spiro-OMeTAD/Au, which yielded an enhanced PCE of 15.49%. Moreover, a PSC with simplified device architecture (FTO/$MAPbI_3$/Spiro-OMeTAD/Au) was successfully fabricated by eliminating the problematic TiO_2 layer in 2017 (Yu et al. 2017). The slow charge extraction at the FTO/perovskite interface was addressed by electrochemically etching the FTO surface to create a porous structure. This surface engineering allowed for a significant increase in photocurrent density while mitigating the hysteretic behavior; thus, a stunning 19.22% efficiency with low hysteresis was achieved. As compared to ETL containing PSCs, the inferior performance of ETL-free PSCs may be due to the severe interfacial charge recombination at TCO/perovskite interface. Huang and his co-workers (Wu et al. 2020) in 2020 worked in this direction: fabricated PSC with a non-annealed, insulating metal oxyhydroxide thin interlayer between TCO and perovskite that facilitated electron tunneling, and suppressed the interfacial charge recombination. This

Table 4.1 Summarized photovoltaic performance of ETL-free PSCs

Year	Device architecture	V_{OC} (V)	J_{SC} (mA.cm^{-2})	FF (%)	PCE (%)	Ref.
2014	ITO/$MAPbI_3$/ Spiro-OMeTAD/Ag	1.01	21.20	63.00	13.50	Liu et al. (2014)
2014	ITO/ITO:Cs_2CO_3/$MAPbI_3$/ Spiro-OMeTAD/Au	1.07	19.90	71.00	15.10	Hu et al. (2014)
2015	FTO/$MAPbI_{3-x}Cl_x$/ Spiro-OMeTAD/Au	1.06	19.76	67.00	14.10	Ke et al. (2015)
2015	ITO/$FAPbI_{3-x}Cl_x$/ Spiro-OMeTAD/Au	1.00	20.29	76.00	15.49	Xu et al. (2015)
2016	FTO/$MAPbI_{3-x}Cl_x$:C_{70}/ Spiro-OMeTAD /Au	1.06	17.40	74.10	13.60	Pascual et al. (2016)
2016	ITO/PEDOT:PSS/$MAPbI_3$/ Ti/Au	0.89	24.38	59.49	12.91	Shi et al. (2016)
2017	FTO/$MAPbI_3$/ Spiro-OMeTAD/Au	1.13	21.45	75.10	18.20	Yu et al. (2017)
2017	FTO/$MAPbI_3$/ Spiro-OMeTAD/Au	1.12	22.81	75.00	19.22	Yu et al. (2017)
2018	ITO/PTAA/$MAPbI_3$/Ti/Au	1.03	22.50	78.00	18.10	Chen et al. (2018)
2018	ITO/$MAPbI_3$/ Spiro-OMeTAD/Au	1.06	23.61	77.79	19.52	Han et al. (2018)
2019	FTO/$MAPbI_3$/ Spiro-OMeTAD/Au	1.08	22.20	77.00	18.50	Liao et al. (2019)
2019	ITO/TAPC/$MAPbI_3$/ Spiro-OMeTAD/Au	1.12	22.07	78.00	19.40	Kong et al. (2019)
2019	FTO/TMAH/$MAPbI_3$/ Spiro-OMeTAD/Au	1.17	23.22	73.00	20.10	Huang et al. (2019)
2020	ITO/MAAc/$MAPbI_3$/ Spiro-OMeTAD/Au	1.15	22.40	82.00	21.00	Li et al. (2020)
2020	FTO/a-NbOH/ $Cs_{0.07}FA_{0.70}MA_{0.23}PbI_3$/ Spiro-OMeTAD/Au	1.16	23.00	79.00	21.10	Wu et al. (2020)
2021	FTO/PPPDE/CsMAFAPb(I, Br)$_3$/Spiro-OMeTAD/Au	1.13	22.80	77.00	19.70	Huang et al. (2021)
2021	ITO/5-AVA/ $FA_{0.9}Cs_{0.1}PbI_{3-x}Cl_x$/ Spiro-OMeTAD/Ag	1.08	23.91	75.00	19.37	Wang et al. (2021)

(continued)

Table 4.1 (continued)

Year	Device architecture	V_{OC} (V)	J_{SC} (mA.cm^{-2})	FF (%)	PCE (%)	Ref.
2022	FTO/NaF/ $(FAPbI_3)_{0.9}(MAPbBr_3)_{0.1}$/ Spiro-OMeTAD/Au	1.13	23.30	76.00	20.00	Sadegh et al. (2022)
2022	ITO/GHJP/Spiro-OMeTAD/ Au	1.09	24.70	76.00	20.55	Gao et al. (2022)
2023	FTO/BDADI/$FAPbI_3$/ Spiro-OMeTAD/Au	1.14	24.68	78.13	22.02	Hui et al. (2023)
2024	FTO/KF/ $(FAPbI_3)_{0.97}(MAPbBr_3)_{0.03}$/ Spiro-OMeTAD/Au	1.16	24.58	80.90	23.04	Deng, et al. (2024)

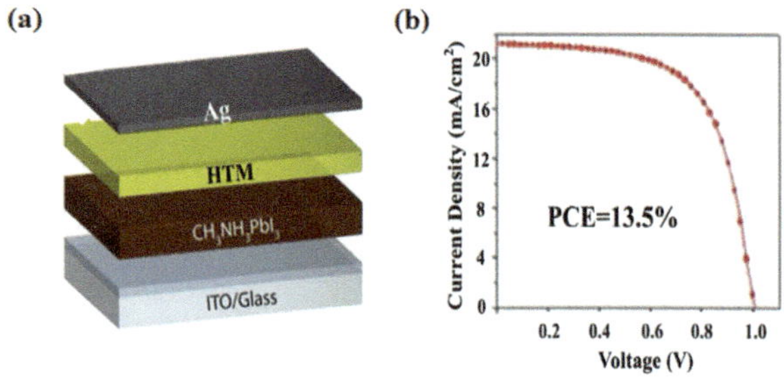

Fig. 4.3 **a** Device architecture, **b** J-V curve of ETL-free PSC fabricated by Liu et al., reproduced with permission from ref. (Liu et al. 2014)

device architecture showed a record efficiency of 21.1% for n–i–p configuration and outperformed the metal oxide ETL-containing PSC counterparts. Significantly, Hui et al. (2023) in 2023 fabricated an ETL-free PSC by in situ formation of a low-dimensional perovskite layer between FTO/perovskite interface, resulting in reduced defect density in perovskite, and improved energy level alignment between perovskite and anode, which facilitated charge carrier transport and collection, and suppressed the carrier recombination. Consequently, the resulting ETL-free PSC with $FAPbI_3$ perovskite material exhibited an impressive PCE of 22.02% under ambient environmental conditions. To the best of our knowledge, the highest PCE in 2024 achieved for ETL-free PSC is 23.04% to date (Deng et al. 2024) as highlighted in Table 4.1, which still lags behind the ETL containing PSC counterparts (26.7% at present) (Best Research-Cell Efficiency Chart 2024).

ETL-free PSCs are expected to have a promising future, and significant research has been made to advance large-scale production, flexible devices, and long-term stability. The energy level mismatch between perovskite and TCO as well as a weak built-in electric field are the primary barriers, resulting in poor charge extraction and significant charge recombination at the TCO/perovskite interface. These challenges can be solved by extending the carrier lifetime in perovskite, which allows for significant electron extraction and by changing interfacial band alignment with an additional layer.

Current Status of HTL-Free PSCs

As compared to ETL, hole transport layer (HTL) is a p-type semiconductor that is particularly useful for extracting holes and blocking electrons from the active layer of the PSCs. High-performance PSCs commonly employ HTLs such as PTAA, NiO_x, and Spiro-OMeTAD. However, some materials are either costly or have complex synthesis procedures that make them unfavorable for low-cost production. As already elucidated previously, despite the superb light absorber, perovskite can efficiently facilitate the transfer of charge carriers owing to its ambipolar nature. As a result, it is feasible to make perovskite photovoltaics without using any HTL. In the history of PSC evolution, the HTL-free design represents a distinct category, demonstrating remarkable stability. Lowering production costs and simplifying the process will increase the device's competitiveness, which is critical for commercialization. Aside from these advantages, the primary disadvantages of HTL-free devices include the usage of metallic electrodes directly on top of the perovskite layer, which decreases device performance. For instance, silver electrodes can induce an energy mismatch with the perovskite layer, as well as the formation of silver halides, which damages the perovskite layer (Lamberti et al. 2021). If Au is employed, it can migrate into the perovskite layer to generate gold halide complexes (Lamberti et al. 2021). Other studies have shown the devices using carbon as the top electrodes having similar work function as Au and are chemically inert to the perovskite layer, protecting it from humidity and so enhancing its longevity (Lamberti et al. 2021).

For the first time in 2012, Etgar et al. (2012) fabricated the HTL-free perovskite and obtained a PCE of 5.5% under 1000 W/m^2 and 7.3% under a lower light intensity of 100 W/m^2. They employed a mesoscopic $MAPbI_3/TiO_2$ heterojunction structure followed by Au deposition as a metal electrode. The employment of Au as an electrode is very expensive and increases the overall device cost. The work function of carbon (− 5.0 eV) closely matches with that of Au (− 5.1 eV) making it an ideal substitute for Au. Therefore, Ku et al. (2013) used the low-cost carbon electrode and surprisingly obtained a PCE of 6.64% for fully printable PSC which is even higher than that of the Au version. In addition, they deposited a ZrO_2 layer as a spacer layer between the TiO_2 and perovskite interface which acted as an insulated layer to prevent short circuit. Even while the performance was not as great as that of devices made using HTL and Au electrode, this design still illustrates the possibility of creating the most affordable photovoltaic devices using low-cost carbon electrode and a printable manufacturing approach (Fig. 4.4).

Furthermore, researchers showed interest in nano-structured solar cells owing to their potential for efficiency enhancement soon as various nanorods, nanosheets (NSs), and nanowires were already used in DSSCs (Laban et al. 2013; Aharon et al. 2014; Shi et al. 2014). Studies have demonstrated that the (001) facets of anatase

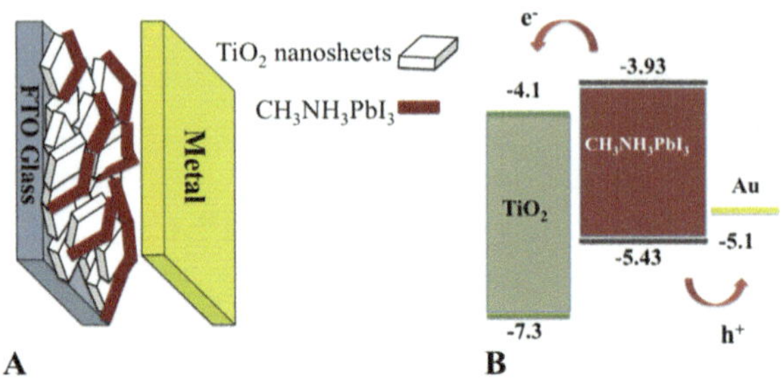

Fig. 4.4 **a** Device structure; **b** Energy level diagram of the HTL-free PSC. (Adapted from Etgar et al. (2012) with permission)

TiO_2 are more reactive than the (101) facets (Ogomi et al. 2014), and it has been suggested that the higher ionic charge on the exposed (001) facets enhances the attachment of light absorbers to the TiO_2 surface, which in turn facilitates the injection of electrons into the conduction band of the oxide (Aharon et al. 2014; Shi et al. 2014). The same group presented TiO_2 nanosheets (NSs) with a high proportion of exposed (001) facets, used as electron collectors in hole-conductor-free mesoscopic heterojunction solar cells with carbon counter electrodes (CEs). Using a two-step sequential method to infiltrate $CH_3NH_3PbI_3$ perovskite, the solar cells with TiO_2 NSs as electron collectors achieved an impressive PCE of up to 10.64% under AM1.5 simulated sunlight of 100 mW cm^{-2} (Rong et al. 2014).

Generally, in normal PSC, we deposit the ETL, absorber layer, and HTL in sequence. In 2016, Senyun et al. (Ye et al. 2016) deposited both perovskite absorber layer and HTL in a single step. They synthesized a mixed precursor solution of $MAPbI_{3-x}Cl_x$ and CuSCN which aided in passivation of trap states along with facilitated charge extraction. Eventually, the resulting device showed a PCE of 18.1% with minimum hysteresis (Ye et al. 2016). Further in the next year, they used the Cu(thiourea)I [Cu(Tu)I] which exhibited better trap passivation and obtained a PCE of 19.9% for inverted PSCs (Ye et al. 2017). Eliminating HTL in the PSCs, on the other hand, typically results in a significant efficiency loss and poor carrier transport owing to the energy-level mismatching at the ITO/perovskite interface. In this direction, Cheng and his workers (Kong et al. 2019) in 2019 introduced an organic monomolecular layer (ML) which increases ITO's effective work function and optimizes energy-level alignment at the interface. The ML-modified HTL-free PSC achieved an overall PCE of 19.4% which was significantly higher than HTL-free PSCs on bare ITO (10.26%). In 2022, Zhou et al. (2022) fabricated HTL-free PSCs using a synergistic doping approach. This method reduces defects and interfacial energy barriers. The HTL-free PSCs achieved a remarkable PCE of 22% with negligible hysteresis.

With the continuous efforts of researchers over the past few years, the PCE of the HTL-free PSC has surpassed 22%, approaching close to HTL counterparts. The efficiency of HTL-free designs has improved over time, with researchers exploring novel electrode materials or modifications to enhance charge extraction and stability, offering HTL-free PSCs a promising avenue for further developments in the field of photovoltaics (Table 4.2).

Table 4.2 Summarized year-wise photovoltaic performance of HTL-free PSC

Year	Device configuration	V_{OC} (V)	J_{SC} ($mA.cm^{-2}$)	FF (%)	PCE (%)	Ref.
2012	FTO/TiO_2 nanosheet/$MAPbI_3$/Au	0.63	16.10	57.00	5.50	Etgar et al. (2012)
2013	FTO/TiO_2/$MAPbI_3$/Au	0.71	18.80	60.00	8.04	Laban et al. (2013)
2013	FTO/TiO_2/ZrO_2/$MAPbI_3$/C	0.87	12.40	61.00	6.64	Ku et al. (2013)
2014	ITO/ZnO/$MAPbI_3$/Ag	1.01	17.20	66.00	11.60	Liu et al. (2014)
2014	FTO/m-TiO_2/m-ZrO_2/$MAPbI_3$/C	0.87	20.10	61.00	10.64	Rong et al. (2014)
2014	FTO/m-TiO_2/m-ZrO_2/$(5\text{-AVA})_xMA_{1-x}PbI_3$/C	0.85	22.80	66.00	12.80	Mei et al. (1979)
2015	FTO/TiO_2/CsBr/$CsPbBr_3$/C	1.26	11.80	72.00	10.70	Liu et al. (2015)
2015	FTO/c-ZnO/$MAPbI_3$/C	0.81	19.98	54.00	8.73	Zhou et al. (2015)
2016	ITO/c-TiO_2/$MAPbI_{3-x}Cl_x$ + (CuSCN)/Au	1.04	22.80	76.00	18.10	Ye et al. (2016)
2016	FTO/c-TiO_2/m-TiO_2/m-Al_2O_3/$MAPbI_3$/C/SWCNTs	1.01	21.26	69.00	14.70	Li et al. (2016)
2016	FTO/m-TiO_2/m-ZrO_2/$MAPbI_{3-x}(BF_4)_x$/C	0.95	18.15	76.00	13.20	Chen et al. (2015)
2017	ITO/$MAPbI_{3-x}Cl_x$ + Cu(Tu)I/C_{60}/BCP/Ag	1.12	22.30	80.00	19.90	Ye et al. (2017)
2017	FTO/m-TiO_2/m-Al_2O_3/$MAPbI_3(SrCl_2)_{0.1}$/C	1.05	20.18	75.00	15.90	Zhang et al. (2017)
2018	FTO/c-TiO_2/m-TiO_2/$MAPbI_3$–SWCNT/CWCNT-C	0.97	22.36	72.00	15.70	Wang et al. (2019)
2018	PEDOT (2 nm) modified-ITO/$MAPbI_{3-x}Cl_x$/C_{60}/BCP/Ag	1.05	22.70	80.00	19.30	Liu et al. (2018)
2018	ITO/$MAPbI_3$:F_4TCNQ/C_{60}/BCP/Cu	1.10	22.70	81.00	20.20	Wu et al. (2018)
2019	FTO/Nb_2O_5/$Cs_{0.99}Rb_{0.01}PbI_2Br$/C	1.24	14.02	69.00	12.00	Guo et al. (2019)
2019	FTO/c-TiO_2/m-TiO_2/$MAPbI_3$/$MAPbI_xBr_{3-x}$/C	1.07	20.80	73.00	16.20	Liu et al. (2019)
2019	ML-ITO/$MAPbI_{3-x}Cl_x$/C_{60}/BCP/Ag	1.12	22.07	79.00	19.42	Kong et al. (2019)
2020	FTO/c-TiO_2/m-TiO_2/$Cs_{0.05}(MA_{0.15}FA_{0.85})_{0.95}Pb(I_{0.86}Br_{0.14})_3$/C	0.99	23.34	71.00	16.41	Meng et al. (2020)
2021	FTO/SnO_2/$MAPbI_3$ (PFTeDA)/C	1.10	23.85	71.80	18.90	Ye et al. (2021)
2021	FTO/TiO_2/$FAPbI_3$ (F5PEAI)/C	1.02	22.25	76.70	17.50	Chen et al. (2021)

(continued)

Table 4.2 (continued)

Year	Device configuration	V_{OC} (V)	J_{SC} (mA.cm^{-2})	FF (%)	PCE (%)	Ref.
2022	ITO/$Cs_{0.05}(FA_{0.92}MA_{0.08})_{0.95}Pb(I_{0.92}Br_{0.08})_3$:Cu(Tu)Cl + TSC/Choline chlorde/C_{60}/BCP/Cu	1.16	23.21	81.70	22.00	Zhou et al. (2022)
2023	FTO/c-TiO_2/m-TiO_2/$Cs_{0.4}FA_{0.6}PbI_3$/Carbon	1.05	23.50	79.10	19.52	Li et al. (2024)
2024	FTO/SnO_2/$MAPbI_3$/Mn_3O_4-C	1.06	23.89	74.80	19.03	Gou et al. (2024)
2024	FTO/SnO_2/$FAPbI_3$/C	1.09	24.66	72.10	19.41	Ma et al. (2024)

Current Status of ETL- and HTL-Free PSCs

Although ETL- and HTL-free PSCs have been extensively investigated, and numerous solutions have been proposed to address the issues associated with these simplest architectures, the PSCs both ETL- and HTL-free have also been reported by some researchers. The reason might be traced back to issues encountered by the simplest device configuration such as charge extraction and perovskite layer deposition. To the best of our knowledge, there are some reported research works on both ETL- and HTL-free PSCs which are summarized in Table 4.3. In 2012, Chen et al. (2012) fabricated a Schottky solar cell based on perovskite semiconductor ($CsSnI_3$) with a device architecture ITO/$CsSnI_3$/Au/Ti. However, the resultant device exhibited a PCE of 0.88% only. Similarly, Zhang et al. (2015) in 2015, fabricated a PSC device with architecture FTO/$MAPbI_{3-x}Cl_x$/Au and investigated the photovoltaic performance of the resulting device. The device with $MAPbI_{3-x}Cl_x$ as perovskite absorber layer yielded the following photovoltaic metrics: V_{oc} = 0.77 V, J_{sc} = 7.11 mA.cm^{-2}, FF = 0.53, and PCE = 3.2%, which showed an improvement in PSC device. Some milestones are noteworthy in this roadmap: using $CsPbBr_3$ as the light absorber, Tang et al. (Duan et al. 2018) fabricated the simplest PSC with architecture, FTO/$CsPbBr_3$/carbon. Due to high energy barriers between valance band of $CsPbBr_3$ and carbon work function as well as between conduction band of $CsPbBr_3$ and FTO, the charge recombination was found to be quite significant, resulting in a low PCE of 2.35%. Further, they enhanced the charge extraction and boosted the PCE to 4.1% by modifying the $CsPbBr_3$/carbon and FTO/$CsPbBr_3$ interfaces using $CsPbBrI_2$ quantum dots and graphene, respectively. Furthermore, Li and his coworkers (Cui et al. 2019) fabricated a CTL-free device by taking into account the p-type perovskite/n-type perovskite homojunction. The fabricated device with architecture FTO/n-type $MAPbI_3$/p-type $MAPbI_3$/Au demonstrated an improved PCE of 8.08%. By including M-FTO (modified FTO with distinct porous structure) and $MAPbI_3$-incorporated NiO nanoparticles in a device architecture, Yin et al. (2019) improved the simplest PSC, showing PCE of 14.65% which was much higher than the PCE of the control device (PCE = 5.71%) that had pure $MAPbI_3$ and pristine FTO (P-FTO). In addition, a PCE of 11.15% (Yin et al. 2019) was achieved by taking only M-FTO in place of P-FTO in the initial device architecture (M-FTO/$MAPbI_3$/Ag). These studies suggest that even the simplest PSCs may be able to attain much greater performance with proper optimization of the perovskite film quality and interfaces. Further research should be done on this type of architecture since it can be designed by the low-cost and simplest fabrication procedure.

Table 4.3 Summarized photovoltaic performance of both ETL- and HTL-free PSCs

Year	Device architecture	V_{OC} (V)	J_{SC} (mA.cm^{-2})	FF (%)	PCE (%)	Ref.
2012	ITO/$CsSnI_3$/Au/Ti	0.42	4.80	0.22	0.88	Chen et al. (2012)
2015	FTO/$MAPbI_{3-x}Cl_x$/Au	0.77	7.11	0.53	3.20	Zhang et al. (2015)
2016	Au/$MAPbI_3$ single crystal/Au	0.82	2.28	–	1.88	Dong et al. (2016)
2018	FTO/$CsPbBr_3$/C	1.05	4.64	0.48	2.35	Duan et al. (2018)
2018	Al/$MAPbI_3$/MoO_3/Au	0.40	5.32	0.24	0.51	Joo Yang et al. (2018)
2018	Au/OMETP/$MAPbI_3$/ClTP/Au	0.55	11.93	0.48	3.11	Lin et al. (2018)
2019	FTO/n-type $MAPbI_3$/p-type $MAPbI_3$/Au	0.80	21.81	0.46	8.08	Cui et al. (2019)
2019	P-FTO/$MAPbI_3$/Ag	0.83	14.11	0.49	5.71	Yin et al. (2019)
2019	M-FTO/$MAPbI_3$/Ag	0.92	18.65	0.65	11.15	Yin et al. (2019)
2019	M-FTO/$MAPbI_3$–NiO/Ag	1.01	19.41	0.75	14.65	Yin et al. (2019)

Challenges and Future Research Directions

The potential of ETL- and HTL-free PSCs is indisputable owing to advantages like reduced cost, increased stability, and simplified architectures. Nevertheless, the challenges that come with eliminating CTLs must be tackled first if the technology is to be pushed in the direction of commercial viability. The foremost challenge concerns with finding the alternate ways to enable efficient charge extraction while stopping the charge recombination. Electrode engineering to create a proper energy alignment and exploring materials that provide passivation at the interface between the electrode and perovskite absorber layer can be the potential avenues for the exploration. The absence of CTLs directly exposes the perovskite layer, making it more vulnerable to environment stressors. Further endeavors are needed in developing the HHP compositions that are less susceptible to degradation. Moreover, optimization of the perovskite material by improving its crystallinity and fabricating a defect-free, high-quality film is of paramount importance for maintaining high PCE, as these defects and grain boundaries trap the carriers and account to non-radiative recombination losses. More research efforts aligned toward temperature-controlled annealing, solvent engineering, and additive engineering can result in the improvement of film quality. The growing efforts in exploring the dopants and molecules that can be introduced to passivate defects and enhance charge transport are the welcome steps in this direction. The fundamental idea behind exploring HTL-free/ ETL-free designs is to scale up the manufacturing processes by eliminating some layers from the perovskite architecture. However, large-scale, large-area fabrication of high-performing perovskite films without ETL/HTL remains a hurdle as solution processing techniques are being employed in the present laboratory-scale methods. It is very difficult to scale up to levels of industrial production while keeping the same uniformity and PCE of the film as in the laboratory. Thus, the methods which show more compatibility towards industrial manufacturing processes such as slot-die coating, spray coating, and blade coating should be investigated more to ensure uniform, defect-free films across a large area to scale up the fabrication processes.

Conclusion

Developing PSCs without ETLs and HTLs is a promising step in the right direction for the coming generations of the photovoltaic technologies. Such simplification of the device structure and architectures can significantly make PSCs more attractive from the commercial perspective by overall lowering the production costs involved and enhancing the long-term stability of these solar cells. However, the challenges regarding extraction of electrons and holes, stability, scalability, and electrode compatibility of these solar cells must be addressed first if some concrete outcomes are to be expected at the commercial level. The use of state-of-the-art fabrication methods, various surface passivation techniques, and novel electrode materials are critically important for optimizing the device performance. Future research must also align in exploring the perovskite materials with enhanced intrinsic stability and charge transport properties. The persistent exploration of the perovskite materials and their device architectures by trying different interfaces and compositions will certainly be a game changer in tackling these challenges. By overcoming some environmental and technical challenges, ETL-free, HTL-free, and both ETL- and HTL-free PSCs could be a transformative technology in the landscape of global renewable energy. In a nutshell, the selection of optimal architecture for any PSC is ultimately determined by the intended objectives and applications while balancing the factors like cost, stability, scalability, and performance. For instance, for large-scale manufacturing where cost-effectiveness is foremost important, the ETL-free PSCs are more suitable due to their simplified fabrication process and lower production costs, whereas, for deployment in extreme conditions, HTL-free PSCs are a better choice because of their stability under environmental stressors. Although fabrication of fully ETL- and HTL-free PSCs is very challenging technologically yet, they hold immense promise for developing highly flexible, lightweight, and ultrathin photovoltaic devices, which are essential for next-generation portable and wearable energy solutions. The future of these PSCs has immense possibilities and potential as their adaptability to different device configurations and architecture, combined with sustained research, will ensure their participation in the evolving commercial photovoltaic landscape as a versatile and promising contender.

References

Aharon S et al (2014) Depletion region effect of highly efficient hole conductor free $CH_3NH_3PbI_3$ perovskite solar cells. Phys Chem Chem Phys 16(22):10512–10518. https://doi.org/10.1039/C4CP00460D

Best Research-Cell Efficiency Chart | photovoltaic research | NREL. Accessed 22 Sept 2024. [Online]. Available: https://www.nrel.gov/pv/cell-efficiency.html

Bhattarai S et al (2021) Carrier transport layer free perovskite solar cell for enhancing the efficiency: a simulation study. Optik (Stuttg) 243:167492. https://doi.org/10.1016/J.IJLEO.2021.167492

Chen Z et al (2012) Schottky solar cells based on $CsSnI_3$ thin-films. Appl Phys Lett 101(9):93901. https://doi.org/10.1063/1.4748888/111837

Chen T et al (2018) Efficient perovskite solar cells with titanium cathode interlayer. Solar RRL 2(11):1800167. https://doi.org/10.1002/SOLR.201800167

Chen X et al (2019) Hole blocking layer-free perovskite solar cells with high efficiencies and stabilities by integrating subwavelength-sized plasmonic alloy nanoparticles. ACS Appl Energy Mater 2(3):2094–2103. https://doi.org/10.1021/ACSAEM.8B02145

Chen X et al (2021) Tailoring the dimensionality of hybrid perovskites in mesoporous carbon electrodes for type-II band alignment and enhanced performance of printable hole-conductor-free perovskite solar cells. Adv Energy Mater 11(18):2100292. https://doi.org/10.1002/AENM.202100292

Chen J et al (2015) Hole-conductor-free fully printable mesoscopic solar cell with mixed-anion perovskite $CH_3NH_3PbI_{(3-x)}$ $(BF_4)_x$. https://doi.org/10.1002/aenm.201502009

Chowdhury TA et al (2023) Stability of perovskite solar cells: issues and prospects. RSC Adv 13(3):1787–1810. https://doi.org/10.1039/D2RA05903G

Cui P et al (2019) Planar p–n homojunction perovskite solar cells with efficiency exceeding 21.3%. Nat Energy 4(2):150–159. https://doi.org/10.1038/s41560-018-0324-8

Deendyal PK et al (2024) Chronological evolution of stability in hybrid halide perovskite solar cells. Solar RRL 8(6):2300957. https://doi.org/10.1002/SOLR.202300957

Deng J et al (2024) Salt-based catalyzer to aid heterogeneous nucleation enabling > 23% efficient electron-transport-layer-free perovskite solar cells. Adv Funct Mater 2408392. https://doi.org/10.1002/ADFM.202408392

Dong Q et al (2016) Lateral-structure single-crystal hybrid perovskite solar cells via piezoelectric poling. Adv Mater 28(14):2816–2821. https://doi.org/10.1002/ADMA.201505244

Duan J et al (2018) Simplified perovskite solar cell with 4.1% efficiency employing inorganic $CsPbBr_3$ as light absorber. Small 14(20):1704443. https://doi.org/10.1002/SMLL.201704443

Etgar L et al (2012) Mesoscopic $CH_3NH_3PbI_3/TiO_2$ heterojunction solar cells. J Am Chem Soc 134(42):17396–17399. https://doi.org/10.1021/JA307789S

Gao L et al (2022) Cascaded band gap design for highly efficient electron transport layer-free perovskite solar cells. Chem Commun 58(47):6749–6752. https://doi.org/10.1039/D2CC01807A

Gou Y et al (2024) Work function tuning of carbon electrode to boost the charge extraction in hole transport layer-free perovskite solar cells. Small 20(38):2403342. https://doi.org/10.1002/SMLL.202403342

Guo Y et al (2019) Efficient and hole-transporting-layer-free $CsPbI_2Br$ planar heterojunction perovskite solar cells through rubidium passivation. Chemsuschem 12(5):983–989. https://doi.org/10.1002/CSSC.201802690

Han Q et al (2018) Carrier dynamics engineering for high-performance electron-transport-layer-free perovskite photovoltaics. Chem 4(10):2405–2417. https://doi.org/10.1016/j.chempr.2018.08.004

Hu Q et al (2014) Engineering of electron-selective contact for perovskite solar cells with efficiency exceeding 15%. ACS Nano 8(10):10161–10167. https://doi.org/10.1021/NN5029828

Huang L et al (2016) Toward revealing the critical role of perovskite coverage in highly efficient electron-transport layer-free perovskite solar cells: an energy band and equivalent circuit model perspective. ACS Appl Mater Interfaces 8(15):9811–9820. https://doi.org/10.1021/ACSAMI.6B00544

Huang C et al (2019) Facile fabrication of highly efficient ETL-free perovskite solar cells with 20% efficiency by defect passivation and interface engineering. Chem Commun 55(19):2777–2780. https://doi.org/10.1039/C9CC00312F

Huang L et al (2019) Simple, robust, and going more efficient: recent advance on electron transport layer-free perovskite solar cells. Adv Energy Mater 9(24):1900248. https://doi.org/10.1002/AENM.201900248

Huang Y et al (2020) Towards simplifying the device structure of high-performance perovskite solar cells. Adv Funct Mater 30(28):2000863. https://doi.org/10.1002/ADFM.202000863

Huang S et al (2021) Outstanding performance of electron-transport-layer-free perovskite solar cells using a novel small-molecule interlayer modified FTO substrate. Chem Eng J 422:130001. https://doi.org/10.1016/J.CEJ.2021.130001

Hui W et al (2023) Stable electron-transport-layer-free perovskite solar cells with over 22% power conversion efficiency. Nano Lett 23(6):2195–2202. https://doi.org/10.1021/ACS.NANOLETT.2C04720

Jiang Q et al (2019) Surface passivation of perovskite film for efficient solar cells. Nat Photonics 13(7):460–466. https://doi.org/10.1038/s41566-019-0398-2

Joo Yang S et al (2018) Visualization and investigation of charge transport in mixed-halide perovskite via lateral-structured photovoltaic devices. Adv Funct Mater 28(41):1804067. https://doi.org/10.1002/ADFM.201804067

Jung HS et al (2015) Perovskite solar cells: from materials to devices. Small 11(1):10–25. https://doi.org/10.1002/SMLL.201402767

Ke W et al (2015) Efficient hole-blocking layer-free planar halide perovskite thin-film solar cells. Nat Commun 6(1):1–7. https://doi.org/10.1038/ncomms7700

Kojima A et al (2009) Organometal halide perovskites as visible-light sensitizers for photovoltaic cells. J Am Chem Soc 131(17):6050–6051. https://doi.org/10.1021/JA809598R

Kong W et al (2019) Organic monomolecular layers enable energy-level matching for efficient hole transporting layer free inverted perovskite solar cells. ACS Nano 13(2):1625–1634. https://doi.org/10.1021/ACSNANO.8B07627

Ku Z et al (2013) Full printable processed mesoscopic $CH_3NH_3PbI_3/TiO_2$ heterojunction solar cells with carbon counter electrode. Sci. Rep. 3(1):1–5. https://doi.org/10.1038/srep03132

Laban WA et al (2013) Depleted hole conductor-free lead halide iodide heterojunction solar cells. Energy Environ Sci 6(11):3249–3253. https://doi.org/10.1039/C3EE42282H

Lamberti F et al (2021) The non-innocent role of hole-transporting materials in perovskite solar cells. Solar RRL 5(10):2100514. https://doi.org/10.1002/SOLR.202100514

Lan Z-R et al (2024) Surface passivation with diaminopropane dihydroiodide for p-i-n perovskite solar cells with over 25% efficiency. Adv Funct Mater 34(12):2312426. https://doi.org/10.1002/ADFM.202312426

Li H et al (2016) 14.7% efficient mesoscopic perovskite solar cells using single walled carbon nanotubes/carbon composite counter electrodes. Nanoscale 8(12):6379–6385. https://doi.org/10.1039/C5NR07347B

Li D et al (2020) In situ interface engineering for highly efficient electron-transport-layer-free perovskite solar cells. Nano Lett 20(8):5799–5806. https://doi.org/10.1021/ACS.NANOLETT.0C01689

Li L et al (2024) Moisture induced secondary crystal growth boosting the efficiency of hole transport layer-free carbon-based perovskite solar cells beyond 19.5%. Adv Funct Mater 34(1):2308428. https://doi.org/10.1002/ADFM.202308428

Liao JF et al (2019) Maze-like halide perovskite films for efficient electron transport layer-free perovskite solar cells. Solar RRL 3(3):1800268. https://doi.org/10.1002/SOLR.201800268

Liao JF et al (2020) Understanding of carrier dynamics, heterojunction merits and device physics: towards designing efficient carrier transport layer-free perovskite solar cells. Chem Soc Rev 49(2):354–381. https://doi.org/10.1039/C8CS01012A

Lin X et al (2018) Effect of grain cluster size on back-contact perovskite solar cells. Adv Funct Mater 28(45):1805098. https://doi.org/10.1002/ADFM.201805098

Liu D et al (2014) Compact layer free perovskite solar cells with 13.5% efficiency. J Am Chem Soc 136(49):17116–17122. https://doi.org/10.1021/JA508758K

Liu L et al (2015) Fully printable mesoscopic perovskite solar cells with organic silane self-assembled monolayer. J Am Chem Soc 137(5):1790–1793. https://doi.org/10.1021/JA5125594

Liu D et al (2018) Ultrathin hole extraction layer for efficient inverted perovskite solar cells. ACS Omega 3(6):6339–6345. https://doi.org/10.1021/ACSOMEGA.8B00741

Liu J et al (2019) In situ growth of perovskite stacking layers for high-efficiency carbon-based hole conductor free perovskite solar cells. J Mater Chem A Mater 7(22):13777–13786. https://doi.org/10.1039/C9TA02772F

Ma J et al (2024) Octylammonium Iodide Induced in-situ healing behavior at perovskite/carbon interface: the slow-release effect caused by carbon black adsorption. Small 20(30):2310196. https://doi.org/10.1002/SMLL.202310196

Mei A et al (2014) A hole-conductor-free, fully printable mesoscopic perovskite solar cell with high stability. Science (1979) 345(6194):295–298. https://doi.org/10.1126/SCIENCE.1254763

Meng F et al (2020) Intermediate-controlled interfacial engineering for stable and highly efficient carbon-based PSCs. ACS Appl Mater Interfaces 12(30):34479–34486. https://doi.org/10.1021/ACSAMI.0C11419

Miyasaka T et al (2020) Perovskite solar cells: can we go organic-free, lead-free, and dopant-free? Adv Energy Mater 10(13):1902500. https://doi.org/10.1002/AENM.201902500

Ogomi Y et al (2014) All-solid perovskite solar cells with HOCO-R-NH_3^+I- anchor-group inserted between porous Titania and perovskite. J Phys Chem C 118(30):16651–16659. https://doi.org/10.1021/JP412627N

Pascual J et al (2016) Electron transport layer-free solar cells based on perovskite-fullerene blend films with enhanced performance and stability. Chemsuschem 9(18):2679–2685. https://doi.org/10.1002/CSSC.201600940

Ravishankar S et al (2018) Influence of charge transport layers on open-circuit voltage and hysteresis in perovskite solar cells. Joule 2(4):788–798. https://doi.org/10.1016/j.joule.2018.02.013

Rong Y et al (2014) Hole-conductor-free mesoscopic $TiO_2/CH_3NH_3PbI_3$ heterojunction solar cells based on anatase nanosheets and carbon counter electrodes. J Phys Chem Lett 5(12):2160–2164. https://doi.org/10.1021/JZ500833Z

Roy P et al (2020) A review on perovskite solar cells: evolution of architecture, fabrication techniques, commercialization issues and status. Sol Energy 198:665–688. https://doi.org/10.1016/J.SOLENER.2020.01.080

Sadegh F et al (2022) Facile NaF treatment achieves 20% efficient ETL-free perovskite solar cells. ACS Appl Mater Interfaces 14(34):38631–38641. https://doi.org/10.1021/ACSAMI.2C06110

Shi J et al (2014) Hole-conductor-free perovskite organic lead iodide heterojunction thin-film solar cells: High efficiency and junction property. Appl Phys Lett 104(6):63901. https://doi.org/10.1063/1.4864638/237078

Shi T et al (2016) Ti/Au cathode for electronic transport material-free organic-inorganic hybrid perovskite solar cells. Sci Rep 6(1):1–6. https://doi.org/10.1038/srep39132

Shin SS et al (2017) Colloidally prepared La-doped $BaSnO_3$ electrodes for efficient, photostable perovskite solar cells. Science (1979) 356(6334):167–171. https://doi.org/10.1126/SCIENCE.AAM6620

Song J et al (2017) Highly efficient and stable low-temperature processed ZnO solar cells with triple cation perovskite absorber. J Mater Chem A Mater 5(26):13439–13447. https://doi.org/10.1039/C7TA03331A

Suresh Kumar N et al (2021) A review on perovskite solar cells (PSCs), materials and applications. J Materiomics 7(5):940–956. https://doi.org/10.1016/J.JMAT.2021.04.002

Wang Y et al (2019) Carbon nanotube bridging method for hole transport layer-free paintable carbon-based perovskite solar cells. ACS Appl Mater Interfaces 11(1):916–923. https://doi.org/10.1021/ACSAMI.8B18530

Wang J et al (2021) Heterojunction engineering and ideal factor optimization toward efficient MINP perovskite solar cells. Adv Energy Mater 11(48):2102724. https://doi.org/10.1002/AENM.202102724

Woong Jung J et al (2015) High-performance semitransparent perovskite solar cells with 10% power conversion efficiency and 25% average visible transmittance based on transparent CuSCN as the hole-transporting material. Adv Energy Mater 5(17):1500486. https://doi.org/10.1002/AENM.201500486

Wu WQ et al (2018) Molecular doping enabled scalable blading of efficient hole-transport-layer-free perovskite solar cells. Nat Commun 9(1):1–8. https://doi.org/10.1038/s41467-018-04028-8

Wu W-Q et al (2020) Suppressing interfacial charge recombination in electron-transport-layer-free perovskite solar cells to give an efficiency exceeding 21 %. Angew Chem 132(47):21166–21173. https://doi.org/10.1002/ANGE.202005680

Xu X et al (2015) Working mechanism for flexible perovskite solar cells with simplified architecture. Nano Lett 15(10):6514–6520. https://doi.org/10.1021/ACS.NANOLETT.5B02126

Ye S et al (2016) A strategy to simplify the preparation process of perovskite solar cells by co-deposition of a hole-conductor and a perovskite layer. Adv Mater 28(43):9648–9654. https://doi.org/10.1002/ADMA.201603850

Ye S et al (2017) A breakthrough efficiency of 19.9% obtained in inverted perovskite solar cells by using an efficient trap state passivator Cu(thiourea)I. J Am Chem Soc 139(22):7504–7512. https://doi.org/10.1021/JACS.7B01439

Ye T et al (2021) Cost-effective high-performance charge-carrier-transport-layer-free perovskite solar cells achieved by suppressing ion migration. ACS Energy Lett 6:3044–3052. https://doi.org/10.1021/ACSENERGYLETT.1C01186

Yin X et al (2019) Minimalist design of efficient, stable perovskite solar cells. ACS Appl Mater Interfaces 11(13):12460–12466. https://doi.org/10.1021/ACSAMI.8B21692

Yu H et al (2017) Outstanding performance of hole-blocking layer-free perovskite solar cell using hierarchically porous fluorine-doped tin oxide substrate. Adv Energy Mater 7(22):1700749. https://doi.org/10.1002/AENM.201700749

Yu H et al (2017) Large grain-based hole-blocking layer-free planar-type perovskite solar cell with best efficiency of 18.20%. ACS Appl Mater Interfaces 9(9):8113–8120. https://doi.org/10.1021/ACSAMI.6B15710

Zhang Y et al (2015) Charge selective contacts, mobile ions and anomalous hysteresis in organic–inorganic perovskite solar cells. Mater Horiz 2(3):315–322. https://doi.org/10.1039/C4MH00238E

Zhang H et al (2017) $SrCl_2$ derived perovskite facilitating a high efficiency of 16% in hole-conductor-free fully printable mesoscopic perovskite solar cells. Adv Mater 29(15):1606608. https://doi.org/10.1002/ADMA.201606608

Zhou H et al (2015) Low-temperature processed and carbon-based $ZnO/CH_3NH_3PbI_3/C$ planar heterojunction perovskite solar cells. J Phys Chem C 119(9):4600–4605. https://doi.org/10.1021/JP512101D

Zhou Z et al (2020) Highly efficient inverted hole-transport-layer-free perovskite solar cells. J Mater Chem A Mater 8(2):503–512. https://doi.org/10.1039/C9TA10694D

Zhou X et al (2022) Highly efficient and stable hole-transport-layer-free inverted perovskite solar cells achieved 22% efficiency through p-type molecular synergistic doping. Nano Energy 104:107988. https://doi.org/10.1016/J.NANOEN.2022.107988

Zhu Q et al (2016) Compact layer free perovskite solar cells with a high-mobility hole-transporting layer. ACS Appl Mater Interfaces 8(4):2652–2657. https://doi.org/10.1021/ACSAMI.5B10555

Chapter 5
Study of the Electrode Materials for Perovskite Solar Cells

Prachi Jain, Sakshi Chaudhary, S. Shankar, and O. P. Thakur

Abstract Currently, the perovskite solar cells (PSCs) have materialized out as a next-gen source of energy in the area of photovoltaic technology. Considering their cheap cost and producing high power, the commercialization of PSCs devices has become crucial. However, addressing stability in these cells has now become a current issue before its commercialization on large scale. A proper choice over deposition of top electrodes on the perovskite layer plays very vital role for maintaining the stability of device over longer period of time. The rear electrode plays vital role in producing stable PSCs. This book chapter reports the study on the comparison of various rear electrodes (metal electrodes, transparent conductive oxides (TMO) based electrodes, conductive polymers, and carbon-based electrodes) on the basis of their cost, efficiency, and stability. This chapter concludes the benefits and limitations of using rear electrodes materials in PSCs for practical device application.

Keywords Perovskite solar cells · Photo-conversion efficiency · Metal electrode · Nickel · Gold · Silver

P. Jain
Department of Physics, Vellore Institute of Technology, Amaravati, Andhra Pradesh 522241, India

S. Chaudhary
Department of Physics, Applied Science Cluster, UPES, Dehradun, Uttarakhand, India

S. Shankar
Functional Materials Research Lab, Department of Physics, ARSD College, Dhaula Kuan, New Delhi, India

P. Jain · O. P. Thakur (✉)
Materials Analysis and Research Laboratory, Department of Physics, Netaji Subhas University of Technology, Sector-3, Dwarka, New Delhi 110078, India
e-mail: opthakur@nsut.ac.in

M. Kumar and A. Raj (eds.), *Perovskite Solar Cells*, Springer Tracts in Electrical and Electronics Engineering, https://doi.org/10.1007/978-981-96-9463-1_5

Introduction

General Outline of Perovskite Solar Cells (PSCs)

Firstly, the photovoltaic SCs were first served to produce electrical energy/current from the energy released by sun. There are many varieties of photovoltaic SCs which have been emerged out to reshape the new environment by modulating renewable sources of energy such as Silicon solar cells, III-V solar cells, organic solar cells, Dye-sensitized solar cells (DSSC), quantum dot sensitized solar cells (QDSCs), Perovskite solar cells (PSCs) (Acik and Darling 2016; Tsujimoto et al. 2012). It has been noteworthy to mention that Si-solar cells were being in use from the longer duration of time as compared to other varieties of solar cells. This is because they are highly non-toxic, enormous availability, 27% conversion efficiency with almost zero degradation and are highly stable against the environment. These qualities made silicon-based solar cells a successful commercialized solar cell device from a long time. However, along with its benefits, there are few limitations are also associated with these Si-solar cells. They need high temperature processing, intensive energy is required for their manufacturing process, and it has relatively large global demand (Wang et al. 2019a). The effective solar cells should include various properties such as continuous or homogenous crystal structure with minimum grain boundaries which would be helpful for fast movement of photo-generated charge carriers. These fundamental properties have been exactly shown by Perovskite solar cells (PSCs). The main fundamental view behind commercializing the PSCs is their ability to withstand with the low production costs and capability to reproduce high quality semiconductors.

Perovskite solar cells (PSC) devices have gained enormous attention after their first demonstration in the year 2009, by Kojima et al. (2009a). These cells are highly researched in all over the world due to its improved photovoltaic (PV) performance and low-cost processing (Chen et al. 2021a; Yoo et al. 2021a). The research on PSCs is mostly driven due to its low cost, advanced optoelectronic properties including tuned bandgap, high value of absorption coefficient, very less recombination rate, elevated mobility of charge carriers (Eperon et al. 2014; Wolf et al. 2014; Stranks et al. 2013; Herz et al. 2017). Within a gap of a decade, the power conversion efficiency (PCE) of PSC devices have proceeded from 3.8% to 25.5% for small area single junction PSCs. While, in perovskite silicon tandem cells, this efficiency has been further risen upto 29%. This is the highest value obtained till now for thin film photovoltaics (Zheng et al. 2020). However, the degradation by external stimuli is still a big objection before commercializing the PSCs. The degradation of PSCs first initialized through the interfaces, which includes both substrates and metal electrodes (Green et al. 2014). The light absorbers based on Perovskites are generally represented as ABO_3, where A is a monovalent cation, mainly Cs^+, formamidinium [FA^+], methylammonium [MA^+], B is a metal cation with valency 2, and X denotes a halide ion like Cl^-, I^-, etc. (Chen et al. 1971). Researchers have used few techniques known as surface passivation or post fabrication treatment (1 1) in order to stabilize the perovskite metal electrode interface. However, very minimal attention has been given to the perovskite and substrate interface since it faces some difficulty in the measurements techniques such as scanning electron microscopy (SEM), Atomic force microscopy (AFM), etc. Although, it has been also unveiled that through the help of compositional or interfacial engineering strategies, the power stability of PSCs has now increased up to more than 1000 h in the presence of sunlight (Liu et al. 2020). The enhancement in the stability of the PSCs can be improved by opting some beneficial techniques such as structural engineering as well as the interfacial engineering across the boundaries, utilizing more stable charge transport materials or by improving the quality of encapsulation devices (Chen et al. 2022). The study of suitable electrode materials for the study of PSCs is quite equally important.

Challenges in the Development of PSCs

However, the PCE of the PSCs have drawn an exponential growth but still PSCs are facing difficulty in their commercialization at the market level. There's still a big challenge to commercialize the perovskite solar cells in spite of producing high PCE with stability over long hours. Considering all these facts, it was noted that role of electrode materials in PSCs also have a significant role in commercializing PSCs. Additionally, the study on electrode materials have been received very much less attention of the researchers. The front electrodes of PSCs are formed of transparent conductive oxides (TCOs) layers such as Fluorine-doped tin oxide (FTO) and Indium-doped tin oxide (ITO). This layer is applied on the front side of PSCs as it forms stable structure even when combined with proper charge transport inter-layers (Liang et al. 2018). In contrast, it has been noted down that the deposition of rear electrodes on the perovskite material's top or there are many limitations associated with the transport layers of charge which includes material selection for electrodes and deposition conditions. Further, many researchers and their groups reported that gold (Au) can be used as most stable rear electrode in PSCs, but it becomes unaffordable at the large scale production. Till now, the Au layer fails to achieve stability if the light absorber layer of perovskite is permeable, the metal ions were intended to disperse into the perovskite layer and will contribute toward degradation in solar cell performance (Lee et al. 2020; Huang et al. 2019). During few recent decades, study on different kind of rear electrode materials such as copper (Cu), nickel (Ni), aluminum (Al), tungsten (W), molybdenum (Mo), noble metals like Au, Ag, some non-metallic electrodes which include TCOs such as FTO, ITO, carbon electrodes, and conductive polymers electrodes have been explored so far. Since, the studies on many common electrode materials like silver nanowires, metal electrodes, carbon electrodes, TCO (transition conductive oxides), etc. have been already reported. The overall device performance is simulated by the properties and morphology of the electrode material. The main work of electrode materials is to collect electrons from ETL (electron transport layer) and holes from HTL (hole transport layer). It has been also studied that resistivity is considered as an intrinsic property which can influence the overall device performance like R_s (series resistance), FF (fill factor), PCE. The selection of suitable electrodes is necessary for the formation of PSCs. Even, the choice of particular electrodes materials also depend upon HTL and ETL. The role of interfacial layer is also crucial for determining the efficiency of PSCs (Shao and Loi 2020; Njema et al. 2024; Tao et al. 2024). In this proposed book chapter, the analysis on the progress of research undertaken for the various electrode materials required for the fabrication of PSCs in terms of their efficiency and stability have been studied in detail. There are three main important factors like efficiency, cost, and stability which are linked with the electrode materials for the PSCs fabrication. Hence, it have been concluded that optimization of all the functional layers of device is mandatory to ensure the best PSC exhibiting longer life time.

Review on Electrode Materials

The price of the electrode materials also plays a crucial role in determining the efficiency of PSCs. Although the analysis on the price of electrode materials is usually very much less studied and always kept at the bottom of the research but in developing PSCs, the price is also very crucial. Figure 5.1 gives the complete overview of the cost of various rear electrodes required to fabricate the PSCs. It has been also explored that non-noble metal such as Cu, Al are bit cheaper than Au and transition conductive oxide. On the other hand, carbon based rear electrode is found to be the cheapest one (besides grapheme and CNTs, which are quite expensive than metals). But Au cannot be employed on a large scale due to its cost issue. Hence, there's an urgent need to find such electrode which can be even used on a larger scale with low cost. To meet this requirement, nickel (Ni), tungsten (W), and molybdenum (Mo) have been proven to be the best alternative for the development of PSCs. In addition, some transition metal oxides like ITO, FTO, Al-doped zinc oxide (AZO), etc. have been also utilized as rear electrodes in developing semi-transparent PSCs. Semi-transparent solar cells are installed as solar windows, thus high price of ITO can be affordable at some specific requirements. The role of sheet resistance of electrodes plays vital role in deciding the transport and charge collection properties of PSCs (Wu et al. 2019; Yang et al. 2020a; Deng et al. 2016; Ku et al. 2015).

The work function of the electrode materials is also necessary for the selection of electrode materials. There should be a resonance of the work function of rear electrode with the fermi energy level of the charge transport material in contact. However, the materials having their work functions lying between -5.1 to -4.1 eV is meant to be suitable for the construction of efficient and flexible PSC (Sun et al. 2020). The efficiency of the PSCs or the transfer of charge between electrode and ETL/HTL usually depend on the electrode's work function. Figure 5.2 exhibit the distribution of different electrode materials on basis of their work-functions. Usually, a barrier gets formed when the work-function of working electrode fails to resonate with the fermi energy of the charge transport layer. If the charge transport layer is electron based or ETL, then LUMO (lowest unoccupied molecular orbital) gets formed. Rather, if this layer is HTL or hole-dominated, then formation of HOMO

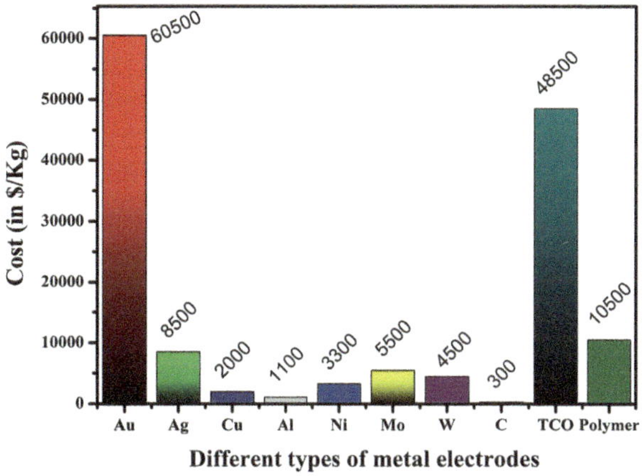

Fig. 5.1 Overview on the costs of different electrodes

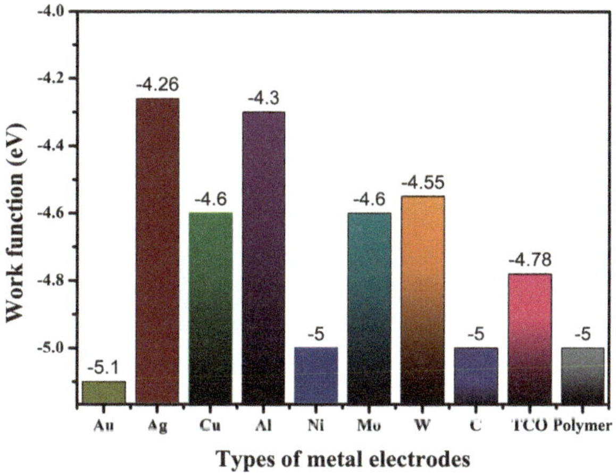

Fig. 5.2 Classification of various rear earth electrodes on the basis of their work functions

layer takes place. The rear electrodes with high optical transmittance are usually preferred for the deposition over perovskite layers. Along with, the robustness of the electrodes is also very necessary for developing PSCs to work for longer duration of time. The various electrodes are compared on the basis of their work-function (eV), sheet resistance (Ω), and their cost per kg (in $), as shown in Table 5.1.

Summarization of the Electrode Materials Utilized for Perovskite Solar Cells (PSCs)

Recent years have seen a surge in interest in perovskite solar cells (PSCs) because of their high efficiency, affordability, and adaptability. The choice of electrode materials,

Table 5.1 Comparison of different types of counter electrodes on the basis of their work-function (eV), R_{sq} (Ω), and their cost per kg (in $)

S. No.	Metal	Work function (eV)	Sheet resistance (R_{sq}) [Ω sq^{-1}]	Cost (in $/per kg)	References
1	Au	− 5.1	< 1	60,500	Yang et al. (2020a)
2	Ag	− 4.26	< 1	8500	Wu et al. (2019)
3	Cu	− 4.6	< 1	2000	Deng et al. (2016)
4	Al	− 4.3	< 1	1100	Raptis et al. (2020)
5	Ni	− 5	1.96	3300	Kim et al. (2018a)
6	Mo	− 4.6	5.1	5500	Jeong et al. (2015)
7	W	− 4.55	10.4	4500	Wang et al. (2017a)
8	C	− 5	6–180	300	Liu et al. (2016)
9	TCO	− 4.78	9.9–15	48,500	Bush et al. (2016)
10	Polymer based	− 5	100	10,500	Jiang et al. (2016b)

which are essential for charge collection and device stability, has a significant impact on the functionality and stability of these cells. In perovskite solar cells, electrode materials—the conductive contacts—are essential to charge collection and device efficiency. Additionally, the stability of the device is influenced by the rear electrodes, which serve as a protective layer.

Key electrode materials used for PSCs include the following:

1. **Front electrodes using Transparent conductive oxides (TCOs)**: One of the most popular TCOs is indium tin oxide (ITO), which has superior electrical conductivity, smoothness, and transparency.

   However, the high price and scarcity of indium have prompted research into substitutes.

   Fluorine-doped Tin Oxide (FTO): FTO is frequently utilized as a transparent electrode in PSCs, particularly in applications that call for high-temperature processing. Even, FTO is more thermally stable than ITO. There is another TCO named AZO (aluminum-doped zinc oxide) which is an inexpensive and also readily available substitute for ITO and FTO. Even though its conductivity is a little lower, it might work for some applications.
2. **Electrodes based on Hole Transport Layer (HTL)**: Basically, this is a layer which is deposited between anode and the photoactive layer of the PSCs to

improvise the overall device performance. They are categorized majorly into 3 kinds: Au electrodes, Ag electrodes, and carbon-based electrodes. Au is a widely used as a back electrode in the PSC. It is expensive but it might permeate the perovskite and compromise stability over time. Whereas, silver (Ag) is used because of its strong conductivity and is less expensive than Au. However, under some circumstances, it is susceptible to oxidation and migration, which may jeopardize stability. Graphene and carbon paste are examples of inexpensive, chemically stable carbon compounds. Metal electrodes can be swapped out for carbon-based electrodes in devices that need more stability and economy.

3. **Electron Transport Layer (ETL) Electrodes**: Titanium dioxide (TiO_2) is the most widely used ETL because of its superior electron mobility and stability. Despite the possibility of contributing to hysteresis, it is frequently utilized in mesoporous and planar PSCs. However, zinc Oxide (ZnO) is an excellent substitute ETL due to its high electron mobility, however, stability may be reduced if ZnO undergoes a chemical reaction with the perovskite layer. Out of all ETLs available, Tin oxide, or SnO_2, has become a viable ETL material due to its chemical stability with perovskites, low-temperature manufacturing compatibility, and strong electron mobility.
4. **Transparent Conductive Polymers**: In certain PSCs, conductive polymers—like PEDOT—are utilized in place of metal oxides such as ETLs or HTLs. They provide adaptability and the possibility of using them in lightweight applications. Conjugated polymers and other organic materials are also being investigated for flexible, inexpensive PSCs. Many improvement have been made but their performance still falls short of that of inorganic materials. NiO_x has demonstrated potential as a stable, affordable HTL that lowers production costs and increases device lifespan.
5. **Electrodes based on Two-dimensional (2D) materials**: Graphene and Additional 2D Substances have gained attention in the advanced PSCs because of their superior mechanical strength, flexibility, and electrical conductivity. To enhance charge transmission, these substances can be used as substitute electrodes or additives. They can increase the efficiency, stability, and adaptability in a variety of applications, such as flexible and wearable electronics and large-scale solar panels.

The applications of several electrode types (including metal electrodes, transparent conductive oxides (TCO), polymers, and carbon-based materials) in the creation of extremely stable, effective PSCs are covered in detail in this suggested book chapter. A summary of the various kinds of electrodes made of metal, inorganic, and organic materials is provided, with particular attention to how they affect the stability and efficiency of the device. Lastly, the difficulties and opportunities in creating stable and effective electrode materials are discussed.

Structure of PSCs

Methylammonium lead iodide ($CH_3NH_3PbI_3$, or $MAPbI_3$) is a widely used material in solar cells. It features a large organic cation represented as, methyl ammonium ($CH_3NH_3^+$), a smaller size of lead (Pb) cation, and halogen anion of I^-. Miyasaka et al. first utilized this structure in solar cells in 2009 (Kojima et al. 2009b). Previously, it was reported that $MAPBX_3$ was adsorbed onto TiO_2 and were utilized in DSSC (dye-sensitized solar cells) devices having electrolytes in liquid form. These configurations achieved efficiency about 3.9% (Kojima et al. 2009b). Lately, it was also observed that these liquid electrolytes reduced the lifespan of PSCs as it got leaked within the cell due to its volatile and corrosive nature. Hence, in 2012, Park and his group, invented advanced technology-based PSCs with perovskite material based photoactive layer in a solid state meso-porous structures PSCs with Spiro-OMeTAD as HTL layer and TiO_2 as a ETL layer (Kim 2012). The most common configuration of PSCs contains mainly 5 components: (a) transparent conductive glass as substrates like FTO, ITO, or tin doped ITO (b) Electron transport layer (ETL) in which n-semiconductors are employed, (c) perovskite materials as light extracting materials, (d) p-type semiconductors as hole transport layers (HTL), and (5) metal layers as back electrodes. After excitation of light on the metal surface, then e^-/hole pairs are prompted and starts migrating toward the interfaces. The separation of charges occurs at interfaces by injecting electrons into ETL_s and holes into HTLs. In n-i-p PSCs, injected holes are transported through HTLs toward the metal layer (Hsiao et al. 2015). On the other hand, electrons are transported through ETLs and go through the external circuit to metal contact. The two configurations are usually employed to fabricate perovskite solar cells (PSCs), these configurations are p-i-n and n-i-p, and perovskite layer is stuffed amidst n-type transport layer (ETL) and p-type transport layer called HTL (Frost et al. 2014; Liu et al. 2015). While, on the basis of structure, the PSCs are mainly categorized into 2 types, named as (1) Mesoporous layered structured based PSC in which a mesoporous layer is incorporated and is further represented for both n-i-p as well as p-i-n architectures. Other type of PSC is, planar structured PSC, it is further shown 2 other architectures named as n-i-p and p-i-n. In the planar one, the plane perovskite layer is incorporated to fabricate the solar cell. The structure for all configurations along with their corresponding energy band diagram is represented in Fig. 5.3. In the perovskite solar cell (PSC), there are few requirements that should be followed by counter electrode.

In recent years, TCOs such as ITO as well as FTO are commonly employed as bottom electrodes in transparent form due to its good conductivity and transmittance. Through high temperature etching, the FTO or ITO is applied on a glass substrate. This process can negatively affect the active layer when employed as a top transparent electrode in developing ST (semi-transparent)/ solar cells (SCs). Beyond conductive oxides, there are other different varieties of transparent top electrodes have been developed over the past two decades, including thin metals, transparent conducting polymers like poly(3,4-ethylenedioxythiophene): poly(styrenesulfonate) (PEDOT:

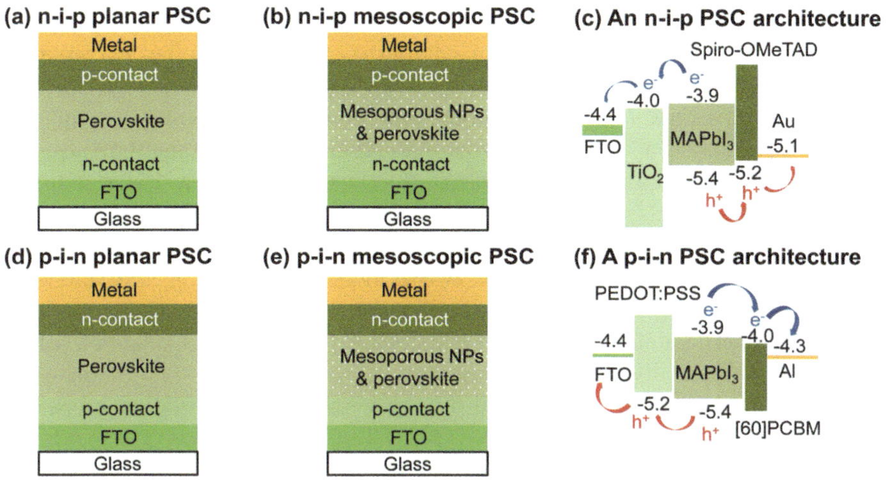

Fig. 5.3 **a** Planar structured based n-i-p PSC, **b** Mesoscopic n-i-p structured PSC, **c** energy band diagram for n-i-p architecture based FTO/TiO_2/$CH_3NH_3PbI_3$/Spiro-OMeTAD/Au **d** Planar n-i-p configuration based PSC, **e** Mesoscopic p-i-n structured PSC, and **f** Energy band diagram for p-i-n configuration based FTO/PEDOT:PSS/$MAPbI_3$/[60]PCBM/Al (*note* reproduced from ref (Wang et al. 2019a) with permission from Royal Society of Chemistry, copyright 2019)

PSS), metal nanowires (NWs), carbon nanotubes (CNTs), and graphene (Li et al. 2015; Salim et al. 2015).

Metal-Based Electrodes

The presence of high conductivity and light reflectivity properties make the metal electrodes as the best choice for preparing perovskite solar cells (PSCs) (Li et al. 2017). Wang et al. have reported the diffusion of iodide ions ejected from $MaPbI_3$ absorber film encompassing through the PCBM ([6,6]-phenyl-C61-butyric acid methyl ester) as ETL layer. The resulted ejected I^- ions reacted with silver (Ag) electrode and formed silver iodide (AgI). This whole process would lead to huge loss of MAI (methylammonium iodide) and finally yielded lead iodide (PbI_2) with lowest conductivity (Li et al. 2017). However, it was unveiled that MAI related materials or other perovskite materials get diffused into PCBM (ETL) layer which would cause high corrosion of the meta electrodes. This is the main reason behind the rapid degradation behind the p-i-n PSCs by employing fullerene-based materials as electron transport layers. The almost same behavior has been shown by p-i-n PSCs. Similar kind of mechanism was also represented by Qi et al., in which they investigated the corrosion phenomena of Ag electrodes and analyzed that the presence of moisture content caused the pinholes in spiro-OMeTAD layer from which the diffusion of species containing iodine into the silver electrode layer and would lead to the formation of AgI as shown in Fig. 5.1 (Akbulatov et al. 2017; Kato et al. 2015). Although performance of gold (Au) electrode is better as compared to silver (Ag) electrode but it generally migrates into the perovskite layer through spiro-OmeTAD even at 70 °C, which might cause non-radiative recombination in the devices (Domanski et al. 2016). The same studies have been performed by Cacovich et al. group, which reported the diffusion of Au and I ions by absorbing the light (Cacovich et al. 2017). Additionally, Shlenskaya et al. also studied that the Au generally forms iodide even at room temperature and could lead to form polyiodide, $[AuI_2]^-$ and $[AuI_4]^-$ (Shlenskaya et al. 2018).

The metals having low costs such as Mo, Al, Cu, Ni have been also utilized as top electrodes for developing perovskite solar cells (PSCs) to replace highly expensive gold and silver electrodes. In case of inverted PSCs, the Al electrodes are found to more suitable than rest of the electrodes due to its high conductivity, low cost, and low work-function (You et al. 2014a). It was also proposed by You et al. which have made first PSC using Al electrode by considering substrate ITO/PEDOT:PSS/$CH_3NH_3PbI_{3-x}Cl_x$/PCBM/Al device structure as shown in Fig. 5.4 (You et al. 2014b). The PSC which was fabricated on rigid substrates achieved maximum power conversion efficiency (PCE) of 11.5%. Whereas, the device fabricated on a flexible substrate achieved maximum PCE of 9.2%. The similar studies have been done by another group in which they showed that how moisture-induced grain growth method is efficient in improving the film quality, mobility of charge carriers, and lifetime. The PSCs grown by this method yielded 17% of PCE with FF near about 0.80.

Now, the main challenge comes is stability of the PSCs devices. For achieving this, Huang et al. added the thin polystyrene layer between perovskite layer and ETL in order to minimize the recombination of charges at cathode. Also, this interlayer polymer film will prevent degradation of perovskite caused by the moisture content (Wang et al. 2016). It was also shown that Al based PSC showed a $V_{oc} = 1.10$ V, $J_{sc} =$

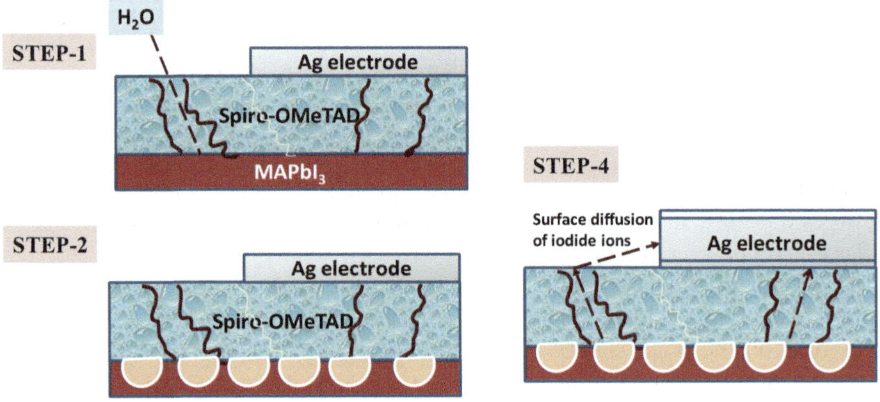

Fig. 5.4 Schematic diagram for the mechanism of the formation of silver iodide layer under top silver electrodes of perovskite solar cells, when reacts with water

22.9 mA/cm^2, and a fill factor of 80.6% with PCE of 20.3%. The efficiency of these electrodes remained for 120 h when exposed to 40% relative humidity (RH) without encapsulation. It was exhibited that inverted p-i-n based $MAPbI_{3-x}Br_x$ with (NDI (naphthalene diimide) -ID) delivered the PCE upto 20.2%. Along with, the NDI-ID (N,N′-Bis(1-indanyl)naphthalene-1,4,5,8-tetracarboxylic diimide) based PSCs have shown the maximum retentivity of 90% for 500 h at 100 °C without encapsulation, only illumination by the sun (Jung et al. 2018). Some buffer layers like MoO_x, CuO_x, NiO_x, VO_x, needs to be encapsulated between HTL and Al electrode to improve the efficiency of PSCs devices. This is done in order to prevent oxidation of aluminum as an energy barrier created between HTL/Al interface due to their mismatch in the energy levels (Li and Gao 2020). To overcome this, Huang and his team conducted the pioneering research using copper (Cu) as the rear electrode in inverted perovskite solar cells (PSCs), achieving promising outcomes (Zhao et al. 2016a).

Gold Electrodes

It has been observed that among all metal electrodes which were earlier utilized as the top most electrode in developing PSC, the gold metal has been considered as a noble metal. The efficiency of the PSCs developed using Au electrode is maximum till reported using any other metal electrodes. It has shown the tremendous performance in PSCs as its wavefunction is well matched with the common hole transport layers (HTLs), Spiro-OMeTAD, or NiO_x, etc. The maximum efficiency of PSCs, nearly equivalent to 25.2% has been noticed using Au electrode with thickness 100 nm. Yoo et al., found that electrode using Au metal is the reliable choice for producing highly efficient PSCs but their cost of installation is quite expensive when considered on large scale (Yoo et al. 2021b; Assi et al. 2021). It has been also discovered that, ultrathin-gold coated based perovskite solar cells can exhibit much improved efficiency % as compared to the transition conductive oxides-based PSCs. It has been also found that, gold electrode in its nano-film form can exhibit much better results. Through thermal evaporation technique and a dry transfer approach, the Au electrode was converted into its nano-porous film type structure. The PCE of 19% was recorded using nano-porous Au film in PSCs. The preparation of Au nanofilm is quite easy as well as a feasible approach to deposit its layer on the PSCs. Whereas, the deposition of Au electrode is quite difficult to achieve high performance PSCs. Utilization of Au electrode in film form boost up its application on an industry level (Yang et al. 2020b).

Silver Electrodes

It is very well known that, Ag is much cheaper than Au metal and hence, it has gained a popularity as back electrode material in PSCs. In case of p-i-n structured

PSCs, the Ag electrodes are mainly used as a cathode as its work-function got well matched with that of LUMO level of commonly used ETLs. The alignment of the work functions resulted in the creation of an ohmic contact. Snaith et al., was the first one who have reported the deposition of Ag electrode for the development of PSCs. The successive enhancement has been also observed in the PSC device performance when Ag powder with particle size of 20 nm was thermally deposited. This is generally because of the boosting up the charge transport process across the ETL/HTL and electrode (Nath et al. 2021). Figure 5.5 establishes the methodology for the deposition of metal (Ag)-based electrodes onto the layer of perovskite for the contriving of PSCs. Currently, Ag electrode was employed as a anode in the device architecture comprising of glass/FTO/TiO_2/Al_2O_3/$CH_3NH_3Pb(X = I,Br)_3$/Spiro-OMeTAD/Ag. Here, the thermal evaporation technique was opted to deposit the silver films. Steirer et al. have currently reported the full solution-based procedure for preparing the PSCs. In which they have fabricated PSC using Ag nanofilm as the back electrode by lamination procedure. The first setep here is, to drop PEDOT:PSS on a flexible PET substrate using spin coater to modify into its film form. Then, the film was annealed at a particular temperature of 120 °C. After that, the inorganic metal halide or $MAPbI_3$ dissolved in antisolvent was deposited on PET substrate layer using spin coating and further annealed at temp 65 °C. On this assembly, the Ag electrode was deposited using Thermal evaporation technique. The PCBM was deposited on the whole assembly using spin coater and then further annealed at 105 °C. In this way, the metal electrode deposited PSC device can be fabricated using thermal evaporation technique as shown in Fig. 5.5. It has been reported that, under relative humidity conditions, these cells were capable to exhibit the PCE of 10.03% as shown by Kim and his co-workers (Kim et al. 2016).

Nanowires of Gold@Silver-Based Electrodes

It has been reported by Liu and his co-workers, that Au coated Ag nanowires based electrodes can be deposited using the spray pyrolysis technique. The device structure would be of the form like, ITO/PTAA:F4-TCNQ/$MAPbI_3$/PC61BM/Au-coated Ag nanowires. This type of architecture exhibited fast transport of charges across the layers. The better stability is shown by those PSCs prepared using spray pyrolysis technique. This improved stability is accounted due to the Au shell which is wrapped around Ag nanowire. This gold shell also acts a separation layer stuffed in between perovskite/HTL and Ag nanowires. It has been reported that silver (Ag) electrodes degrade more quickly than gold (Au) when in direct contact with the perovskite layer. Additionally, devices made with Ag nanowires as top electrodes have shown a noticeable drop in efficiency within just 10 min. However, comparatively the efficiency of the PSCs devices fabricated through Au-coating Ag NWs have been risen to 85% of the original efficiency of silver electrode even after exposed for 7 days in air. The PSCs developed using Au coating have intended to exhibit the maximum

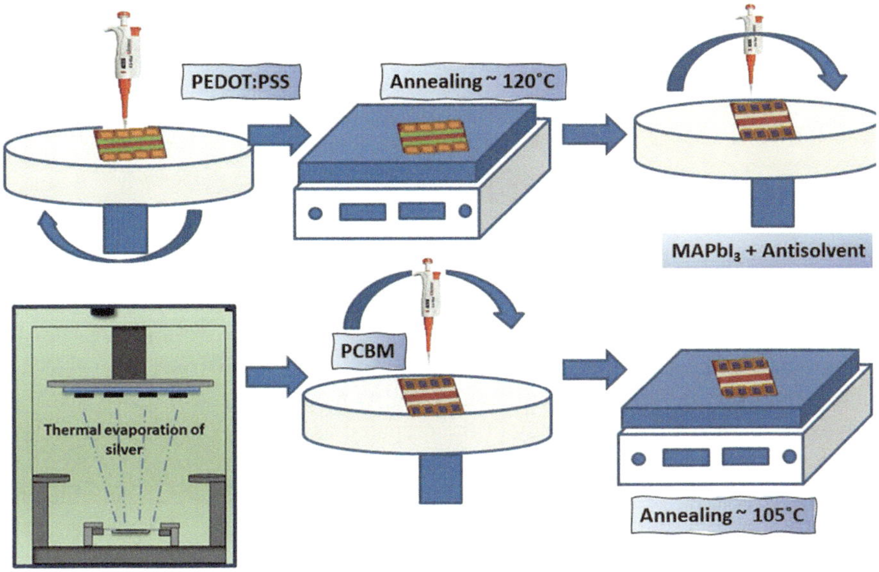

Fig. 5.5 Detailed Stepwise representation for preparing the metal (Ag) electrode based perovskite solar cells (PSCs)

power conversion efficiency of 11% with an open circuit voltage of 0.99 V, and a current density o 18.5 mA/cm^2, with a FF (fill factor) of 64.3% (Paetzold et al. 2015).

Aluminum Electrodes

It has been unveiled that, till date, among all the metal-based electrodes reported for developing PSCs, the Al metal has been found to be the most commonly used, as well as the cheaper one, easier to afford, kind of electrode material. Another main reason behind using Al electrode in PSC, is that its work-function is well matched with that of perovskite layer, which further supports the fast movement of charge carrier across the layers. Many groups have studied Al electrode in many architectures like, Glass/ITO/PEDOT:PSS/$MAPbI_3$/PCBM/Al devices (Nawaz et al. 2016). Li et al. have studied the devices fabricated on a flexible substrate with p-i-n structure like, PET/Ag-mesh/PH1000/PEDOT:PSS (35 nm)/$MAPbI_3$ (B280 nm)/ PCBM/Al (100 nm), and this whole assembly detected PCE of 14%, with Al at top electrode. Using several solution processed techniques, the You et al. have performed the architecture of the assembly, Glass/FTO/NiO_x_NP/$MAPbI_3$/ZnO NP/Al which exhibited a maximum value of efficiency around 16.1% (Li et al. 2016). When compared to polymer-based transport layers, the improved stability has been noted down when metal oxides-based layers were incorporated in place of them, as specified by You et al. (You et al. 2016).

Nevertheless, there are a lot of degradation issues associated with Al electrodes when exposed to air and moisture, this degradation caused oxidation which further reduce the working lifespan of PSCs (Zhao et al. 2016b).

Al-Ag Alloy-Based Electrodes

It has been very well researched by many researchers groups and they have found that, metal alloys based electrodes can show the better efficiencies as compared to single-single metal electrode. Some metal alloys are believed to exhibit the promising results when embedded as a top electrode in PSCs (Jiang et al. 2016a; Luo et al. 2015). Jiang group's dealt with the formation of Ag:Al metal alloy electrode based PSC, they have observed that efficiency of the prepared assembly is found to be 11.75% which is slight better than the value of the efficiency recorded for single Ag metal (11.45%). Also, the efficiency of electrode in its alloy form yielded better efficiency when compared to efficiency of Al electrode, which is around 7.95% (Jiang et al. 2016a). This high value of efficiency for alloy system has been recorded. As this assembly has resulted the construction of AlO_x passivation layer among PCBM and Ag:Al electrode. The formation of this passivation layer also suppresses the diffusion of Ag ions along with the suppression in the penetration of moisture ions. This also caused the strong adhesion or contact between Ag:Al electrode and the active layer. A required ohmic contact gets formed at the PCBM/electrode interface due to the configuration of AlO_x (passivation layer).

Cu Electrodes

It has been shown by Wang et al. that there are some low-cost non precious metals like Cu, Ni, Mo, etc. are also available which have the capability to replace Ag metal films in PSCs. There was no corrosion found in these non-precious metals (Wang et al. 2017a). In the electronic industry level, copper electrodes are widely utilized on a large scale because of the inclusion of its extremely good electrical conductivity along with its low cost, and it also perceives minimum oxidation resistance. The study on copper-based electrodes for developing PSCs devices was first studied by Huang et al. They have utilized copper electrode in inverted structured PSC, which have embarked good stability as compared to other electrodes. They even explored that Cu didn't decay even when put in direct contact with the $MAPbI_3$. Under mild thermal aging situations, the diffusion of Cu ions was not observed in perovskites (Zhao et al. 2016a). ITO/PTAA (poly[bis(4-phenyl)(2,4,6-trimethylphenyl)]amine)/perovskite/indene-C60 bis-adduct/C60/bathocuproine (BCP)/Cu PSC device has been fabricated, where it has been observed by the group that, no damage or loss has been caused to C60/BCP layer from thermal evaporation of copper (Deng et al. 2016). Finally, it has been discovered that Cu electrode when combined with $FMAPbI_3$, then it gave the final efficiency of 18.3%. This efficiency is quite very high as compared to when other metal electrodes were directly deposited onto films of perovskites. Even after exposing metal films in air or moisture for some hours or let suppose for 20 h, the films became lusterless. In spite of the fact that, the Cu film maintained its "cuprous red" color even after exposing it for more than 20 h (Deng et al. 2016). The unsealed copper electrodes engrossed in MAPbI3 or $FA_{0.4}MA_{0.6}PbI_3$ devices could nurture the device's performance with almost negligible power losses, which could even last for 20–30 days in the environment which contain 30–60% RH (relative humidity) values. More recently, Chen et al. have recorded a highly efficient/ stable PSCs devices with PCE around 23.2% in blade laminated p-i-n structured PSC by keeping a BHC (benzylhydrazine hydrochloride) as a reducing agent, which is a remarkable value reported till now (Chen et al. 2021b). This device was observed to sustain operational stability for more than 1000 h with nearly zero degradation. Additionally, the Cu paste and its inks are readily available commercially which allows the easy manufacturing of PSCs with Cu electrodes deposited as the top electrodes (Zhao et al. 2016a; Shao et al. 2015).

Nickel Electrodes

As compared to all other metal electrodes, the Ni electrodes received less attention in comparison with Al, Ag, and Cu electrodes for PSCs fabrication. Jiang et al. first reported the studies on the deposition of Ni electrode on the perovskite layer (Jiang et al. 2014). They have declared a nickel film coating done onto the spiro-OMETAD layer via sputtering technique to develop a regular mesoporous PSC device. This device exhibited a maximum PCE of 10.4% and could be reused again by washing. Hence, the Ni electrode proved to be the least expensive metal electrode which can be utilized for the fabrication of PSCs. After that, Ku and his co-workers produced a Ni-etched mesoporous PSC device delivering PCE of around 13.6%, which could be recycled and still preserved an efficiency of 12% (Ku et al. 2015). It has been observed that, mesoporous structured nickel film with 2 μ m thickness has shown a better conductivity value along with the smooth working of PSCs devices. After that, Bryant and his group, reported that a nickel electrode based ST-PSC obtained a PCE of 15.7%, which is high as considered to the Ni-based PSCs devices reported till now. Ni electrodes can yield PCE values comparable to that of Au electrode, and to that being low cost. Hence, the Ni electrode can be used as a best alternative to gold electrodes in developing PSCs. Table 5.1 summarizes the different metal electrodes on the basis of their work-function, sheet resistance, and their cost in $ per kg. Frequent metal electrodes utilized as electrodes for fabricating PSCs (Wu et al. 2019; Yang et al. 2020a; Deng et al. 2016; Wang et al. 2017a; Raptis et al. 2020; Kim et al. 2018a; Jeong et al. 2015; Liu et al. 2016; Bush et al. 2016; Jiang et al. 2016b).

Transparent Conductive Oxides-Based Electrodes

In optoelectronic devices, the transparent conductive oxides (TCO) are generally employed in the place of bottom electrode for the transmission of light. The traditional electrodes like FTO, ITO, IZO, AZO, IWO are being used for the fabrication of PSCs (Dou et al. 2017; Weerasinghe et al. 2015; Pisoni et al. 2018; Kim et al. 2018b). As compared to the above listed TCOs based electrodes, the FTO requires high manufacturing temperature above 350 °C to be deposited on a polymer substrate (Bae et al. 2006; Kawashima et al. 2004). Whereas, the IZO could be even deposited at 100 °C, and other remaining electrodes such as ITO, AZO, IWO require only room temperature for their deposition on the substrate (Taylor et al. 2008; Park et al. 2017). Dou et al. investigated that AZO encounters very strong reaction with the perovskite films leading toward the decay in the performance of PSCs (Pisoni et al. 2018; Park et al. 2017). However, it has been observed that IZO has a lower sheet resistance and can be prepared at even low temperature, and it has also improved the overall PSC's performance. Whereas, due to the presence of higher optical transmittance, the ITO based PSCs exhibit better short circuit current (J_{sc}) as compared to other transition conductive oxides. Despite of exhibiting lower sheet resistance by IZO, they have got failed to fabricate flexible and stable PSCs devices because they got broken down when bending at a smaller radius. More recently, it has been found by Kim et al, that the IWO films prepared via arc plasma ion plating method at room temperature tend to replace ITO in PSCs due to its high PCE of 11.33% which is quite high as compared to ITO based PSCs devices which are having the PCE of 10.6% (Kim et al. 2018b). Additionally, the IWO films persist higher work-function near 4.85 eV which is high as compared to ITO films having work-function of 4.65 eV. The IWO (due to its high value of work-function) attributed to its higher optical transmittance, mainly in IR region.

Many researchers have worked on several ways for increasing the net output power conversion efficiency (PCE) of PSCs using TCOs based electrodes. FTO has been accepted worldwide as the most efficient TCO based electrode for developing PSCs due to its some attractive features which include high chemical stability against perovskite films, low manufacturing temperature, better optical transmittance, etc. Moreover, the best commercial PSCs have been fabricated using ITO/polymer substrates, which records the maximum PCE of more than 20%. Despite having extensive usage of ITO in developing PSCs, there are several other issues which prevent ITO electrodes for developing idea of PSCs, some of them are, (1) ideal PSCs are required to maintain almost 90% PCE even after underwent bending about 1000 cycles at a small radius of 2–4 mm. (2) Also, the ITO electrodes lack flexibility and easily breakable which degrades the efficiency of PSCs (Zhang et al. 2020). The occurrence of cracks or breakage in FTO films can yield to high sheet resistance, which would lead toward device degradation. Reducing the thickness of ITO films can improve the flexibility of PSCs along with the less breakage or cracks on ITO. Liu and his group exhibited, ITO based PSCs can exhibit better mechanical stability even with a radius around 4 mm. The recent literature proved that, embedding a thin silver (Ag) layer on the bottom side of ITO film will enhance its flexibility, will improve efficiency (11.01%), and also produce very minimal value of sheet resistance, around 7 Ω/sq. The diagram for this particular phenomena is shown in Fig. 5.6a (Poorkazem et al. 2015). Since, it has been overlooked that transition metal oxides readily react with iodide present in the perovskite and hence decaying the overall performance of PSC. So, to overcome this issue, McGehee et al. developed inverted semi-transparent ITO based PSCs by utilizing a solution of AZO nanoparticles (~50 nm) as a buffer layer and hole-blocking layer, as represented in Fig. 5.6b (Bush et al. 2016). It has been unveiled from Fig. 5.6c, the ITO films fabricated through direct DC sputtering formed an amorphous phase at room temperature, and the films processed through this technique exhibit enormous value of R_s with least transmittance, coarse surface with lacking in flexibility. Whereas, Kim et al. synthesized the ITO films on PET substrates using vertical plasma arc plating technique, exhibiting very low sheet resistance around 15.75 Ω/sq with net transmittance of 85.8% as shown in Fig. 5.6d (Kim et al. 2018b). The prepared ITO films persist a smooth surface with almost minimal value of resistance. All the enlisted literature which was done on TCO based electrodes for developing PSCs along with their % efficiency has been tabulated in Table 5.2. (Xu et al. 2022).

Nonetheless, ITO has been widely accepted as one of the most popular TCO electrode which can be utilized for developing PSC due to its outstanding electrical and optical properties, when considered at bottom electrode. However, utilizing ITO as the top electrode is quite difficult task due to its complex and damaging procedure of deposition, which cannot be commercialized on a wider scale.

Polymer Electrodes

It has been observed that under different bending conditions, a severe breakage or crack occurred on the ITO layer for a curvature radius less than 10 mm. Hence, to eliminate this problem, the PEDOT:PSS (Poly(3,4-ethyldioxythiophene) polystyrene sulphonate) is employed in fabricating the top most layer of the PSCs. Being one of the most widely used conductive polymer, which sometimes also work as an alternative to ITO due to its high mechanical properties, extremely good flexibility, high optical transmittance, solution processability, etc. PEDOT:PSS can exhibit high conductivity upto 1000 S/m and can be utilized as a hole transport layer (HTL) due to its workfunction (5 eV) similar to that of Au. In the recent times, the PEDOT:PSS is employed as one of the most commonly as well as the commercialized conductive polymers which can be even used as a buffer layer in PSCs. Another advantage behind the utilization of PEDOT:PSS is that, it perceives low sheet resistance along with high transparency in the visible region and hence it keeps a great potential for its usability as a top electrode in the fabrication of PSCs. Zhou et al. proposed the usage of PEDOT:PSS as a top electrode in PSCs when embedded in a configuration like glass/FTO/c-or m $TiO_2/MAPbI_3$/Spiro-OMeTAD/PEDOT:PSS, as shown

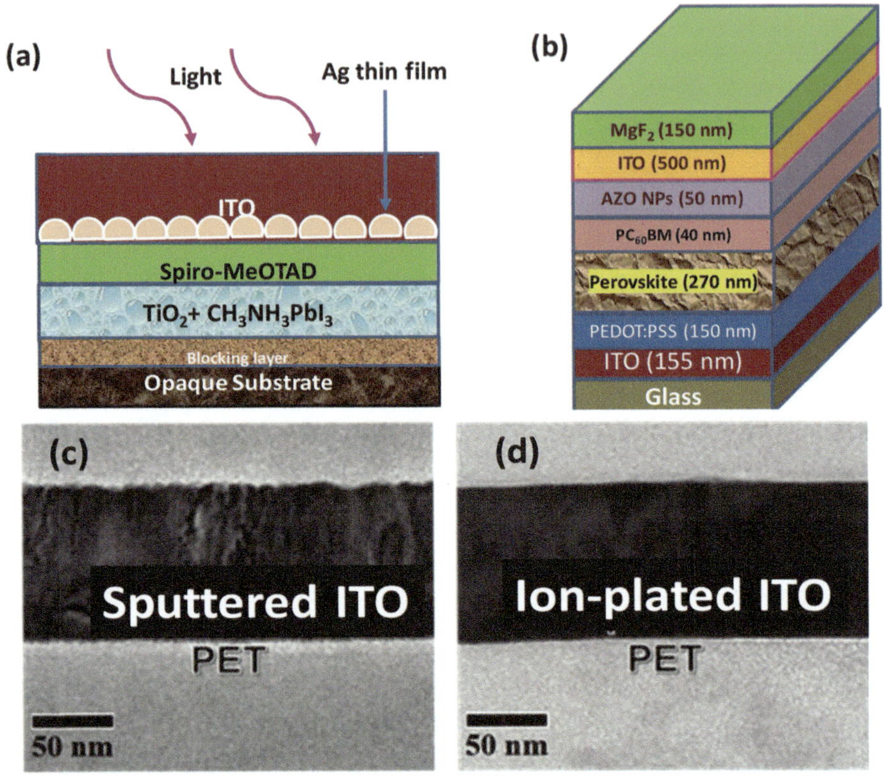

Fig. 5.6 **a** PSCs with sputtered ITO top electrode, **b** Schematic representation of the SC having double-layered HTL, **c** Images near cross-section for the sputtered ITO film, **d** and ion-plated ITO film (*note* reproduced from ref (Kim et al. 2018a) with permission from Royal Society of Chemistry, copyright 2018)

in Fig. 5.7a (Jiang et al. 2015). Additionally, using different antireflection coatings on PEDOT:PSS electrode, it enables PSCs emerging out different colors and increase device complexity. The efficiency of these prepared PSCs using PEDOT:PSS electrode can reach upto 13–15%. Also, Shi et al. developed the flexible and easily detachable PSCs device based on PEDOT:PSS electrode as shown in Fig. 5.7b (Wang et al. 2017b). The commercially used PSCs using PEDOT:PSS electrode is also represented in Fig. 5.8.

Carbon-Based Electrodes

Carbon electrode perovskite solar cell has great potential in commercial application based on its low cost, superior stability, and facile fabrication process.

Table 5.2 Summarized results of efficiency % obtained by various TCOs based electrodes

Electrode/Substrate	Sheet resistance	Flexibility of the device	Power conversion efficiency (PCE) (%)	References
ITO/PET	10–15	2000 tension only cycles at radius (r = 5 mm)	21.0	Dong et al. (2021); Burst et al. (2014)
ITO/glass	10	–	24.4	Xiong et al. (2022)
ITO/willow glass	12	Bending at radius in cm scale	19.72	Burst et al. (2014); Dai et al.)2020)
ITO/PEN	14	2000 bending cycles at r = 10 mm	21.10	Yang et al. (2021a)
PEDOT:PSS/PET	24	5000 bending cycles acquired at r = 3 mm	19.0	Hu et al. (2019)
PEDOT:PSS/PET	234.3	10,000 bending cycles at r = 0.5 mm	17.03	Lee et al. (2019)
Ag grids/PDMS/PET	7.9	1000 bending cycles at r = 6 mm	18.49	Yang et al. (2021b)
Graphene (top)/PDMS	396 ± 23	1000 bending cycles at r = 4 mm	15	Heo et al. (2018)
AgNWs/graphene/polymer	8.06	–	10.419	Dong et al. (2016)
MoO_3/graphene/PEN	552.0	5000 bending cycles at r = 2 mm	16.8	Yoon et al. (2017)

Carbon Black

The thermal breakdown of carbon-containing materials in the absence of adequate air can produce carbon black. The graphite layer is present within the carbon black structural diagram (as shown in Fig. 5.9a), and some groups (–COOH and –C–OH) are present near the edge (Boehm 1994; Hazra et al. 2019). The carbon black particles, which were around 15 mm in size, were shaped like spheres (Fig. 5.9b). The bulk specific gravity of granular carbon black was 0.35 to 0.4, that of powdered carbon black was 0.04 to 0.08, and the specific gravity of carbon black was 1.8 to 1.9 (Ku et al. 2013). It was quite huge in specific surface area and was in powder form. Carbon black has a WF of 5.05 eV, which makes it effective in removing holes from the perovskite layer. Nonetheless, hot treatment had a major impact on the carbon electrode has to be sintered at high temperature (400–500 °C). Since graphite was more conductive than carbon black. Researchers frequently combined the two materials in certain ratio, as listed in Table 5.3. The PCE reached its greatest point

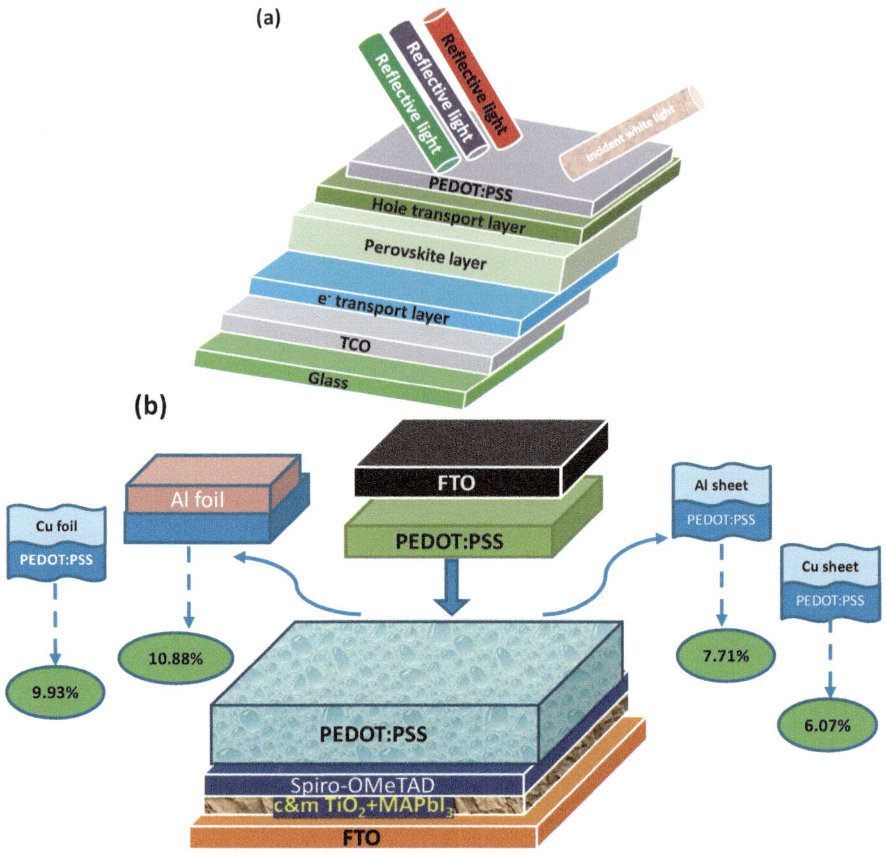

Fig. 5.7 **a** Architecture of cells with conductive polymer PEDOT:PSS as the top electrode prepared using transfer printing [figure has been redrawn considering ref (Chen et al. 2022), **b** Stacking structured PSCs with various electrode substrates and the corresponding PCEs

at 13.53% when graphite to carbon black ratio was 3 to 1. It could be an indication of how well graphite and carbon black work.

Ku et al. employed graphite/carbon black as the PSC electrode (Ku et al. 2013). Then, by gathering candle soot, Wei et al. used pure carbon black as the black electrode for C-PSCs (Wei et al. 2015a). This carbon black has a very porous structure. During the device preparation process, the carbon electrode was also printed on the prefabricated perovskite sheet. Finally, the poor conductivity of this carbon black allowed the cell to attain a PCE of 2.6%. Following that, they decided to use commercial carbon black as the black electrode, which significantly increased the device PCE till it reached 11.0%. Yang et al. had conducted similar studies (Wei et al. 2014). Because carbon black is so tiny, it had to be combined with methylammonium iodide (MAI) to create carbon ink before carbon electrodes could be deposited using inkjet printing technology. The PbI_2 precursor layer was covered with carbon

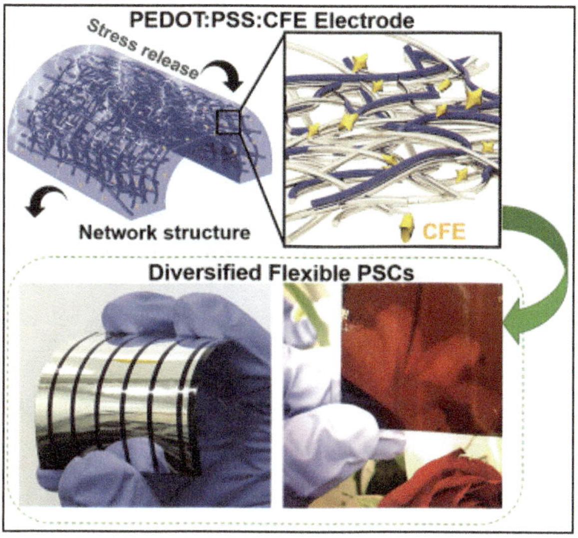

Fig. 5.8 Commercially available polymer electrodes-based PSCs [under open access license ref Hu et al. (2019)

ink, which caused MAI to react with PbI_2 and crystallize into a cubic perovskite phase. This immobilized and embedded the carbon black, producing carbon electrons. It improved the C-PSC black contact, which sped up carrier migration and decreased charge recombination. Consequently, a PCE of 11.6% was attained (Huang et al. 2018). Between the carbon black electrode and perovskite sheet, Gong et al. have recently added an intermediary layer of carbon black (Gong et al. 2020). The structure of the perovskite /carbon black/carbon electrode was used to create the device, and carbon black was used to make hole extraction easier. To produce a high-quality perovskite film, they used a one step spin-coating technique combined with a preheating procedure (Bai et al. 2018; Guo et al. 2019; Wang et al. 2019b; Fan et al. 2019; Nam et al. 2017). Fig. 5.9c–d show the device setup and band alignment (Gong et al. 2020). The production process for the carbon black interlayer, carbon black electrode, and perovskite layer are shown in Fig. 5.9e. The carbon black interlayer acted as a transition layer between the carbon rear of the electrode and perovskite, speeding up the extraction of holes. The outcome was a reported PCE of 13.13% for the device (Gong et al. 2020). To enhance the perovskite/ carbon interface, Chu et al. created a novel kind of carbon black electrode using nanoscale carbon black and carbon fiber. Using double-layer carbon electrodes (conductive and coherent layers), HTL-free PSCs were produced, raising the PCE to 14.1% (Chu et al. 2020). Furthermore, Wei et al. combined carbon black and graphite at low temperatures to create a thermoplastic carbon sheet for the flexible HTL-free PSC black electrode (Wei et al. 2015a). The ideal interface contact between the perovskite and carbon film was achieved with the following parameters: hot pressure (0.25 MPa) and a carbon film composition (3:1).

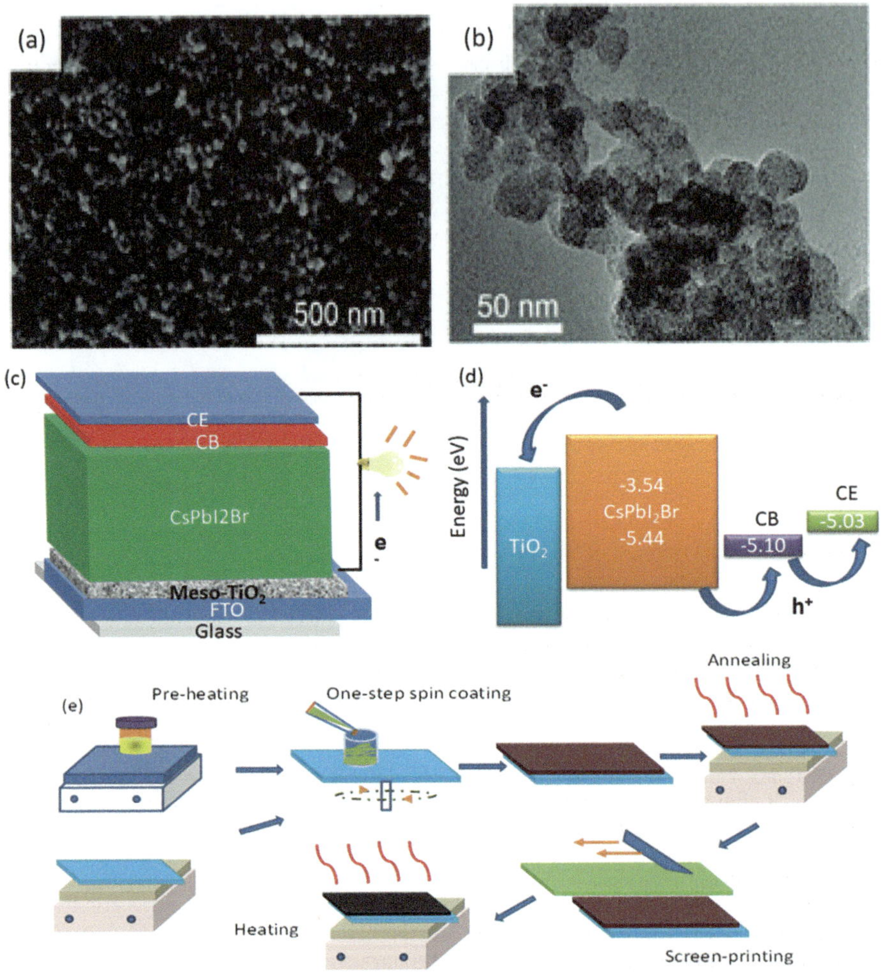

Fig. 5.9 **a** SEM and **b** TEM images of carbon black nanoparticles [under open access license, ref (Luo et al. 2017). **c** Architecture of a carbon black interlayer-based PSC. **d** Energy band alignment of a CB interlayer-based PSC. **e** Schematic of the fabrication process of C-PSCs

Table 5.3 Comparison of perovskite solar cell characteristics under standard 1 sun illumination using various elemental compositions (Cheng et al. 2016; Wei et al. 2015a)

Electrode composition	V_{oc} (V)	J_{sc} (mA/cm^2)	FF	PCE (%)
Carbon black	0.87	20.2	0.56	9.8
Graphite	0.97	18.8	0.57	10.27
Graphite/CBNPs (5:1)	0.98	20	0.58	11.43
Graphite/CBNPs (3:1)	1	21.3	0.63	13.53
MWCNT	0.87	17.94	0.51	8.04
Carbon/MWCNT	0.93	21.3	0.59	11.6

Graphene

Graphene, a sheet of carbon atoms arranged in a honeycomb lattice structure, is one of the carbon materials with a wide range of unique characteristics (as shown in Fig. 5.10a). Graphene's excellent mechanical flexibility, high charge mobility (1020 cm^2v^{-1} s^{-1}), strong electrical conductivity, and last but not least, its superior flexibility and light transmittance make it appropriate for use as a transparent electrode in flexible PSCs. However, the device structure was either planar or mesoporous, and to treat TiO_2 and create thick layers, a high temperature was required. In addition to consuming time and energy, the method made it more difficult to build flexible electronics. ETL and HTL, which are positioned at the anode and carbon electrodes, respectively, gathered photogenerated electrons and holes in the inverted (p-i-n) configuration (Babu et al. 2020). Here, the carbon electrode has a high degree of transparency within the visible and infrared spectrum. Consequently, graphene is the best option.

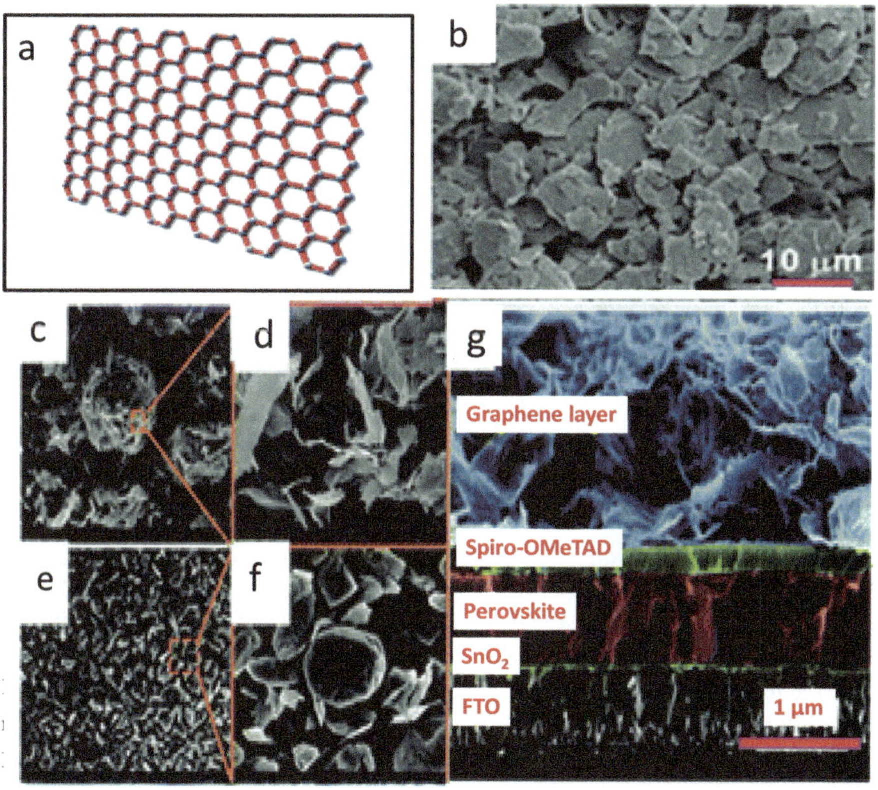

Fig. 5.10 **a** Schematic of graphene. **b** SEM image of graphene. SEM images at low and high magnifications of **c** and **d** UG and **e** and **f** bulk graphite, **g** Cross-sectional SEM image of G-PSC

Figure 5.10a structure of Graphene nanosheet, (b) displays the graphene SEM picture [reproduced from ref (Wei et al. 2015b) with permission from Royal Society of Chemistry, copyright 2015,], (c–f) displays the SEM images of bulk graphite and ultra-thin graphite [reproduced from ref (Duan et al. 2017) with permission from Elsevier, copyright 2017], (g) Cross-sectional SEM image of the graphene-based PSC structure [reproduced from ref (Zhang et al. 2019) with permission from Royal Society of Chemistry, copyright 2019].

Figure 5.10b displays the graphene SEM picture. Figure 5.10c–f displays the SEM images of bulk graphite and ultra-thin graphite. Graphene was originally used by You et al. as a transparent black electrode in inverted PSCs (You et al. 2015). After 1 sun illumination, the PCE of the front electrode side (FTO) and rear electrode (graphite side) were 12.02% and 11.6%, respectively. Transparent conductive oxide was frequently substituted with graphene as the front electrode in inverted PSCs (Sung et al. 2016). High-efficiency TCO-free PSCs were created employing graphene transparent anodes and organic hole transport materials poly (3,4-ethylene dioxythiophene): poly (styrene sulfonate) (PEDOT: PSS). The maximum PCE of 17.1% was attained by adding MoO_3 and fine-tuning the thickness of the MoO_3 layer between the graphene anode and PEDOT: PSS. Simultaneously, the PCE of the device with an ITO transparent electrode was 18.8% (Sung et al. 2016; Heo et al. 2017; Ishikawa et al. 2019). These studies found that PCE vitally depended on the number of graphene layers (Mazzotta et al. 2018). The single-layer constructed device PCE was 9.18%, but the two-layer graphene device had the highest performance, with a validated PCE of 12.37. Zhang et al. reported that graphene-based PSCs exhibit a PCE of 18.65% (Zhang et al. 2019). The optimized unpackaged device retained 90% of its initial PCE after aging for 1000 h at a high temperature of 850 °C. However, the increased graphene layers resulted in the deterioration of PCE (PCE for three and four layers were 11.45% and 11.27%, respectively). A large quantity of layers reduced the transmittance of electrodes and decreased the amounts of photons reaching the absorber layer.

Carbon Nanotube

With superior optoelectronics capabilities equivalent to graphene, carbon nanotubes were one-dimensional cylinders coiled with graphene. Since carbon nanotubes have amazing optical characteristics and flexibility, they were included in the list of potential materials for flexible PSC electrodes. These materials are included in the list of potential materials for flexible PSC electrodes. These materials included both single-walled and multi-walled carbon. The SEM images of CNTs are shown in Fig. 5.11a. Luo et al. first described cross-stacked carbon nanotube film in their work (Luo et al. 2018). Later, they employed a SnO_2-coated carbon nanotube film as the electrode in flexible inverted PSCs (Luo et al. 2017). Flexible inverted PSCs with SnO_2@CSCNT cathodes demonstrated notably better photovoltaic performance as compared to devices without SnO_2 coating. According to the findings, SnO_2@CSCNT was a favorable cathode material for PSC operation over the long term. The CSCNT film typical SEM picture is shown in the Fig. 5.11b. To limit the contact with BP and demonstrate the maximal doping impact, Jeon et al. devised an ex situ triflic acid (TFMS) vapor doping approach (Jeon et al. 2020). For conventional M-PSCs, the PCE would drop with an increase in the concentration of 2,2,7,7-tetracis (N, N-di p-methoxyphenylamine)-9,90-spirobifluorene (Spiro-MeTAD) in HTL (Fig. 5.11c). It worked like this: a higher HTL concentration led to a thicker Spiro-MeOTAD layer, which in turn reduced the hole transport capacity, and vice versa. It did not, however, occur in the case of CNT lamination PSCs as, before spiro MeOTAD restricting hole extraction, CNTs had already extracted holes in the interface (Fig. 5.11d). When Spiro-MeOTAD and optimized carbon nanotubes were combined (Fig. 5.11e), the best PCE of C-PSCs was 18.8%, greater than that of M-PSCs (Fig. 5.11f). (Jeon et al. 2020).

This resulted from the enhanced electrical conductivity achieved by doping TFMS and the enhanced hole selectivity of carbon nanotubes. In conclusion, due to its excellent flexibility, light transmittance, and superior conductivity, graphene performs well as a transparent electrode from an application standpoint (Sung et al. 2016). Because of their special structure that facilitates hole extraction, carbon nanotubes show good results in counter electrodes (Aitola et al. 2016), whereas carbon black can expand the pores in carbon paste, encourage charge transfer, and enhance interface contact

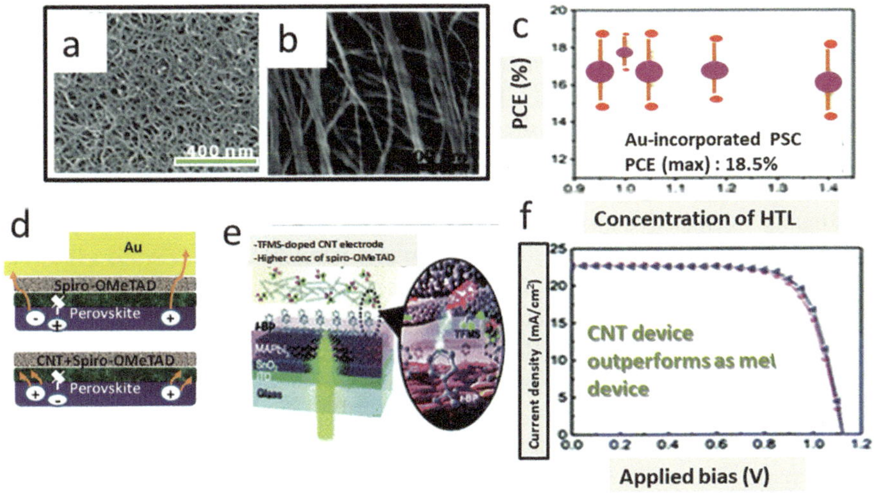

Fig. 5.11 **a** SEM image of CNTs [reproduced from ref (Wei et al. 2015b) with permission from Royal Society of Chemistry, copyright 2015], **b** Typical SEM image of the CSCNT film [reproduced from ref (Luo et al. 2018) with permission from Wiley, copyright 2018], **c** M-PSCs with different Spiro-OMeTAD concentrations. **d** Depiction of the charge flow in M-PSCs (above) and CNT-based PSCs (below). **e** Schematic of the optimized TFMS-doped CNT electrode PSC device. **f** J–V curves of the Au-based PSCs (red circle) and the optimized TFMS-doped CNT-PSCs (blue triangle) [reproduced from ref (Jeon et al. 2020) with permission from Royal Society of Chemistry, copyright 2020]

(Gong et al. 2020). Consequently, it makes sense to assume that the PCE could eventually catch up to that of traditional M-PSC nanotubes (MWCNTs) if carbon materials are used as the primary electrode material. SWCNTs typically have a diameter of 1.5 mm (Wei et al. 2015b). The charge transfers more quickly through the tube as it passes because the wall could be twisted. CNTs exhibited a work-function value of 4.95 eV and a band gap that varies along with changes in diameter (Mohammed 2020). Additionally, it has been demonstrated that the mechanical characteristics of CNTs are very advantageous for the use of flexible PSCs (Ma et al. 2020). When compared to the carbon black/graphite composite electrodes high treatment temperature (above 450 °C), CNT films could be incorporated into SCs without the need for sintering.

Conclusion

In this book chapter, the progress of research in developing the electrodes materials for perovskite solar cells (PSCs) targeting on the prime features of different electrodes, comparison of the efficiency and stability of the different electrodes has been studied successfully. The comparison on the efficiency, stability, and cost of all electrodes materials which can be used for fabricating the perovskite solar cells (PSCs) have been deeply studied in detail. The deep analysis on the research progress in the development of such electrode materials, focusing on the application features of different electrodes and their impact on the efficiency and stability of the devices have been completed successfully. The "golden triangle" parameters (i.e., efficiency, stability, cost) associated with the use of different electrode materials have been compared. Among all available metal electrodes, Cu is arguably one of the most promising metal electrodes because of its exceptional electrical conductivity and reasonably inexpensive cost. Although, an additional research is still required to meet the desired efficiency and stability requirements, Ni, Mo, and W also showed a lot of promising results. Although achieving competitively high PCEs of C-PSCs is still a significant problem, carbon-based materials are known for their low cost and exceptional stability, as well as their close Fermi energy levels to Au. Additionally, new back electrode material types like TCOs and conductive polymers have recently shown promising results in semi-transparent and tandem perovskite solar cells. We have outlined current developments in the creation of PSC rear electrode materials in this paper. In order to obtain high efficiency, long lifetime, and an acceptable product cost, certain parameters such energy level alignment, stability, conductivity, and availability should be carefully examined, even though a variety of rear electrode materials have been used successfully in PSCs.

References

Acik M, Darling SB (2016) Graphene in perovskite solar cells: device design, characterization and implementation. J Mater Chem A 4(17):6185–6235

Aitola K, Sveinbjörnsson K, Correa-Baena JP, Kaskela A, Abate A, Tian Y et al (2016) Carbon nanotube-based hybrid hole-transporting material and selective contact for high efficiency perovskite solar cells. Energy Environ Sci 9(2):461–466

Akbulatov AF, Frolova LA, Griffin MP, Gearba IR, Dolocan A, Vanden Bout DA et al (2017) Effect of Electron-transport material on light-induced degradation of inverted planar junction perovskite solar cells. Adv Energy Mater 7(19):1700476

Assi AA, Saleh WR, Mohajerani E (2021) Effect of metals (Au, Ag, and Ni) as cathode electrode on perovskite solar cells. In: IOP conference series: earth and environmental science, vol 722, no 1. IOP Publishing, p 012019

Babu V, Fuentes Pineda R, Ahmad T, Alvarez AO, Castriotta LA, Di Carlo A et al (2020) Improved stability of inverted and flexible perovskite solar cells with carbon electrode. ACS Appl Energy Mater 3(6):5126–5134

Bae JW, Lee SW, Yeom GY (2006) Doped-fluorine on electrical and optical properties of tin oxide films grown by ozone-assisted thermal CVD. J Electrochem Soc 154(1):D34

Bai D, Bian H, Jin Z, Wang H, Meng L, Wang Q, Liu SF (2018) Temperature-assisted crystallization for inorganic $CsPbI_2Br$ perovskite solar cells to attain high stabilized efficiency 14.81%. Nano Energy 52:408–415

Boehm HP (1994) Some aspects of the surface chemistry of carbon blacks and other carbons. Carbon 32(5):759–769

Burst JM, Rance WL, Meysing DM, Wolden CA, Metzger WK, Garner SM, et al (2014) Performance of transparent conductors on flexible glass and plastic substrates for thin film photovoltaics. In: 2014 IEEE 40th photovoltaic specialist conference (PVSC). IEEE, pp 1589–1592

Bush KA, Bailie CD, Chen Y, Bowring AR, Wang W, Ma W et al (2016) Thermal and environmental stability of semi-transparent perovskite solar cells for tandems enabled by a solution-processed nanoparticle buffer layer and sputtered ITO electrode. Adv Mater 28(20):3937–3943

Cacovich S, Cinà L, Matteocci F, Divitini G, Midgley PA, Di Carlo A, Ducati C (2017) Gold and iodine diffusion in large area perovskite solar cells under illumination. Nanoscale 9(14):4700–4706

Chen W, Li X, Li Y, Li Y (1971) Energy Environ Sci 2020:13

Chen R, Zhang W, Guan X, Raza H, Zhang S, Zhang Y et al (2022) Rear electrode materials for perovskite solar cells. Adv Func Mater 32(26):2200651

Chen S, Dai X, Xu S, Jiao H, Zhao L, Huang J (2021) Science 373:902; Jeong J, Kim M, Seo J, Lu H, Ahlawat P, Mishra A, Yang Y, Hope MA, Eickemeyer FT, Kim M (2021) Nature 592:381

Chen S, Xiao X, Gu H, Huang J (2021b) Iodine reduction for reproducible and high performance perovskite solar cells and modules. Sci Adv 7:eabe8130

Cheng N, Liu P, Qi F, Xiao Y, Yu W, Yu Z (2016) Multi-walled carbon nanotubes act as charge transport channel to boost the efficiency of hole transport material free perovskite solar cells. J Power Sources 332:24–29

Chu QQ, Sun Z, Ding B, Moon KS, Yang GJ, Wong CP (2020) Greatly enhanced power conversion efficiency of hole-transport-layer-free perovskite solar cell via coherent interfaces of perovskite and carbon layers. Nano Energy 77:105110

Dai X, Deng Y, Van Brackle CH, Chen S, Rudd PN, Xiao X et al (2020) Scalable fabrication of efficient perovskite solar modules on flexible glass substrates. Adv Energy Mater 10(1):1903108

De Wolf S et al (2014) Organometallic halide perovskites: sharp optical absorption edge and its relation to photovoltaic performance. J Phys Chem Lett 5:1035–1039

Deng Y, Dong Q, Bi C, Yuan Y, Huang J (2016) Air-stable, efficient mixed-cation perovskite solar cells with Cu electrode by scalable fabrication of active layer. Adv Energy Mater 6(11)

Domanski K, Correa-Baena JP, Mine N, Nazeeruddin MK, Abate A, Saliba M et al (2016) Not all that glitters is gold: metal-migration-induced degradation in perovskite solar cells. ACS Nano 10(6):6306–6314

Dong H, Wu Z, Jiang Y, Liu W, Li X, Jiao B et al (2016) A flexible and thin graphene/silver nanowires/polymer hybrid transparent electrode for optoelectronic devices. ACS Appl Mater Interfaces 8(45):31212–31221

Dong Q, Chen M, Liu Y, Eickemeyer FT, Zhao W, Dai Z et al (2021) Flexible perovskite solar cells with simultaneously improved efficiency, operational stability, and mechanical reliability. Joule 5(6):1587–1601

Dou B, Miller EM, Christians JA, Sanehira EM, Klein TR, Barnes FS et al (2017) High-performance flexible perovskite solar cells on ultrathin glass: implications of the TCO. J Phys Chem Lett 8(19):4960–4966

Duan M, Rong Y, Mei A, Hu Y, Sheng Y, Guan Y, Han H (2017) Efficient hole-conductor-free, fully printable mesoscopic perovskite solar cells with carbon electrode based on ultrathin graphite. Carbon 120:71–76

Eperon GE, Stranks SD, Menelaou C, Johnston MB, Herz LM, Snaith HJ (2014) Formamidinium lead trihalide: a broadly tunable perovskite for efficient planar heterojunction solar cells. Energy Environ Sci 7(3):982–988

Fan Y, Fang J, Chang X, Tang MC, Barrit D, Xu Z et al (2019) Scalable ambient fabrication of high-performance $CsPbI_2Br$ solar cells. Joule 3(10):2485–2502

Frost JM, Butler KT, Brivio F, Hendon CH, Van Schilfgaarde M, Walsh A (2014) Atomistic origins of high-performance in hybrid halide perovskite solar cells. Nano Lett 14(5):2584–2590

Gong S, Li H, Chen Z, Shou C, Huang M, Yang S (2020) $CsPbI_2Br$ perovskite solar cells based on carbon black-containing counter electrodes. ACS Appl Mater Interfaces 12(31):34882–34889

Green MA, Ho-Baillie A, Snaith HJ (2014) The emergence of perovskite solar cells. Nat Photonics 8(7):506–514

Guo Y, Yin X, Liu J, Que W (2019) Highly efficient $CsPbIBr_2$ perovskite solar cells with efficiency over 9.8% fabricated using a preheating-assisted spin-coating method. J Mater Chem A 7(32):19008–19016

Hazra SK, Ghosh S, Nandi TK (2019) Photo-thermal conversion characteristics of carbon black-ethylene glycol nanofluids for applications in direct absorption solar collectors. Appl Therm Eng 163:114402

Heo JH, Shin DH, Kim S, Jang MH, Lee MH, Seo SW et al (2017) Highly efficient $CH_3NH_3PbI_3$ perovskite solar cells prepared by $AuCl_3$-doped graphene transparent conducting electrodes. Chem Eng J 323:153–159

Heo JH, Shin DH, Lee ML, Kang MG, Im SH (2018) Efficient organic–inorganic hybrid flexible perovskite solar cells prepared by lamination of polytriarylamine/$CH_3NH_3PbI_3$/anodized Ti metal substrate and graphene/PDMS transparent electrode substrate. ACS Appl Mater Interfaces 10(37):31413–31421

Herz LM et al (2017) Charge-carrier mobilities in metal halide perovskites: fundamental mechanisms and limits. ACS Energy Lett 2:1539–1548

Hsiao YC, Wu T, Li M, Liu Q, Qin W, Hu B (2015) Fundamental physics behind high-efficiency organo-metal halide perovskite solar cells. J Mater Chem A 3(30):15372–15385

Hu X, Meng X, Zhang L, Zhang Y, Cai Z, Huang Z et al (2019) A mechanically robust conducting polymer network electrode for efficient flexible perovskite solar cells. Joule 3(9):2205–2218

Huang Z, Proppe AH, Tan H, Saidaminov MI, Tan F, Mei A, Tan C-S, Wei M, Hou Y, Han H (2019) ACS Energy Lett 4:1521

Huang B, Kong G, Esfahani EN, Chen S, Li Q, Yu J et al (2018) Ferroic domains regulate photocurrent in single-crystalline $CH_3NH_3PbI_3$ films self-grown on FTO/TiO_2 substrate. NPJ Quantum Mater 3(1):30

Ishikawa R, Watanabe S, Yamazaki S, Oya T, Tsuboi N (2019) Perovskite/graphene solar cells without a hole-transport layer. ACS Appl Energy Mater 2(1):171–175

Jeon I, Shawky A, Seo S, Qian Y, Anisimov A, Kauppinen EI et al (2020) Carbon nanotubes to outperform metal electrodes in perovskite solar cells via dopant engineering and hole-selectivity enhancement. J Mater Chem A 8(22):11141–11147

Jeong I, Kim HJ, Lee BS, Son HJ, Kim JY, Lee DK et al (2015) Highly efficient perovskite solar cells based on mechanically durable molybdenum cathode. Nano Energy 17:131–139

Jiang Q, Sheng X, Shi B, Feng X, Xu T (2014) Nickel-cathoded perovskite solar cells. J Phys Chem C 118(45):25878–25883

Jiang F, Liu T, Zeng S, Zhao Q, Min X, Li Z et al (2015) Metal electrode–free perovskite solar cells with transfer-laminated conducting polymer electrode. Opt Express 23(3):A83–A91

Jiang Z, Chen X, Lin X, Jia X, Wang J, Pan L et al (2016a) Amazing stable open-circuit voltage in perovskite solar cells using AgAl alloy electrode. Sol Energy Mater Sol Cells 146:35–43

Jiang Y, Luo B, Jiang F, Jiang F, Fuentes-Hernandez C, Liu T et al (2016b) Efficient colorful perovskite solar cells using a top polymer electrode simultaneously as spectrally selective antireflection coating. Nano Lett 16(12):7829–7835

Jung SK, Heo JH, Lee DW, Lee SC, Lee SH, Yoon W et al (2018) Nonfullerene electron transporting material based on naphthalene diimide small molecule for highly stable perovskite solar cells with efficiency exceeding 20%. Adv Func Mater 28(20):1800346

Kato Y, Ono LK, Lee MV, Wang S, Raga SR, Qi Y (2015) Silver iodide formation in methyl ammonium lead iodide perovskite solar cells with silver top electrodes. Adv Mater Interfaces 2(13):1500195

Kawashima T, Ezure T, Okada K, Matsui H, Goto K, Tanabe N (2004) FTO/ITO double-layered transparent conductive oxide for dye-sensitized solar cells. J Photochem Photobiol, A 164(1–3):199–202

Kim H-S, Lee C-R, Im J-H, Lee K-B, Moehl T, Marchioro A, Moon S-J et al (2012) Lead iodide perovskite sensitized all-solid-state submicron thin film mesoscopic solar cell with efficiency exceeding 9%. Sci Rep 2(1):591

Kim A, Lee H, Kwon HC, Jung HS, Park NG, Jeong S, Moon J (2016) Fully solution-processed transparent electrodes based on silver nanowire composites for perovskite solar cells. Nanoscale 8(12):6308–6316

Kim JH, Seok HJ, Seo HJ, Seong TY, Heo JH, Lim SH et al (2018a) Flexible ITO films with atomically flat surfaces for high performance flexible perovskite solar cells. Nanoscale 10(44):20587–20598

Kim JG, Na SI, Kim HK (2018b) Flexible and transparent IWO films prepared by plasma arc ion plating for flexible perovskite solar cells. AIP Adv 8(10)

Kojima A, Teshima K, Shirai Y, Miyasaka T (2009a) Organometal halide perovskites as visible-light sensitizers for photovoltaic cells. J Am Chem Soc 131:6050–6051

Kojima A, Teshima K, Shirai Y, Miyasaka T (2009b) Organometal halide perovskites as visible-light sensitizers for photovoltaic cells. J Am Chem Soc 131(17):6050–6051

Ku Z, Rong Y, Xu M, Liu T, Han H (2013) Full printable processed mesoscopic $CH_3NH_3PbI_3$/TiO_2 heterojunction solar cells with carbon counter electrode. Sci Rep 3(1):3132

Ku Z, Xia X, Shen H, Tiep NH, Fan HJ (2015) A mesoporous nickel counter electrode for printable and reusable perovskite solar cells. Nanoscale 7(32):13363–13368

Lee G, Kim MC, Choi YW, Ahn N, Jang J, Yoon J et al (2019) Ultra-flexible perovskite solar cells with crumpling durability: toward a wearable power source. Energy Environ Sci 12(10):3182–3191

Lee J, Singh S, Kim S, Baik S (2020) Carbon 157:731

Li GR, Gao XP (2020) Low-cost counter-electrode materials for dye-sensitized and perovskite solar cells. Adv Mater 32(3):1806478

Li MH, Shen PS, Wang KC, Guo TF, Chen P (2015) Inorganic p-type contact materials for perovskite-based solar cells. J Mater Chem A 3(17):9011–9019

Li Y, Meng L, Yang Y, Xu G, Hong Z, Chen Q et al (2016) High-efficiency robust perovskite solar cells on ultrathin flexible substrates. Nat Commun 7(1):10214

Li J, Dong Q, Li N, Wang L (2017) Direct evidence of ion diffusion for the silver-electrode-induced thermal degradation of inverted perovskite solar cells. Adv Energy Mater 7(14):1602922

Liang L, Cai Y, Li X, Nazeeruddin MK, Gao P (2018) Nano Energy 52:211

Liu Y, Hong Z, Chen Q, Chang W, Zhou H, Song TB et al (2015) Integrated perovskite/bulk-heterojunction toward efficient solar cells. Nano Lett 15(1):662–668

Liu Z, Shi T, Tang Z, Sun B, Liao G (2016) Using a low-temperature carbon electrode for preparing hole-conductor-free perovskite heterojunction solar cells under high relative humidity. Nanoscale 8(13):7017–7023

Liu Z et al (2020) A holistic approach to interface stabilization for efficient perovskite solar modules with over 2000-hour operational stability. Nat Energy 5:596–604

Luo Y, Chen X, Zhang C, Li J, Shi J, Sun Z et al (2015) Ag Al alloy electrode for efficient perovskite solar cells. RSC Adv 5(69):56037–56044

Luo Q, Ma H, Hao F, Hou Q, Ren J, Wu L et al (2017) Carbon nanotube based inverted flexible perovskite solar cells with all-inorganic charge contacts. Adv Func Mater 27(42):1703068

Luo Q, Ma H, Hou Q, Li Y, Ren J, Dai X et al (2018) All-carbon-electrode-based endurable flexible perovskite solar cells. Adv Func Mater 28(11):1706777

Ma C, Clark S, Liu Z, Liang L, Firdaus Y, Tao R et al (2020) Solution-processed mixed-dimensional hybrid perovskite/carbon nanotube electronics. ACS Nano 14(4):3969–3979

Mazzotta G, Dollmann M, Habisreutinger SN, Christoforo MG, Wang Z, Snaith HJ et al (2018) Solubilization of carbon nanotubes with ethylene-vinyl acetate for solution-processed conductive films and charge extraction layers in perovskite solar cells. ACS Appl Mater Interfaces 11(1):1185–1191

Mohammed MK (2020) High-performance hole conductor-free perovskite solar cell using a carbon nanotube counter electrode. RSC Adv 10(59):35831–35839

Nam JK, Jung MS, Chai SU, Choi YJ, Kim D, Park JH (2017) Unveiling the crystal formation of cesium lead mixed-halide perovskites for efficient and stable solar cells. J Phys Chem Lett 8(13):2936–2940

Nath B, Ramamurthy PC, Mahapatra DR, Hegde G (2021) Effect of top electrode using silver nano powder on the performance of perovskite solar cells. In: 2021 IEEE 48th photovoltaic specialists conference (PVSC). IEEE, pp 0326–0329

Nawaz A, Erdinc AK, Gultekin B, Tayyib M, Zafer C, Wang K, Akram MN (2016) Morphology study of inverted planar heterojunction perovskite solar cells in sequential deposition. Int J Energy Power Eng 10:922–926

Njema GG, Kibet JK, Ngari SM (2024) A review of interface engineering characteristics for high performance perovskite solar cells. Meas: Energy 100005

Paetzold UW, Qiu W, Finger F, Poortmans J, Cheyns D (2015) Nanophotonic front electrodes for perovskite solar cells. Appl Phys Lett 106(17)

Park JI, Heo JH, Park SH, Hong KI, Jeong HG, Im SH, Kim HK (2017) Highly flexible InSnO electrodes on thin colourless polyimide substrate for high-performance flexible $CH_3NH_3PbI_3$ perovskite solar cells. J Power Sources 341:340–347

Pisoni S, Fu F, Widmer R, Carron R, Moser T, Groening O et al (2018) Impact of interlayer application on band bending for improved electron extraction for efficient flexible perovskite mini-modules. Nano Energy 49:300–307

Poorkazem K, Liu D, Kelly TL (2015) Fatigue resistance of a flexible, efficient, and metal oxide-free perovskite solar cell. J Mater Chem A 3(17):9241–9248

Raptis D, Stoichkov V, Meroni SM, Pockett A, Worsley CA, Carnie M et al (2020) Enhancing fully printable mesoscopic perovskite solar cell performance using integrated metallic grids to improve carbon electrode conductivity. Curr Appl Phys 20(5):619–627

Salim T, Sun S, Abe Y, Krishna A, Grimsdale AC, Lam YM (2015) Perovskite-based solar cells: impact of morphology and device architecture on device performance. J Mater Chem A 3(17):8943–8969

Shao S, Loi MA (2020) The role of the interfaces in perovskite solar cells. Adv Mater Interfaces 7(1):1901469

Shao Y, Wang Q, Dong Q, Yuan Y, Huang J (2015) Vacuum-free laminated top electrode with conductive tapes for scalable manufacturing of efficient perovskite solar cells. Nano Energy 16:47–53
Shlenskaya NN, Belich NA, Grätzel M, Goodilin EA, Tarasov AB (2018) Light-induced reactivity of gold and hybrid perovskite as a new possible degradation mechanism in perovskite solar cells. J Mater Chem A 6(4):1780–1786
Stranks SD et al (2013) Electron-hole diffusion lengths exceeding 1 micrometer in an organometal trihalide perovskite absorber. Science 342:341–344
Sun X, Deng X, Li Z, Xiong B, Zhong C, Zhu Z et al (2020) Dopant-free cross conjugated hole-transporting polymers for highly efficient perovskite solar cells. Adv Sci 7(13):1903331
Sung H, Ahn N, Jang MS, Lee JK, Yoon H, Park NG, Choi M (2016) Transparent conductive oxide-free graphene-based perovskite solar cells with over 17% efficiency. Adv Energy Mater 6(3)
Tao L, Huang Y, Ding B, Wang H, Tang J, Zhang S et al (2024) Interfacial toughening for high-efficiency perovskite solar modules. Mater Today Energy 101611
Taylor MP, Readey DW, van Hest MF, Teplin CW, Alleman JL, Dabney MS et al (2008) The remarkable thermal stability of amorphous In-Zn-O transparent conductors. Adv Func Mater 18(20):3169–3178
Tsujimoto Y, Yamaura K, Takayama-Muromachi E (2012) Oxyfluoride chemistry of layered perovskite compounds. Appl Sci 2(1):206–219
Wang L, Li GR, Zhao Q, Gao XP (2017a) Non-precious transition metals as counter electrode of perovskite solar cells. Energy Storage Mater 7:40–47
Wang S, Hou K, Xing Y, Dong Q, Wang K, Lv Y, Shi Y (2017b) Flexibly assembled and readily detachable photovoltaics. Energy Environ Sci 10(10):2117–2123
Wang B, Iocozzia J, Zhang M, Ye M, Yan S, Jin H et al (2019a) The charge carrier dynamics, efficiency and stability of two-dimensional material-based perovskite solar cells. Chem Soc Rev 48(18):4854–4891
Wang Z, Liu X, Lin Y, Liao Y, Wei Q, Chen H et al (2019b) Hot-substrate deposition of all-inorganic perovskite films for low-temperature processed high-efficiency solar cells. J Mater Chem A 7(6):2773–2779
Wang Q, Dong Q, Li T, Gruverman A, Huang J (2016) Thin insulating tunneling contacts for efficient and water-resistant perovskite solar cells. Adv Mater 28(31)
Weerasinghe HC, Dkhissi Y, Scully AD, Caruso RA, Cheng YB (2015) Encapsulation for improving the lifetime of flexible perovskite solar cells. Nano Energy 18:118–125
Wei Z, Yan K, Chen H, Yi Y, Zhang T, Long X et al (2014) Cost-efficient clamping solar cells using candle soot for hole extraction from ambipolar perovskites. Energy Environ Sci 7(10):3326–3333
Wei H, Xiao J, Yang Y, Lv S, Shi J, Xu X et al (2015a) Free-standing flexible carbon electrode for highly efficient hole-conductor-free perovskite solar cells. Carbon 93:861–868
Wei Z, Chen H, Yan K, Zheng X, Yang S (2015b) Hysteresis-free multi-walled carbon nanotube-based perovskite solar cells with a high fill factor. J Mater Chem A 3(48):24226–24231
Wu S, Chen R, Zhang S, Babu BH, Yue Y, Zhu H et al (2019) A chemically inert bismuth interlayer enhances long-term stability of inverted perovskite solar cells. Nat Commun 10(1):1161
Xiong Z, Chen X, Zhang B, Odunmbaku GO, Ou Z, Guo B et al (2022) Simultaneous interfacial modification and crystallization control by biguanide hydrochloride for stable perovskite solar cells with PCE of 24.4%. Adv Mater 34(8):2106118
Xu Y, Lin Z, Wei W, Hao Y, Liu S, Ouyang J, Chang J (2022) Recent progress of electrode materials for flexible perovskite solar cells. Nano-Micro Lett 14(1):117
Yang Z, Babu BH, Wu S, Liu T, Fang S, Xiong Z et al (2020a) Review on practical interface engineering of perovskite solar cells: from efficiency to stability. Solar Rrl 4(2):1900257
Yang F, Liu J, Lu Z, Dai P, Nakamura T, Wang S et al (2020b) Recycled utilization of a nanoporous Au electrode for reduced fabrication cost of perovskite solar cells. Adv Sci 7(6):1902474

Yang L, Xiong Q, Li Y, Gao P, Xu B, Lin H et al (2021a) Artemisinin-passivated mixed-cation perovskite films for durable flexible perovskite solar cells with over 21% efficiency. J Mater Chem A 9(3):1574–1582

Yang Y, Min F, Qiao Y, Li Z, Vogelbacher F, Liu Z et al (2021b) Embossed transparent electrodes assembled by bubble templates for efficient flexible perovskite solar cells. Nano Energy 89:106384

Yoo JJ, Seo G, Chua MR, Park TG, Lu Y, Rotermund F, Kim Y-K, Moon CS, Jeon NJ, Correa-Baena J-P (2021a) Nature 590:587

Yoo JJ, Seo G, Chua MR, Park TG, Lu Y, Rotermund F et al (2021b) Efficient perovskite solar cells via improved carrier management. Nature 590(7847):587–593

Yoon J, Sung H, Lee G, Cho W, Ahn N, Jung HS, Choi M (2017) Superflexible, high-efficiency perovskite solar cells utilizing graphene electrodes: towards future foldable power sources. Energy Environ Sci 10(1):337–345

You J, Hong Z, Yang Y, Chen Q, Cai M, Song TB et al (2014a) Low-temperature solution-processed perovskite solar cells with high efficiency and flexibility. ACS Nano 8(2):1674–1680

You J, Yang YM, Hong Z, Song TB, Meng L, Liu Y et al (2014b) Moisture assisted perovskite film growth for high performance solar cells. Appl Phys Lett 105(18)

You P, Liu Z, Tai Q, Liu S, Yan F (2015) Efficient semitransparent perovskite solar cells with graphene electrodes. Adv Mater 27(24):3632–3638

You J, Meng L, Song TB, Guo TF, Yang Y, Chang WH et al (2016) Improved air stability of perovskite solar cells via solution-processed metal oxide transport layers. Nat Nanotechnol 11(1):75–81

Zhang C, Wang S, Zhang H, Feng Y, Tian W, Yan Y et al (2019) Efficient stable graphene-based perovskite solar cells with high flexibility in device assembling via modular architecture design. Energy Environ Sci 12(12):3585–3594

Zhang J, Zhang W, Cheng HM, Silva SRP (2020) Critical review of recent progress of flexible perovskite solar cells. Mater Today 39:66–88

Zhao J, Zheng X, Deng Y, Li T, Shao Y, Gruverman A et al (2016a) Is Cu a stable electrode material in hybrid perovskite solar cells for a 30-year lifetime? Energy Environ Sci 9(12):3650–3656

Zhao L, Kerner RA, Xiao Z, Lin YL, Lee KM, Schwartz J, Rand BP (2016b) Redox chemistry dominates the degradation and decomposition of metal halide perovskite optoelectronic devices. ACS Energy Lett 1(3):595–602

Zheng X et al (2020) Managing grains and interfaces via ligand anchoring enables 22.3%-efficiency inverted perovskite solar cells. Nat Energy 5:131–140

Chapter 6
Scaling up Perovskite Photovoltaic from Lab to Commercial Market: Progress and Challenges

Stacy A. Lynrah, Ph. Nonglen Meitei, and P. Chinnamuthu

Abstract The recent developments in perovskite photovoltaics, a cutting-edge technology praised for its high efficiency and low production expenses, have seen considerable progress. However, using perovskite solar cells on a large-scale faces several challenges, including lowering production costs, producing large-area panels, maintaining high production rates, maintaining environmental concern, and achieving high efficiency in converting sunlight to energy. Taking all these into considerations, this chapter delves into the advancements in material science, manufacturing methodologies, and device performance while pinpointing key challenges like stability, environmental impact, long-term durability, and reproducibility. It also gives a clear look at the current state of perovskite photovoltaics and offers ideas for overcoming the challenges to make them commercially viable. The chapter highlights the market potential of perovskite photovoltaics by detailing both the progress made and the ongoing challenges. It also outlines a roadmap for forthcoming research and development efforts in this field.

Keywords Perovskite · Challenges · Commercial market · Fabrication method · Lead-free perovskite

S. A. Lynrah (✉)
Electronics and Communication Engineering Department, National Institute of Technology Calicut, Kozhikode 673601, Kerala, India
e-mail: stacy@nitc.ac.in

Ph. N. Meitei
Centre for Nanotechnology, Indian Institute of Technology Guwahati, Guwahati 781039, Assam, India

P. Chinnamuthu
Electronics and Communication Engineering Department, National Institute of Technology Nagaland, Dimapur, Nagaland, India

M. Kumar and A. Raj (eds.), *Perovskite Solar Cells*, Springer Tracts in Electrical and Electronics Engineering, https://doi.org/10.1007/978-981-96-9463-1_6

Overview of Perovskite Photovoltaics

In recent years, advancements in perovskite solar cells (PSCs) have led to significant improvements, positioning them as one of the most promising developments in the field of solar energy. They are an exciting new technology in solar power, known for their high efficiency, affordability, and adaptability. Unlike traditional solar cells, which primarily utilize single-layer silicon semiconductors, PSCs are constructed using perovskite-structured compounds that can be incorporated with other materials, like silicon, to enhance light absorption across a broader spectrum. Furthermore, the manufacturing of PSCs is less resource-intensive and can be accomplished using simpler techniques like solution processing, which reduces production costs and allows for the creation of thinner, lighter, and more flexible solar panels. These benefits enable their integration into various applications such as building-integrated solar panels, wearable gadgets, portable electronics, and near-space applications (Ahmad et al. 2021; Roy et al. 2022; Zhang et al. 2024). The discovery of the mineral perovskite marks the beginning of the development of perovskite materials, $CaTiO_3$, by Gustav Rose in 1839. However, the application of perovskites in solar cells did not gain traction until the early twenty-first century. The initial breakthrough came in 2009 when Tsutomu Miyasaka et al. reported the first PSCs with 3.8% efficiency. They developed a PSC using a perovskite material as the optical absorber layer and a liquid electrolyte. This early version of the $MAPbI_3$ PSCs accomplished a conversion efficiency of 3.8%, which was quite promising for its time (Miyasaka and Jena 2022). Yet, it faced stability challenges, since the liquid electrolytes used in those early perovskite solar cells were often corrosive, degrading the perovskite material over time. By 2012, researchers such as Kim et al. had demonstrated significant improvements in device architecture and material processing, achieving power conversion efficiencies (PCEs) 9.7%. To make perovskite solar cells more stable and efficient, researchers replaced liquid electrolytes—which could leak, degrade, or be sensitive to moisture—with solid hole conductors. This change aims to create more durable cells that perform well for longer, improving their commercial potential (Kim et al. 2012). However, the latest world record was set by LONGi Green Energy Technology Co., Ltd. (referred to as "LONGi") in 2024, crystalline silicon-perovskite tandem solar cells achieving an impressive efficiency of 34.6%. This surpasses the previous record of 33.7%, which was set by King Abdullah University of Science & Technology (KAUST) although there is lagging in making it widely accessible. The optical bandgap of perovskite materials, generally ranging from 1.5 to 2.5 eV, offers significant tunability. Thus, their tuning bandgap can be readily adjusted by varying the halide composition, such as mixing iodide with bromide or bromide with chloride. This flexibility in bandgap tuning enables the development of multicolor light-emitting devices and allows customization for applications that demand specific bandgap properties (Ou et al. 2019). The charge carrier mobility is ~ 10 to 45 cm^2/Vs, however, these values can vary depending on the materials, method of deposition, structures, etc. (Kober-Czerny et al. 2022).

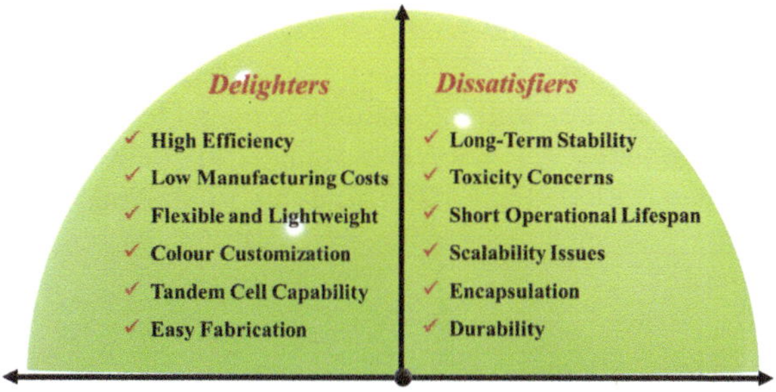

Fig. 6.1 Generalized delighters and dissatisfiers of PSCs

Shifting from laboratory-scale research to commercial-scale production has been a critical focus in recent years. PSCs have been started in labs and impressed scientists with their ability to turn sunlight into electricity, moving from experimental models to practical, mass-produced products involves many challenges and opportunities that need careful handling. The path from innovation to practical application is often challenging, and PSCs are no different. There is a complex process of scaling up PSCs, and the progress done so far is focusing on the key scientific, engineering, and economic factors that affect their shift from the lab to real-world use. Efforts have been directed toward understanding and mitigating the challenges associated with scaling up PSCs, including uniformity in large-area film deposition, long-term operational stability, and the development of cost-effective manufacturing processes (Yang et al. 2024; Zhang and Park 2024). Partnerships among academia, industry, and government entities have had a profound impact in enhancing the commercial potential of PSCs. While PSCs offer exciting benefits, especially in efficiency, flexibility, and cost-effectiveness, yet there are still key challenges that need to be addressed, particularly regarding stability, environmental impact, and long-term reliability as shown in Fig. 6.1.

Challenges in Scaling Up

Scaling up the manufacturing of PSCs involves addressing several critical challenges that must be overcome to transition from laboratory-scale research to large-scale industrial production. Maintaining high efficiencies while scaling up manufacturing is crucial for widespread deployment. Scalable fabrication techniques like sheet-to-sheet and roll-to-roll processing are being investigated, however, inconsistencies in testing protocols and a shortage of comprehensive field data have restricted the ability to assess performance across different devices. Overall, there are intrinsic and extrinsic factors affecting the stability of the perovskite, which should be paid attention to during the preparation of perovskite (Kore et al. 2024; Li et al. 2024).

This stability can be understood through several key factors like the Goldschmidt Tolerance Factor (T). The Goldschmidt tolerance factor is a critical parameter that helps predict the perovskite structure's stability. It is defined based on the ionic sizes or ratio of the bond length of the A, B, and X (A–X or B–X) components in the general formula ABX_3. When tolerance factor is ~ 1, it indicates a stable or perfect cubic structure, while deviations can lead to distortions or phase transitions i.e., if $t < 1$ octahedral distortion is evaluated. In alkali metal halide perovskites, formability is when $0.813 < t < 1.107$. (Yu et al. 2022; Chen et al. 2018; Raj et al. 2021, 2022).

$$t = \frac{r_a + r_x}{\sqrt{2}(r_b + r_x)} \tag{6.1}$$

The ionic radii of the cations at the A and B sites and the anions at the X site are denoted as r_a, r_b, and r_x, respectively. Variations in cation size can induce distortions in the crystal lattice, influencing the performance of PSCs. Another parameter is the octahedral factor which is the ratio of the ionic radius of the divalent cation (r_b) to the radius of the anion (r_x) given in Eq. (6.2). It determines the structural distortions and phase stability (Hussain et al. 2018)

$$\mu = \frac{r_b}{r_x} \tag{6.2}$$

Moreover, the size of the cations and anions also influences the arrangement and structural stability of the perovskite lattice. The ability of the organic cation (such as methylammonium) to fit into the octahedral cavities created by the B cation and halides is also crucial for maintaining structural integrity (Wang et al. 2017). The stability of perovskites is also linked to their decomposition pathways, which can vary significantly based on composition. The decomposition can lead to different gaseous products depending on the stability of the material under thermal or environmental stress. The Gibbs energy changes associated with these reactions provide insight into relative stabilities among different perovskite compositions. The tendency for certain perovskites to phase separate into more stable compounds can be composition-dependent. The creation of domains with distinct electrical characteristics resulting from phase separation may affect the efficiency of PSCs (Senocrate et al. 2019). Other factors that affect perovskite stability include ion accumulation at interfaces, hysteresis, and poor encapsulation allowing moisture and oxygen absorption. Addressing these intrinsic and extrinsic degradation factors through strategies like compositional engineering, interfacial modifications, and device encapsulation is crucial to improving the efficiency and long-term stability of PSCs. The thermodynamic stability of perovskites is significantly influenced by their composition, which includes the type and ratio of cations and anions (Zhang et al. 2024).

Transitioning from lab-scale to industrial-scale production necessitates the development of large-area processing methods that align with industry standards. Current lab techniques often rely on glove-box environments with techniques such as spin coating, which are not feasible for large-scale production (Vesce et al. 2024). Developing industry-compatible methods that can produce high-quality PSCs efficiently and cost-effectively is a major challenge. Scalable deposition methods such as blade coating, slot-die coating, and other ambient air techniques are being explored to facilitate this transition and ensure uniformity across larger surfaces (Chen et al. 2021; Du et al. 2020).

The existence of lead in perovskite materials raises health and environmental concerns, complicating the path to commercialization. Addressing these issues is essential for public acceptance and regulatory compliance. Researchers are investigating alternative materials that could replace lead without compromising efficiency, as well as methods to encapsulate or sequester lead in finished products. Efforts to develop lead-free PSCs have been ongoing, with researchers exploring alternative materials such as tin-based and bismuth-based perovskites. These materials have the potential to replace lead without compromising efficiency (Wang et al. 2021; Kwon et al. 2023). However, they come with their challenges, such as stability and efficiency, which researchers are actively working to overcome. This section will be thoroughly addressed in the upcoming section on environmental considerations.

Advancements in Manufacturing Techniques

Scaling up perovskite solar cells is a significant challenge, particularly in developing scalable deposition techniques. Researchers have utilized various methods for producing the perovskite absorber layer, with the primary approaches being solution-based and vapor-based techniques. Solution-based techniques commonly include spin coating, spray coating, inkjet printing, screen printing, and slot die coating, while vapor-based techniques are mainly categorized as physical vapor deposition (PVD) and chemical vapor deposition (CVD). Additionally, other techniques, such as the sol–gel method and electrodeposition, are also employed for the deposition of perovskite film. Each method has distinct advantages and limitations. Table 6.1 shows the summary of common perovskite thin film deposition processes. In contrast to solution-based deposition, vapor-based deposition allows for the creation of multilayer films over large areas. This method enables easy adjustment of charge collection at interfaces and the production of uniform, high-purity films. However, vapor-based techniques require significant initial investment and a controlled environment, including vacuum conditions and precise temperatures. Despite yielding high-quality films, vapor-based techniques are hampered by lower throughput and encounter challenges in producing large-area films due to the difficulty in maintaining consistency and uniformity over expansive areas (Abzieher et al. 2024). Table 6.1 shows different techniques used for developing PSCs. Each deposition technique has its unique advantages and challenges regarding scalability in the production of perovskite solar cells (Abzieher et al. 2024; Wang et al. 2016a; Bishop et al. 2020; Lu et al. 2021; Duarte and Andrade 2024). The choice of method will depend on balancing efficiency, cost-effectiveness, and the ability to maintain high-quality film properties across larger areas.

Table 6.1 Different techniques used for manufacturing PSCs

Method	Advantages	Disadvantages	Scalability on large scale
Spin coating	Simplest and cost-effective	Slow processing speed, material wastage	No
Spray coating	Simple, cost-effective, high-quality film, reproducibility	Required series of processing parameters	Yes
Inkjet printing	Simple, low-cost and reproducibility	Slow process and nozzle clogging problem	Yes
Slot-die coating	Quality film, reproducibility, continuous production capability	High maintenance cost	Yes
Vapor deposition	Reproducibility, good quality film	High-cost setup, low production	Yes

In order to transition from lab-scale to large-scale fabrication, roll-to-roll (R2R) offers a promising and straightforward path. The R2R method is a continuous manufacturing technique that involves the processing of flexible substrates through a series of rollers. This process facilitates efficient and high-throughput fabrication with uniform layers of perovskite materials over large areas. Its adaptable nature allows for integrating various deposition techniques, such as slot-die coating and spray coating (Li 2021; Beynon et al. 2023).

The hybrid approach integrates R2R processing with vapor-based evaporation techniques to enhance the scale of the fabrication process. This method allows for the fabrication of the perovskite film using R2R techniques, while vapor-based methods are employed for the deposition of the top contact. However, this approach presents certain drawbacks, including time constraints, cost implications, and production rate limitations. Parvazian and Watson (2024) have documented the use of carbon as the top electrode for the PSCs, effectively bypassing the need for costly metals such as gold and silver. This advancement holds the potential to greatly improve the scalability and sustainability of the hybrid model for large-scale production.

Spin coating excels in producing high-quality films at small scales but struggles with scalability. Spray coating offers speed and flexibility but faces challenges in achieving uniformity. Slot-die coating combines the benefits of high throughput and material efficiency while requiring careful calibration. Spray coating and slot-die coating also show promise but may not achieve the same level of throughput and efficiency as R2R when fully optimized. R2R manufacturing stands out as a highly effective method for industrial-scale production of PSCs due to its high throughput, material efficiency, and adaptability. As research continues to refine this technology and address existing challenges, R2R could play an essential role in advancing the commercialization of PSCs, making them a viable option for widespread renewable energy applications. Tackling the issues related to process optimization and environmental stability will be essential for unlocking the full potential of R2R technology in commercializing PSCs. As advancements continue in this area, R2R could play a pivotal role in advancing renewable energy solutions globally.

Materials Innovation

To design PSCs with the highest possible PCE, a deep understanding of the material's fundamental properties is essential. This includes thoroughly investigating and comprehending the intrinsic characteristics of perovskite materials, such as their crystal structure, electronic properties, defect dynamics, and interfaces with other materials. By doing so, mitigating factors that limit efficiency, such as defects that trap charge carriers or interfaces that cause energy losses can be identified. This enables the optimization of material composition and fabrication processes, leading to significant improvements in the overall performance and stability of PSCs.

The first PSCs were made using a specific type of perovskite material known as methylammonium lead halide ($MAPbI_3$), a hybrid organic–inorganic halide perovskite material (HOIHPM). Hybrid organic–inorganic halide perovskite materials represent a groundbreaking development in the field of photovoltaics and optoelectronics. One of the most notable advantages of HOIHPMs is their impressive PCE, which has reached around 29.3%. This level of performance is comparable to traditional silicon photovoltaics, which have seen limited improvements in efficiency over the past 15 years. Their high efficiency positions them as a practicable substitute to silicon-based solar cells and has attracted significant interest from research institutes and start-ups, such as Oxford PV, which are pioneering efforts to commercialize these materials. This indicates a growing market interest and the potential for HOIHPMs to serve as a key factor in shaping the future of solar energy. (Jonathan et al. 2022). Their unique crystal structure allows them to efficiently absorb sunlight and convert it into electricity with high power conversion efficiencies that rival or exceed traditional silicon solar cells. The crystal structure of $MAPbI_3$ is characterized by the presence of methylammonium (CH_3NH_3) cations at the A-site, lead (Pb) cations at the B-site, and iodide (I) anions at the X-site. This groundbreaking work was reported by Tsutomu Miyasaka and colleagues in 2009, marking a significant milestone in the field of photovoltaics. The introduction of methylammonium lead iodide ($MAPbI_3$) into solar cell technology was revolutionary due to its exceptional light absorption properties, long carrier diffusion lengths and high charge carrier mobility (Kojima et al. 2009).

More investigation has been done by C. Momblona et al. on the performance of $MAPbI_3$ PSCs, focusing on the impact of varying active layer thicknesses ranging from 300 to 900 nm and highlighting the optimal solar cell efficiency. They found out that a PCE of 12.7%, is achieved with a perovskite layer thickness of ~ 300 nm.

The study emphasizes the role of built-in potential in enhancing charge carrier separation and discusses how increasing layer thickness can improve short-circuit current density (J_{sc}) while potentially decreasing fill factor (FF).

With the appropriate modifications to the device architecture, particularly in the electron-blocking layer, high efficiencies can be maintained even at greater thicknesses. Thicker perovskite layers can absorb more sunlight, leading to a higher generation of photo-induced charge carriers, which is essential for enhancing the J_{sc}. However, this must be balanced with the charge carrier diffusion lengths; if the layer is too thick, there is an increased risk of recombination losses, where generated charge carriers recombine before reaching the electrodes. The study highlights that the long charge carrier diffusion lengths in methylammonium lead iodide perovskite allow for effective charge transport even in thicker layers, which is beneficial for maintaining efficiency. Additionally, the thickness affects the FF and open-circuit voltage (V_{oc}) of the solar cells, as thicker layers may lead to a decrease in FF due to increased series resistance and potential recombination losses. Various light intensities—100%, 50%, 20%, 10%, and 1%—were radiated onto the thick devices, resulting in FF of 48.7, 52.6, 55.9, 58.6, and 63.4, respectively. These results indicate that the FF increases as the light intensity decreases, with the highest FF of 63.4 observed at 1% light intensity (Momblona et al. 2014).

Although $MAPbI_3$ became the cornerstone of early perovskite research due to its ability to absorb light and transport charges efficiently, their hysteresis effect which is attributed to unbalanced carrier mobilities needs to be mitigated to address this issue. This led to the emergence of Formamidinium Lead Iodide ($FAPbI_3$), a perovskite material that shares many similarities with $MAPbI_3$ but with a key advantage: its slightly lower bandgap (~1.48 eV compared to $MAPbI_3$'s ~ 1.55 eV), which brings it closer to the ideal range for solar energy absorption. $FAPbI_3$ not only promises higher efficiency as shown in Fig. 6.2 due to its superior infrared absorption but also shows enhanced thermal stability, making it more suited to outdoor applications. As research progressed, $FAPbI_3$ became a vital addition to the perovskite family, offering an alternative path toward more efficient and durable solar cells.

Figure 6.3 provides a detailed comparison of the PCE of $FAPbI_3$ and $MAPbI_3$ solar cells, highlighting the superior performance of $FAPbI_3$ in photovoltaic applications achieving higher PCE values over time, reflecting advancements in material processing and device architecture. Additionally, eliminating lattice stress and stabilizing the pure α-$FAPbI_3$ phase plays a major role in enhancing the performance and longevity of PSCs. A stable α-$FAPbI_3$ phase exhibits superior electronic properties, including higher charge carrier mobility and longer diffusion lengths, which reduce recombination losses and enhance PCE. The favorable optical properties and stability of the α-$FAPbI_3$ phase, including its optimal band gap for light absorption, allow for effective utilization of the solar spectrum, enabling configurations that can exceed the Shockley–Queisser efficiency limit of single-junction cells (Zheng et al. 2022).

Parallel to these advancements, Methylammonium Lead Bromide ($MAPbBr_3$) emerged as another important variant, albeit with a wider bandgap (~2.3 eV). Solar devices utilizing these films have demonstrated an average power PCE of 8.7%, derived from forward and reverse scans. The devices achieved an open-circuit voltage

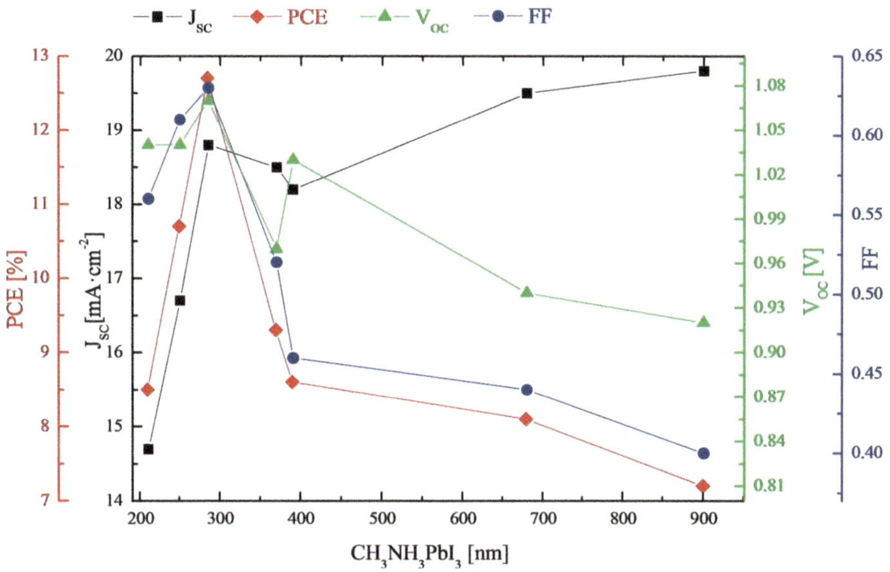

Fig. 6.2 The relationship between J_{sc}, V_{OC}, FF, and PCE values and perovskite thickness [Image reproduced from ref. (Momblona et al. 2014) (Open access agreement)]

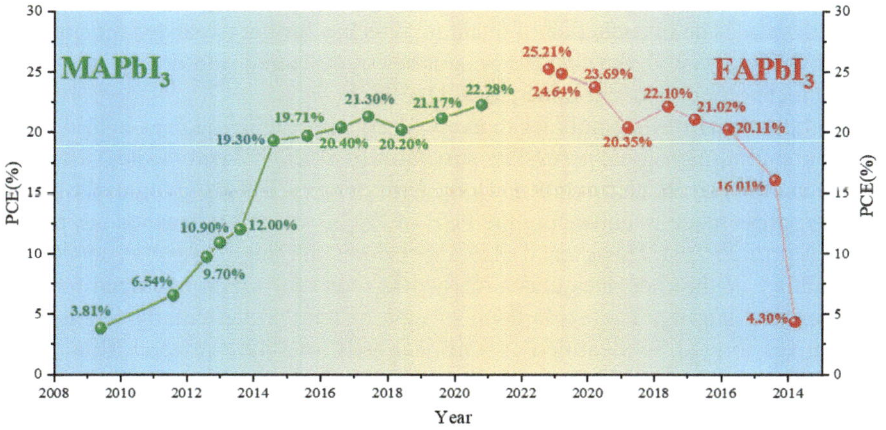

Fig. 6.3 Advancements in perovskite solar cells using MA-based and FA-based materials [Image reproduced from ref. (Zheng et al. 2022) (Open access agreement)]

(VOC) of 1.45 V, a short-circuit current density (JSC) of 9.75 mA/cm^2 and a fill factor of 61.5% (Sheng et al. 2015). Although this wider bandgap limits its use in single-junction solar cells, $MAPbBr_3$ has found its place in tandem solar cells, by pairing with lower-bandgap materials to create multi-junction structures that capture a wider solar spectrum. This combination leads to higher overall efficiency, as $MAPbBr_3$

absorbs light from the higher-energy portion of the spectrum while the other material focuses on the lower-energy part.

The pursuit of more stable and efficient perovskites didn't stop there. Researchers soon turned their attention to all-inorganic perovskites like Cesium Lead Halides ($CsPbX_3$), specifically Cesium Lead Iodide ($CsPbI_3$) and Cesium Lead Bromide ($CsPbBr_3$). These materials stand out due to their improved stability, as the inorganic cesium cation replaces the organic methylammonium or formamidinium cations, which are prone to degradation. Cesium-based perovskites, particularly $CsPbI_3$, offer the potential for enhanced thermal and environmental robustness. However, they face challenges in maintaining their perovskite phase, especially at room temperature, where they tend to revert to a less-optimal, non-perovskite phase. Researchers are continuously working on stabilizing these materials, hoping to combine their inherent stability with the high efficiencies seen in organic–inorganic hybrid perovskites (Eaton et al. 2016; Yue et al. 2019; Wang et al. 2024).

In addition to these bulk perovskite materials, recent advancements have introduced perovskite quantum dots into the mix. Quantum dots, which are nanometer-sized particles, have unique properties due to quantum confinement effects that allow for proper tuning over their optical properties. Through size modification of these quantum dots, researchers can fine-tune their bandgaps, enabling them to absorb different wavelengths of light more efficiently. This ability to manipulate light absorption makes quantum dots highly valuable in the design of high efficiency, multi-functional solar cells, as well as in light-emitting diodes (LEDs) and other optoelectronic devices. The introduction of quantum dots has further expanded the horizons of perovskite research, pushing the boundaries of efficiency and flexibility in solar energy technology (Chi and Banerjee 2021, 2022).

As the field of PSCs continues to evolve, new approaches such as hybrid materials and multidimensional perovskite structures (e.g., 2D/3D perovskites) are being investigated to elevate both output and long-term stability. It also highlights the significant progress made in enhancing the PCE of PSCs, which has reached impressive levels (up to 25.7%). However, it also underscores the ongoing issues concerning the stability of these devices under real-world scenario, which is crucial for their commercial viability. The exploration of new materials, particularly low-toxicity alternatives to lead, is identified as a critical area for future research (Elangovan et al. 2024).

Despite these advantages, HOIHPM faces significant challenges, particularly concerning stability, reproducibility, and toxicity that need to be overcome for successful commercialization. The material is prone to degradation under moisture and oxygen, leading to the formation of non-photovoltaically active phases. The instability of mixed-halide perovskites under illumination can limit achievable voltages and have implications for device performance and lifetime. Ion migration and phase segregation in the photo absorber materials can also lead to stability issues under solar cell operation, illumination, or reverse bias conditions. Additionally, advancements in fabrication techniques, the use of passivation layers, have been developed to produce high-quality structures, which are crucial for achieving optimal device performance (Vijayan et al. 2020).

Coming to metal oxides and their outstanding electronic characteristics, particularly their high electron mobility and appropriate energy bandgap, serve as excellent electron transport layers (ETLs) in PSCs, enabling efficient charge extraction while reducing energy losses, which ultimately enhances the overall performance and particularly the stability of the devices. In the context of PSCs, this means that electrons produced in the perovskite layer can quickly move to the cathode without significant delay. When light is absorbed by the perovskite material, it generates electron–hole pairs (excitons). For the solar cell to function effectively, efficient charge transport relies on high electron mobility, as it enables electrons to move swiftly from the perovskite layer to the cathode, reducing recombination losses. The rapid transport of electrons is vital for sustaining high photocurrent and maximizing the overall efficiency of PSCs. The energy bandgap determines the light wavelength ranges where a material can absorb and convert into electrical energy. For optimal performance, the energy bandgap of the electron transport layer should align well with the absorption spectrum of the perovskite material ensuring that the maximum amount of light energy can be utilized for charge generation (Chander and Tripathi 2022). Another significant factor is the proper alignment for facilitating effective charge extraction; it reduces the energy barrier for electron transfer while preventing holes from migrating into the electron transport layer. Achieving proper energy level alignment involves a combination of understanding the electronic properties of the materials, adjusting work functions, optimizing interfaces, and selecting appropriate materials. This selective transport is vital for maintaining charge balance within the solar cell, which directly influences the V_{oc} and overall power conversion efficiency. The ability of metal oxides to be engineered for optimal energy alignment makes them suitable candidates for ETL in perovskite solar cells (Soonmin et al. 2023).

Metal oxide ETLs are essential components in PSCs, significantly influencing their performance through various mechanisms. By efficiently extracting and transporting electrons to the cathode, the ETL minimizes the recombination of these charge carriers (largely dependent on alignment), which is a major loss mechanism in solar cells. Additionally, metal oxides like titanium dioxide (TiO_2) and zinc oxide (ZnO) exhibit high electron mobility, enabling fast electron movement and preventing charge build-up at the interface, further decreasing the likelihood of recombination. Valadi K et al. provide a comprehensive overview of various metal oxides used as ETLs in PSCs, highlighting their key properties and characteristics. A lower Conduction Band Minimum (CBM) value typically correlates with better electron extraction capability, which is crucial for device performance. The CBM values range from − 4.5 eV (for WO_3) to − 3.65 eV (for $SrTiO_3$), with most materials having values close to − 4 eV, suitable for a wide range of optoelectronic applications. The materials listed have bandgaps ranging from 2.6 eV to 4.0 eV, indicating a variety of optical absorption characteristics. For example, WO_3, with a bandgap of 2.6–3.1 eV, has a relatively narrow bandgap, making it suitable for applications where higher light absorption is beneficial, while other materials such as $BaSnO_3$ (3.1 eV) and ZnO (3.3 eV) are more transparent. ZnO and SnO_2 exhibit notably high electron mobility (200 $cm^2\ V^{-1}\ s^{-1}$ and 250 $cm^2\ V^{-1}\ s^{-1}$, respectively), making them promising candidates for high-speed electronic and optoelectronic devices. The refractive indices of the materials in this table range from 1.95 (for WO_3) to 2.5 (for TiO_2), with these values affecting the design of devices, particularly in optimizing light absorption and minimizing reflection losses. In summary, it highlights key properties—conduction band minimum, bandgap, bulk mobility, and refractive index—of several metal oxides, each of which has its unique advantages and limitations. These parameters are critical in determining the suitability of these materials for different types of devices, influencing efficiency, performance, and overall design considerations (Valadi et al. 2021). Overall, by improving charge extraction, ensuring proper energy alignment, providing stability, and minimizing recombination losses, metal oxide ETLs significantly enhance the efficiency and performance of PSCs, making them a promising technology for sustainable energy solutions.

In recent years, research efforts have intensified toward creating environmentally friendly lead-free PSCs, focusing on removing lead from the perovskite layers. To reduce the amount of Pb in perovskites and develop non-toxic metal halide perovskites, two common strategies are employed. One involves partially substituting lead with non-toxic metal cations, while the other replaces lead entirely with alternative cations such as tin, germanium, titanium, silver, bismuth, or antimony (Fig. 6.4). Lead-based halide perovskites, such as methylammonium lead iodide ($MAPbI_3$), have demonstrated exceptional efficiency in converting solar energy into electricity, with efficiencies reaching over 23%. However, the presence of lead poses serious health risks due to its toxicity and potential for bioaccumulation. Lead exposure is linked to various health issues, including neurological damage, particularly in children. As a result, the commercialization of lead-based PSCs is hindered by regulatory concerns and public perception regarding safety and environmental impact (Ke and Kanatzidis 2019).

Figure 6.5 provides a comparison of the potential environmental consequences of replacing lead (Pb) with various alternatives. While Pb itself is toxic, some of the lead-free alternatives (especially silver, indium, and tellurium) have higher overall environmental costs in other categories like energy demand, acidification, and eutrophication. This highlights the trade-offs involved in choosing alternative materials for PSCs (Ke and Kanatzidis 2019; Schileo and Grancini 2021).

Among the aforementioned, tin (Sn^{2+}), a divalent metal shares similarities with lead (Pb^{2+}) in terms of electronic configuration and effective ionic radius, and therefore it is the most widely used. The ionic radius of Sn^{2+} is approximately 1.35 Å, compared to 1.49 Å for Pb^{2+}, allowing Sn-based perovskites to exhibit comparable optoelectronic properties, such as photoabsorption and charge carrier mobility. Sn-based perovskites typically have a direct optical bandgap varying from 1.2 to 1.4 eV, which is advantageous for solar energy absorption. Additionally, they demonstrate higher charge carrier mobility, reported between 10^2–10^3 cm^2/Vs outperforming many lead-based counterparts (Wang et al. 2021). Sn-based perovskites, particularly those with the composition $CsSnI_3$, have demonstrated considerable potential in achieving high PCE. Recent studies have reported efficiencies exceeding 10%, with some advancements reaching up to 14% in mixed tin–lead formulations. Furthermore, innovative synthesis techniques such as solvent modulation and controlled processing conditions have been employed to improve film morphology and reduce

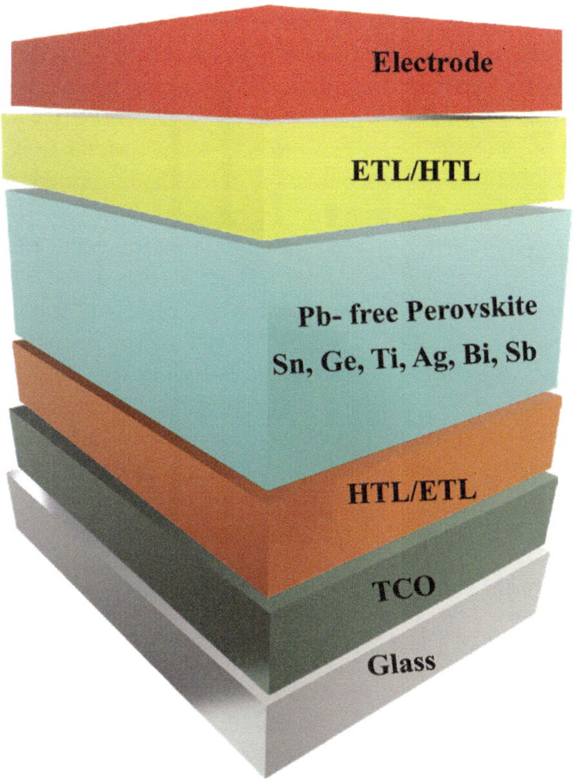

Fig. 6.4 Structure of Pb-free perovskite solar cells with alternative materials

defect densities in tin-based perovskites (Cao and Yan 2021). Understanding defect chemistry is essential for overcoming common misconceptions about the instability of tin-based PSCs, and despite the advancements, several challenges remain in the development of tin-based PSCs. The oxidation of Sn^{2+} to Sn^{4+} continues to be a significant hurdle, necessitating ongoing research into protective strategies and alternative formulations that minimize this effect (Fan et al. 2022). Additionally, while mixed tin–lead perovskites show promise for achieving higher efficiencies, balancing lead content to mitigate toxicity while maintaining performance is a critical area for future exploration. Up to now, a significant number of studies have focused on Sn-based PSCs. Performance Metrics of various Lead-Free PSCs has been discussed by Ahmed et al. listing the photovoltaic parameters in these studies. Using of different lead-free absorber and adapting the different device structure has an effect on the parameters as well as PCE of the PSCs overall (Ahmed et al. 2024).

Germanium (Ge) based perovskites typically adopt the general formula ABX_3 and with its similar ns^2 electronic structure to lead (Pb) and tin (Sn), allow for enhanced light absorption efficiency and improved carrier diffusion lengths in the perovskite structure. Notably, Ge has a smaller ionic radius than lead and tin, which can enhance the packing efficiency and stability of the perovskite structure. Ge-based

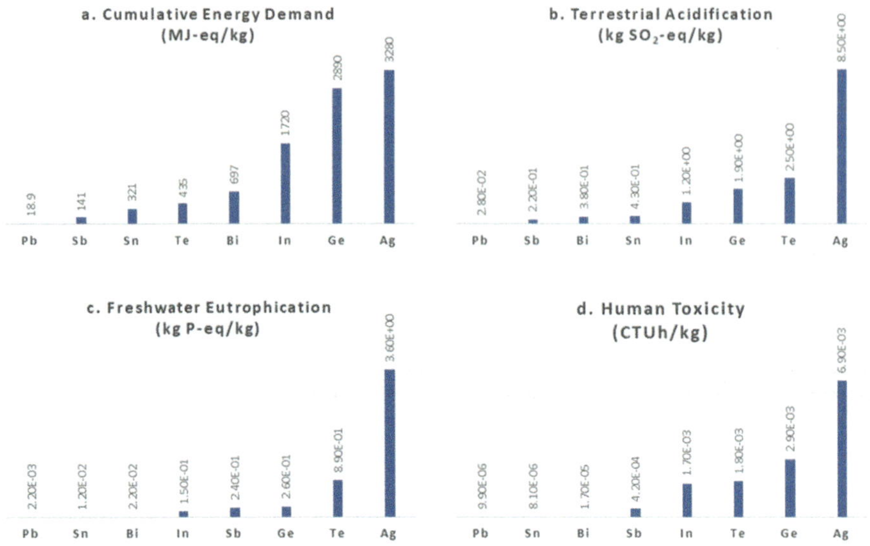

Fig. 6.5 Environmental impact comparison of lead (Pb) and alternative elements for Pb-free perovskites, covering **a** cumulative energy demand, **b** terrestrial acidification, **c** freshwater eutrophication, and **d** human toxicity (cancer and non-cancer) [Image from ref. (Schileo and Grancini 2021) Copyright 2020. Reproduced with permission from Royal Society of Chemistry and based on data from Nuss et al. [ref. (Nuss and Eckelman 2014) (Open access agreement)]

perovskites, such as $MAGeI_3$, $CsGeI_3$, and $FAGeI_3$, have demonstrated remarkable thermal stability, remaining stable at temperatures up to 150 °C. The bandgaps of these materials can be finely tuned by altering the A-site cations, enabling adjustments close to the optimal bandgap of conventional lead halide perovskites, typically around 1.5 eV. For instance, $CsGeI_3$ exhibits an optical bandgap of 1.63 eV, while other $AGeI_3$ compounds generally show wider bandgap values of 2.0 eV or more. This tunability, combined with their rhombohedral crystal structure at room temperature, positions Ge-based perovskites as viable candidates for high-performance, environmentally friendly solar cells. Researchers have highlighted the potential of Ge to replace Pb in perovskite materials, paving the way for advancements in solar technology that prioritize both efficiency and environmental safety (Aktas et al. 2022; Yang et al. 2021).

Mingzhen Liu discusses the advantages of efficient planar heterojunction PSCs, particularly those produced through vapor deposition methods. A high absorption coefficient and long charge carrier diffusion lengths allow for the effective absorption of a broad spectrum of sunlight, even with ultra-thin films, typically around 500 nm thick. Such thin films are advantageous because they reduce material costs while maintaining high efficiency levels, making them competitive with traditional silicon-based solar cells (Liu et al. 2013). Materials innovation has been pivotal in the rapid advancement of PSCs, driving improvements in efficiency, stability, and scalability. By optimizing the perovskite composition, researchers have achieved higher power conversion efficiencies while tailoring the material properties to enhance device longevity. Innovations such as the incorporation of metal oxides for charge transport layers, development of lead-free alternatives, and the exploration of hybrid organic–inorganic structures have addressed key challenges, including environmental concerns and degradation under operational conditions.

Economic and Environmental Considerations

Perovskite materials are inherently unstable, particularly under environmental stressors. Parameters such as moisture, oxygen, heat, and light exposure can deteriorate the perovskite layer. This instability can result in a loss of efficiency and performance over time, making it challenging to maintain the operational lifespan of PSCs. PSCs encounter several key challenges related to stability that impact their long-term performance and commercial viability. One major issue is compositional instability, where variations in the materials used can lead to degradation over time, resulting in a loss of efficiency. Additionally, ion migration within the perovskite structure can create defects and recombination centers, further compromising the cell's performance. Environmental factors also play a significant role; PSCs are sensitive to moisture, oxygen, and UV light, all of which can cause substantial degradation. Temperature fluctuations present another challenge, as elevated temperatures can accelerate the degradation processes, leading to faster performance declines. Moreover, the regular day-night cycles that solar cells experience can cause ion redistribution, potentially creating new defects that affect stability. Lastly, the lack of standardized testing protocols for PSCs means that their long-term stability under real-world scenarios is not fully understood, which can hinder their development and adoption in the market (Jošt et al. 2020). Moreover, the environmental impact of PSCs is a growing concern, particularly regarding the potential release of toxic lead during degradation. This has prompted investigations into lead-free alternatives and recycling methods to mitigate environmental risks associated with PSCs (Ali et al. 2023). So based on the aforementioned points the PSCs were thoroughly researched though mostly degradation over time can lead to a significant decrease in the PCE of PSCs. Even small reductions in efficiency can impact the overall energy yield, making them less competitive compared to established solar technologies. The operational lifetime of PSCs is a crucial metric for commercial viability. For commercial applications, a target lifespan of 20 years or more is necessary. How the environmental factors affect the PSCs or the life span overall are explained in detail in the next section.

Moisture has a significant negative impact on PSCs through various mechanisms. One major effect is hydrolysis, in which water exposure causes the perovskite material to decompose into non-perovskite phases. This transformation results in the loss of the desirable crystalline structure that is essential for efficient light absorption and charge transport, ultimately reducing the PCE of the solar cells. As the perovskite phase degrades, the material's ability to effectively convert sunlight into electricity is compromised (Urbina 2020). The perovskite layer, often composed of materials like $MAPbI_3$, is particularly susceptible to moisture. When exposed to water, the perovskite can undergo hydrolysis, resulting in the formation of byproducts such as PbI_2, CH_3NH_2, and HI. Additionally, moisture can induce the formation of hydrated phases, which may initially enhance crystallization but ultimately compromise the structural integrity of the perovskite layer if moisture levels become excessive. The hydrogen bonding between the organic cations and inorganic units is disrupted by moisture, further affecting the stability of the perovskite structure. Bhushan K Pore et al. reveal that when $MAPbI_3$ is exposed to moisture, it degrades and changes color from black to yellow, transforming into lead iodide (PbI_2) and releasing gases like methylamine and hydrogen iodide. Moisture can cause the formation of hydrated phases of the perovskite, which may further decompose into starting compounds and other hydrated forms. Despite these challenges, advancements like inverted PSCs have shown high power conversion efficiencies and maintained over 95% of their initial efficiency even after extensive damp-heat testing for over 1000 h. In summary, while moisture can enhance some aspects of perovskite crystallization, excessive exposure ultimately compromises the efficiency and stability of PSCs.

Figure 6.6 illustrates the intricate relationship between moisture and the stability of triple cation PSCs, specifically focusing on the composition $Cs_{0\cdot05}FA_{0\cdot85}MA_{0\cdot15}Pb(I_{0\cdot85}Br_{0\cdot15})_3$. The figure depicts how water molecules interact with the grain boundaries of the perovskite crystals, forming a hydrated phase that can enhance the potential energy difference between individual crystals and their boundaries, potentially improving solar cell performance. It highlights the effects of constant heating during the annealing process, where partially hydrated phases can revert to the original perovskite structure, but some samples may degrade, resulting in byproducts such as CsI, CH_3NH_2, HI, PbI_2, and H_2O. This indicates that while a minimal amount of moisture can be beneficial for crystallization, excessive humidity—illustrated by the degradation observed at 50% humidity—accelerates the breakdown of the perovskite material, leading to the significant formation of degradation products. Additionally, the figure include a schematic representation of the crystal structure of $FAPbI_3$ before and after the incorporation of TMS^+ (trimethylsilyl), demonstrating how such additives can modify the crystal structure and enhance solar cell performance, as evidenced by the accompanying JV (current–voltage) plots. Overall, Fig. 6.5 encapsulates the delicate balance required in moisture management to optimize the stability and efficiency of perovskite solar cells, emphasizing the dual role of moisture as both a potential enhancer and a detrimental factor (Kore et al. 2024).

As the temperature increases, perovskite materials, particularly those based on $MAPbI_3$, tend to decompose. The crystal structure of perovskite materials is affected by temperature, with different phases (α, β, and γ) appearing depending on the crystallization temperature. For instance, at 100 °C, an intermediate phase between the tetragonal and cubic phases emerges, leading to a loss in device performance (Mesquita et al. 2019). Q. Meng et al. suggest his decomposition often begins with the formation of lead iodide (PbI_2) at elevated temperatures. At temperatures around 200 °C, the perovskite films can become predominantly PbI_2, leading to a complete

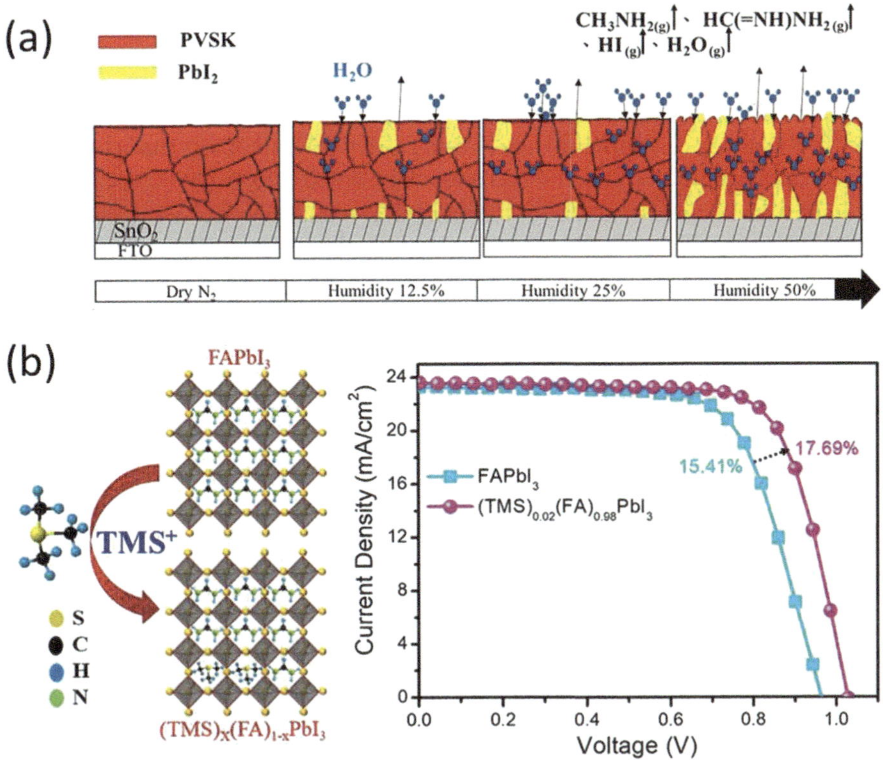

Fig. 6.6 **a** demonstrates the influence of moisture on the stability and degradation of triple cation perovskite solar cells ($Cs_{0.05}FA_{0.85}MA_{0.15}Pb(I_{0.85}Br_{0.15})_3$). It illustrates how water interacts with grain boundaries, forming hydrated phases that can enhance or degrade performance, particularly under varying humidity levels. **b** The schematic also shows the effects of TMS^+ incorporation on the crystal structure and solar cell efficiency, highlighting the importance of moisture management for optimal stability [Image reproduced from ref. (Kore et al. 2024) (Open access agreement)]

loss of the perovskite structure at temperatures around 250 °C. This thermal instability is a critical challenge for the commercial viability of PSCs. The efficiency of PSCs generally decreases with increasing temperature. The study indicates that as the heat treatment temperature rises, the J_{SC} and V_{OC} decline, resulting in a linear reduction in PCE. For example, the external quantum efficiency drops significantly from 80% at room temperature to as low as 9% at 250 °C. This reduction is attributed to the degradation of the perovskite material and the increase in defect density, which negatively impacts charge carrier mobility and recombination rates. Higher temperatures lead to an increase in defect density within the perovskite films, which can rise from approximately 3.87×10^{17} cm^{-3} to 9.03×10^{17} cm^{-3} as the temperature increases. Additionally, the absorption edge of the material redshifts, resulting in the reduction of the band gap from 1.569 eV to 1.508 eV. (Meng et al. 2021). While high temperatures can lead to thermal decomposition and efficiency loss, controlled heat treatment at moderate temperatures can enhance device performance by passivating defects. Understanding these temperature effects is essential for improving the thermal stability and commercial viability of PSCs. Elevated temperatures can also accelerate degradation processes, with faster degradation observed under open-circuit conditions due to increased charge carrier concentrations that create additional recombination centers. Furthermore, temperature impacts the material properties of the solar cells, such as the charge transport characteristics of materials like Spiro-OMeTAD. In real-world scenarios, solar modules can reach temperatures well above the standard testing conditions of 25 °C, often exceeding 70 °C, leading to performance drops that laboratory tests may not fully capture (Jošt et al. 2020).

Among all the PSCs, the most significant outcome is the promising radiation stability exhibited by the flexible PSCs. Flexible PSCs demonstrate superior tolerance to radiation compared to traditional glass-based solar cells and commercially available crystalline silicon (c-Si) solar cells. This enhanced stability is crucial for their potential use in harsh space environments, where radiation exposure is a significant concern. Flexible PSCs experience minimal performance degradation even after exposure to low-energy proton radiation, specifically at 100 keV. The findings show only small changes in important output characteristics, suggesting that these solar cells can maintain their efficiency under conditions that simulate several years of exposure in space. This resilience is further underscored by the absence of color center formation, a common issue in glass-based solar cells when exposed to radiation. The lack of such defects in flexible PSCs is a notable advantage, contributing to their overall radiation resistance (Malinkiewicz et al. 2020). Strategies have been established to minimize this light loss, like designing efficient light trapping structures, since these structures help to trap more light into the cells, thereby reducing light loss. Using of a better indium tin oxide (ITO) layer was suggested to reduce parasitic absorption. This improvement in the transparent conducting oxide layer allows for more effective light transmission into the active layer, further enhancing the overall efficiency of the solar cells. The advantages of ray optics-based light trapping mechanisms, which can effectively capture a broad spectrum of sunlight while maintaining cost efficiency. This approach is particularly beneficial for perovskite solar cells, allowing for enhanced light absorption without the complexities associated with wave optics (Wang et al. 2016b).

Oxygen: A key challenge in PSCs is the interaction between the perovskite layer and water molecules, as well as oxygen. This interaction is inevitable and occurs even during the cell assembly and testing process. When they react with oxygen, it leads to the formation of defects and the degradation of the crystalline structure. This reaction can create non-radiative recombination centers, where they act as traps reducing the number of free charge carriers overall. This in turn reduces the charge carrier lifetime and overall efficiency. The primary effects of oxygen exposure is the initiation of oxidation reactions. In addition to direct oxidation, the presence of oxygen can facilitate the formation of reactive oxygen species (ROS) when combined with light exposure. These ROS can further exacerbate degradation processes, leading to a more rapid decline in the performance of the solar cells. The interaction between oxygen and light can create a feedback loop where the degradation accelerates under operational conditions, significantly shortening the lifespan of the devices (Urbina 2020). To mitigate the adverse effects of oxygen, researchers are exploring various strategies, including the use of protective coatings and encapsulation materials that can shield perovskite layers from oxidative environments. By improving the stability of perovskite materials against oxygen exposure, it is possible to enhance the overall performance and longevity of perovskite solar cells, making them more competitive in the renewable energy market.

Additionally, UV radiation can induce photodegradation in perovskite materials. Prolonged exposure to UV light can break chemical bonds within the perovskite structure, leading to the formation of defects and reduced efficiency. When PSCs are exposed to different UV radiations UV-B (310–317 nm) and UV-A (360–380 nm). UV-B exposure is especially harmful, causing a reduction in PCE of over 50% after more than 1700 h of sunlight. In contrast, UV-A exposure did not produce similar degradation effects. The degradation was observed across various architectures with different ETL, including SnO_2, TiO_2, nanoparticle-TiO_2, and electron-beam TiO_2, indicating that the detrimental effects of UV-B are widespread regardless of the specific layer used (Farooq et al. 2018). Degradation occurs under both high and low illumination intensities of UV light, suggesting that the degradation mechanism is robust and not solely dependent on light intensity (Lang et al. 2018).

Future Directions and Outlook

One of the foremost challenges in scaling up PSC is improving their stability under real-world conditions. Transitioning from lab-scale production to industrial-scale manufacturing poses significant challenges. Although PSCs have shown remarkable efficiency in the lab, they often degrade under environmental conditions such as moisture, oxygen, heat, and UV light. Additionally, the intrinsic stability of the perovskite materials themselves must be addressed through innovations in material chemistry, such as developing more robust perovskite compositions or incorporating additives that enhance resilience in real-world conditions. Future directions should prioritize the development of scalable deposition techniques that maintain the high quality of perovskite films while ensuring cost-effectiveness. Developing advanced encapsulation techniques to shield the cells from external factors is essential for enhancing PSC stability. Innovations in interface engineering and hybrid structures could significantly improve the durability of PSCs, boosting their viability for prolonged use. Techniques such as roll-to-roll processing and inkjet printing should be optimized for large-area applications. Additionally, establishing standardized protocols for quality control during manufacturing will be essential to ensure consistency and reliability in PSC performance. The integration of PSCs with established solar technologies, such as silicon-based cells in tandem configurations, presents an exciting opportunity for enhancing overall efficiency. Optimizing bandgap tuning in perovskites is crucial for enabling efficient coupling with other materials and should be a key area of focus in ongoing research. This integration not only improves power conversion efficiencies but also allows for innovative applications in building-integrated photovoltaics (BIPV) and portable energy solutions. Given ongoing concerns about the toxicity of lead-based perovskites, it is essential to prioritize the development of environmentally friendly materials in the next phase of research. Research into lead-free perovskite formulations that achieve similar efficiencies is critical. Moreover, strategies for recycling and managing end-of-life perovskite modules will be vital for promoting sustainability within the industry. To address the stability challenges faced by PSCs, particularly for commercial applications, a multifaceted approach is necessary. One of the primary issues is compositional instability, where variations in the materials used can lead to degradation over time. To combat this, researchers are focusing on optimizing the composition of perovskite materials. This involves exploring mixed halides or alternative materials that exhibit greater stability and are less sensitive to environmental factors. By developing formulations that enhance compositional stability, the longevity and efficiency of PSCs can be significantly improved. Another critical challenge is the sensitivity of PSCs to environmental conditions such as moisture, oxygen, and UV light. Effective encapsulation techniques are essential to protect these cells from such harmful elements. Advanced encapsulation materials and methods can create a robust barrier that minimizes exposure to moisture and oxygen, thereby prolonging the lifespan of the solar cells. This protective layer is crucial for maintaining performance in real-world conditions. Temperature management is also vital, as elevated temperatures can accelerate degradation processes. Implementing thermal management strategies can help maintain optimal operating temperatures for perovskite solar cells. This could involve designing solar modules that dissipate heat more effectively or utilizing materials that are less sensitive to temperature fluctuations. By controlling the thermal environment, the performance and stability of the cells can be enhanced. Ion migration is another significant issue that can lead to the formation of defects within the perovskite structure. To mitigate this, research is being directed toward developing materials that minimize ion migration. This may include using additives or alternative hole-transporting materials that stabilize the perovskite layer and limit the movement of ions. By addressing ion migration, the overall stability and efficiency of the solar cells can be improved. Furthermore, the establishment of standardized testing protocols is crucial for assessing the long-term stability of PSCs. Current testing often does not fully replicate real-world conditions, which can lead to an incomplete understanding of their performance. By creating testing protocols that simulate various environmental conditions, including day-night cycles and temperature variations, researchers can gain better insights into the durability of these cells. This understanding is essential for ensuring that perovskite solar cells meet commercial standards. In addition to these strategies, hybrid approaches that combine perovskite materials with established technologies, such as silicon, can enhance overall efficiency and stability. Tandem solar cells that leverage the strengths of both materials can be particularly effective in improving performance and making perovskite technology more attractive for commercial applications. Conducting extensive field testing under diverse environmental conditions is also important. This real-world data collection can provide valuable insights into the performance and degradation mechanisms of perovskite solar cells, guiding further improvements and optimizations. Finally, ensuring that PSCs comply with industry standards and certifications can enhance their credibility and acceptance in the market. Adhering to safety, performance, and durability standards set by regulatory bodies is essential for fostering trust among consumers and investors. By implementing these comprehensive strategies, the stability and reliability of PSCs can be significantly enhanced, paving the way for their successful commercialization and enabling them to compete effectively with established solar technologies.

While significant strides have been made in developing perovskite solar cells, continued efforts are necessary to address stability, scalability, environmental impact, and integration with existing technologies. To bridge the gap between research and commercialization, fostering collaborations between academia, industry, and regulatory bodies will be essential. Such partnerships can facilitate knowledge transfer, resource sharing, and joint ventures aimed at overcoming technical challenges. Engaging with policymakers to establish supportive frameworks for perovskite technology can also accelerate its adoption in the market.

Conclusion

The journey of scaling up PSCs from laboratory research to commercial markets has been marked by significant progress and persistent challenges. As this chapter concludes, it is evident that while the remarkable efficiency and cost-effectiveness of PSCs present a compelling case for their adoption in the photovoltaic industry, several critical obstacles must be addressed to facilitate their successful commercialization. In a matter of years, the PCE of PSCs has surged dramatically, rising from 3.8% to 34.6% making PSCs highly competitive with traditional silicon-based solar cells. While the potential for PSCs in the commercial market is immense, realizing this potential requires concerted efforts across research, industry collaboration, and regulatory frameworks. Innovations in material science, and advanced fabrication techniques, have improved the optoelectronic properties of perovskites, enhancing their light absorption and charge transport capabilities. Moreover, scalable deposition methods have been developed, allowing for the production of large-area films suitable for commercial applications. By addressing stability concerns and scaling manufacturing processes effectively, PSCs can transition from promising laboratory results to viable solutions in the global energy landscape. To overcome these challenges, a multi-faceted approach is essential. Continued research into novel materials that enhance stability while maintaining efficiency is crucial. Additionally, developing advanced encapsulation techniques and exploring hybrid structures that combine perovskites with established photovoltaic technologies could provide pathways to improve durability and performance. In conclusion, while the potential for perovskite solar cells in the commercial market is immense, realizing this potential requires concerted efforts across research, industry collaboration, and regulatory frameworks. By addressing stability concerns and scaling manufacturing processes effectively, PSCs can transition from promising laboratory results to viable solution in the global energy landscape. The future of perovskite photovoltaics holds promise not only for enhancing renewable energy production but also for contributing to sustainable development goals worldwide.

References

Abzieher T, Moore DT, Roß M, Albrecht S, Silvia J, Tan H et al (2024) Vapor phase deposition of perovskite photovoltaics: short track to commercialization? Energy Environ Sci 17(5):1645–1663. https://doi.org/10.1039/D3EE03273F

Ahmad S, Kazim S, Grätzel M (eds) (2021) Perovskite solar cells: materials, processes, and devices. Wiley. ISBN: 978-3-527-82580-6

Ahmed S, Gondal MA, Alzahrani AS, Parvaz M, Ahmed A, Hussain S (2024) Recent trends and challenges in lead-free perovskite solar cells: a critical review. ACS Appl Energy Mater 7(4):1382–1397. https://doi.org/10.1021/acsaem.3c02327

Aktas E, Rajamanickam N, Pascual J, Hu S, Aldamasy MH, Di Girolamo D et al (2022) Challenges and strategies toward long-term stability of lead-free tin-based perovskite solar cells. Commun Mater 3(1):104. https://doi.org/10.1038/s43246-022-00327-2

Ali MU, Mo H, Li Y, Djurišić AB (2023) Outdoor stability testing of perovskite solar cells: necessary step toward real-life applications. APL Energy 1(2). https://doi.org/10.1063/5.0155845

Beynon D, Parvazian E, Hooper K, McGettrick J, Patidar R, Dunlop T et al (2023) All-printed roll-to-roll perovskite photovoltaics enabled by solution-processed carbon electrode. Adv Mater 35(16):2208561. https://doi.org/10.1002/adma.202208561

Bishop JE, Smith JA, Lidzey DG (2020) Development of spray-coated perovskite solar cells. ACS Appl Mater Interfaces 12(43):48237–48245. https://doi.org/10.1021/acsami.0c14540

Cao J, Yan F (2021) Recent progress in tin-based perovskite solar cells. Energy Environ Sci 14(3):1286–1325. https://doi.org/10.1039/D0EE04007J

Chander S, Tripathi SK (2022) Recent advancement in efficient metal oxide-based flexible perovskite solar cells: a short review. Mater Adv 3(19):7198–7211; Cao Z, Li C, Deng X, Wang S, Yuan Y, Chen Y et al (2020) Metal oxide alternatives for efficient electron transport in perovskite solar cells: beyond TiO_2 and SnO_2. J Mater Chem A 8(38):19768–19787. https://doi.org/10.1039/D2MA00700B

Chen Y, Zhang L, Zhang Y, Gao H, Yan H (2018) Large-area perovskite solar cells—a review of recent progress and issues. RSC Adv 8(19):10489–10508. https://doi.org/10.1039/C8RA00384J

Chen S, Dai X, Xu S, Jiao H, Zhao L, Huang J (2021) Stabilizing perovskite-substrate interfaces for high-performance perovskite modules. Science 373(6557):902–907. https://doi.org/10.1126/science.abi6323

Chi W, Banerjee SK (2021) Development of perovskite solar cells by incorporating quantum dots. Chem Eng J 426:131588. https://doi.org/10.1016/j.cej.2021.131588

Chi W, Banerjee SK (2022) Application of perovskite quantum dots as an absorber in perovskite solar cells. Angew Chem Int Ed 61(9):e202112412. https://doi.org/10.1002/anie.202112412

Du M, Zhu X, Wang L, Wang H, Feng J, Jiang X et al (2020) High-pressure nitrogen-extraction and effective passivation to attain highest large-area perovskite solar module efficiency. Adv Mater 32(47):2004979. https://doi.org/10.1002/adma.202004979

Duarte VC, Andrade L (2024) Recent advancements on slot-die coating of perovskite solar cells: the lab-to-fab optimisation process. Energies 17(16):3896. https://doi.org/10.3390/en17163896

Eaton SW, Lai M, Gibson NA, Wong AB, Dou L, Ma J et al (2016) Lasing in robust cesium lead halide perovskite nanowires. Proc Natl Acad Sci 113(8):1993–1998. https://doi.org/10.1073/pnas.1600789113

Elangovan NK, Kannadasan R, Beenarani BB, Alsharif MH, Kim MK, Inamul ZH (2024) Recent developments in perovskite materials, fabrication techniques, band gap engineering, and the stability of perovskite solar cells. Energy Rep 11:1171–1190. https://doi.org/10.1016/j.egyr.2023.12.068

Fan Y, Wu Y, Xu Y, Li W, Zhou H, Zhang X (2022) Situation and perspectives on tin-based perovskite solar cells. Sustainability 14(24):16603. https://doi.org/10.3390/su142416603

Farooq A, Hossain IM, Moghadamzadeh S, Schwenzer JA, Abzieher T, Richards BS et al (2018) Spectral dependence of degradation under ultraviolet light in perovskite solar cells. ACS Appl Mater Interfaces 10(26):21985–21990. https://doi.org/10.1021/acsami.8b03024

Hussain I, Tran HP, Jaksik J, Moore J, Islam N, Uddin MJ (2018) Functional materials, device architecture, and flexibility of perovskite solar cell. Emergent Mater 1:133–154. https://doi.org/10.1007/s42247-018-0013-1

Jonathan L, Diguna LJ, Samy O, Muqoyyanah M, Abu Bakar S, Birowosuto MD, El Moutaouakil A (2022) Hybrid organic–inorganic perovskite halide materials for photovoltaics towards their commercialization. Polymers 14(5):1059. https://doi.org/10.3390/polym14051059

Jošt M, Lipovšek B, Glažar B, Al-Ashouri A, Brecl K, Matič G et al (2020) Perovskite solar cells go outdoors: field testing and temperature effects on energy yield. Adv Energy Mater 10(25):2000454. https://doi.org/10.1002/aenm.202000454

Ke W, Kanatzidis MG (2019) Prospects for low-toxicity lead-free perovskite solar cells. Nat Commun 10(1):965. https://doi.org/10.1038/s41467-019-08918-3

Kim HS, Lee CR, Im JH, Lee KB, Moehl T, Marchioro A et al (2012) Lead iodide perovskite sensitized all-solid-state submicron thin film mesoscopic solar cell with efficiency exceeding 9%. Sci Rep 2(1):591. https://doi.org/10.1038/srep00591

Kober-Czerny M, Motti SG, Holzhey P, Wenger B, Lim J, Herz LM, Snaith HJ (2022) Excellent long-range charge-carrier mobility in 2D perovskites. Adv Func Mater 32(36):2203064. https://doi.org/10.1002/adfm.202203064

Kojima A, Teshima K, Shirai Y, Miyasaka T (2009) Organometal halide perovskites as visible-light sensitizers for photovoltaic cells. J Am Chem Soc 131(17):6050–6051. https://doi.org/10.1021/ja809598r

Kore BP, Jamshidi M, Gardner JM (2024) The impact of moisture on the stability and degradation of perovskites in solar cells. Mater Adv. https://doi.org/10.1039/D3MA00828B

Kwon N, Lee J, Ko MJ, Kim YY, Seo J (2023) Recent progress of eco-friendly manufacturing process of efficient perovskite solar cells. Nano Convergence 10(1):28. https://doi.org/10.1186/s40580-023-00375-5

Lang F, Shargaieva O, Brus VV, Neitzert HC, Rappich J, Nickel NH (2018) Influence of radiation on the properties and the stability of hybrid perovskites. Adv Mater 30(3):1702905. https://doi.org/10.1002/adma.201702905

Li C (2021) Roll-to-roll compatible methods and outlook for perovskite solar cells. In: E3S web of conferences, vol 242. EDP Sciences, p 01005. https://doi.org/10.1051/e3sconf/202124201005

Li B, Li S, Gong J, Wu X, Li Z, Gao D et al (2024) Fundamental understanding of stability for halide perovskite photovoltaics: the importance of interfaces. Chem 10(1):35–47. https://doi.org/10.1016/j.chempr.2023.09.002

Liu M, Johnston MB, Snaith HJ (2013) Efficient planar heterojunction perovskite solar cells by vapour deposition. Nature 501(7467):395–398. https://doi.org/10.1038/nature12509

Lu D, Zhang W, Kloo L, Belova L (2021) Inkjet-printed electron transport layers for perovskite solar cells. Materials 14(24):7525. https://doi.org/10.3390/ma14247525

Malinkiewicz O, Imaizumi M, Sapkota SB, Ohshima T, Öz S (2020) Radiation effects on the performance of flexible perovskite solar cells for space applications. Emergent Mater 3:9–14. https://doi.org/10.1007/s42247-020-00071-8

Meng Q, Chen Y, Xiao YY, Sun J, Zhang X, Han CB et al (2021) Effect of temperature on the performance of perovskite solar cells. J Mater Sci: Mater Electron 32:12784–12792. https://doi.org/10.1007/s10854-020-03029-y

Mesquita I, Andrade L, Mendes A (2019) Temperature impact on perovskite solar cells under operation. Chemsuschem 12(10):2186–2194. https://doi.org/10.1002/cssc.201802899

Miyasaka T, Jena AK (2022) Research background and recent progress of perovskite photovoltaics. Perovskite photovoltaics and optoelectronics: from fundamentals to advanced applications, pp 1–60. https://doi.org/10.1002/9783527826391.ch1

Momblona C, Malinkiewicz O, Roldán-Carmona C, Soriano A, Gil-Escrig L, Bandiello E et al (2014) Efficient methylammonium lead iodide perovskite solar cells with active layers from 300 to 900 nm. Apl Mater 2(8). https://doi.org/10.1063/1.4890056

Nuss P, Eckelman MJ (2014) Life cycle assessment of metals: a scientific synthesis. PLoS ONE 9(7):e101298. https://doi.org/10.1371/journal.pone.0101298

Ou Q, Bao X, Zhang Y, Shao H, Xing G, Li X et al (2019) Band structure engineering in metal halide perovskite nanostructures for optoelectronic applications. Nano Mater Sci 1(4):268–287. https://doi.org/10.1016/j.nanoms.2019.10.004

Parvazian E, Watson T (2024) The roll-to-roll revolution to tackle the industrial leap for perovskite solar cells. Nat Commun 15:3983. https://doi.org/10.1038/s41467-024-48518-4

Raj A, Kumar M, Anshul A (2021) Recent advancement in inorganic-organic electron transport layers in perovskite solar cell: current status and future outlook. Mater Today Chem 22:100595. https://doi.org/10.1016/j.mtchem.2021.100595

Raj A, Kumar M, Anshul A (2022) Recent progress in cesium-based lead-free halide double perovskite materials for photovoltaic applications. Phys Status Solidi (A) 219(22):2200425. https://doi.org/10.1002/pssa.202200425

Roy P, Ghosh A, Barclay F, Khare A, Cuce E (2022) Perovskite solar cells: a review of the recent advances. Coatings 12(8):1089. https://doi.org/10.3390/coatings12081089

Schileo G, Grancini G (2021) Lead or no lead? Availability, toxicity, sustainability and environmental impact of lead-free perovskite solar cells. J Mater Chem C 9(1):67–76. https://doi.org/10.1039/D0TC04552G

Senocrate A, Kim GY, Grätzel M, Maier J (2019) Thermochemical stability of hybrid halide perovskites. ACS Energy Lett 4(12):2859–2870. https://doi.org/10.1021/acsenergylett.9b01605

Sheng R, Ho-Baillie A, Huang S, Chen S, Wen X, Hao X, Green MA (2015) Methylammonium lead bromide perovskite-based solar cells by vapor-assisted deposition. J Phys Chem C 119(7):3545–3549. https://doi.org/10.1021/jp512936z

Soonmin H, Hardani Nandi P, Mwankemwa BS, Malevu TD, Malik MI (2023) Overview on different types of solar cells: an update. Appl Sci 13(4):2051. https://doi.org/10.3390/app13042051

Urbina A (2020) The balance between efficiency, stability and environmental impacts in perovskite solar cells: a review. J Phys: Energy 2(2):022001. https://doi.org/10.1088/2515-7655/ab5eee

Valadi K, Gharibi S, Taheri-Ledari R, Akin S, Maleki A, Shalan AE (2021) Metal oxide electron transport materials for perovskite solar cells: a review. Environ Chem Lett 19(3):2185–2207. https://doi.org/10.1007/s10311-020-01171-x

Vesce L, Stefanelli M, Rossi F, Castriotta LA, Basosi R, Parisi ML (2024) Perovskite solar cell technology scaling-up: eco-efficient and industrially compatible sub-module manufacturing by fully ambient air slot-die/blade meniscus coating. Prog Photovoltaics Res Appl 32(2):115–129. https://doi.org/10.1002/pip.3741

Vijayan A, Johansson MB, Svanström S, Cappel UB, Rensmo H, Boschloo G (2020) Simple method for efficient slot-die coating of $MAPbI_3$ perovskite thin films in ambient air conditions. ACS Appl Energy Mater 3(5):4331–4337. https://doi.org/10.1021/acsaem.0c00039

Wang G, Liu D, Xiang J, Zhou D, Alameh K, Ding B, Song Q (2016a) Efficient perovskite solar cell fabricated in ambient air using one-step spin-coating. RSC Adv 6(49):43299–43303. https://doi.org/10.1039/C6RA05893K

Wang DL, Cui HJ, Hou GJ, Zhu ZG, Yan QB, Su G (2016b) Highly efficient light management for perovskite solar cells. Sci Rep 6(1):18922. https://doi.org/10.1038/srep18922

Wang Z, Shi Z, Li T, Chen Y, Huang W (2017) Stability of perovskite solar cells: a prospective on the substitution of the A cation and X anion. Angew Chem Int Ed 56(5):1190–1212. https://doi.org/10.1002/anie.201603694

Wang M, Wang W, Ma B, Shen W, Liu L, Cao K et al (2021) Lead-free perovskite materials for solar cells. Nano-Micro Lett 13:1–36. https://doi.org/10.1007/s40820-020-00578-z

Wang X, Gao Y, Guo J, Wu M (2024) Synthesis of VS_4 and VO on carbon nanofibers for back interface modification in HTL free CsPbBr 3-based perovskite solar cells. Carbon 230:119566. https://doi.org/10.1016/j.carbon.2024.119566

Yang D, Zhang G, Lai R, Cheng Y, Lian Y, Rao M et al (2021) Germanium-lead perovskite light-emitting diodes. Nat Commun 12(1):4295. https://doi.org/10.1038/s41467-021-24616-5

Yang C, Hu W, Liu J, Han C, Gao Q, Mei A et al (2024) Achievements, challenges, and future prospects for industrialization of perovskite solar cells. Light: Sci Appl 13(1):227. https://doi.org/10.1038/s41377-024-01461-x

Yu Y, Xia J, Liang Y (2022) Basic understanding of perovskite solar cells and passivation mechanism. AIP Adv 12(5). https://doi.org/10.1063/5.0058441

Yue M, Su J, Zhao P, Lin Z, Zhang J, Chang J, Hao Y (2019) Optimizing the performance of $CsPbI_3$-based perovskite solar cells via doping a ZnO electron transport layer coupled with interface engineering. Nano-Micro Lett 11:1–14. https://doi.org/10.1007/s40820-019-0320-y

Zhang C, Park NG (2024) Materials and methods for cost-effective fabrication of perovskite photovoltaic devices. Commun Mater 5:194. https://doi.org/10.1038/s43246-024-00636-8

Zhang L, Wang Y, Meng X, Zhang J, Wu P, Wang M et al (2024) The issues on the commercialization of perovskite solar cells. Mater Futures 3(2):022101. https://doi.org/10.1088/2752-5724/ad37cf

Zheng Z, Wang S, Hu Y, Rong Y, Mei A, Han H (2022) Development of formamidinium lead iodide-based perovskite solar cells: efficiency and stability. Chem Sci 13(8):2167–2183. https://doi.org/10.1039/D1SC04769H

Chapter 7
Stability of Perovskite Solar Cell: A Real Challenge

Ipsita Mohanty and Udai P. Singh

Abstract Perovskite solar cells (PSCs) have attracted global attention due to their remarkable efficiency gains, increasing from 3.8% in 2009 to 26% today, driven by the exceptional properties of perovskite materials such as excellent light absorption, high charge-carrier mobility, and long lifetimes. Metal halide perovskites, including methyl ammonium lead iodide, offer the potential for low-cost, scalable solar technologies. However, the primary challenge lies in their instability and degradation, particularly under moisture and environmental exposure, limiting their operational lifespan to well below the 10-year requirement for commercial viability. The soft, ionic nature of perovskite materials complicates efforts to enhance long-term stability. Current research focuses on structural modifications, advanced charge transport materials, and improved encapsulation techniques to address these challenges. This chapter examines key factors influencing PSC stability, highlights recent advancements, and explores future directions, emphasizing encapsulation's pivotal role in achieving durable and commercially viable devices while inspiring further innovation in this transformative field.

Keywords Perovskite · PSCs · Stability · Efficiency · Degradation

Introduction

Non-renewable fossil fuels have been the primary contributor of reliable energy supply worldwide. But with increase in population, global energy demand is also in consistent rise. However, the depletion of these finite energy sources, coupled with their associated environmental challenges, has raised significant concerns. So, a growing shift to alternative energy sources, especially renewable energy is

I. Mohanty
Department of Physics, Dharanidhar University, Keonjhar, Odisha, India

U. P. Singh (✉)
School of Electronics Engineering, Kalinga Institute of Industrial Technology (KIIT) Deemed to be University, Bhubaneswar, Odisha, India
e-mail: singhup@kiit.ac.in

M. Kumar and A. Raj (eds.), *Perovskite Solar Cells*, Springer Tracts in Electrical and Electronics Engineering, https://doi.org/10.1007/978-981-96-9463-1_7

highly necessary. Within this framework, solar energy stands out as a crucial renewable resource due to its abundance, sustainability, and inherent purity. Additionally, harnessing solar energy poses no threat to the environment (Celik et al. 2017). Among the diverse array of solar technologies, photovoltaics (PV) has garnered significant attention.

Numerous types of solar cells have been engineered till date, but the silicon solar cells remain the most prevalent due to its impressive power conversion efficiency (PCE) and their durability. Despite this, these cells lack behind because of their long fabrication processes and high production costs. In recent times, PSCs have gained recognition as a promising alternative to the traditional silicon solar cells (Cho et al. 2017). Organic–inorganic perovskites have become a leading option among the next generation solar cells, due to its remarkable certified PCE exceeding 25% (https://www.nrel.gov/pv/cell-efficiency.html). Since official certification began in 2013, these materials have consistently set new records, indicating that it may only be inevitable before they surpass the current world record PCE of 26.8%, held by conventional crystalline silicon (Si) hetero-structure cells (Ahn and Choi 2024).

To achieve the commercialization of perovskite-based solar devices across various applications, the most significant issue is the material's low stability. This instability is particularly concerning for the cells that operate in energy-harvesting modes, as these materials tend to degrade much more rapidly when exposed to light illumination and electrical bias. To start with, the lead-based PSCs featuring $CH_3NH_3PbI_3$ have achieved an impressive power conversion efficiency of 25.7% (Celik et al. 2017). However, these halide perovskites face a critical challenge: inherent instability, as they degrade rapidly when exposed to humid environments due to the susceptibility of their organic components to decomposition (Mohanty et al. 2024). Despite numerous theories and strategies aimed at mitigating the degradation of $CH_3NH_3PbI_3$, a definitive solution remains elusive. The organic component of the material, methyl ammonium ($CH_3NH_3^+$), is particularly susceptible to degradation under humid conditions, leading to the breakdown of the perovskite structure and the formation of unwanted by-products like lead iodide (PbI_2). This degradation process significantly reduces the material's efficiency and lifetime. Additionally, the soft, ionic nature of the perovskite lattice makes it prone to phase transitions, which further destabilizes the material under environmental stress. This has been well explained in the later section of this chapter. Thus, with such difficulties, achieving stable, durable $CH_3NH_3PbI_3$-based devices remains a key focus of research, with efforts directed toward improving encapsulation, modifying the perovskite structure, and using alternative materials to enhance its resistance to degradation (Ava et al. 2019).

Similarly, the stability of methyl ammonium tin iodide ($CH_3NH_3SnI_3$) poses significant challenges, similar to its lead-based counterpart, $CH_3NH_3PbI_3$, but with additional complexities due to the chemical nature of tin. $CH_3NH_3SnI_3$ is highly susceptible to oxidation, where tin (Sn^{2+}) readily oxidizes to tin (Sn^{4+}) in the presence of oxygen and moisture, leading to rapid degradation of the perovskite structure (Mohanty et al. 2023). This oxidative instability accelerates the material's decomposition and limits its operational lifespan. Furthermore, like $CH_3NH_3PbI_3$, the $CH_3NH_3^+$ component in $CH_3NH_3SnI_3$ is prone to degradation under humid conditions, compounding the stability issues. Thermal instability and phase transitions further exacerbate the material's vulnerability. Efforts to improve the stability of $CH_3NH_3SnI_3$ focus on incorporating additives, alloying with other cations, enhancing encapsulation techniques, and developing alternative organic or inorganic components to mitigate its sensitivity to environmental factors and oxidation (Ayaydah et al. 2023).

This chapter focuses on the complex factors affecting PSC stability, strategies for improvement, recent advancements, and future prospects. It prioritises on the critical facets of PSC stability. The hurdles posed by environmental factors such as oxygen, moisture, temperature, and light remain unresolved. Emphasis is placed on the importance of solar cell encapsulation, which is crucial for future commercialization efforts. This chapter aims to furnish comprehensive insight on the stability issues and challenges associated with PSCs, with the goal of attracting young researchers to this emerging field. Through ongoing innovation and dedication to solving stability issues, the future of perovskite solar cells looks promising, paving the way for their integration into the global energy landscape.

Factors Affecting Stability

Environmental Factors

During the fabrication of PSCs, factors like moisture, oxygen, temperature, and UV phonons can explicitly impact the stability of the perovskite layer, which is schematically represented in Fig. 7.1. Predominantly, organo-metal halides are extremely sensitive to moisture, as excessive water can impair perovskite's crystallinity, leading to degradation of their structure and performance. Niu et al. reviewed the chemical stability of PSCs. In this review, the authors have thoroughly studied the issue of stability considering the most-widely used perovskite $CH_3NH_3PbI_3$. According to reports, the moisture, oxygen, and UV radiation are the critical contributors to the degradation process of perovskite materials. These environmental factors accelerate the breakdown of the perovskite layer, compromising its structural integrity and reducing overall device stability (Niu et al. 2015). The degradation of stability follows the following reactions:

$$CH_3NH_3PbI_3(s) \leftrightarrow PbI_2(s) + CH_3NH_3I\ (aq) \quad (7.1)$$

$$CH_3NH_3I\ (aq) \leftrightarrow CH_3NH_2(aq) + HI\ (aq) \quad (7.2)$$

$$4HI(aq) + O_2(g) \leftrightarrow 2I_2(s) + 2H_2O\ (l) \quad (7.3)$$

$$2HI(aq) \leftrightarrow H_2(g) + I_2(s) \quad (7.4)$$

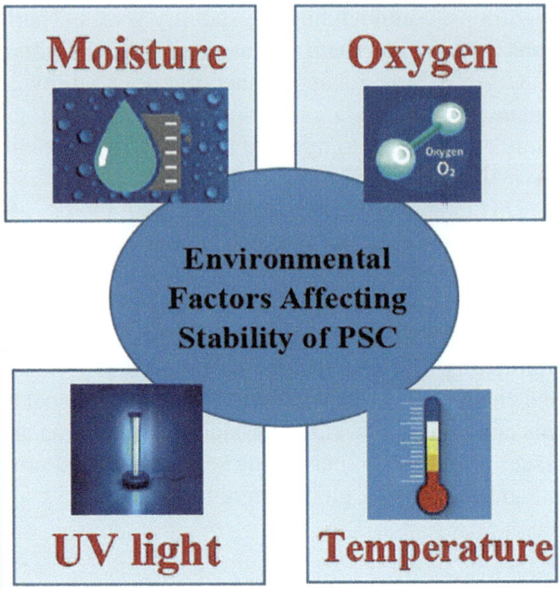

Fig. 7.1 Environmental factors affecting stability of PSC

The reaction (7.1) represents the synthesis and decomposition process of $CH_3NH_3PbI_3$. The equilibrium of reaction (7.2) results in the coexistence of CH_3NH_3I, CH_3NH_2, and HI within the film. There exist two primary pathways for the degradation of HI in the subsequent step. The first method involves a redox reaction facilitated by oxygen (reaction 7.3). The second method is a photochemical reaction, where HI decomposes into H_2 and I_2 when exposed to UV radiation (reaction 7.4). The consumption of HI through reactions (7.3) and (7.4) propels the entire degradation process forward. Other types of perovskites, such as $CH_3NH_3PbBr_3$ and $CH_3NH_3PbI_2Cl$, undergo analogous reactions, with the only difference being the substitution of reactants with the corresponding chemicals (Khalid and Mallick 2023).

Considering the fact that, organic–inorganic halide perovskites are highly vulnerable to moisture and oxygen, majority of the manufacturing procedure needs to be conducted in a glove box filled with inert gas. When these constructed models are tested in peripheral circumstances, substantial degradation has been commonly monitored. Seok et al. reported that $CH_3NH_3PbI_3$ began to decompose at 55% humidity, evidenced by a noticeable colour transition from dark brown to yellow. This degradation leads to a substantial drop in efficiency, limiting the potential for outdoor applications of PSCs (Noh et al. 2013).

Apart from moisture, oxygen, humidity, and stability of PSCs are also greatly affected by UV light exposure. Reports reveal that on exposure to UV light, the stability of PSCs result in a significant degradation in PCE. Farooq et al. (2018) systematically studied the stability of PSCs based on $CH_3NH_3PbI_3$, exposing them to (i) 310–317 (UV-B range) and (ii) 360–380 nm (UV-A range), under accelerated conditions. The key findings of this article highlight that UV-B light (310–317 nm) has a harmful effect on the stability of PSCs, leading to more than 50% degradation in efficiency after over 1700 sun hours of exposure. Photo degradation occurs under both high and low intensities of UV-B light, but is not observed under UV-A light (360–380 nm) exposure (Khalid and Mallick 2023). Similarly, Ito et al. also reported the degradation of $CH_3NH_3PbI_3$ to PbI_2 after exposure to light for 12 h. The degradation reactions are as follows (Niu et al. 2015; Ito et al. 2014):

$$2\,I^- \leftrightarrow I_2 + 2\,e^- \tag{7.5}$$

$$2\,CH_3NH_3^+ \leftrightarrow 3\,CH_3NH_2 \uparrow + 3\,H^+ \tag{7.6}$$

$$I^- + I_2 + 3\,H^+ + 2\,e^- \leftrightarrow 3\,HI \uparrow \tag{7.7}$$

Krishnan et al. have reviewed the key factors affecting the stability of PSCs. The authors reviewed that solution-processed $MAPbI_3$ remains unaffected by exposure to O_2; however, its degradation in ambient air or moisture is primarily attributed to the presence of H_2O. Also, trace amounts of H_2O can act as an n-dopant, altering the energy levels of $MAPbI_3$. The proposed degradation pathway involves the release of NH_3 and HI as volatile by-products, leaving behind PbI_2, making the decomposition reaction irreversible. Studies indicate that moisture combined with light accelerates the decomposition of $MAPbI_3$ into PbI_2, a process that cannot be reversed. Under sunlight, $MAPbI_3$ readily forms residues like PbI_2 and I_2, whereas in the dark, it tends to form reversible hydrates. Exposure to 60% relative humidity for 18 h results in irreversible decomposition under light but forms stable hydrates in darkness. While stability tests in the dark are a standard practice, evaluating stability under illumination is crucial to understanding the material's operational efficiency and degradation mechanisms, as exposure to moisture and light leads to permanent breakdown (Krishnan et al. 2019).

Intrinsic Material Properties

Apart from environmental factors, intrinsic material properties like crystal structure, phase stability and ion migration also have an essential part in the stability issue of PSCs, as shown in Fig. 7.2. Generally, perovskite has cubic structure along with ABX_3 crystal structure, where A = monovalent cation, B = bivalent cation, and X = monovalent halide. The structural stability of PSCs reflects the ability of their crystal structure to withstand various external factors. Achieving crystal symmetry in halide perovskites depends on maintaining an appropriate tolerance factor. This tolerance factor supplies a tentative approximation of both the stability and potential deformities within the crystal structure, playing a crucial influence on deciding its overall durability (Khalid and Mallick 2023). The Goldschmidt tolerance factor τ is utilized to compute the level of stability of PSCs, which is defined as:

$$\tau = \frac{R_A + R_B}{\sqrt{2}(R_B + R_X)} \tag{7.8}$$

where R_A, R_B, and R_X denote the ionic radii of A, B, and X ions, respectively. A stable perovskite crystal structure is typically achieved when the tolerance factor falls among 0.7 and 1.0. If the tolerance factor exceeds this range, lattice distortion occurs, which can lead to instability in the device (Elangovan et al. 2024). The widely used $MAPbI_3$, holds a tolerance factor (τ) close to 0.9, allowing it to crystallize in the photoactive black perovskite phase. In contrast, $CsPbI_3$, with a smaller τ value of around 0.85, and $FAPbI_3$, with a higher τ value of about 1, prefer to solidify into the photo-inactive δ phase at room temperature (Rao et al. 2021).

The nature of phase transition under several circumstances also affects the stability of PSCs. Stoumpos et al. experimented on $MAPbI_3$ to study the phase transition of perovskite, and it has been observed that perovskite exhibits a tetragonal phase at 373 K. This suggests stability among the tetragonal and cubic phases of perovskite materials. The phase of perovskites varies with temperature, and they exhibit a stable orthorhombic phase as the temperature drops lower than 100 K. Different phases emerge as temperature changes, influencing the overall stability of the material (Stoumpos et al. 2013). As the temperature rises to 160 K, the perovskite transitions from the stable orthorhombic phase to a tetragonal phase (Kawamura et al. 2002). As the temperature rises further to 330 K, the tetragonal phase is substituted by the cubic phase. This temperature-dependent phase transition serves a decisive role in establishing the structural properties of perovskite materials (Khalid and Mallick 2023; Whitfield et al. 2016).

Perovskite crystals, being ionic in nature and inherently soft, have been widely observed to exhibit phenomena such as ion migration, segregation, mixed electronic-ionic conduction, and partial amorphization, particularly in Pb-based halide perovskites (Liu et al. 2021; Guerrero et al. 2016). These phenomena are considered major contributors to the instability of perovskite materials and are also linked to abnormal photocurrent–voltage (*J*–*V*) behaviours, like the light-soaking effect and hysteresis (Ahn and Choi 2024; Lopez-Varo et al. 2018). According to

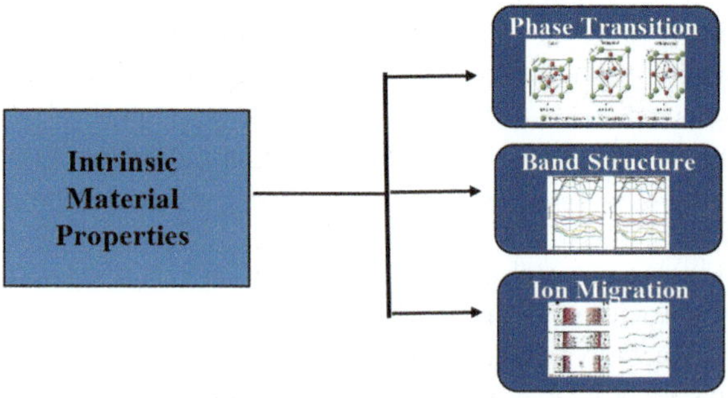

Fig. 7.2 Intrinsic material properties

reports, the presence of a space-charge zone at the PSC interface has been demonstrated and he observed hysteresis has been influenced by both the scan rate and scan direction. As the device performance is evaluated based on the JV curve characteristics, hysteresis can lead to misinterpretation of performance metrics (Zarazua et al. 2016; Belisle et al. 2017). Due to ion migration at grain boundaries, the photocurrent has been degraded that overall degrades the stability of PSC (Zhang et al. 2020).

Strategies for Enhancing Stability

Material Engineering

Despite the considerable challenges, PSCs are still hailed as the future of photovoltaic (PV) technology. Extensive research and development efforts are focused on bringing this vision a reality, with scientists and companies aiming to improve efficiency and stability, extend lifespan, and substitute toxic substances with reliable alternatives. To upgrade the stability of PSCs, material engineering plays a crucial role. This includes chemical engineering of the perovskite layer, the use of additives, and encapsulation techniques (Xu et al. 2016).

Chemical engineering of perovskite requires essential modification of X-site anions and A-site cations. To form a stable perovskite structure, the size of A cation is highly important as the ionic radii of A, B, and X in ABX_3 is often used to access the distortion of the MX_6 octahedron. Specifically, a smaller ionic radius for X tends to promote the formation of cubic structures. The A-site cation must fit within the void created by four corner-sharing MX_6 octahedral. The structural and optoelectronic differences between $MAPbI_3$ and $FAPbI_3$ are likely caused by the variation in the ionic radii of their constituent ions. This structural instability in orthorhombic perovskites is possible to be mitigated by adjusting τ through suitable substitutions, as the τ serves as a valuable indicator of crystal structure stability and distortion (Berhe et al. 2016).

Encapsulation of PSCs is another crucial strategy aimed at enhancing their stability and reliability, addressing their inherent vulnerability to environmental stressors such as humidity, light, and temperature fluctuations. Effective encapsulation protects the delicate perovskite materials from degradation, thereby prolonging their operational lifespan and improving performance efficiency. Recent advancements include an encapsulation strategy that is compatible with industrial applications, solvent-less, and strain-free based on a viscoelastic (semi-solid)/highly viscous (liquid) polyolefin, such as homo polymer PIB. The viscoelasticity of polyisobutylene (PIB) plays a key role in minimizing thermomechanical stresses, both during the encapsulation process and in response to temperature fluctuations during accelerated aging tests. When used without edge sealants, PSCs encapsulated with PIB-based materials successfully endured multiple accelerated aging results, incorporating ISOS-D1 preconditioning (240 h), ISOS-D2 (over 1000 h at 85 °C), ISOS-L1 light soaking

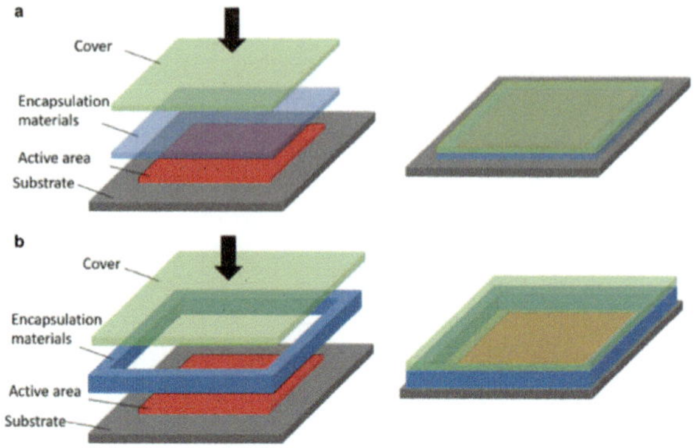

Fig. 7.3 **a** Surface and **b** edge encapsulation in PSC (Dipta et al. 2024)

(over 1000 h), a customized thermal shock test (200 cycles), and a modified humidity freeze test (10 cycles). Remarkably, these devices preserved beyond 80% of their initial PCE throughout the tests (Mariani et al. 2024). In recent years, significant advancements have been made in encapsulation and lead containment techniques in lead-based PSCs. Dipta et al. have extensively reviewed various encapsulation techniques addressing lead leakage and stability concerns in PSCs. These techniques include hydrophobic, lead-absorbing ionogel encapsulants, epoxy resin-based self-healing polymers, on-device lead-absorbing encapsulants, thin-film encapsulants, and mixed resin-based encapsulation. According to the reports, the ideal encapsulant for PSCs should be a hydrophobic, transparent insulator that efficiently blocks moisture and oxygen while preventing lead leakage. The hydrophobic nature of the material helps to repel water, a major cause of perovskite degradation, thereby preserving the stability and efficiency of the cells. Transparency ensures that the encapsulant does not hinder light absorption, which is essential for maintaining high photovoltaic performance. Additionally, the encapsulant must act as a barrier to prevent lead from leaching out, addressing environmental and health concerns associated with lead-based perovskite materials (Dipta et al. 2024). Li et al. summarized and highlighted standard test procedures for PSCs that can be used to compare results across different labs. Figure 7.3 demonstrates the contrast between edge-sealed and surface-sealed encapsulants. Encapsulants can be used to cover the surface, the edges, or both the active areas of devices. Significant research indicates that sealing the perovskite provides the best results for device stability and longevity (Li et al. 2021a).

Interface Engineering

Interface engineering has emerged as a productive technique for attaining prolonged stability of PSCs (Li and Liu 2017). For instance, adding a surplus layer separating the perovskite layer and the charge transport layers can help prevent adverse chemical reactions while ensuring efficient carrier transport at the interface (Xie et al. 2016). Xie et al. added designed and synthesized a fullerene derivative $C_{60}NH_2$ with an amino-terminated group, and used it as a passivation layer between TiO_2 and perovskite in planar perovskite solar cells to improve the quality of the perovskite film, electron transfer efficiency, photovoltaic performance, and stability of the devices (Li et al. 2021a). In this work, [6,6]-Phenyl-C_{61}-butyric acid 2-((2-(dimethylamino)ethyl)(methyl)amino) ethyl ester (PCBDAN) has been employed as an interlayer to enhance the performance of polymer-based devices. Figure 7.4 shows the SEM micrographs of the devices incorporating PCBDAN reveal intact perovskite and PCBM/PCBDAN layers. It has been observed that in devices without PCBDAN, damage to the active layers and direct contact between various layers lead to a reduction in shunt resistance and an escalation in series resistance, suggesting interface deterioration.

Utilization of different types of hole transport layer (HTL) and electron transport layer (ETL) is one of the major interface engineering techniques to normalize the stability issue. According to reports, the stability of the HTL and ETL in PSCs is crucial, as their instability can directly affect the stability of the perovskite layer. The peak efficiencies in the PSCs have been attained with organic transport layers. However, due to their hygroscopic and acidic nature, these organic layers can absorb moisture from the environment, making them highly susceptible to degradation. Once these organic layers degrade, water can penetrate the perovskite layer, leading to its degradation. Replacing organic charge transport layers with inorganic materials has been shown to enhance PSC stability, as inorganic materials offer higher resistance to moisture. Despite the advantages of inorganic charge transport layers, such as their low cost and excellent stability, they generally exhibit lower efficiency compared to their organic counterparts (Byranvand et al. 2019).

Interface engineering also include the methodology of waterproofing the tunnelling layer in order to stabilize the perovskite layer. Waterproofing of tunnelling layer involves addition of a hydrophobic contact layer that will be acting as a barrier between the interface of perovskite and charge transport layers (Li and Liu 2017). The additional layer is to be applied on the top of the absorber layer that will ensure it does not negatively impact charge transport and protect the perovskite layer as an insulating tunnelling contact (Byranvand et al. 2019). Cross-linking and passivation is another technique of interface engineering. In chemistry, cross-linking denotes to the process of connecting molecular chains through chemical bonds, facilitated

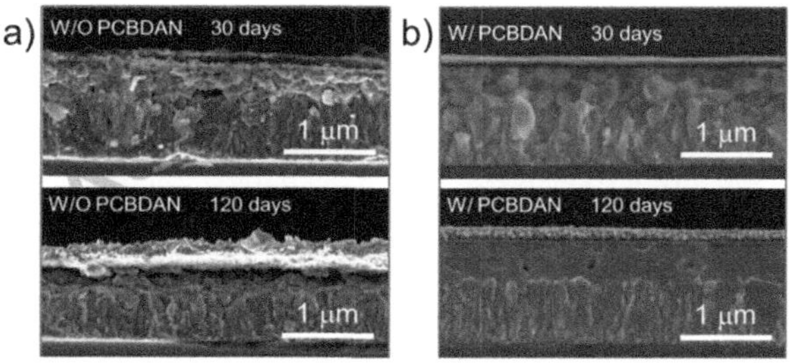

Fig. 7.4 SEM cross-sectional micrographs of devices **a** without and **b** with PCBDAN stored in air-condition with 45% (Li et al. 2021a)

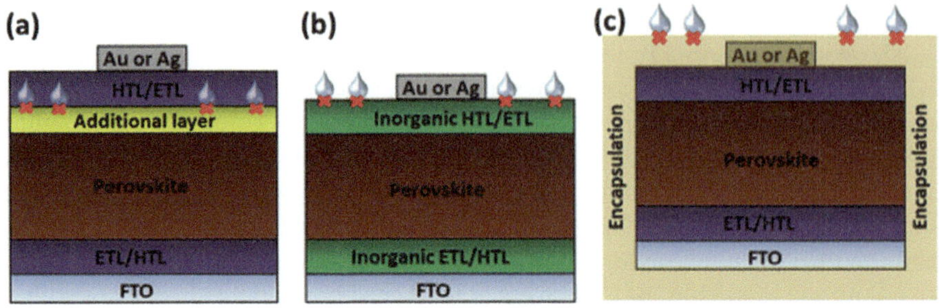

Fig. 7.5 Interface engineering (Byranvand et al. 2019)

by factors like light, heat, or mechanical force. Cross-linking can help stabilize the interfacial surfaces of devices, improving both morphology of the film and device stability. Various inorganic materials, polymers, or small molecules are utilized as supplementary layers or interfacial cross-linking agents in PSCs to enhance their long-lasting stability (Byranvand et al. 2019). Byranvand et al. have reviewed the prevailing approaches for improving the stability of PSCs involve transforming either the attributes of the perovskite material or the interfaces within the device. Figure 7.5 depicts the schematic representation of waterproofing, use of supplementary layer or an inorganic HTL and encapsulation to enhance moisture stability of PSCs (Byranvand et al. 2019).

Dong et al. reported the deposition of ultrathin compact aluminium oxide (Al_2O_3) as an interface modification method by atomic layer deposition (ALD) to stabilize $CH_3NH_3PbI_3$. Ultrathin ALD-Al_2O_3 acts as a waterproof layer, isolating the $CH_3NH_3PbI_3$ layers from moisture and thereby significantly enhancing the device's stability, while minimally affecting performance through the tunnelling effect. Notably, the PCE of the ALD-Al_2O_3-based cell sustained approximately 90% of its initial value after 24 days of storage in air (Dong et al. 2015).

Recent Advancements in Stability Enhancements

A brief summary of the recent advancements in the field of enhancing the stability of PSCs has been tabulated in Table 7.1. The table highlights various PSC structures, displaying a range of materials and strategies aimed at improving both stability and efficiency. It compares different HTL, ETL, and additives, illustrating the impact of material choices on device performance. Ye et al. have used a copper-based HTL on an inverted PSC to strengthen the device's stability and performance. The fabrication of high-grade $CH_3NH_3PbI_3$ perovskite films on a CuSCN layer using a one-step fast deposition-crystallization method has proven advantageous. This method produces films with reduced surface roughness and lower interface contact resistance compared to the conventional two-step deposition process, resulting in improved film quality and better device performance. This process yielded an efficiency of 16.6% and the stability of the PSC remains nearly unchanged, demonstrating excellent long-term durability (Ye et al. 2015). Stability enhancements are achieved through methods such as iodine confinement, surface modifications, and the incorporation of quantum dots (QDs), with several devices maintaining over 90% of their original efficiency for up to 1000 h or more. Yang et al. incorporated poly(2-vinylpyridine) into perovskite films in order to boost the stability of PSCs by confining iodine. The incorporation of P2VP effectively inhibits the diffusion and volatilization of iodine in perovskite solar cells, suppressing perovskite decomposition and significantly improving device stability. PSCs utilizing P2VP achieved an efficiency of 22.97% and retained 90.9% of their initial efficiency after being subjected to 85 °C for 750 h, showcasing enhanced durability and performance under thermal stress (Yang et al. 2023). Notably, devices using advanced strategies like multifunctional molecules or surface modifications exhibit both high efficiency and long-term stability, making them strong candidates for commercialization. Thus, the table summarises significant progress in material engineering and device optimization, demonstrating promising pathways toward commercial viability for PSCs.

Conclusion

In conclusion, the transition from non-renewable energy sources to sustainable alternatives like solar power is vital for addressing the growing global energy demand and mitigating environmental concerns. PSCs have become a promising alternative to conventional Si-based photovoltaics due to their impressive efficiency and potential for lower production costs. However, their commercialization faces significant challenges, primarily related to their stability. This chapter has explored the key factors influencing the stability of PSCs, including environmental degradation, intrinsic material properties, and the impact of ion migration. It has also highlighted strategies to enhance stability, such as material and interface engineering, and encapsulation techniques. Through these advancements, the stability of PSCs has seen considerable improvements, as demonstrated by devices retaining high efficiency even under prolonged stress conditions. Ongoing research focused on addressing these stability challenges will play a crucial role in the future integration of PSCs into the global energy landscape, ultimately contributing to the widespread adoption of solar technology.

Table 7.1 Recent advancements in stability enhancements

Sl No.	Device structure	Implementation	Stability	PCE (%)	References
1	ITO/CuSCN/$CH_3NH_3PbI_3$/C_{60}/BCP/Ag	CuSCN as HTL	Almost invariable, 28 h. Stored in ambient air in dark environment	16.6	Ye et al. (2015)
2	FTO/SnO_2/$(FAPbI_3)_{0.85}(MAPbBr_3)_{0.15}$/spiro-OMeTAD/Ag	SnO_2 as ETL and mixed halide perovskite	Stable for 30 days, when stored in dry air	18.4	Baena et al. (2015)
3	ITO/NiO_X/$MAPb(I_{1-x}Br_x)_3$ ($x = 0.3$)/PCBM/Al	Doped Perovskite and inverted PSC	> 80%, 40 days, Stored in nitrogen glove box in dark environment	15.35	Mali et al. (2016)
4	ITO/CoO_X/$CH_3NH_3PbI_3$/PCBM/Ag	CoO_X as HTL	> 90%, 1000 h. Stored in nitrogen glove box in dark environment	14.5	Shalan et al. (2016)
5	FTO/WO_X/$CH_3NH_3PbI_{3-x}Cl_x$/spiro-OMeTAD/Au	WO_3 as ETL	20%, 10 h. Stored in AM 1.5G illumination	10	Gheno et al. (2017)
6	ITO/P3CT-N/$Cs_{0.1}FA_{0.9}PbI_3$/PCBM/Ag	Iodine confinement	> 90%, 1000 h	22.97	Yang et al. (2023)
7	ITO/P3CT-N/$MAPbI_3$/PCBM/C_{60}/TPBi/Cu	Addition of multifunctional additive of dipentaerythritol pentaacrylate (DPPA) into perovskite films	> 90%, 1000 h	22	Li et al. (2021b)
8	FTO/c-TiO_2/$MAPbI_3$/Spiro-OMeTAD/Au	Isopropyl alcohol (IPA) treatment on formamidinium iodide (FAI) solution	92% MPPT, 98% dark 600 h	23.5	Xin et al. (2020)

(continued)

Table 7.1 (continued)

Sl No.	Device structure	Implementation	Stability	PCE (%)	References
9	FTO/ETL/$MAPbI_3$/DT + MoS_2 modification/ Spiro-OMeTAD/Au	Surface modification using 1-dodecanethiol (DT) and MoS_2 nanosheets	Retained 80% after 2200 h in dark	18.9	Karimipour et al. (2021)
10	FTO/TiO_2/$FAMAPbI_3$/ $CsPbX_3$ QDs/ SpiroMeOTAD/Au	Use of an interface layer in the form of α-$CsPbI_3$ quantum dots (Q Ds)	> 82%, 30 days	18.56	Liu et al. (2018)
11	ITO/$Cs_{0.06}MA_{0.14}$ $FA_{0.8}PbI_3$/PCBM/Ag	Addition of a multifunctional molecule, 4-guanidinobenzoic acid hydrochloride (GBAC) to form of H-bond-bridged intermediate	Thermal stability with a T_{98} lifetime beyond 1000 h. Stored in an N_2-filled glovebox	24.8	Li et al. (2023)
12	ITO/SnO_2/(KCl, KMS, KFSI or KTFSI)/PCBM/Ag	Interface stabilization strategy leveraging the synergistic effects of fluorine and sulfonyl functional groups	Excellent stability	24.17	Gong et al. (2023)
13	ITO/SnO_2/FAI/MABr/ Spiro-OMeTAD/Ag	Prevention of iodide ion migration from the perovskite	> 88% after 1920 h of continuous illumination	24.6	Shen et al. (2024)
14	ITO/SnO_2/$MAPbI_3$ + $NaHCO_3$/Spiro-OMeTAD/ Ag	Sodium bicarbonate ($NaHCO_3$) is used to induce the formation of porous PbI_2	> 90%, Storage in ambient air (30–40% relative humidity) for 2160 h	24	Du et al. (2023)
15	FTO/PEDOT:PSS/$MAPbI_3$ + DLBA/Spiro-OMeTAD/ Ag	Use of carbonyl functional materials	Long-term stability	23.20	Xu et al. (2023)

Acknowledgements The author would like to acknowledge the financial support received from DST_FIST program (SR/FST/ET-1/2021/862)

References

Ahn N, Choi M (2024) Towards long-term stable perovskite solar cells: degradation mechanisms and stabilization techniques. Adv Sci 11(4):2306110

Ava TT, Al Mamun A, Marsillac S, Namkoong G (2019) A review: thermal stability of methylammonium lead halide based perovskite solar cells. Appl Sci 9(1):188

Ayaydah W, Raddad E, Hawash Z (2023) Sn-based perovskite solar cells towards high stability and performance. Micromachines 14(4):806

Baena JPC, Steier L, Tress W, Saliba M, Neutzner S, Matsui T, Hagfeldt A (2015) Highly efficient planar perovskite solar cells through band alignment engineering. Energy Environ Sci 8(10):2928–2934

Belisle RA, Nguyen WH, Bowring AR, Calado P, Li X, Irvine SJ et al (2017) Interpretation of inverted photocurrent transients in organic lead halide perovskite solar cells: proof of the field screening by mobile ions and determination of the space charge layer widths. Energy Environ Sci 10(1):192–204

Berhe TA, Su WN, Chen CH, Pan CJ, Cheng JH, Chen HM, Hwang BJ (2016) Organometal halide perovskite solar cells: degradation and stability. Energy Environ Sci 9(2):323–356

Byranvand MM, Kharat AN, Taghavinia N (2019) Moisture stability in nanostructured perovskite solar cells. Mater Lett 237:356–360

Celik I, Phillips AB, Song Z, Yan Y, Ellingson RJ, Heben MJ, Apul D (2017) Environmental analysis of perovskites and other relevant solar cell technologies in a tandem configuration. Energy Environ Sci 10(9):1874–1884

Cho KT, Paek S, Grancini G, Roldán-Carmona C, Gao P, Lee Y, Nazeeruddin MK (2017) Highly efficient perovskite solar cells with a compositionally engineered perovskite/hole transporting material interface. Energy Environ Sci 10(2):621–627

Dipta SS, Rahim MA, Uddin A (2024) Encapsulating perovskite solar cells for long-term stability and prevention of lead toxicity. Appl Phys Rev 11(2)

Dong X, Fang X, Lv M, Lin B, Zhang S, Ding J, Yuan N (2015) Improvement of the humidity stability of organic–inorganic perovskite solar cells using ultrathin Al_2O_3 layers prepared by atomic layer deposition. J Mater Chem A 3(10):5360–5367

Du Y, Wang Y, Wu J, Chen Q, Deng C, Ji R, Lan Z (2023) $NaHCO_3$-induced porous PbI_2 enabling efficient and stable perovskite solar cells. InfoMat 5(6):e12431

Elangovan NK, Kannadasan R, Beenarani BB, Alsharif MH, Kim MK, Inamul ZH (2024) Recent developments in perovskite materials, fabrication techniques, band gap engineering, and the stability of perovskite solar cells. Energy Rep 11:1171–1190

Farooq A, Hossain IM, Moghadamzadeh S, Schwenzer JA, Abzieher T, Richards BS, Paetzold UW (2018) Spectral dependence of degradation under ultraviolet light in perovskite solar cells. ACS Appl Mater Interfaces 10(26):21985–21990

Gheno A, Pham TTT, Di Bin C, Bouclé J, Ratier B, Vedraine S (2017) Printable WO_3 electron transporting layer for perovskite solar cells: influence on device performance and stability. Sol Energy Mater Sol Cells 161:347–354

Gong C, Zhang C, Zhuang Q, Li H, Yang H, Chen J, Zang Z (2023) Stabilizing buried interface via synergistic effect of fluorine and sulfonyl functional groups toward efficient and stable perovskite solar cells. Nano-Micro Lett 15(1):17

Guerrero A, Garcia-Belmonte G, Mora-Sero I, Bisquert J, Kang YS, Jacobsson TJ, Hagfeldt A (2016) Properties of contact and bulk impedances in hybrid lead halide perovskite solar cells including inductive loop elements. J Phys Chem C 120(15):8023–8032

Ito S, Tanaka S, Manabe K, Nishino H (2014) Effects of surface blocking layer of Sb2S3 on nanocrystalline TiO_2 for $CH_3NH_3PbI_3$ perovskite solar cells. J Phys Chem C 118(30):16995–17000

Karimipour M, Khazraei S, Kim BJ, Boschloo G, Johansson EM (2021) Efficiency and stability enhancement of perovskite solar cells utilizing a thiol ligand and MoS_2 (100) nanosheet surface modification. ACS Appl Energy Mater 4(12):14080–14092

Kawamura Y, Mashiyama H, Hasebe K (2002) Structural study on cubic-tetragonal transition of $CH_3NH_3PbI_3$. J Phys Soc Jpn 71(7):1694–1697
Khalid M, Mallick TK (2023) Stability and performance enhancement of perovskite solar cells: a review. Energies 16(10):4031
Krishnan U, Kaur M, Kumar M, Kumar A (2019) Factors affecting the stability of perovskite solar cells: a comprehensive review. J Photon Energy 9(2):021001–021001
Li F, Liu M (2017) Recent efficient strategies for improving the moisture stability of perovskite solar cells. J Mater Chem A 5(30):15447–15459
Li J, Xia R, Qi W, Zhou X, Cheng J, Chen Y, Zhang X (2021a) Encapsulation of perovskite solar cells for enhanced stability: structures, materials and characterization. J Power Sources 485:229313
Li X, Ke S, Feng X, Zhao X, Zhang W, Fang J (2021b) Enhancing the stability of perovskite solar cells through cross-linkable and hydrogen bonding multifunctional additives. J Mater Chem A 9(21):12684–12689
Li F, Deng X, Shi Z, Wu S, Zeng Z, Wang D, Jen AKY (2023) Hydrogen-bond-bridged intermediate for perovskite solar cells with enhanced efficiency and stability. Nat Photon 17(6):478–484
Liu C, Hu M, Zhou X, Wu J, Zhang L, Kong W, Cheng C (2018) Efficiency and stability enhancement of perovskite solar cells by introducing $CsPbI_3$ quantum dots as an interface engineering layer. NPG Asia Mater 10(6):552–561
Liu J, Hu M, Dai Z, Que W, Padture NP, Zhou Y (2021) Correlations between electrochemical ion migration and anomalous device behaviors in perovskite solar cells. ACS Energy Lett 6(3):1003–1014
Lopez-Varo P, Jiménez-Tejada JA, García-Rosell M, Ravishankar S, Garcia-Belmonte G, Bisquert J, Almora O (2018) Device physics of hybrid perovskite solar cells: theory and experiment. Adv Energy Mater 8(14):1702772
Mali SS, Kim H, Shim SE, Hong CK (2016) A solution processed nanostructured p-type NiO electrode for efficient inverted perovskite solar cells. Nanoscale 8(46):19189–19194
Mariani P, Molina-García MÁ, Barichello J, Zappia MI, Magliano E, Castriotta LA, Bonaccorso F (2024) Low-temperature strain-free encapsulation for perovskite solar cells and modules passing multifaceted accelerated ageing tests. Nat Commun 15(1):4552
Mohanty I, Singh UP, Jana S, Mangal S (2024) Influence of annealing on stability and regeneration of degraded $CH_3NH_3PbI_3$ thin films. Appl Phys A 130(2):143
Mohanty I, Mangal S, Jana S, Singh UP (2023) Synthesis and characterization of methylammonium tin iodide ($CH_3NH_3SnI_3$) absorber layer for photovoltaic applications. Mater Today Proc
Niu G, Guo X, Wang L (2015) Review of recent progress in chemical stability of perovskite solar cells. J Mater Chem A 3(17):8970–8980
Noh JH, Im SH, Heo JH, Mandal TN, Seok SI (2013) Chemical management for colorful, efficient, and stable inorganic–organic hybrid nanostructured solar cells. Nano Lett 13(4):1764–1769
NREL. Best research-cell efficiency chart. https://www.nrel.gov/pv/cell-efficiency.html
Rao MK, Sangeetha DN, Selvakumar M, Sudhakar YN, Mahesha MG (2021) Review on persistent challenges of perovskite solar cells' stability. Sol Energy 218:469–491
Shalan AE, Oshikiri T, Narra S, Elshanawany MM, Ueno K, Wu HP, Misawa H (2016) Cobalt oxide (CoO_x) as an efficient hole-extracting layer for high-performance inverted planar perovskite solar cells. ACS Appl Mater Interfaces 8(49):33592–33600
Shen Z, Han Q, Luo X, Shen Y, Wang Y, Yuan Y, Han L (2024) Efficient and stable perovskite solar cells with regulated depletion region. Nat Photon 1–8
Stoumpos CC, Malliakas CD, Kanatzidis MG (2013) Semiconducting tin and lead iodide perovskites with organic cations: phase transitions, high mobilities, and near-infrared photoluminescent properties. Inorg Chem 52(15):9019–9038
Whitfield PS, Herron N, Guise WE, Page K, Cheng YQ, Milas I, Crawford MK (2016) Structures, phase transitions and tricritical behavior of the hybrid perovskite methyl ammonium lead iodide. Sci Rep 6(1):35685

Xie J, Yu X, Sun X, Huang J, Zhang Y, Lei M, Yang D (2016) Improved performance and air stability of planar perovskite solar cells via interfacial engineering using a fullerene amine interlayer. Nano Energy 28:330–337

Xin C, Zhang J, Zhou X, Ma L, Hou F, Shi B, Zhang X (2020) Defects healing in two-step deposited perovskite solar cells via formamidinium iodide compensation. ACS Appl Energy Mater 3(4):3318–3327

Xu T, Chen L, Guo Z, Ma T (2016) Strategic improvement of the long-term stability of perovskite materials and perovskite solar cells. Phys Chem Chem Phys 18(39):27026–27050

Xu Y, Guo X, Lin Z, Wang Q, Su J, Zhang J, Chang J (2023) Perovskite films regulation via hydrogen-bonded polymer network for efficient and stable perovskite solar cells. Angew Chem 135(33):e202306229

Yang H, Li X, Guo X, Lu C, Yuan H, Liu A, Fang J (2023) Enhancing the stability of perovskite solar cells through an iodine confining strategy. ACS Energy Lett 8(9):3793–3799

Ye S, Sun W, Li Y, Yan W, Peng H, Bian Z, Huang C (2015) CuSCN-based inverted planar perovskite solar cell with an average PCE of 15.6%. Nano Lett 15(6):3723–3728

Zarazua I, Bisquert J, Garcia-Belmonte G (2016) Light-induced space-charge accumulation zone as photovoltaic mechanism in perovskite solar cells. J Phys Chem Lett 7(3):525–528

Zhang T, Hu C, Yang S (2020) Ion migration: a “double-edged sword” for halide-perovskite-based electronic devices. Small Methods 4(5):1900552

Chapter 8
Lead-Free Multiferroic Materials for Perovskite-Based Solar Cells: Recent Advancements and Future Insights

Monika Srivastava and R. C. Singh

Abstract The increasing energy demand and the environmental crisis caused due to the unconventional means to fulfil these demands have urged the human race to develop certain technologies based on the utilisation of renewable and sustainable energy sources. The cleanest source of renewable energy is solar energy; this suggests that the photovoltaic application is one of the most promising and dependable way to overcome the energy needs of our planet Earth. However, the solar cells generated on the principle of light to electricity conversion is still lacking an efficient, stable and low-cost materials system for this application. The PSCs formed using perovskite materials, having high absorption coefficient, low binding energy, tunable band gap and exceptionable structural and optical properties, have obtained an efficiency from 3.8 to 25% in a very short span of time, but still there are certain factors which are detrimental in its commercialisation, i.e. stability and toxicity due to lead. Many researchers are focussing on fabricating lead-free stable perovskite solar cells. There has been a number of research work in the last decade incorporating multiferroics as a light absorber in perovskite solar cells. This chapter showcases the properties of perovskite multiferroics, advancement in the field of multiferroics based lead free perovskite solar cells, their challenges and future insights.

Keywords Perovskite · Multiferroics · Solar cells · Photo conversion efficiency · Solar energy

Introduction

The word ‘Multiferroic’ was introduced by Hans Schmidth in 1964, which basically is named as such due coexistence of two ferroic order ferroelectricity and ferromagnetism in the same material (Schmid 1994) or in words multiferroic are the

M. Srivastava · R. C. Singh (✉)
Department of Physics, Centre of Excellence in Solar Cell & Renewable Energy, Sharda University, Greater Noida, India
e-mail: rcsingh@sharda.ac.in

M. Kumar and A. Raj (eds.), *Perovskite Solar Cells*, Springer Tracts in Electrical and Electronics Engineering, https://doi.org/10.1007/978-981-96-9463-1_8

materials which primarily shows two or three primary ferroic properties. Therefore, they find their application in fabrication of they have novel functional devices utilising the new material properties formed by the combination of ne multiple ferroic orders (Schmid 1994; Kim et al. 2023). The ferroelectricity and the ferromagnetic materials are different but in multiferroics they exist in single phase. There are two main reasons which signifies its technological importance, i.e. first, the magnetic bit being controlled by the electric bit and second, their coupling leads to generation of remarkable properties which are not present in their individual state. This could be understanding as that if in analysing a magnetic bit if electric voltage is used instead of a fluctuating electric current, then this reduces the loss of energy as electric current requires large build ups. Thus, these multiferroics are the future of efficient energy data storage.

However, the existence of multiferroics at room temperature is the most important challenge. This is due to the opposing effect of ferroic orders on each other. But the multifield coupling is very effective in multiferroics which enhances their properties to build high efficiency multibit devices which operate at room temperature (Manipatruni et al. 2018; Ramesh and Spaldin 2007). One such multibit coupling leads to the formation of multiferroic perovskites. These materials exhibit some remarkable properties such as high optical absorption coefficient, tunable band gap, long diffusion length, and high photoluminescence properties (Pathak et al. 2015; Aharon and Etgar 2016; Li et al. 2022; Huang et al. 2022; Wang et al. 2016a, 2016b; Lanzetta et al. 2017; Mao et al. 2017).

In the recent years, the use of oxide based multiferroics have gained tremendous importance in solar cell technologies. They have been successful to prove themselves as a versatile substitute to overcome all the existing challenges in the field of photovoltaics. The solar cells work on the three basic phenomena, i.e. charge formation, charge separation and charge transport. The oxide-based multiferroics has an extraordinary property of no-central symmetric structure that enables a suitable separation path for the excitons created by the activation through light. These oxide-based multiferroics are stable under ambient conditions, also it has a cost-effective processing and deposition techniques, which shows that it is a suitable candidate for being used in photovoltaic application. There are a number of multiferroics suitable photovoltaic application such as ($PbZrTiO_3$, $BiFeO_3$, $PbTiO_3$, $BaTiO_3$) that have garnered the attention of researchers (Raj et al. 2024; Lu et al. 2017; Wang et al. 2019, 2003a). BFO has proved to be the most suitable multiferroic for photovoltaic purposes due to its bandgap of 2.7 eV. In addition, it is also considered to be one of the best multiferroic as it simultaneously exhibits antiferromagnetic and ferroelectric nature. Researchers considers it as a potential material for next generation solar cell devices (Wang et al. 2003a). In some of the researches, the study of ferromagnetic and ferroelectric properties of the organic–inorganic lead halide perovskite ($MAPbI_3$) has also been done by investigating orders of electric dipole. The dipoles with polar cations reduced the recombination rate of charge carriers. Some of the researchers focus on the multiferroic TMO ($TbMnO_3$). Structurally, it belongs to the perovskite class which is the most favourable structure to tune in preferential properties such as superconductivity, ferroelectricity, and ferromagnetism. However, both these properties in a single perovskite are very scarce to be found. The ideal perovskite has a formula ABO_3, where *A* and *B* are cations of valency one and two, respectively, and O is the anion which bonds with *A* and *B*. The ABO_3 crystal structure has two cations *A* and *B* of different chemical valences and sizes, respectively, while the oxygen cation bonds with both cations. *A* and *B* can, however, be formed as mixed cations of different valences which helps the perovskites to be tuned in according to the application of the material. The small cation *B* is surrounded by the 6 nearest neighbours of oxygen in octahedron structure, while the large cation *A* is surrounded by maximum 12 oxygen anions. The coordination is stable under the Goldsmith tolerance factor t (Huang et al. 2022; Wang et al. 2016a; Lanzetta et al. 2017). However, the t factor is the responsible factor for the variation in the lattice structure which is further influenced by the other physical parameters—temperature, pressure, etc. The ferroelectricity in perovskites is governed by the soft mode theory which elaborates that during cooling the polar phonon modes becomes softer and goes to negligible frequency t the curie temperature where the transformation from ideal cubic to lower ferroelectric structure takes place (Mao et al. 2017; Wang et al. 2016b). This results in the breaking of centrosymmetry. The magnetic order is governed by the quantum mechanical exchange interactions—both type of interactions, i.e. double and super exchange. The ferromagnetism is, however, induced by the double exchange interactions which is also the most promising magnetic order for multiferroic materials (Wang et al. 2019, 2003a). However, these magnetic order of ferromagnets tends to destroy ferroelectricity and enhance the conductivity through itinerant electrons. Therefore, additional care must be taken for ferroelectricity of magnetic materials.

Multiferroics and Its Properties

As described earlier, multiferroics are the materials which exist in at least order state of ferroelectric and ferromagnetic. Therefore, any given ferromagnet exhibits both piezomagnetic as well as piezoelectric properties. Such type of ferromagnets are the special type of material known as magneto-elastic-coupled multiferroics. There are a number of applications based on these properties of multiferroics. In the multiferroics, the magnetic order originates due to the presence magnetic moments which originate due to the orbital motion of electrons around the nucleus. However, in such materials, particularly in solids, partially filled d and f shells contribute to the magnetic order. As it is known that the f-shell elements are characterised as the rare earths or actinides which are surrounded by the valence electrons, therefore, the ordering is confined below 50 K temperature. For the d-shell materials, the effective magnetic moments are found for Mn, Fe, Ni, and Co. In this case, magnetic order for even high temperatures can occur due to strong direct exchange interactions. Therefore, multiferroic behaviour is shown in materials exhibiting indirect exchange interactions such as oxides, whereas the ligands of transition materials govern the magnetic interactions (Buurma et al. 2016). For ferroelectricity, there are two groups of elements. The first group of elements are titanates, zirconates, and niobates. However, the second group is based on the lone pair configuration which is found in heavy elements such as Pb and Bi. There are certain terminologies used for ferroics:

(a) Ferroelectrics are the materials which has spontaneous polarisation and switches polarisation with respect to the applied electric field.
(b) Pyroelectrics are also the materials which exhibit spontaneous polarisation and ferroelectrics are the subgroup of pyroelectrics. The only difference is that to show symmetry inversion for polarisation the pyroelectrics require the electric breakdown field (Buurma et al. 2016).
(c) Proper ferroelectrics are materials in which ferroelectric ordering is caused by phase transition of primary mode. The responsible mode can be traced from the change in symmetry through the transition.
(d) Improper ferroelectrics are materials in which factor responsible for the ferroelectric order is second-order parameter instead of the first-order parameter. In this case, the polarisation is typically less than 1 $\mu C/cm^2$.
(e) Antiferroelectrics are one in which the local electric dipoles arranged in antiparallel fashion. They cannot be polarised, though they have antiparallel electric ordering.
(f) Ferromagnetics are the materials that possess stable spontaneous magnetisation and can be switched hysterically by an applied magnetic field.
(g) Antiferromagnetic materials have such magnetic moments in ordered form that they are in equal and opposite to each other which causes them to cancel each other completely within each magnetic unit cell.
(h) Ferroelastic materials are such materials in which the deformations are stable and spontaneous and thus can be switched hysterically by an applied stress.
(i) Ferrotoroidic materials have an order parameter of magnetisation and polarisation in spontaneous and switchable form.
(j) Ferrimagnetic materials have incomplete cancellation of magnetic moment in such a way that the net magnetisation can be introduced by an applied magnetic field.

Time and Space Symmetry of Multiferroics

Maxwell was among the first ones to demonstrate the link between electric and magnetic fields with their spatial gradient but what if the coupling between these two is addressed in solids? Pierre Curie proposed the magnetic electric coupling based on the symmetry considerations of the crystal. The multiferroic properties are determined by the symmetry studies related to the space and time inversion. For example, as shown in Fig. 8.1, the time reversal of the symmetry operation leads to the change in the sign of magnetisation while the electric polarisation vector remains invariant. For a spatial symmetry operation, the magnetisation remains invariant while there is a reversal in the electric polarisation. Thus, we can conclude that both

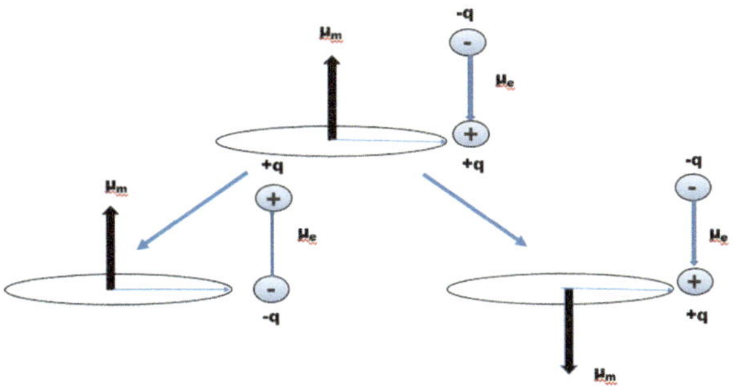

Fig. 8.1 Space and time inversion symmetry

the magnetic materials and dielectrics break the time symmetry and spatial symmetry, respectively.

Therefore, the electric- and magnetic-order states present in the multiferroics breaks both the time and space inversion symmetry. If we examine the microscopic origins of magnetic and electric phenomena, this symmetry violation could be understood. As the electron orbits around the nucleus in the atom, it generates an orbital magnetic moment μ_m. If the electric current generated corresponding to the orbital motion of the electrons is I for a cross-sectional area A, then the magnetic moment is given by the expression,

$$\mu_\mathrm{m} = I.A = \frac{\mathrm{d}Q}{\mathrm{d}t}A \tag{8.1}$$

Applying the time inversion symmetry, results in the magnetic moment reversal as,

$$\mu_{m'} = I \cdot A = -\frac{\mathrm{d}Q}{\mathrm{d}t} \cdot \mathrm{A} = -\mu_m \tag{8.2}$$

Now for a spatial symmetry, lets first consider a system of two charges—q and q separated by the distance r, then the electric dipole moment is given by,

$$\mu_\mathrm{e} = q \cdot r \tag{8.3}$$

For the spatial inversion of $r \to r'$,

$$\mu_{\mathrm{e}'} = -q \cdot r = \mu_\mathrm{e} \tag{8.4}$$

Therefore, multiferroics exhibits violation of space and time inversion symmetry.

Types of Multiferroics

These are the homogeneous compounds in which there is coexistence of electric and magnetic order states coexist. Due to this coexistence of electric- and magnetic-order state, strict symmetry conditions are required. For this, electric-order phase of the crystals should possess spontaneous polarisation and it must be invariant under time symmetry, similarly crystals with magnetic order must possess spontaneous magnetisation and should be symmetric under space symmetry (Vopson 2015; Shubnikov 1951). According to Shubnikov point groups, there are such 122 crystallographic point groups in which 31 exists in electric-order state and 31 exist in magnetic-order state. However, there are only 13-point groups which coexists both in electric- and magnetic-order state. These 13 crystallographic point groups allow the multiferroic phase. Since the ferroics are electrically ordered states, most of the single-phase multiferroics, as shown in Table 8.1, are dielectric oxides. However, the single-phase multiferroics do not occur abundantly in nature and do not hold the ease of processing it artificially. This could possibly be linked to the symmetry restrictions. The single-phase multiferroics are further classified as into (a) Type 1 Multiferroics and (b) Type 2 Multiferroics. This classification is mainly based on the mechanism of electrical polarisation. Type 1 multiferroics are the materials in which the magnetic and electric states exist in the same phase of a single compound but are independent of each other due to variation in their microscopic origin, while type 2 multiferroics has strong magneto-electric coupling as the ferroelectric order in these materials are induced by magnetism. There is a wide categorisation of type 1 and type 2 multiferroics. The materials of ABX_3 crystal class are known as the perovskites, which has octahedral BX_6^- structure with interstitial A cation as shown in Fig. 8.2. This structure class is similar to a number of ferroelectric materials such as $BaTiO_3$. The ferroelectric order is contributed by the lone pair on the A site, such as Bi^{+3} and Pb^{+2}, while magnetism is generated by the transition metals ion present at the B-site. Such as Fe^{+3} or Mn^{+3} (Buurma et al. 2016). A tunable choice of materials can lead to the multiferroicity properties in a material. Among the multiferroics $BiFeO_3$ is the most relevant multiferroic. Its ferroelectric properties are characterised by the lone pair of Bi^{3+} while the magnetic orders are characterised by unpaired spin of Fe. It has a large ferroelectric polarisation (60 μC/ cm^2) (Wang et al. 2003b) and has a ferroelectric transition temperature of 1143 K. while antiferromagnetic ordering sets in at a temperature below 643 K. Since in this case ferroelectricity and magnetic orders originates independently, magneto-electric coupling is weak in this case. In researches, there has been other types of crystal class, $EuTiO_3$ known as quantum paraelectrics. Quantum paraelectrics are the materials in which transition in to the ferroelectric order is reduced below temperature of 0 K. Since at this temperature even the molecular motion or thermal vibrations stops, the quantum effect is predominant (Katsufuji and Takagi 2001). There is another categorisation of multiferroics which originated due to charge ordering in transition metal compounds. Ferroelectricity is caused if, after charge ordering, both sites and bonds turn out to be inequivalent. Such type of mechanism is observed in $Pr_{1/2}Ca_{1/2}MnO_3$ (Efremov et al. 2004).

Table 8.1 Performance table of multiferroics-based perovskite solar cells

S. No.	Lead-free multiferroics	Photovoltaic properties			References
		E_g (eV)	V_{oc} (Volts)	J_{sc} (mA/cm^2)	
1	$BiFeO_3$	2.3–2.8	16	0.12	Chu et al. (2007)
2	$BaTiO_3$	3.2	–	–	Koch et al. (1975)
3	$LiNbO_3$	4.0	–	–	Lu et al. (2011)
4	15%Ca-BFO	1.85	0.46	2 × 10–5	Bharathi et al. (2013)
5	Pr Doped BFO NT	2.27 V	0.21	8.9×10^{-7}	Pei et al. (2020)
6	BFO BCO*	2.10–2.36	1.2	0.013	Chakrabartty et al. (2014)
7	BCFMO	2.41	0.29	0.261	Pei et al. (2020)
8	poly-BFO	1.98–2.24	0.1	2×10^{-7}	Lu et al. (2017)
9	KBNNO	1.1–3.8	0.7×10^{-3}	0.1×10^{-3}	Grinberg et al. (2013)
10	BFCO	1.4–3.2	1.12	12	Mahammedi et al. (2024)

The type 1 multiferroic magnetic materials also exhibit geometric multiferroicity which is caused due to the change in ionic bonds resulting in polar distortion causing geometric multiferroicity. In such type of materials, the magnetic order occurs at a low temperature < 120 K (Efremov et al. 2004). In type 2 multiferroics, the ferroelectricity is in conjunction with a spiralling magnetic phase just like a cycloid. Some of the examples are $TbMnO_3$, $Ni_3V_2O_6$, and $MnWO_4$. Under the temperature of 41 K, the magnetic structure of $TbMnO_3$ resembles a spin waves causing all the spins pointing in the same direction only exhibiting the antiferromagnetic phase with zero magnetic moment. Below the temperature of 28 K, tip of the Mn spins curve out like a cycloid. This spiral magnetic moment is given the name of magnetic frustration for the well-known spin–flop transition in an antiferromagnet and its effect on a cycloid, caused by an external field H, are similar to that in a magnetic field. In this case, the sublattice moments lie in a plane perpendicular to the field, thus allowing their movement in the direction of the field with less energy exchange between neighbouring spins (Khomskii 2009; Liu and Yang 2017).

Multiferroic Perovskites

There are two types of multiferroic perovskites: *A*-site driven and *B*-site driven. The *A*-site driven multiferroic perovskites are basically formed by the ferroelectric order of lone pair of *A* site. One such example is $BaTiO_3$ (Cohen 1992). These types of

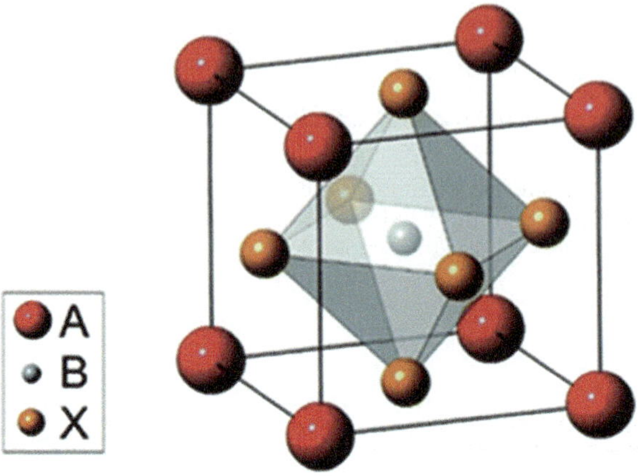

Fig. 8.2 Structure of ABX_3 perovskite material

ferroelectric materials can inculcate by the replacement of *B*-site magnetic replacement such as $BiFeO_3$ and $BiMnO_3$ (Wang et al. 2003c; Hill and Rabe 1999). Since BFO is characterised by the large value of polarisation, the focus of the researchers shifted to the *A*-site-driven multiferroics (Hill and Rabe 1999; Brown et al. 1968; Bibes and Barthélémy 2008; Cheong and Mostovoy 2007; Mostovoy 2006; Kenzelmann and Harris 2008; Kenzelmann et al. 2005; Aliouane et al. 2006; Yamasaki et al. 2007; Fennie 2008; Hill 2002; Catalan and Scott 2009; Herrero-Albillos et al. 2010; Ramazanoglu et al. 2011). Some worked on the colloidal spin modulation, while others on antiferromagnetic state and magnetoelectric coupling. The multiferroic is marked with the canted ferromagnetic order as one of the great achievements; however, the bulk multiferroics does not exhibit magnetic order due to the cyclic modulation in spin of magnetic moments, which is considered as least favourable properties when it comes to application of multiferroics. However, the required application is found in thin film of BFO. Though, experimentally the ferromagnetic property can even be induced in BFO by Fe 3C reduction and even by size effect (Eerenstein et al. 2005; Albrecht et al. 2010). On the theoretical front, the spin canting effect explains the ferromagnetic property on the basis of first principal methods, which is the result of oxygen octahedra tilting (Albrecht et al. 2010). The canted ferromagnetism could exist or induced in the BFO thin film by a number of methods, i.e. by polarised small-angle neutron scattering method (Yang et al. 2014), or by pseudo-collinear antiferromagnetic ordering, as it was found that the epitaxial strain plays an important role on magnetic properties of BFO (Sando et al. 2013, 2014). Also in recent studies, the existence of spin glass state in BFO has been investigated which is the base of existence of short range ferromagnetism in materials such as BFO (Singh et al. 2008a, 2008b). There are certain studies which focusses as to how the polar quantities balance the magnetic orders of a multiferroic material. In one of the studies, it was found that the magnetic plane could be rotated by 71 different orientation of polarisation (Bhattacharjee et al. 2014), even the spin waves can be are easily controlled by the electric field (Rovillain et al. 2010). In the theoretical simulations based on first principle, it was shown that magnetic chirality vector could be switched by 180 polarisation reorientation (Bhattacharjee et al. 2014), and similarly the magnetism of multiferroic BFO can be modified by electric field (Yang et al. 2014). In the similar way, the effect of magnetic field on polarisation as given in a single-crystal BFO, the spin order is governed by the magnetic field (Tokunaga et al. 2010). Several studies also claim that as the strain increases, the curie temperature decreases and thus the spin modulation is destroyed by high value of strain (Sando et al. 2014). Thus, it is evident that epitaxial strain is the source to modify or tune magnetic order as well as ferroelectricity. These results are of great importance because of the applications of thin films in the spintronics devices. Another A-site-driven multiferroic is BMO which was first theoretically predicted by Hill and Rabe (Hill and Rabe 1999). Unlike BFO, BMO exhibits ferromagnetic order below temperature 110 K (Chi et al. 2007a) or 100 K (Goian et al. 2012), though the ferroelectricity of BMO is always questionable due to its centrosymmetry (Sando et al. 2014; Bhattacharjee et al. 2014; Yang et al. 2014; Tokunaga et al. 2010). However, there are still some researches which illustrates the polarisation values around 0.04–0.06 mC/cm^2 for BMO ceramics (Grizalez et al. 2007; Luca et al. 2013) also it is reported that there is a remnant polarisation of 9–16 mC/cm^2 in BMO film with $SrTiO_3$. It is stated generally that thin films of BMO have large polarisation (Grizalez et al. 2007; Luca et al. 2013; Son and Shin 2008; Hatt and Spaldin 2009). Some speculated that it could be due to epitaxial strain but later it was proved that ferroelectricity cannot be induced by epitaxial strain (Guennou et al. 2014). Later on, as demonstrated by the Raman spectroscopy, X-ray diffraction BMO can have ferroelectricity at a pressure range between 37 and 53GPa (McGuire et al. 1966). The crystal structure of BFO and ETO is shown in Fig. 8.3.

One of the *B*-site-driven multiferroics is $EuTiO_3$ which is a perovskite multiferroic also known as ETO. It has a cubic space group, and it behaves as a *G*-type antiferromagnet material below 5.3 K (Chien et al. 1974; Bessas et al. 2013; Rabe et al. 2007). The centrosymmetry of ETO rules out any possibility of ferroelectricity in it. In the theoretical studies, it is found to have three phases of antiferrodistortion. This transition was found at 282 K owing to specific heat measurement and at 235 K due to synchrotron X-ray diffraction studies (Rushchanskii et al. 2012; Bussmann-Holder

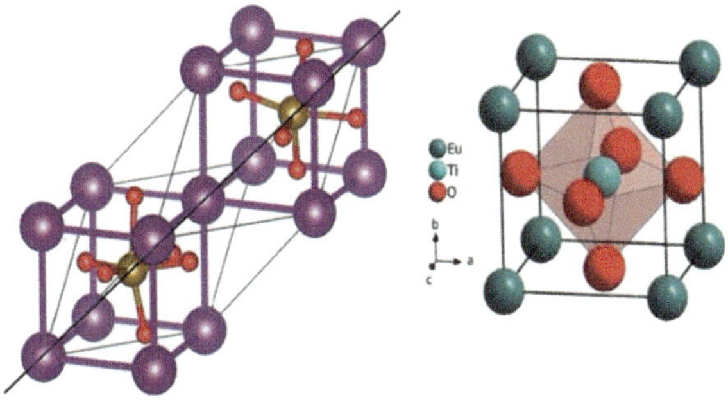

Fig. 8.3 Crystal structure of BFO and ETO

et al. 2011; Allieta et al. 2012). The ETO exhibits large ferroelectricity and ferromagnetism under the effect of biaxial compressive strain (Hemberger et al. 2007). Also ferroelectricity can be induced in ETO by adding dopant to it such as the formation of the compounds $Eu_{0.8}Y_{0.2}MnO_3$ (Morozovska et al. 2013) and $Eu_xSr_{1-x}MnO_3$ after doping ETO with Mn shows ferroelectricity (Morozovska et al. 2013; Jiang and Wu 2003). There has been reports of strong magnetoelectric coupling in ETO by varying dielectric properties under the magnetic field (Jiang and Wu 2003; Wu and Shen 2005; Wu et al. 2004; Shvartsman et al. 2010; Ryan et al. 2013; Glinchuk et al. 2014). According to the Landau-Ginzburg-Devonshire theory, it was found that the magnetic ground state can be changed to ferromagnetic state in the compression film (Azuma et al. 2005; Nechache et al. 2012). There is existence of double-perovskite multiferroics such Bi_2NiMnO_6 (Azuma et al. 2005), Bi_2FeCrO_6 (Glinchuk et al. 2014), Lu_2MnCoO_6 (Azuma et al. 2005), and Y2NiMnO6 (Nechache et al. 2012) which have rock-salt-type arrangement of B-site cations in a double-perovskite unit cell. In these double perovskites, the polar and magnetic-order transition state occur at a very low temperature except for Bi_2FeCrO_6, which occurs at room temperature in the form of thin film (Su et al. 2015), while the relaxor ferroelectrics, i.e. $Pb(Fe_{2/3}W_{1/3})O_3$ (PFW), $PbFe_{0.5}Nb_{0.5}O_3$ (PFN), and $PbFe_{0.5}Ta_{0.5}O_3$ (PFT), have relatively high temperature of polar and magnetic ordering. For example, the polar and magnetic ordering takes place at different temperatures such as 150 K (our 370 K (Martinez et al. 2010; Sil et al. 2020), and 240 K (Rumaiz et al. 2007), respectively. Among the above compounds, PFW has the highest magnetic ordering temperature and is a promising achieving room temperature multiferroicity. These results are based on the experiments of differential scanning calorimetry (DSC), Raman spectra, and dielectric measurement of both ceramics and single crystals. However, there are also the halide perovskites of formula $MAPbI_3$, which shows numerous remarkable properties such tunable bandgap, long diffusion length of charge carriers, and also photoluminescence, etc. A number of studies have been conducted on the ferroelectric and magnetic properties of hybrid halide perovskites, which demonstrates the halide vacancy sites (Nechache et al. 2012; Yáñez-Vilar et al. 2011). In recent years, a number of studies have been conducted to study the multiferroic properties of methyl ammonium lead halide perovskite material, which concludes that a due to low symmetry of the organic cations, the overall symmetry of the structure remains unstable. There are certain perovskites which are termed as appropriate and inappropriate. The appropriate ones are those which are relaxor ferroelectrics. These types of perovskites have maximum application in the energy devices, for example $BiFeO_3$, which are characterised with low dielectric loss, low remanent polarisation, and high saturation polarisation, which are key properties for energy storage application (Su et al. 2015; Sil et al. 2020).

Lead-Free Perovskites Multiferroics

There are a number of multiferroics which fulfil the criteria that they are free from the toxic lead material in their structural composition. The removal of lead not only addresses the environmental issues but also opens a lot of opportunities for the synthesis of multifunctional materials by the use of rational structural form. Now a days, oxide-based perovskites which exhibit the multiferroic properties shows a prospective application in the photovoltaic application. These oxide-based perovskites have a non-centrosymmetric structure which contributes in the easy formation of excitons, i.e. the charge carriers aiding strongly in the solar cell application. In addition, these oxide-based multiferroic perovskites are stable at room temperature and is easily processible and adhere to lost cost fabrication techniques for photovoltaic applications (Jeong et al. 2011; Martinez et al. 2010; Sil et al. 2020). The main multiferroics which are used for the photovoltaic applications are BFO ($BiFeO_3$), $BaTiO_3$ which also shows a narrow bandgap of ~ 2.7 eV (Rumaiz et al. 2007; Rashad et al. 2016). BFO has been proven as the most suitable multiferroics to be used in solar cell devices. It has rhombohedral structure and G-Type magnetic ordering with sufficiently high ferroelectricity of the order of (90–100 μC/ cm^2). In the recent researches, there has been significant focus on lead free $BaTiO_3$, it has high dielectric constant, low toxicity, low dielectric loss and have remarkable thermal stability. However, these properties can be further refined by tuning the doping concentration and drain size (Lu et al. 2017; Deka and Cho 2021). It also displays a unique property of transforming in to five different phases at different temperatures such as rhombohedral phase for $T < -90$ °C, orthorhombic phase for -90 °C $< T <$ 5 °C, tetragonal phase for 5 °C $< T <$ 120 °C, cubic phase for 120 °C $< T <$ 1450 °C and hexagonal phase for further higher temperatures. In addition, the phase change is not only limited to the variation in temperature, instead it occurs by varying the dopant concentration. There are certain reports of doped $BATiO_3$ which shows excellent structural ability, enhanced ferroelectric, and magnetic-order states. Mehmood et al., doped $BaTiO_3$ with La and Ni substituting the Ba site as shown in Fig. 8.4.

This works for improving the dielectric, polarisation, and multiferroic properties (Wang et al. 2019; Zhao et al. 2004). Doping is also done at the Ti site of $BaTiO_3$ which causes the system to go to the tetragonal phase along with the hexagonal phase with adopting conc of Mn > 7.5% (Kinoshita and Yamaji 1976). If $BaTiO_3$ is doped with Gd or Fe, then there is coexistence of orthorhombic phase with the tetragonal phase. Experimentally, it is proved that the multiferroic nature of $BaTiO_3$ enhances with the doping of Mn, Fe, Ni, and Cr. In several studies, it is reported that doping of $BaTiO_3$ with transition metal ions (Mn, Fe, Ni, and Cr) induces magnetism in $BaTiO_3$, thus altering its nature into multiferroics (Kinoshita and Yamaji 1976; Alkathy and Raju 2018; Lukacs et al. 2020; Rani et al. 2015). There are certain double-perovskite multiferroics which are also known as complex perovskites such as Bi_2NiMnO_6, Bi_2FeCrO_6, Lu_2MnCoO_6, and Y_2NiMnO_6—in which one of the sublattices is magnetic and the other is ferroelectric. For these types of materials, the

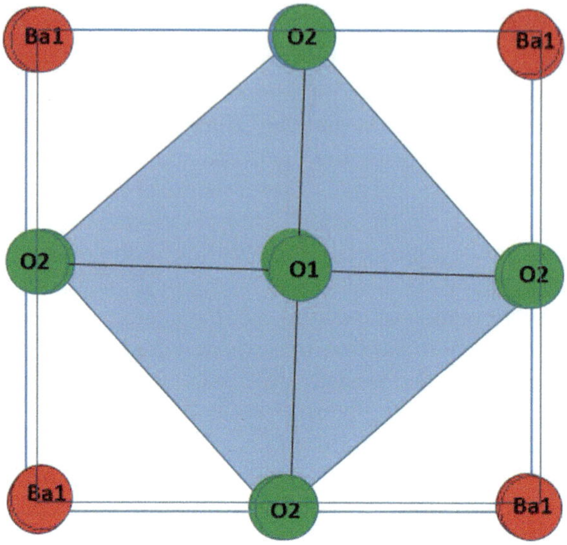

Fig. 8.4 Crystal structure of $BaTiO_3$

transition temperature is very low both for the polar and magnetic phases. However, there are certain double perovskites known as relaxor ferroelectrics—they have high transition temperature for polar and magnetic-order state, such as $Pb(Fe_{2/3}W_{1/3})O_3$ (PFW), $PbFe_{0.5}Nb_{0.5}O_3$ (PFN), and $PbFe_{0.5}Ta_{0.5}O_3$ (PFT), but they are not of our concern as they are lead-based (Azuma et al. 2005; Yáñez-Vilar et al. 2011; Su et al. 2015; Jeong et al. 2011; Martinez et al. 2010; Zhang et al. 2012; Dang et al. 2012, 2011; Blinc et al. 2007; Yang et al. 2010; Fridkin and Popov 1978; Brody and Crowne 1975; Belinicher and Sturman 1980; Seidel et al. 2011; Cao et al. 2010).

Photovoltaic Application of Lead-Free Perovskite Multiferroics

The photovoltaic phenomenon in the ferroelectrics is quite different than the conventional photovoltaic processes. In the conventional solar cell, the band gap of the semiconductor to be used in the photovoltaic application is in the visible range. The photovoltaic process is governed by three phenomena, i.e. charge formation, charge separation, and charge transportation. When the *p*-type and *n*-type materials are brought in contact with one another, they form a space charge region known as the depletion region at the junction. This depletion region helps in charge separation towards their respective electrodes. As the sun illuminates the light-absorbing material of the solar cell, excitons are created and depending on the difference of the quasi-fermi level of *p*- and *n*-type semiconductors, photovoltage is generated giving rise to the photocurrent (Fig. 8.4). The so-developed solar cells depend on the bandgap of the semiconductor material (Jeong et al. 2011; Martinez et al. 2010; Blinc et al. 2007; Yang et al. 2010). However, in the case of ferroelectric materials, a single material is involved in the formation of charge carriers instead of the *p*-type and *n*-type material. With sun illumination, the charges (electron hole pair) get polarised due to formation of the dipoles. The non-centrosymmetry of these materials helps in generating an internal electric field within the material which retains the charge separation of electron and holes—thus causing a significantly high open circuit voltage but the efficiency obtained by ferroelectric photovoltaics (Fig. 8.5) is nowhere close to that of p–n junction-based photovoltaics. For the single-crystal materials such as $LiNbO_3$, $BaTiO_3$ exhibits bulk photovoltaic effect. There are different means to explain this bulk photovoltaic effect such shift current, asymmetric potential well, asymmetric charges, etc. in the simplest model of asymmetric charge carriers the charges flow randomly under drift and diffusion processes, thus constituting the photo current (Fridkin and Popov 1978). However, the shift current model is the most promising model which gives large photovoltage. In this model, the ferroelectric materials are considered as the current source as their conductivity is quite low.

The non-centro-symmetry in the FE crystal maintains the charge separation and provide them momentum to maintain steady-state current (Fig. 8.6). These initial theoretical models that the PV properties of the multiferroics is affected by the polarisation, space charge, and defects and impurities; but according to the first principle calculation, the photo current is independent of material polarisation (Brody and Crowne 1975).

Fig. 8.5 Photovoltaic effect

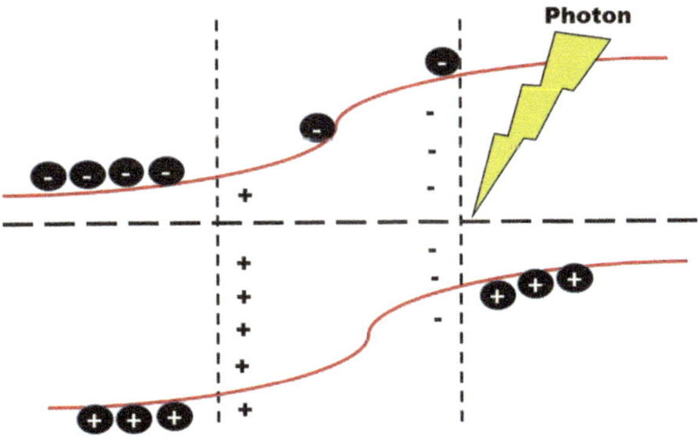

Fig. 8.6 **a** Schematic diagram of multiferroic perovskite solar cell, **b** polarisation effect of multiferroics

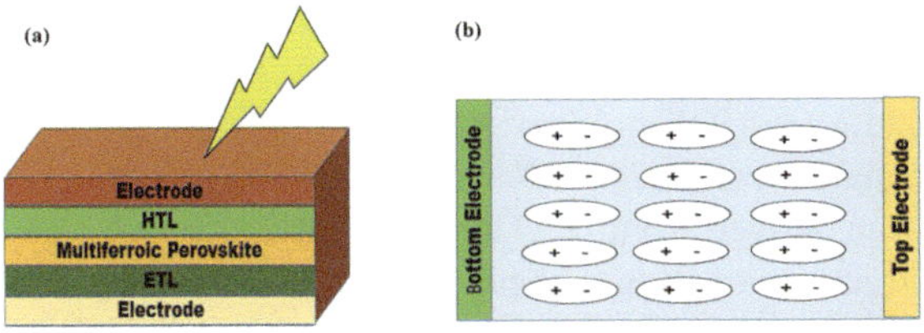

At the interface of FE material and the metal electrode produces an electric field which causes Schottky junction. The local electric field maintains the charge generation and separation in the solar cell. In metal–FE-metal interface, Schottky junction is formed and magnitude of the photo voltage at the junction is very small. The local electric field is developed as a difference between the upper- and lower-metal–FE interface, it is observed that the greater this difference the larger is the value of photo current (Seidel et al. 2011; Cao et al. 2010). For specifically the multiferroics, spin property of the electrons is also realised known as spin photovoltaics. In addition to the electric polarisation, the magnetic spin plays a major role in such materials. These multiferroics have a comparatively low bandgap energy than the ferroelectric material. The multiferroics with low band gap energy facilitates the absorption of light in the visible to near IR region covering the major portion of solar spectra.

Performance of Multiferroic-Based Perovskite Solar Cells

The multiferroics have created a boom in the field of photovoltaics due to ferroelectric and magnetic properties and also due to reduced bandgap energy. New materials are formed by doping and tuning of bandgap to improve the photovoltaic properties of the multiferroics based solar cells. BFO is among the most promising multiferroics. However, its band gap is almost double of what is required in solar cells. This limitation is removed by strain engineering through uniaxial compression. This compressive stress not only reduces the band gap but also transforms it from indirect band gap to the direct band gap material, thus improving the optical absorption. The performance of some BFO-based PVs are as follows: Machado et al., worked on the doping of BFO with Co with 10 and 30% doping concentration. However, it was found that as the concentration of Co increases, the band gap also increases and the one with lower concentration delivers high photo current (Brody and Crowne 1975). This may be due to the increase leakage current with higher concentration of Co. Zhang et al. prepared a solar cell with La-doped BFO and achieved an efficiency of 5.62%, and open circuit voltage (V_{OC}) of 1.74 V with a configuration of FTO/ETLs/La-doped BFO/Au. In the same study, PSCs with SnO_2 and ZnO substituting TiO_2 shows a PCE of 5.45% and 4.11%, respectively, with a V_{OC} of 1.69 V and 1.23 V (Belinicher and Sturman 1980). With a co doping of Ca and Mn to BFO causes an improved photovoltaic performance and improved remnant polarisation (Seidel et al. 2011). With a heterostructure following the pattern FTO/TiO_2/absorber/Spiro-OMeTAD/Au, a lead-free PSC is fabricated based on the absorber Bi0.80La0.20$Fe_{1-x}Ti_xO_3$ (x = 0.0) which attains a PCE = 5.61%, V_{OC} = 0.91 V and a fill factor (FF) = 55.55%, and J_{sc} = 10.87 mA/cm^2 with optimised thickness of absorber and operating temperature (Raj et al. 2024; Pei et al. 2020). A recent multiferroic $KBiFe_2O_5$ (KBFO) has been derived from BFO and have alternate layers of octahedral BO_6 and tetrahedral structures of BO_4. The presence of this tetrahedral layer helps in reducing band gap and in enhancing the photovoltaic properties. The theoretical studies predicted the efficiency of 30% as compared to that 7% for BFO (Zhang et al. 2013). It generates a low photocurrent in the range of μA, while it has a significantly high photo voltage of 8.8 V.

Lead-free all Bismuth based all inorganic Bi_2FeCrO_6-based (BFCO) perovskite solar cell is a metastable multiferroic with tunable band gap, high absorption coefficient and high carrier mobilities. Due to these properties, BFCO finds its potential use in thin-film perovskite solar cells. The BFCO-based PSC was studied and analysed under SCAPS 1-D and under optimised temperature, thickness, doping concentration, achieved room temperature stable PSC with a photo conversion efficiency of 7% with $V_{oc} = 1.12$ V, $J_{sc} = 12$ mA/cm^2, and fill factor = 53.7% (Mahammedi et al. 2024). The band gap compatibilities of ETL, HTL, and BFCO and the J-V characteristic of the BFCO-based PSC are shown in Fig. 8.7 a and b. In one of the studies, BFO is used as photo anode in a dye-sensitised solar cell (DSSC). This reduces the electron traps and increases the conductivity of the electron transport layer. Such type of solar cell is capable of generating a PCE of 0.13% (Kumar et al. 2021; Jalaja and Dutta 2015; Chu et al. 2007; Koch et al. 1975; Lu et al. 2011). In one of the studies when the BFO was co-doped with Mn and Ca, considerable enhancement in the photovoltaic performance of the device was witnessed as shown in Fig. 8.8, with a $V_{oc} = -0.11$ V and $J_{sc} = 0.08$ mA/cm^2 owing to the increase in polarisation and narrowing of the bandgap. This work illustrates that co-doping BFO with Ca and Mn is an effective technique to enhance the photovoltaic performance of BFO-based PSCs (Pei et al. 2020).

Under the theoretical work, a sandwich structure device is projected under SCAPS simulation Inorganic lead-free double perovskite La_2NiMnO_6 (LNMO) with a common hole transport layer CuI and varying electron transport layers. Various electron transport layers were varied such as TiO_2, C60, PCBM, and ZnO to find a suitable front electrode for device application. After careful simulation, it was found that C_{60} and TiO_2 are the most suitable ETLs for the double-perovskite La_2NiMnO_6 to be used in device structure (Raj et al. 2021).

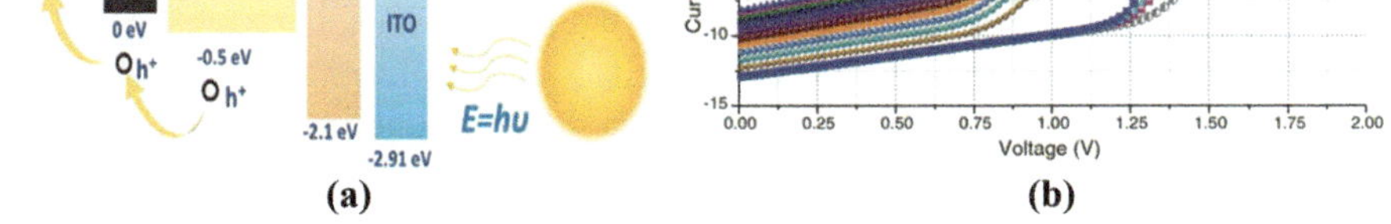

Fig. 8.7 **a** Band gap diagram of BFCO-based PSC, **b** J-V characteristic of BFCO PSC Mahammedi et al. (2024)

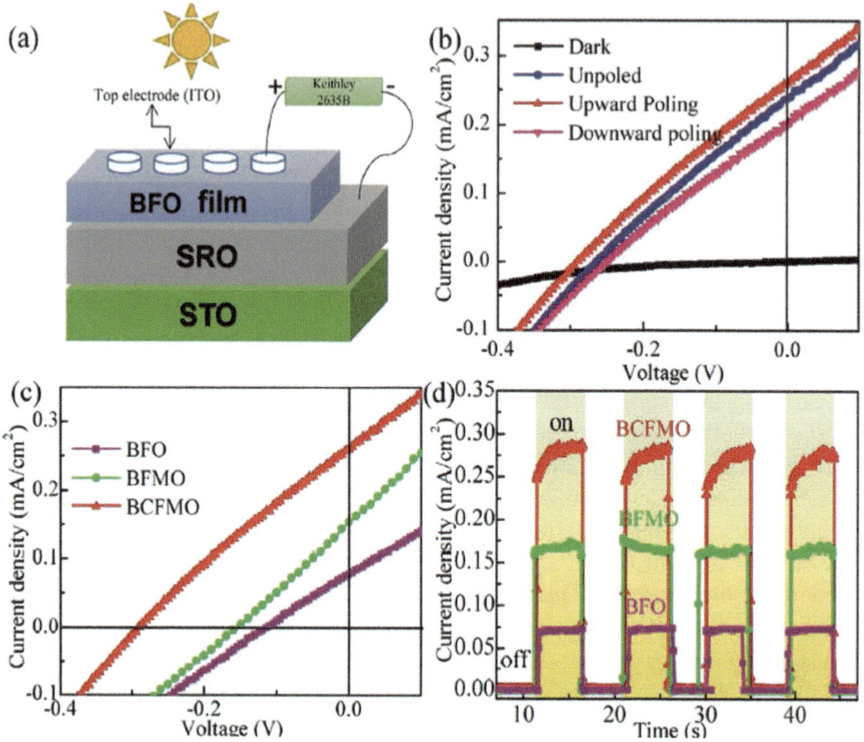

Fig. 8.8 **a** Represents ITO/BCFMO/SRO device in sandwich structure, **b** J-V characteristics of the device in the dark unpoled state, **c** J-V characteristics of the device in the upward polarised state, **d** time-dependent response of current density for doped BFO (Pei et al. 2020)

Challenges and Future Prospects

Indeed, without any discussion, the multiferroic perovskites has paved the way for the next-generation solar cells and has proved itself to be a promising field of research. Of course, it will have taken its own time to give considerably good results. Multiferroic perovskite solar cells are capable of generating excellent photovoltage, but the PCE achieved is quite low comparable to the existing solar cell devices. The low efficiency and low photo current are due to the high energy band gap of multiferroic perovskites. The potential application of the multiferroics in solar cell include multiple functionalities and has numerous merits over the conventional photovoltaics. It provides advantage of tuning its composition to enhance the photovoltaic properties. The technical and financial challenges of commercial solar cells are addressed by recent advancements in the fields of oxide ferroelectric and multiferroics. The next generation of solar cells no longer need to build up a p–n junction, hence bandgap has very little effect on how efficient these solar cells are. Beyond its efficiency and capacity, this class is attractive for future research due to the possibility of coupling between PV and ferroelectric/ ferromagnetic properties.

Conclusion

This chapter introduces recent researches on perovskite multiferroics with application photovoltaics for both theoretical and experimental approaches. In this chapter, we reach to certain conclusions. First, it discusses about the ordering of multiferroics which shows both ferroelectric and magnetic-order state. This chapter also guides us to the findings that the strain engineering is a tool to reduce the bandgap of material to the desirable value. It can be also be concluded that strain engineering behaves as an important tool to attain multiferroicity and also about the importance of magneto-electric coupling in the multiferroics. It also informs us that the ferroelectric polarisation and the magnetic spin helps in charge generation and their separation which is an important phenomenon in photovoltaics. In some cases, it is found that doping improves the performance of the solar cells while in other, due to the leakage current owing to high concentration of the doping the efficiency decreases considerably low. When it comes to photovoltaic application, the existence of ferroelectricity and magnetism is very difficult to coexist, and therefore multiferroics are rare to find. They are wide bandgap materials; therefore, effective solar absorption is less in these materials. Therefore, it is important to reduce the bandgap of multiferroic perovskite material so that it works efficiently in solar cells. Therefore, modification is required in the octahedral sites to allow the band gap to be lowered. For the multiferroics, it is expected that under one sun illumination, it achieves an efficiency up to 20% for a band gap of 1.5 eV. However, the efficiencies of perovskite multiferroic solar cell has not reached this efficiency till now. It is expected that the efficiencies could improve if the intermediate bands could be inserted into multiferroic bandgap, resulting in a multigap semiconductor.

References

Aharon S, Etgar L (2016) Two dimensional organometal halide perovskite nanorods with tunable optical properties. Nano Lett 16(5):3230–3235

Albrecht D, Lisenkov S, Ren W, Rahmedov D, Kornev IA, Bellaiche L (2010) Ferromagnetism in multiferroic $BiFeO_3$ films: a first-principles-based study. Phys Rev B Condensed Matter Mater Phys 81(14):140401

Aliouane N, Argyriou DN, Strempfer J, Zegkinoglou I, Landsgesell S, Zimmermann MV (2006) Field-induced linear magnetoelastic coupling in multiferroic $TbMnO_3$. Phys Rev B Condensed Matter Mater Phys 73(2):020102

Alkathy MS, Raju KJ (2018) Onset of multiferroicity in nickel and lithium co-substituted barium titanate ceramics. J Magn Magn Mater 452:40–47

Allieta M, Scavini M, Spalek LJ, Scagnoli V, Walker HC, Panagopoulos C, Saxena SS, Katsufuji T, Mazzoli C (2012) Role of intrinsic disorder in the structural phase transition of magnetoelectric $EuTiO_3$. Phys Rev B Condensed Matter Mater Phys, 85(18):184107

Azuma M, Takata K, Saito T, Ishiwata S, Shimakawa Y, Takano M (2005) Designed ferromagnetic, ferroelectric Bi_2NiMnO_6. J Am Chem Soc 127(24):8889–8892

Belinicher VI, Sturman BI (1980) The photogalvanic effect in media lacking a center of symmetry. Soviet Physics Uspekhi 23(3):199

Bessas D, Rushchanskii KZ, Kachlik M, Disch S, Gourdon O, Bednarcik J, Maca K, Sergueev I, Kamba S, Ležaić M, Hermann RP (2013) Lattice instabilities in bulk $EuTiO_3$. Phys Rev B Condensed Matter Mater Phys 88(14):144308

Bharathi KK, Lee W-M, Sung JH, Lim JS, Kim SJ, Chu K, Park JW, Song JH, Jo M-H, Yang C-H (2013) Appl Phys Lett 102:012908

Bhattacharjee S, Rahmedov D, Wang D, Íniguez J, Bellaiche L (2014) Ultrafast switching of the electric polarization and magnetic chirality in $BiFeO_3$ by an electric field. Phys Rev Lett 112(14):147601

Bibes M, Barthélémy A (2008) Towards a magnetoelectric memory. Nat Mater 7(6):425–426

Blinc R, Cevc P, Zorko A, Holc J, Kosec M, Trontelj Z, Pirnat J, Dalal N, Ramachandran V, Krzystek J (2007) Electron paramagnetic resonance of magnetoelectric Pb $(Fe_{1/2}Nb_{1/2})O_3$. J Appl Phys 101(3)

Brody PS, Crowne F (1975) Mechanism for the high voltage photovoltaic effect in ceramic ferroelectrics. J Electron Mater 4(5):955–971

Brown WF Jr, Hornreich RM, Shtrikman S (1968) Upper bound on the magnetoelectric susceptibility. Phys Rev 168(2):574

Bussmann-Holder A, Köhler J, Kremer RK, Law JM (2011) Relation between structural instabilities in $EuTiO_3$ and $SrTiO_3$. Phys Rev B Condensed Matter Mater Phys 83(21):212102

Buurma AJC, Blake GR, Palstra TTM, Adem U (2016) Multiferroic materials: physics and properties

Cao D, Xu J, Fang L, Dong W, Zheng F, Shen M (2010) Interface effect on the photocurrent: a comparative study on Pt sandwiched $(Bi_{3.7}Nd_{0.3})$ Ti_3O_{12} and Pb $(Zr_{0.2}Ti_{0.8})$ O_3 films. Appl Phy Lett 96(19)

Catalan G, Scott JF (2009) Physics and applications of bismuth ferrite. Adv Mater 21(24):2463–2485

Chakrabartty J, Nechache R, Li S, Nicklaus M, Ruediger A, Rosei F (2014) Photovoltaic properties of multiferroic $BiFeO_3/BiCrO_3$ heterostructures. J Am Ceram Soc 97(6):1837–1840

Cheong SW, Mostovoy M (2007) Multiferroics: a magnetic twist for ferroelectricity. Nat Mater 6(1):13–20

Chi ZH, Yang H, Feng SM, Li FY, Yu RC, Jin CQ (2007a) Room-temperature ferroelectric polarization in multiferroic $BiMnO_3$. J Magn Magn Mater 310(2):e358–e360

Chi ZH, Yang H, Feng SM et al (2007) Room-temperature ferroelectric polarization in multiferroic $BiMnO_3$. In: Proceedings of the 17th international conference on magnetism the international conference on magnetism, vol 310, no (2, Part 2), pp e358–e360

Chien CL, DeBenedetti S, Barros FDS (1974) Magnetic properties of $EuTiO_3$, Eu_2TiO_4, and $Eu_3Ti_2O_7$. Phys Rev B 10(9):3913

Chu Y-H, Cruz MP, Yang C-H, Martin LW, Yang P-L, Zhang J-X, Lee K, Yu P, Chen L-Q, Ramesh R (2007) Adv Mater 19:2662

Cohen RE (1992) Origin of ferroelectricity in perovskite oxides. Nature 358(6382):136–138

Dang NV, Thanh TD, Hong LV, Lam VD (2011) Structural, optical and magnetic properties of polycrystalline $BaTi_{1-x}FexO_3$ ceramics. J Appl Phys 110(4)

Dang NV, Thanh TD, Lam VD, Hong LV (2012) Structural phase separation and optical and magnetic properties of $BaTi_{1-x}MnxO_3$ multiferroics. J Appl Phys 111(11)

Deka B, Cho KH (2021) $BiFeO_3$-based relaxor ferroelectrics for energy storage: progress and prospects. Materials 14(23):7188

Dos Santos AM, Parashar S, Raju AR, Zhao YS, Cheetham AK, Rao CNR (2002) Evidence for the likely occurrence of magnetoferroelectricity in the simple perovskite, $BiMnO_3$. Solid State Commun 122(1–2):49–52

Eerenstein W, Morrison FD, Dho J, Blamire MG, Scott JF, Mathur ND, Wang J, Scholl A, Zheng H, Ogale SB, Viehland D (2005) Technical comment abstracts. Science 307(5713):1203–1204

Efremov DV, Van Den Brink J, Khomskii DI (2004) Bond-versus site-centred ordering and possible ferroelectricity in manganites. Nat Mater 3(12):853–856

Fennie CJ (2008) Ferroelectrically induced weak ferromagnetism by design. Phys Rev Lett 100(16):167203.39

Fridkin VM, Popov B (1978) Anomalous photovoltaic effect in ferroelectrics. Sov Phys Uspekhi 21(12):981

Glinchuk MD, Eliseev EA, Gu Y, Chen LQ, Gopalan V, Morozovska AN (2014) Electric-field induced ferromagnetic phase in paraelectric antiferromagnets. Phys Rev B 89(1):014112

Goian V, Kamba S, Savinov M, Nuzhnyy D, Borodavka F, Vaněk P, Belik AA (2012) Absence of ferroelectricity in $BiMnO_3$ ceramics. J Appl Phys 112(7)

Grinberg I, West DV, Torres M, Gou G, Stein DM, Wu L, Chen G, Gallo EM, Akbashev ER, Davies PK, Spanier JE, Rappe AM (2013) Nature 503:509

Grizalez M, Martinez E, Caicedo J et al (2007) Occurrence of ferroelectricity in epitaxial BiMnO3 thinfilms. Papers CLACSA XIII, Colombia 2007, 2008 39(11):1308–1310

Guennou M, Bouvier P, Toulemonde P, Darie C, Goujon C, Bordet P, Hanfland M, Kreisel J (2014) Jahn-Teller, polarity, and insulator-to-metal transition in $BiMnO_3$ at high pressure. Phys Rev Lett 112(7):075501

Hatt AJ, Spaldin NA (2009) Strain effects on the electric polarization of $BiMnO_3$. Euro Phys J B 71:435–437

Hemberger J, Schrettle F, Pimenov A, Lunkenheimer P, Ivanov VY, Mukhin AA, Balbashov AM, Loidl A (2007) Multiferroic phases of $Eu_{1-x}Y_xMnO_3$. Phys Rev B Condensed Matter Mater Phys 75(3):035118

Herrero-Albillos J, Catalan G, Rodriguez-Velamazan JA, Viret M, Colson D, Scott JF (2010) Neutron diffraction study of the $BiFeO_3$ spin cycloid at low temperature. J Phys Condens Matter 22(25):256001

Hill NA (2002) Density functional studies of multiferroic magnetoelectrics. Annu Rev Mater Res 32(1):1–37

Hill NA, Rabe KM (1999) First-principles investigation of ferromagnetism and ferroelectricity in bismuth manganite. Phys Rev B 59(13):8759

Huang W, Zhang X, Li Y, Zhou Y, Chen X, Li X, Wu F, Hong M, Luo J, Zhao S (2022) A hybrid halide perovskite birefringent crystal. Angew Chem Int Ed 61(22):e202202746

Jalaja MA, Dutta S (2015) Ferroelectrics and multiferroics for next generation photovoltaics. Adv Mater Lett 6(7):568–584

Jeong IK, Ahn JS, Kim BG, Yoon S, Singh SP, Pandey D (2011) Short-and medium-range structure of multiferroic Pb $(Fe_{1/2}Nb_{1/2})O_3$ studied using neutron total scattering analysis. Phys Rev B Condensed Matter Mater Phys 83(6):064108

Jiang Q, Wu H (2003) Magnetic influence on frequency of soft-phonon mode in incipient ferroelectric $EuTiO_3$. J Appl Phys 93(4):2121–2125

Katsufuji T, Takagi H (2001) Coupling between magnetism and dielectric properties in quantum paraelectric $EuTiO_3$. Phys Rev B 64(5):054415

Kenzelmann FM, Harris AB (2008) Comment on ferroelectricity in spiral magnets. Phys Rev Lett 100(8):089701

Kenzelmann M, Harris AB, Jonas S, Broholm C, Schefer J, Kim SB, Zhang CL, Cheong FSW, Vajk OP, Lynn JW (2005) Magnetic inversion symmetry breaking and ferroelectricity in $TbMnO_3$. Phys Rev Lett 95(8):087206

Khan GG, Das R, Mukherjee N, Mandal K (2012) Phys Status Solidi RRL 6:312

Khomskii D (2009) Classifying multiferroics: mechanisms and effects. Physics 2:20

Kim YS, Jin Z, Park MW, Jeon HC, Lim JY (2023) Multiferroic-field coupling in ultrathin nanofilm halide perovskite at room temperature. Mater Today Phys 35:101109
Kinoshita K, Yamaji A (1976) Grain-size effects on dielectric properties in barium titanate ceramics. J Appl Phys 47(1):371–373
Koch WTH, Munser R, Ruppel W, Würfel P (1975) Solid State Commun 17:847
Kumar AM, Peter IJ, Ramachandran K, Mayandi J, Jayakumar K (2021) Influence of Al-Cu doping on the efficiency of $BiFeO_3$ based perovskite solar cell (PSC). Mater Today Proc 35:62–65
Lanzetta L, Marin-Beloqui JM, Sanchez-Molina I, Ding D, Haque SA (2017) Two-dimensional organic tin halide perovskites with tunable visible emission and their use in light-emitting devices. ACS Energy Lett 2(7):1662–1668
Lebeugle D, Colson D, Forget A, Viret M, Bataille AM, Gukasov A (2008) Electric-field-induced spin flop in $BiFeO_3$ single crystals at room temperature. Phys Rev Lett 100(22):227602
Li M, Zhang X, Xiong Z, Li Y, Zhou Y, Chen X, Song Y, Hong M, Luo J, Zhao S (2022) A hybrid antiperovskite with strong linear and second-order nonlinear optical responses. Angew Chem 134(42):e202211151
Liu H, Yang X (2017) A brief review on perovskite multiferroics. Ferroelectrics 507(1):69–85
Lu Z, Li P, Wan JG, Huang Z, Tian G, Pan D, Fan Z, Gao X, Liu JM (2017) Controllable photovoltaic effect of microarray derived from epitaxial tetragonal $BiFeO_3$ films. ACS Appl Mater Interfaces 9(32):27284–27289
Lu Z, Zhao K, Li X (2011) Ferroelectrics-physical effects. InTech Lallart M (eds), pp 1–666
De Luca GM, Preziosi D, Chiarella F, Di Capua R, Gariglio S, Lettieri S, Salluzzo M (2013) Ferromagnetism and ferroelectricity in epitaxial $BiMnO_3$ ultra-thin films. Appl Phys Lett 103(6)
Lukacs VA, Stanculescu R, Curecheriu L, Ciomaga CE, Horchidan N, Cioclea C, Mitoseriu L (2020) Structural and functional properties of BaTiO3 porous ceramics produced by using pollen as sacrificial template. Ceram Int 46(1):523–530
Machado P, Scigaj M, Gazquez J, Rueda E, Sánchez-Díaz A, Fina I, Gibert-Roca M, Puig T, Obradors X, Campoy-Quiles M, Coll M (2019) Band gap tuning of solution-processed ferroelectric perovskite $BiFe_{1-x}$ Co_x O_3 Thin Films. Chem Mater 31(3):947–954
Mahammedi NA, Benameur A, Gueffaf H, Merabet B, Ozkendir OM, Sato SI (2024) Investigating a Pb-free nip perovskite solar cell with BFCO absorber using SCAPS-1D. Optik 302:171659
Manipatruni S, Nikonov DE, Young IA (2018) Beyond CMOS computing with spin and polarization. Nat Phys 14(4):338–343
Mao L, Wu Y, Stoumpos CC, Traore B, Katan C, Even J, Wasielewski MR, Kanatzidis MG (2017) Tunable white-light emission in single-cation-templated three-layered 2D perovskites $(CH_3CH_2NH_3)_4Pb_3Br_{10-x}Cl_x$. J Am Chem Soc 139(34):11956–11963
Martinez R, Palai R, Huhtinen H, Liu J, Scott JF, Katiyar RS (2010) Nanoscale ordering and multiferroic behavior in Pb $(Fe_{1/2}Ta_{1/2})O_3$. Phys Rev B Condensed Matter Mater Phys 82(13):134104
McGuire TR, Shafer MW, Joenk RJ, Alperin HA, Pickart SJ (1966) Magnetic structure of $EuTiO_3$. J Appl Phys 37(3):981–982
Montanari E, Calestani G, Righi L, Gilioli E, Bolzoni F, Knight KS, Radaelli PG (2007) Structural anomalies at the magnetic transition in centrosymmetric $BiMnO_3$. Phys Rev B Condensed Matter Mater Phys 75(22):220101
Morozovska AN, Gu Y, Khist VV, Glinchuk MD, Chen LQ, Gopalan V, Eliseev EA (2013) Low-symmetry monoclinic ferroelectric phase stabilized by oxygen octahedra rotations in strained $Eu_xSr_{1-x}TiO_3$ thin films. Phys Rev B Condensed Matter Mater Phys 87(13):134102
Mostovoy M (2006) Ferroelectricity in spiral magnets. Phys Rev Lett 96(6):067601
Nechache R, Harnagea C, Pignolet A (2012) Multiferroic properties—structure relationships in epitaxial Bi_2FeCrO_6 thin films: recent developments. J Phys Condens Matter 24(9):096001
Pathak S, Sakai N, Wisnivesky Rocca Rivarola F, Stranks SD, Liu J, Eperon GE, Ducati C, Wojciechowski K, Griffiths JT, Haghighirad AA, Pellaroque A (2015) Perovskite crystals for tunable white light emission. Chem Mater 27(23):8066–8075

Pei W, Chen J, You D, Zhang Q, Li M, Lu Y, Fu Z, He Y (2020) Enhanced photovoltaic effect in Ca and Mn co-doped $BiFeO_3$ epitaxial thin films. Appl Surf Sci 530:147194
Rabe KM, Ahn CH, Triscone JM (eds) (2007) Physics of ferroelectrics: a modern perspective, vol 105. Springer
Raj A, Anshul A, Tuli V, Singh PK, Singh RC, Kumar M (2021) Effect of appropriate ETL on quantum efficiency of double perovskite La_2NiMnO_6 based solar cell device via SCAPS simulation. Mater Today Proc 47:1656–1659
Raj A, Sharma S, Singh DV, Kumar A, Chourasia RK, Siqueiros JM, Herrera OR, Anshul A, Kumar M (2024) Photovoltaic energy conversion in multiferroic perovskite absorber-based devices via experiment and theoretical calculations. Physica B 673:415504
Ramazanoglu M, Ratcliff W, Choi YJ, Lee S, Cheong SW, Kiryukhin V (2011) Temperature-dependent properties of the magnetic order in single-crystal $BiFeO_3$. Phys Rev B Condensed Matter Mater Phys 83(17):174434
Ramesh R, Spaldin NA (2007) Multiferroics: progress and prospects in thin films. Nat Mater 6(1):21–29
Rani A, Kolte J, Gopalan P (2015) Phase formation, microstructure, electrical and magnetic properties of Mn substituted barium titanate. Ceram Int 41(10):14057–14063
Rashad MM, Elseman AM, Hassan AM (2016) Facile synthesis, characterization and structural evolution of nanorods single-crystalline $(C_4H_9NH_3)_2PbI_2X_2$ mixed halide organometal perovskite for solar cell application. Optik 127(20):9775–9787
Rovillain P, De Sousa R, Gallais YT, Sacuto A, Méasson MA, Colson D, Forget A, Bibes M, Barthélémy A, Cazayous M (2010) Electric-field control of spin waves at room temperature in multiferroic $BiFeO_3$. Nat Mater 9(12):975–979
Rumaiz AK, Ali B, Ceylan A, Boggs M, Beebe T, Shah SI (2007) Experimental studies on vacancy induced ferromagnetism in undoped TiO2. Solid State Commun 144(7–8):334–338
Rushchanskii KZ, Spaldin NA, Ležaić M (2012) First-principles prediction of oxygen octahedral rotations in perovskite-structure $EuTiO_3$. Phys Rev B Condensed Matter Mater Phys 85(10):104109
Ryan PJ, Kim JW, Birol T, Thompson P, Lee JH, Ke X, Normile PS, Karapetrova E, Schiffer P, Brown SD, Fennie CJ (2013) Reversible control of magnetic interactions by electric field in a single-phase material. Nat Commun 4(1):1334
Sando D, Agbelele A, Rahmedov D, Liu J, Rovillain P, Toulouse C, Infante IC, Pyatakov AP, Fusil S, Jacquet E, Carrétéro C (2013) Crafting the magnonic and spintronic response of $BiFeO_3$ films by epitaxial strain. Nat Mater 12(7):641–646
Sando D, Agbelele A, Daumont C, Rahmedov D, Ren W, Infante IC, Lisenkov S, Prosandeev S, Fusil S, Jacquet E, Carrétéro C (2014) Control of ferroelectricity and magnetism in multi-ferroic $BiFeO_3$ by epitaxial strain. Philos Trans R Soc A Math Phys Eng Sci 372(2009):20120438
Schafranek R, Payan S, Maglione M, Klein A, Barrier height at (Ba, Sr) TiO_3/Pt interfaces studied by photoemission. Phys Rev B (19)
Schmid H (1994) Multi-ferroic magnetoelectrics. Ferroelectrics 162(1):317–338
Seidel J, Fu D, Yang SY, Alarcón-Lladó E, Wu J, Ramesh R, Ager JW III (2011) Efficient photovoltaic current generation at ferroelectric domain walls. Phys Rev Lett 107(12):126805
Shubnikov AV (1951) Symmetry and antisymmetry of finite figures. USSR Acad, Moscow
Shvartsman, V.V., Borisov, P., Kleemann, W., Kamba, S. and Katsufuji, T., 2010. Large off-diagonal magnetoelectric coupling in the quantum paraelectric antiferromagnet $EuTiO_3$. Phys Rev B Condensed Matter Mater Phys 81(6):064426
Sil S, Luitel H, Dhar J, Chakrabarti M, Ray PP, Bandyopadhyay B, Sanyal D (2020) Defect induced room temperature ferromagnetism in methylammonium lead iodide perovskite. Phys Lett A 384(14):126278
Singh MK, Katiyar RS, Prellier W, Scott JF (2008b) The Almeida-Thouless line in $BiFeO_3$: is bismuth ferrite a mean field spin glass. J Phys Condens Matter 21(4):042202
Singh MK, Prellier W, Singh MP, Katiyar RS, Scott JF (2008) Spin-glass transition in single-crystal $BiFeO_3$. Phys Rev B Condensed Matter Mater Phys 77(14):144403

Son JY, Shin YH (2008) Multiferroic $BiMnO_3$ thin films with double $SrTiO_3$ buffer layers. Appl Phys Lett 93(6)

Sosnowska I, Przeniosło R (2011) Low-temperature evolution of the modulated magnetic structure in the ferroelectric antiferromagnet $BiFeO_3$. Phys Rev B Condensed Matter Mater Phys 84(14):144404

Su J, Yang ZZ, Lu XM, Zhang JT, Gu L, Lu CJ, Li QC, Liu JM, Zhu JS (2015) Magnetism-driven ferroelectricity in double perovskite Y_2NiMnO_6. ACS Appl Mater Interfaces 7(24):13260–13265

Tokunaga M, Azuma M, Shimakawa Y (2010) High-field study of strong magnetoelectric coupling in single-domain crystals of $BiFeO_3$. J Phys Soc Jpn 79(6):064713

Vopson MM (2015) Fundamentals of multiferroic materials and their possible applications. Crit Rev Solid State Mater Sci 40(4):223–250

Wang J et al (2003b) Epitaxial $BiFeO_3$ multiferroic thin film heterostructures. Science 299(5613):1719–1722

Wang J, Scholl A, Zheng H, Ogale SB, Viehland D, Schlom DG, Spaldin NA, Rabe KM, Wuttig M, Mohaddes L, Neaton J (2005) Response to comment on Epitaxial $BiFeO_3$ multiferroic thin film heterostructures. Science 307(5713):1203–1203

Wang KH, Wu L, Li L, Yao HB, Qian HS, Yu SH (2016a) Large-scale synthesis of highly luminescent perovskite-related $CsPb_2Br_5$ nanoplatelets and their fast anion exchange. Angew Chem 128(29):8468–8472

Wang N, Cheng L, Ge R, Zhang S, Miao Y, Zou W, Yi C, Sun Y, Cao Y, Yang R, Wei Y (2016b) Perovskite light-emitting diodes based on solution-processed self-organized multiple quantum wells. Nat Photonics 10(11):699–704

Wang MZ, Feng HJ, Qian CX, He J, Feng J, Cao YH, Yang K, Deng ZY, Yang Z, Yao X, Zhou J (2019) $PbTiO_3$ as electron-selective layer for high-efficiency perovskite solar cells: enhanced electron extraction via tunable ferroelectric polarization. Adv Func Mater 29(1):1806427

Wang JBNJ, Neaton JB, Zheng H, Nagarajan V, Ogale SB, Liu B, Viehland D, Vaithyanathan V, Schlom DG, Waghmare UV, Spaldin NA (2003) Epitaxial $BiFeO_3$ multiferroic thin film heterostructures. Science 299(5613):1719–1722

Wang JBNJ, Neaton JB, Zheng H, Nagarajan V, Ogale SB, Liu B, Viehland D, Vaithyanathan V, Schlom DG, Waghmare UV, Spaldin NA (2003) Epitaxial $BiFeO_3$ multiferroic thin film heterostructures. Science 299(5613):1719–1722

Wu H, Shen WZ (2005) Magnetoelectric effect in perovskite quantum paraelectric $EuTiO_3$. Solid State Commun 133(8):487–491

Wu H, Jiang Q, Shen WZ (2004) A possible coupling mechanism between magnetism and dielectric properties in $EuTiO_3$ within the framework of the transverse-field Ising model. Phys Lett A 330(5):358–364

Yamasaki Y, Sagayama H, Goto T, Matsuura M, Hirota K, Arima T, Tokura Y (2007) Electric control of spin helicity in a magnetic ferroelectric. Phys Rev Lett 98(14):147204.38

Yáñez-Vilar S, Mun ED, Zapf VS, Ueland BG, Gardner JS, Thompson JD, Singleton J, Sánchez-Andújar M, Mira J, Biskup N, Señarís-Rodríguez MA (2011) Multiferroic behavior in the double-perovskite Lu_2MnCoO_6. Phys Rev B Condensed Matter Mater Phys 84(13):134427

Yang SY, Seidel J, Byrnes SJ, Shafer P, Yang CH, Rossell MD, Yu P, Chu YH, Scott JF, Ager Iii JW, Martin LW (2010) Above-bandgap voltages from ferroelectric photovoltaic devices. Nat Nanotechnol 5(2):143–147

Yang Y, Íñiguez J, Mao AJ, Bellaiche L (2014) Prediction of a novel magnetoelectric switching mechanism in multiferroics. Phys Rev Lett 112(5):057202

Zhang G, Wu H, Li G, Huang Q, Yang C, Huang F, Liao F, Lin J (2013) New high T c multiferroics $KBiFe_2O_5$ with narrow band gap and promising photovoltaic effect. Sci Rep 3(1):1265

Zhang Y, Sun H, Yang C, Su H, Liu X (2019) Modulating photovoltaic conversion efficiency of $BiFeO_3$-based ferroelectric films by the introduction of electron transport layers. ACS Appl Energy Mater 2(8):5540–5546

Zhang P, Grinting D, Yu SC, Nghia NX, Dang NV, Lam VD (2012) Influences of annealing temperature on structural characterization and magnetic properties of Mn-doped $BaTiO_3$ ceramics. J Appl Phys 112(1)
Zhao Z, Buscaglia V, Viviani M, Buscaglia MT, Mitoseriu L, Testino A, Nygren M, Johnsson M, Nanni P (2004) Grain-size effects on the ferroelectric behavior of dense nanocrystalline $BaTiO_3$ ceramics. Phys Rev B Condens Matter Mater Phys 70(2):024107

Chapter 9
Latest Updates on Materials, Device Designs, and Performance Optimization in Low-Bandgap Perovskite Solar Cells

Fency Sunny, Linda Maria Varghese, Nandakumar Kalarikkal, and Subila Kurukkal Balakrishnan

Abstract Halide perovskites have attracted widespread attention over the past two decades owing to excellent charge carrier dynamics and photovoltaic properties. The conversion efficiency of solar cells based on these materials has reached up to 26% in the past years making them the leading materials in photovoltaics today. Enhanced performance and diversification into different types of solar cells have opened new interest in low-bandgap perovskite-based solar cells. This chapter explores the latest developments in the area of low-bandgap perovskite solar cells detailing the different materials involved, the device architecture, and the factors determining performance enhancement. The constituent elements involved in perovskite engineering is discussed, including formamidinium-based perovskites, cesium-based perovskites, lead–tin double perovskites, and mixed halide perovskites. The device design strategies are elaborated covering sequential deposition, additive engineering, interfacial and bandgap engineering, and tandem configurations. Finally, the performance enhancement approaches and optimization schemes involving controlling the defect density, improving charge extraction, reducing recombination losses, and creating multi-junction arrangements are reviewed further to provide a holistic impression on the latest updates on low-bandgap perovskite solar cells.

F. Sunny
International and Inter University Centre for Nanoscience and Nanotechnology, Mahatma Gandhi University, Kottayam, Kerala, India

L. M. Varghese · S. K. Balakrishnan (✉)
School of Chemical Sciences, Mahatma Gandhi University, Kottayam, Kerala, India
e-mail: subila@mgu.ac.in

N. Kalarikkal
International and Inter University Centre for Ultrafast Studies, Mahatma Gandhi University, Kottayam, Kerala, India

School of Pure and Applied Physics, Mahatma Gandhi University, Kottayam, Kerala, India

M. Kumar and A. Raj (eds.), *Perovskite Solar Cells*, Springer Tracts in Electrical and Electronics Engineering, https://doi.org/10.1007/978-981-96-9463-1_9

Introduction

Halide perovskites have garnered worldwide attention in the past decade owing to the exceptional photophysical properties. Combined with the low-cost solution processing methods and the tunabilty of bandgap, halide perovskites provide suitable alternative over other thin film solar cells (Raj et al. 2021, 2022, 2023). Solar cells based on perovskites have achieved power conversion efficiencies (PCS) comparable to silicon solar cells. The direct bandgap present in these materials makes them excellent absorbers in the entire solar spectrum from visible to near infrared (NIR) (Ibrahim et al. 2024).

Improving efficiency and extending the usable solar spectrum remains the prominent reason in exploring low-bandgap materials for photovoltaics (Ibrahim et al. 2024). Apart from visible light absorption many variants of perovskites show high near infrared absorption making them suitable candidates for low-bandgap solar cells. Perovskites with bandgap from 1.2 to 1.8 eV can be considered as low-bandgap materials while those from 1.8 to 3 eV are classified as wide bandgap materials. The bandgap vs efficiency of low-bandgap perovskite solar cells is shown in Fig. 9.1. The graph shows the efficiency achieved so far for devices with different bandgap with a maximum efficiency of 20.03% for 1.38 eV bandgap material.

The correlation between efficiency and bandgap lies in the amplified utilization of the solar spectrum in low-bandgap material-based solar cells. Different halide perovskites, including formamidinium-based, cesium-based, lead- and tin-based, and mixed anion perovskites have all shown suitable low-bandgap that can be utilized in solar cells. However, these materials face numerous challenges that obstruct commercialization. In addition to the environmental stability issues that plague halide perovskites, there is also phase instability that actively disrupt long lifetime of these devices. Overcoming these challenges using various device and performance optimization strategy is required for translational applications (Alam and Ashraf 2020; Tonui et al. 2018).

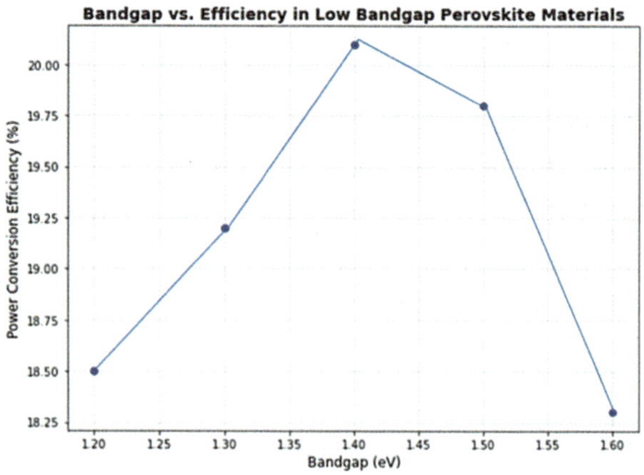

Fig. 9.1 Bandgap versus power conversion efficiency of low-bandgap perovskite solar cells (Ibrahim et al. 2024). Copyright 2024. Reproduced with permission from Taylor & Francis

This chapter focuses on the various low-bandgap materials used in perovskite solar cells and the updates in the various device design and performance optimization strategies for efficient low-bandgap solar cells. The current development in various device design approaches including sequential deposition, additive engineering, interfacial and bandgap engineering, and tandem configurations are detailed to provide a pathway to efficient device fabrication. The performance optimization schemes for these low-bandgap materials including reducing defect density and recombination losses along with tactics for improving charge extraction are also elucidated to direct productive research in the field.

Low-Bandgap Perovskites

Formamidinium (FA)-Based Perovskites

Ever since first reported by Koh et al. in 2014, FA-based perovskites have been hailed as the alternative near infrared (NIR) solar cell due to the low bandgap of the material. The proposed device showed high open circuit voltage combined with a favourable fill factor of 68.7%. The power conversion efficiency (PCE) of this material was observed to be 4.3%, indicating the viability of using this sensitizer in low-bandgap solar cells (Koh et al. 2014).

Computational studies have investigated the efficiency of $FASnI_3$ as a solar cell absorber. It was observed the low dielectric constant of $FASnI_3$ leads to greater electric field strength which reduced the charge carrier recombination leading to better power conversion efficiency (PCE). The study calculated a PCE of 17.33% for the proposed material (Shukla et al. 2021). The efficiency of FA-based perovskite solar cells have been increased via cesium cation alloying. $Cs_xFA_{1-x}PbI_3$ was synthesised showing better thermal stability, optimal bandgap and improved charge mobility and carrier lifetime. The conversion efficiency was found to be 19.85% for the mesoscopic device synthesised via the absorber material (Prochowicz et al. 2019). CsCl was added into the precursor solution of FAI and PbI in DMF/DMSO leading to an increase in the grain size and film quality as shown in Fig. 9.2.

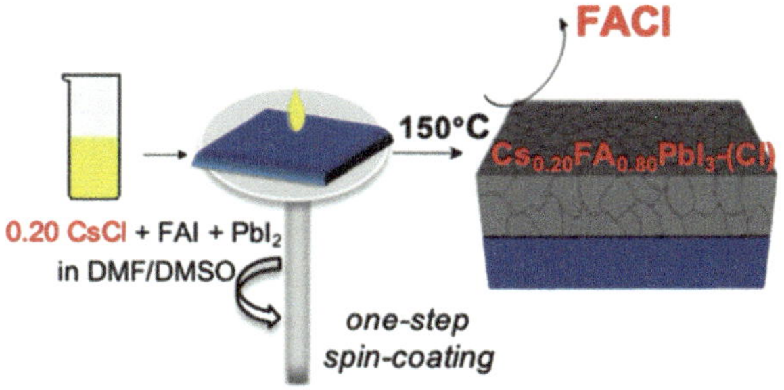

Fig. 9.2 Schematic representation showing the one-step synthesis of $Cs_xFA_{1-x}PbI_3$ low-bandgap perovskite (Prochowicz et al. 2019). Copyright@2019 Reproduced with permission from American Chemical Society

Methylammonium (MA) has also been utilized in FA-based perovskites to improve the conversion efficiency. Polycrystalline $MA_{0.03}FA_{0.97}Pb(I_{0.97}Br_{0.03})_3$ films have been fabricated showing improved PCE. Incorporation of an inorganic cation Cs has further the average efficiency up to 20.3% and improved the photostability leading to a retention of 90% of initial PCE value (Li et al. 2018).

Cesium-Based Perovskites

Organic–inorganic hybrid perovskites such as methylammonium (MA)-based halide perovskites have been on the forefront of low-bandgap materials for tandem solar cell application. However, the instability of the organic molecule in the presence of light and heat has led to an increased interest in inorganic alternatives such as cesium-based low-bandgap perovskites. Replacing MA with cesium has shown formation of thermally and optically stable halide absorber material with bandgap of 1.9 eV and conversion efficiency of 6.5% (Beal et al. 2016).

Cesium-based iodide perovskites have been synthesised to be efficient low-bandgap counterpart. Mixing cesium and formamidinium in iodide-only perovskite has been seen to tune the bandgap from 1.5 to 1.8 eV (Werner et al. 2018). $CsPbI_2Br$ perovskite has achieved a planar architecture device having a PCE of 9.8% with enhanced stability—thermal and optical—compared to the organic–inorganic perovskite counterparts (Sutton et al. 2016).

Lead–Tin-Mixed Perovskites

Reduced toxicity, high absorptivity, and current matching with wide bandgap counterparts in tandem solar cells have catapulted Pb–Sn-mixed perovskites to be good candidates in low-bandgap photovoltaics (Wang et al. 2023). Combining $FASnI_3$ and $MAPbI_3$ was found to have great results in power conversion with PCE of 15% (Liao et al. 2016). Instability of these perovskites due to easy oxidation of tin proves to be the greatest challenge in furthering research with these materials. Prasanna et al. outlined an effective way of passivation by form compact thin films that were large-grained. They also made use of indium tin oxide as the hole transport layer in the cell leading to a thermally and environmentally stable low-bandgap solar cell. They also observed that keeping the tin content below 50% and capping the cell with oxide such as indium zinc oxide helped in increasing the PCE and stability of Pb–Sn solar cells (Prasanna et al. 2019).

Carrier recombination happening at grain boundaries was observed via dark current density to be a prominent reason for low power conversion efficiency in mixed Pb–Sn-based low-bandgap solar cells. Introduction of Br was seen to passivate the grain boundaries leading to PCE greater than 19%. Incorporating the halide anion increased the fill factor making uniform films and lowered the open circuit voltage deficit improving the overall performance of the solar cell (Li et al. 2019).

Mixed Anion Perovskites

Halide exchange as always proved to be the most common method for bandgap tuning in halide perovskites. The valance and conduction band in perovskites have been primarily consisting of the halide anion and the bivalent cation orbitals (Sunny et al. 2024). Therefore, the tuning of bandgap, and thereby the optical properties, can be effectively tuned via replacing halide anions or mixed anion perovskites. Introducing more than one halide ion in the perovskite structure leads to lattice distortion provide effective strain and beneficial defects to alter the optoelectronic properties.

Mixing the anions in Pb–Sn perovskites have shown the formation of tunable narrow bandgap solar cells from 1.92 to 1.38 eV. $CsPb_{1-x}Sn_xI_2Br$ is predicted as an ideal perovskite with favourable bandgap, but previously was demonstrated unable to convert to perovskite phase by thermal annealing. G Chen et al. successfully prepared a series of $CsPb_{1-x}Sn_xI_2Br$ perovskites with tunable bandgaps from 1.92 to 1.38 eV for the first time via low annealing temperature (60 °C) (Chen et al. 2021). Mixed bromo-iodide perovskite-based solar cells was found to have a PCE of 9.8% achieved by stabilizing the efficient black phase and by tuning the bandgap (Sutton et al. 2016).

Device Design Strategies

Optimization of low-bandgap solar cells lies equally in finding the perfect materials as in the device design. Different strategies have been utilized to construct high-performance low-bandgap solar cells. The proper synthesis of the absorber layer plays an important role in producing efficient and stable low-bandgap solar cells. Utilizing appropriate deposition techniques including sequential deposition and additive engineering along with controlling the interface between the different layers of the solar cell and forming tandem configurations helps in fabricating effectual low-bandgap solar cell.

Sequential Deposition

Mixed anion halide perovskites form a prominent element of low-bandgap solar cells. The absorbers for the cells have been fabricated using sequential deposition. Better quality films leads to stable performance brought on by the larger grain size and reduced defects in the layer. Sequential deposition has been employed to facilitate better bandgap tuning via mixed halide perovskite synthesis. $MA_{0.03}FA_{0.97}Pb(I_{0.97}Br_{0.03})_3$ was synthesised via sequential deposition exhibiting record PCE. The absorber layer was further improved via the introduction of Cs through the sequential addition of CsCl precursor along with PbI resulting in a PCE of 20.3%. The process of sequential deposition helps in forming larger continuous grains. This commences from the lowered rate of nucleation on CsCl containing PbI_2 film on the addition of FAI (Li et al. 2018).

Low-bandgap solar cells using $FAPbI_3$ was synthesised using the sequential deposition method, wherein a thin film of PbI_2 spin coated onto mesoporous TiO_2 on FTO coated glass was dipped into a concentrated solution of FAI. The absorber layer thus fabricated was cleaned, and the hole transport layer and the electrodes were deposited onto it forming the final device (Fig. 9.3a). The device showed an extended absorption of light and a conversion efficiency of 4.3% (Koh et al. 2014).

Inverted structure low-bandgap solar cells have been fabricated using two-step sequential deposition method. The ease of fabrication and increased repeatability makes it a promising tool. In this cell, the bottom organic hole-selective layer was treated with the binary modulation system composed of a polyelectrolyte and an ammonium salt leading to refinement of the interfaces. This optimized cell exhibited a PCE of 23.4% and improved thermal and operational stability (Chen et al. 2023).

Additive Engineering

Improving the lifetime of perovskite solar cells via long-term stability is one of the prominent challenges in perovskite solar cell commercialization. One way to achieve this is by additive engineering. Additive engineering involves the addition of organic or inorganic compounds that helps in improving the grain size and thereby the overall performance of the cells. Different functional groups have been recognized as effective additives in halide perovskite solar cells (Pereyra et al. 2021).

In low-bandgap solar cells, additive engineering helps in improved heterostructure construction, control of film morphology, surface and/or grain boundary passivation, and gradient-distribution of film formation and carrier dynamics modulation (Wang et al. 2023). In Sn-based perovskites, the need for additive engineering arises from the oxidation of Sn^{2+} which is detrimental to the performance of the device and from rapid crystallization that occurs while film formation. Addition of molecules such as SnF_2 has proved to be helpful in controlling the defect formation and degradation (Dey et al. 2021).

Chloride addition has proven to be an effectual morphology modulator for low-bandgap iodide perovskites thin films. Chloride perovskites have also been used as templates for iodide perovskite synthesis owing to the volatility of chlorine leading to the formation of large grained iodide perovskites upon annealing at appropriate temperatures. MACl added into $FaPbI_3$ perovskite film via antisolvent method

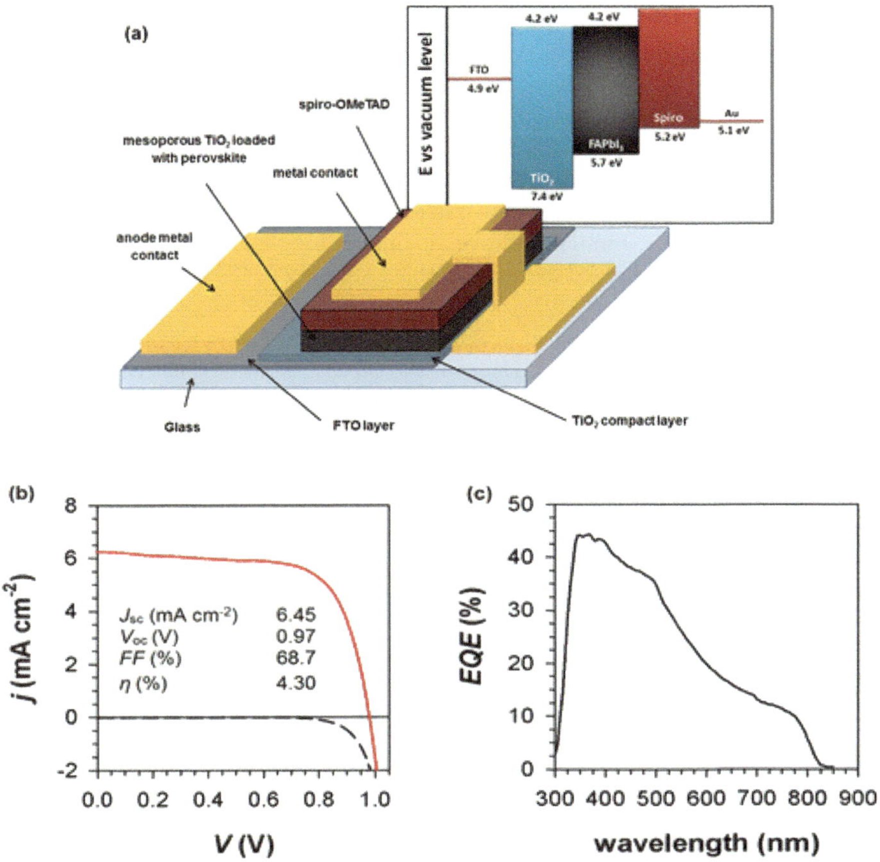

Fig. 9.3 **a** Schematic representation of the device design of the solar cell. Inset shows the energy level of the different layers in the cell. **b** The current–voltage curve for the $FAPbI_3$ device under AM1.5G illumination and under dark conditions. **c** EQE (external quantum efficiency) of the solar cell (Koh et al. 2014). Copyright@2014 Reproduced with permission from American Chemical Society

resulted in increasing the crystal size of $FAPbI_3$ and thereby the PCE of the device to 23.08% (Kim et al. 2019).

Molecule additives including organic solvents such as DMSO or methylamines have been utilized to solvate precursors to form better quality films. Highly crystalline $FAPbI_3$ was synthesised through intramolecular exchange method brought about by the interaction of FAI (MABr) solution on a pre-deposited PbI_2/DMSO film. The intermediate phase formed via DMSO addition helps in the formation of improved absorber layer resulting in a device with PCE 20.2% (Yang et al. 2015).

Phase destabilization is the major challenge faced in FA- and Cs-based low-bandgap perovskites. At room temperature, $FAPbI_3$ has greater affinity to be in the yellow non-perovskite phase rather than in the black phase. Several additives have been employed to stabilize the perovskite including cations, addition of hydroiodic acid, phenylethylammonium (PEA) iodide, etc. Surface functionalization via long-chain alkyl and/or aromatic ammonium cations have also been employed for phase stabilization. Long-chain ammonium cations helps in phase stabilization via the decrease in the nanocrystal size and thereby the available surface, brought about by surface functionalization. Addition of PEA to form alternate 2D perovskite layer in $FAPbI_3$ grain boundaries has also shown to increase the moisture and environmental stability of these films. Similar additives including long-chain ammonium cations and PEA have been employed to increase phase stability in Cs-based perovskites as well (Liu et al. 2020).

Interfacial and Bandgap Engineering

Bandgap Engineering

Bandgap engineering is one of the important strategies to optimize the performance of the perovskite solar cell. By altering the bandgap, it is possible to enhance the light absorption, and minimize the recombination losses tailor the material suitable for photovoltaic applications. The optimization focuses on balancing the bandgap to stabilize and improve the broad spectral absorption along with improving the crystallinity and tuning composition and grain boundaries. In 2014, Kanatzidis and Snaith fabricated a solar cell of bandgap of 1.3 eV, $MASnI_3$ which have significantly lower bandgap than benchmark $MAPbI_3$ of 1.55 eV bandgap. This red shift in bandgap enables broader absorption towards infrared photons, critical for enhancing the single-junction solar cell. Along with that, incorporation of Br into $MASnI_{1-x}Br_x$ led to the modification of bandgap and optimization of composition, and achieved a power conversion efficiency (PCE) of 5.37% (Noel et al. 2014; Hao et al. 2014).

Further studies revealed that $FASnI_3$ exhibits better performance than $MASnI_3$ in both performance and stability. But when we discuss about the Sn-based solar cells, its oxidation is needed to address properly and researchers improved the properties by introducing SnF_2, an additive that act as reducing agent to impede the oxidation of Sn^{2+} to Sn^{4+}, thus preserving the desired bandgap. This approach resulted in a higher PCE of 6.22% (Lee et al. 2016a). Ke et al. introduced hollow-structured $MASnI_3$ and $FASnI_3$ with ethylenediammonium (EDA) as an additive to obtain low-bandgap high PCE. Further bandgap tuning through Br improved crystallinity and stability (Ke et al. 2017). To solve the intrinsic challenges faced with pure Sn-based perovskites, high oxidation rate and doping densities research are shifted to mixed Sn–Pb perovskites, which provided more control over bandgap tuning. Hayase et al. introduced $MASn_{0.5}Pb_{0.5}I_3$ with a bandgap of ~ 1.17 eV with initial PCE of 4% and subsequently raised the efficiency to 15.6% with different modifications (Ogomi et al. 2014). Additional efforts by Yang et al. optimize the incorporation of Br in mixed Sn–Pb perovskites, forming $MAPb0.5Sn_{0.5}(I_{0.8}Br_{0.2})_3$ with an ideal bandgap of 1.35 eV for single-junction solar cells (Yang et al. 2017).

Low-bandgap PSCs using FA/Cs perovskites have demonstrated bandgaps around 1.2–1.3 eV, showing enhanced thermal and photostability. Devices such as $MA_{1-x}FA_xPb_{0.7}Sn_{0.25}I_3$ achieved a PCE of 14.2% with a bandgap of 1.33 eV (Yang et al. 2016). $(FASnI_3)_{0.6}(MAPbI_3)_{0.4}$ fabricated by combining $FASnI_3$ and $MAPbI_3$ precursor solutions led to the production solar cell with a bandgap of 1.25 eV, yielding PCEs up to 17.5% (Zhao et al. 2017). Eperon et al. developed solar cells with a bandgap of 1.2 eV and a PCE of 14.8% with composition $FA_{0.75}Cs_{0.25}Sn_{0.5}Pb_{0.5}I_3$ (Eperon et al. 2016). Leijtens et al. pushed the PCE to 15.6% by incorporating MACl and formic acid into the precursor solution, while Padture and co-workers achieved a single-junction bandgap of ~ 1.3 eV by optimized composition with PCEs up to 15.8% (Leijtens et al. 2018; Zong et al. 2017). Mixed Sn–Ge compositions in $FA_{0.75}MA_{0.25}Sn_{1-x}Ge_xI_3$ achieve a bandgap range of 1.40 eV, and $CsPb_{0.4}Sn_{0.6}I_{2.4}Br_{0.6}$ exhibit a bandgap of 1.35 eV with efficiency improvements up to 12.34%. These advances highlight the potential of tailored perovskite materials for optimized photovoltaic performance (Ito et al. 2018; Lee et al. 2020; Song et al. 2019). Table 9.1 shows examples of solar cells with low-bandgap perovskite.

Interfacial Engineering

Interfacial engineering plays a crucial role in the stability and performance of solar cell by alleviates the charge recombination at the interface and improves the performance of photovoltaic cell. A key aspect of interfacial engineering lies in the surface passivation techniques that passivating surface defects and grain boundaries which are the primary reason for the non-radiative recombination. Interface modification using guanidinium thiocyanate (GuaSCN) in Sn–Pb perovskites $(FASnI_3)_{0.6}(MAPbI_3)_{0.4}$ forms a 2D layer at grain boundaries that reduces the defect density and improves stability. Along with that, it improves the carrier life time and block oxygen infiltration result a PCE of 23.1% with a bandgap of 1.23 eV (Tong et al. 2019). Similarly, Sargent and co-workers introduce phenethylammonium iodide (PEAI) into $Cs_{0.1}MA_{0.2}FA_{0.7}Pb_{0.5}Sn_{0.5}I_3$ PSCs effectively passivate the surface defects via ligand-anchoring interaction by which non-radiative carrier recombination was suppressed and leading to a high efficiency of 23.7% (Wei et al. 2020). Similarly, Jianga and co-workers applied the thin-layer phenethylammonium iodide (PEAI) into MA-free $FA_{0.83}Cs_{0.17}Pb_{0.5}Sn_{0.5}I_3$ PSC and showed a lifespan 20 times longer than the MA-containing PSC, with a bandgap of 1.49 eV (Jiang et al. 2022). By the incorporation of phenyl-C_{61}-butyric acid methyl ester (PCBM) into the Sn–Pb perovskite $FA_{0.5}MA_{0.5}Sn_{0.5}Pb_{0.5}I_3$ (FAMA), a conduction band spike structure improved the charge extraction and reduced recombination rates, increasing PCEs to 17.6% with a bandgap of 1.25 eV (Song et al. 2019).

Furthermore, alloying strategies such as the use of formamidinium thiocyanate (FASCN) and methylammonium thiocyanate (MASCN) into the perovskite $FAPbI_3$ have proven effective in stabilizing the α-phase of $FAPbI_3$ at room temperature. This method improved the stability against humidity-induced degradation along with lowering of bandgap from 2.43 eV (δ-$FAPbI_3$) to 1.56 eV (α-$FAPbI_3$) with PCE of

Table 9.1 Summary of low-Eg single-junction perovskite solar cells reported in the recent years

Type	Absorber	Eg (eV)	VOC (V)	JSC (mA/cm^2)	FF (%)	PCE (%)
n-i-p	$MASn_{0.75}Pb_{0.25}I_{3-x}Cl_x$	1.25	0.457	19.63	57.2	5.13
p-i-n	$MA_{0.9}Cs_{0.1}Pb_{0.5}Sn_{0.5}I_3$	1.28	0.7	23.32	61.7	10.07
p-i-n	$MASn_{0.5}Pb_{0.5}I_3$	1.28	0.69	22.8	65	10.24
p-i-n	$MASn_{0.1}Pb_{0.9}I_3$	1.31	0.78	19	67.1	10.25
p-i-n	$MASn_{0.5}Pb_{0.5}I_3$	1.18	0.75	26.3	68.8	13.6
p-i-n	$MAPb0.75Sn_{0.25}I_3$	1.3	0.74	23.8	78.6	13.9
p-i-n	$MA_{0.5}FA_{0.5}Pb_{0.75}Sn_{0.25}I_3$	1.33	0.78	23.03	79	14.19
p-i-n	$FA_{0.75}Cs_{0.25}Sn_{0.5}Pb_{0.5}I_3$	1.2	0.74	26.7	71	14.8
p-i-n	$(FASnI_3)_{0.6}(MAPbI_3)_{0.4}$	1.24	0.79	26.86	70.6	15.08
p-i-n	$MASn_{0.25}Pb_{0.75}I_3$	1.26	0.77	24	66	12.08
p-i-n	$MAPb0.75Sn_{0.25}I_3$	1.3	0.73	23.5	79	13.9
p-i-n	$MA_{0.5}FA_{0.5}Pb_{0.5}Sn_{0.5}I_3$	1.2	0.78	25.69	70	14.01
p-i-n	$MASn_{0.25}Pb_{0.75}I_3$	1.3	0.8	26.2	72.5	15.2
p-i-n	$MAPb0.5Sn_{0.5}I_3$	1.22	0.89	26.8	71	15.61
p-i-n	$(FASnI_3)_{0.6}(MAPbI_3)_{0.4}$	1.25	0.85	28.5	72.5	17.6
p-i-n	$MAPb0.5Sn_{0.5}(I_{0.8}Br_{0.2})_3$	1.35	0.9	25.9	75	17.63
p-i-n	$(FASnI_3)_{0.6}(MAPbI_3)_{0.4}$	1.24	0.83	27.3	78.5	17.8
n-i-p	$MASn_{0.74}Pb_{0.26}I_3$	1.28	0.79	25.5	69	14.04
p-i-n	$(t\text{-}BA)_2(FA_{0.85}Cs_{0.15})_4(Pb_{0.6}Sn_{0.4})_nI_{16}$	1.24	0.7	24.2	63	10.6
p-i-n	$MASn_{0.5}Pb_{0.5}I_xCl_{3-x}$	1.2	0.7	24.7	71	12.3
p-i-n	$MAPb0.4Sn_{0.6}I_{2.6}Br_{0.4}$	1.25	0.8	23.38	72	13.4
p-i-n	$FA_{0.7}MA_{0.3}Sn_{0.3}Pb_{0.7}I_3$	1.26	0.78	23.6	78	13.6
p-i-n	$(FAPbI_3)_{0.7}(CsSnI_3)_{0.3}$	1.3	0.7	26.3	79.7	14.7
p-i-n	$FAPb_{0.7}Sn_{0.3}I_3$/PEAI	1.33	0.78	26.46	79	16.26
p-i-n	$FAPb_{0.75}Sn_{0.25}I_3$	1.36	0.81	28.23	75	17.25
p-i-n	$FA_{0.5}MA_{0.5}Sn_{0.5}Pb_{0.5}I_3$	1.25	0.75	30.29	74	17.6
p-i-n	$(FASnI_3)_{0.6}(MAPbI_3)_{0.4}$	1.25	0.86	27.92	75	18.03
p-i-n	$(FASnI_3)_{0.6}(MAPbI_3)_{0.4-x}(MAPbCl_3)_x$	1.25	0.84	29	74.5	18.1
p-i-n	$(FASnI_3)_{0.6}(MAPbI_3)_{0.34}(MAPbBr_3)_{0.06}$	1.27	0.88	28.7	74.6	19.03
p-i-n	$(FAPb_{0.6}Sn_{0.4}I_3)_{0.85}(MAPb0.6Sn_{0.4}Br_3)_{0.15}$	1.28	0.87	26.45	79.1	18.2
p-i-n	$Cs_{0.3}FA_{0.7}Sn_{0.3}Pb_{0.7}I_3$	1.34	0.78	29.1	79.9	18.3
p-i-n	$FA_{0.75}Cs_{0.25}Sn_{0.4}Pb_{0.6}I_3$	1.29	0.72	30.2	74	15.4
p-i-n	$FA_{0.83}Cs_{0.17}Pb(I_{0.6}Br_{0.4})_3$	1.25	0.85	28.6	72	17.6
p-i-n	$FA_{0.85}MA_{0.1}Cs_{0.05}Sn_{0.5}Pb_{0.5}I_3$	1.28	0.86	29.8	79.1	20.4
p-i-n	$FA_{0.75}Cs_{0.25}Sn_{0.5}Pb_{0.5}I_3$	1.25	0.74	31.7	75.4	16.5

19.09% (Chen et al. 2020). In the recent research, Albrecht and co-workers modified the low-bandgap perovskite cell $Cs_{0.05}(FA_{0.77}MA_{0.23})_{0.95}Pb(I_{0.77}Br_{0.23})_3$ with self-assembled, methyl-substituted carbazole monolayer as the hole-selective layer exhibited high PCE of 29.15%, in ambient air without encapsulation (Al-Ashouri et al. 2020). Since (NiO_x) is found as a promising hole-transport layer for perovskite solar cells, the researchers are keep trying to passivate the defects at the NiO_x/perovskite (bandgap of 1.5 eV) interface originated from Ni $\geq 3^+$ states due to a nickel deficiency at the NiO_x surfaces. Woif and co-workers focused on that and used a dye molecule N719 to passivate the surface traps and chemical reaction at the NiO_x/perovskite interface and to increase power conversion efficiency (PCE) from 17.3 to 20.4% (Zhumagali et al. 2021).

Novel low-bandgap S,N-heteroacene-based oligothiophenes were developed as hole-transport materials (HTMs) by Zhaoning and co-workers who achieved a bandgap tuned to 1.49 eV and a power conversion efficiencies (PCEs) of 9.5–10.5% in $CH_3NH_3PbI_3$-based PSCs, compared to devices without HTMs of PCE 7.6% (Pereyra et al. 2021; Qin et al. 2014). Additionally, a cascade-type electron extraction design using a $ZnO/SnO_2/C_{60}$-SAM tri-layer in perovskite $FA_{0.5}MA_{0.5}Sn_{0.5}Pb_{0.5}I_3$ PSCs improved PCE of 13.8% with a bandgap of 1.3 eV. This method effectively engineered the energy-level alignment of electron transport layers, reduces the interfacial recombination and minimize voltage loss, which resulted in an improved PCE of perovskite solar cell. The charge recombination and extraction properties of the devices with different ETL were schematically represented in Fig. 9.4 (Liu et al. 2019). In 2022, Yuanyuan and co-workers achieved effective Sn(IV)-defects passivation in $CsPb_{0.6}Sn_{0.4}I_3$ IHPs by suppressing the oxidation using acetylhydrazine-incorporated antioxidative solution system. This delivery enhanced PCEs with 1.4 eV bandgap (Hu et al. 2022). These approaches demonstrate the crucial role of bandgap tuning and interface engineering in optimizing PSC efficiency.

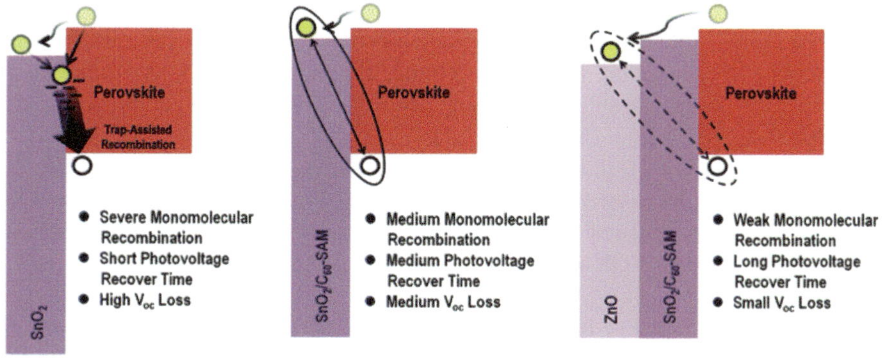

Fig. 9.4 Schematic representation of the charge recombination situations in cases of different ETLs in low-bandgap $FA_{0.5}MA_{0.5}Sn_{0.5}Pb_{0.5}I_3$ PSCs (Liu et al. 2019). Copyright 2019. Reproduced with permission from Royal Society of Chemistry

Tandem Configurations

Multifunction tandem solar cells are designed in such a way that semiconductor absorber layers with different bandgaps are stacked together to reduce energy losses from thermalization of photoexcited carriers. The most common tandem cells are double-junction configuration—consisting of a top cell of high bandgap for high energy photons and low-bandgap bottom cell for low energy photons. Four-terminal (4-T) and two-terminal (2-T) architecture are the two types of fabrication method for double-junction tandem cells. In a 4-T tandem design, the two sub cells are either stacked by mechanical or optical splitter method. In a 4-T configuration, each sub cell has its own electrodes and can operate at its maximum power point, as they are only optically coupled, not electrically connected. However, 4-T design has drawbacks like optical losses and high manufacturing costs as four sub cells required four electrodes. In the case of 2-T tandem architecture, the two sub cells are monolithically integrated on the same substrate. Here, the sub cells are connected both optically and electrically by an interconnecting layer (IL), either a heavily doped tunnel junction or a transparent conductive oxide (TCO). Since the device shares a common transparent front electrode and an opaque back electrode, the cost of production is low compared to 4-T architecture; however, it presents fabrication challenges related to the processing of the IL and the second cell, including cell polarity, substrate roughness, deposition temperature, and solvent compatibility makes them less attractive than4-T architecture (Song et al. 2019).

Four-Terminal Perovskite/CIGS-Based Solar Cells

When considering four-terminal (4T) cells, a different scenario emerges, where the engineering of material becomes a critical research focus. Unlike monolithic configurations, numerous studies have investigated 4T perovskite/CIGS solar cells. Notably, Shen and co-workers developed 4T cells with perovskite/CIGS by strategically integrating transparent conducting electrodes (TCEs) consisted of a dense TiO_2 layer on an ITO bottom electrode with a multi-cation perovskite absorber ($Cs_{0.1}Rb_{0.05}FA_{0.75}MA_{0.15}PbI_{1.8}Br_{1.2}$). When stacked with a CIGS bottom cell (bandgap 1.13 eV), the overall tandem efficiency of the cell reached 23.4%, with stability against oxygen-induced degradation (Shen et al. 2018). Bailie and his team used silver nanowires (AgNWs) as a back contact material in perovskite sub cells and paired with CIGS to achieve a efficiency of 18.6 (Bailie et al. 2015). In the similar way AgNWs in a $MAPbI_3$/CIGS device were employed by Lee et al. though the efficiency remained at 10% due to complex processing and solvent interactions (Lee et al. 2016b). Long with that Gharibzadeh and co-workers explored the 2D/3D perovskite heterostructure by combining ($FA_{0.83}Cs_{0.17}Pb(I_{1-X}Br_X)_3$) bulk structure with a 2D passivation agent (n-butylammonium bromide) and combined with CIGS bottom cells (bandgap 1.13 eV, PCE 21.2%) obtain overall tandem efficiency of 23.4% (Gharibzadeh et al. 2020).

Four-Terminal Perovskite–Perovskite Tandem Solar Cells

Compared to existing tandem technologies, perovskite tandems offer substantial design and processing flexibility, while maintaining a low thermal budget for fabrication, the bandgaps of both the upper and lower cells can be finely tuned to optimize the performance. In the beginning stages of this technology, Li et al. constructed a 4T TSC using perovskite materials ($MAPbBr_3$ on top of $MAPbI_3$), with a composite HTL of carbon nanotubes and PMMA, and a transparent gold grid electrode. Due to the limited NIR transmission the efficiency of the solar cell was restricted to 9.46% (Li et al. 2016a). Jen and colleagues later achieved a 4T all-perovskite TSC with a PCE of 19.08% by combining $MA_{0.5}FA_{0.5}Pb_{0.75}Sn_{0.25}I_3$ PSCs with a small bandgap (1.33 eV). NIR response of perovskite are improved by increasing absorber layer thickness and grain size, resulting in a Sn–Pb PSC with a validated PCE of 17.01%, obtained by the studies of Zhao and co-workers (Zhao et al. 2017). Yang et al. demonstrated the perovskite tandem solar cell using a 1.55 eV $MAPbI_3$ top cell, and a 1.33 eV $MA_{0.5}FA_{0.5}Pb_{0.75}Sn_{0.25}I_3$ bottom cell, achieving a combined power conversion efficiency (PCE) of 19.1%.The thin layers of $MAPbI_3$ allowed sufficient photons to reach the bottom cell to result the high efficiency (Yang et al. 2016). Eperon et al. improved this design using a 1.6 eV $FA_{0.83}Cs_{0.17}Pb(I_{0.83}Br_{0.17})_3$ top cell and a 1.2 eV $FA_{0.75}Cs_{0.25}Sn_{0.5}Pb_{0.5}I_3$ bottom cell, achieving a total PCE of 20.3% (Eperon et al. 2016). The researchers found another tandem solar cell by combining 1.58 eV $FA_{0.3}MA_{0.7}PbI_3$ with 1.25 eV $(FASnI_3)_{0.6}(MAPbI_3)_{0.47}$ with efficiency of 21.2% (Song et al. 2019).

Two-Terminal All Perovskite Tandem Solar Cells

In the case of 2-terminal (2-T) perovskite tandem solar cell, the interlayer (IL) play a crucial role as it not only connects two sub cells but also protects the underlying sub cell from the possible damage, by the processing of the second cell. The Zhou and co-workers were the first team who introduced the concept of monolithic 2-T al-perovskite tandem solar cells by connecting two methylammonium lead iodide ($MAPbI_3$) cells with organic ILs (Jiang et al. 2016). Heo and Im also fabricated 2-T $MAPbBr_3$-$MAPbI_3$ tandem solar cells with a bandgap of 1.5 eV off bottom cell and achieve a PCE of 10.8% but limited the current density to lower value due to the physical contact between the sub cells (Heo and Im 2016). In the later years, McGehee, Snaith, and co-workers introduced a low-bandgap mixed tin–lead (Sn–Pb) perovskite tandem cell featured a 1.8 eV $FA_{0.83}Cs_{0.17}Pb(I_{0.5}Br_{0.5})_3$ top cell and a 1.2 eV $FA_{0.75}Cs_{0.25}Sn_{0.5}Pb_{0.5}I_3$ bottom cell, connected by a zinc–tin oxide (ZTO)/SnO_2/indium tin oxide (ITO) interlayer. The ITO layer protects the first sub cell during the solution deposition of the second sub cell and also provides electrical and optical connection between the sub cells, resulting in a PCE of 16.9%. Following more optimization, the researchers achieved a PCE of 19.1% with the use of a more stable top cell composition, $FA_{0.83}Cs_{0.17}Pb(I_{0.7}Br_{0.3})_3$, and improved the quality of the bottom cell using methylammonium chloride (MACl) and formic acid with $FA_{0.75}Cs_{0.25}S_{0.25}Pb_{0.5}I_3$ (Leijtens et al. 2018). Jen and co-workers reported 2-T all perovskite tandem cells with impressive PCE of 18.4% using $MA_{0.9}Cs_{0.1}Pb(I_{0.6}Br_{0.4})_3$ (1.82 eV) as the top cell and $MASn_{0.5}Pb_{0.5}I_3$ (1.22 eV) as the bottom cell, which is connected by sputtered ITO recombination layer (Rajagopal et al. 2017). In the similar way, researchers demonstrated a 2-T tandem cells with a 1.75 eV $FA_{0.8}Cs_{0.2}Pb(I_{0.7}Br_{0.3})_3$ top cell and a 1.25 eV $(FASnI_3)_{0.6}(MAPbI_3)_{0.4}$ bottom cell connected by vacuum-processed ILs of MoO_3/Ag/ITO. The constructed cell shows a PCE of 21% and significant performance improvements achieved by the use of a bulk-passivation strategy which increase grain size, carrier mobility, and suppress trap-assisted recombination in the low-Eg perovskite film. Additionally, the tandem devices exhibited decent operational stability, maintaining over 85% of the initial PCE after 80 h of continuous operation (Zhao et al. 2018). Choy and co-workers offer an alternative route to integrating perovskite sub cells using an ultrathin Cu/Au thermionic emission-based interlayer and achieved an efficiency of 17.9%. In addition to solution-processed all-perovskite solar cells, the vapour-deposited perovskite solar cells are also possible; Bolink and co-workers employed a combination of solution-processed wide-bandgap $FA_{0.75}Cs_{0.15}Pb(I_{0.3}Br_{0.7})_3$ top cells and vacuum-deposited $MAPbI_3$ bottom cells achieves a champion PCE of 18.1% (Forgács et al. 2017).

Performance Optimization Approaches

Reducing Defect Density

For achieving high efficiency and long-term stability, reducing defect density in low-bandgap perovskite is essential. This can be accompanied by various strategies like material engineering, surface passivation, processing optimization and optimization of environment for fabrications. These approaches ultimately addressing different types of defects contributes to minimizing the defect density and improving the performance and durability of perovskite solar cell. With respect to defects, its properties such as defect density, defect position and electron, hole capture cross-section in the perovskite layers are needs to be addressed. Due to high Shockley–Read–Hall recombination rate, the deep defect nearer to the centre of perovskite bandgap are identified as most harmful, and this high in defect density reduces PCE (Taheri et al. 2021). Bulk defects and interfacial defects are two types of defects that observed in the perovskite materials. Unlike crystalline solar cell, which manufactured by well-established semiconductor processes, perovskites are commonly fabricated by solution-based methods like spin-coating, blade-coating, spray-coating or vapour-based techniques which results some advantages and disadvantages to the solar cell (Ding et al. 2019). The main advantage as well as the disadvantages of this process is the rapid crystal growth, often necessitating post-treatment annealing to remove residual solvents. This creates a wide range of defects due to imperfect lattice alignments such as point defects, impurity atoms, dislocations, grain boundaries, and precipitates. These defects can create trap states within the bandgap and centre for recombination sites (Ono et al. 2020).

Different Types of Defects and Their Impact

To enhance the efficiency of PSCs, a comprehensive knowledge of the types, densities, and energy position of these defects are essential. Through the experimental studies, researchers identified various defects like Halide vacancies: Halide vacancies are the most frequently reported defects. Halide vacancies, such as Cl^-, Br^-, and I^-, can expose under-coordinated Pb^{2+} atoms, which are highly reactive and susceptible to oxidation, leading to rapid degradation (Duan and Uddin 2022).

Anti-sites and Halide Excess: Defects such as negatively charged Pb-I anti-sites like PbI_3^- or halide excess are the next most common, defect observed in studies. These defects can cause localized distortions in the perovskite lattice, contributing to the formation of non-radiative recombination centres (Ono et al. 2020).

Cation Vacancies and Mobile Ions: Cation vacancies (e.g., Cs^+, MA^+, FA^+) and mobile or volatile species such as iodine or MA^+ ions are also prevalent. These defects creating new defect sites that adversely affect the stability and efficiency of PSCs and can migrate under an electric field or during thermal annealing (Dunfield et al. 2020).

Moisture-Sensitive Defects at Grain Boundaries: There are different types of defects at grain boundaries are reported but moisture-sensitive defects at grain boundaries play a critical role in triggering degradation processes, particularly under environmental stress. The presence of interfaces, such as those between the perovskite layer and adjacent layers such as ETL and HTL, further complicates the defect landscape (Castro-Méndez et al. 2019).

Strategies for Reducing Defect Density

Effective passivation strategies are essential to mitigate the detrimental effects of defects in perovskite solar cells. Passivating the molecules with Lewis acid and Lewis base are found to be an effective way as they can simultaneously neutralize both positively and negatively charged defects. For example, Nenon and co-workers proposed the idea of passivating the material relative to its softness or hardness of the Lewis base character. From their findings, it is clear that passivating coordinated Pb^{2+} ions, which exhibit a relatively soft Lewis acid nature, are found to be effective with softer Lewis bases like alkylphosphonates, fluorinated carboxylates, and sulfonates rather than harder Lewis bases, such as alkylcarboxylates, carbonates, and nitrates (Nenon et al. 2018). However, the surface appeared to be charge-free, and halides located at grain boundaries remain mobile under electric fields are needed to address properly.

Adding additives like dimethyl sulfoxide (DMSO), N-methyl-2-pyrrolidone (NMP), and 1,8-diiodooctane (DIO) to the perovskite precursor solutions, which form complexes with PbI_2 can slow down perovskite crystallization kinetics and improve film crystallinity (Song et al. 2019; Liang et al. 2014). To stabilize the perovskite structure, strengthening the interaction between A-site cations (such as MA^+, FA^+, Cs^+) and $[PbX_6]^{4-}$ (X = I, Br) octahedra has been shown to help to some extent (Yang et al. 2018). However, additional research is required to develop strategies that make these interactions stronger and more stable under operational conditions. Perovskite composition can be optimized using mixed cation or anion formulations, which enhance crystal quality and reduce defects. Surface passivation techniques like introducing organic or inorganic layers in the interface of perovskites reduce trap states and protect against degradation which also improves charge transport and reduces recombination losses. Processing optimization, through solvent selection, anti-solvent techniques, and post-treatment processes, helps achieve uniform crystal growth and controlling environmental conditions like humidity and temperature during fabrication minimizes defect formation and improves film quality.

To improve the performance and stability of perovskite solar cells, it is crucial to manage a variety of defect types through different strategies. Some issues are addressed recently but ongoing research is needed to further understand defect dynamics and develop materials and methods that mitigate their impact, ensuring the long-term viability of perovskite-based photovoltaic technologies.

Improving Charge Extraction

Low-bandgap perovskite is a promising candidate for solar cells due to their ability to absorb a broader range of the solar spectrum, offering higher theoretical efficiencies. Despite their potential, these cells often face challenges related to its charge extraction as it directly affects the power conversion efficiency. Improving charge extraction requires careful control of materials and interfaces to minimize non-radiative recombination losses and maximize carrier transport efficiency. The efficiency of charge extraction in perovskite solar cells is dependent on various factors like carrier mobility, carrier life time, and interfacial engineering (Taheri et al. 2021). High carrier mobility is crucial for ensuring charges move quickly through the perovskite material to the respective electrodes without recombination. Longer carrier lifetimes allow more time for charges to reach the interfaces before they recombine and proper alignment of energy levels and charge transport layers between the perovskite minimizes energy barriers that impede charge transfer. In low-bandgap perovskites, compared to the lead-based perovskite, Sn-based perovskites tend to have lower carrier mobility and shorter lifetime (Yang et al. 2016). This arises due to the higher density of trap states caused by defects in the perovskite crystal structure, which lead to non-radiative recombination in the material. Both Pb and Sn perovskites are highly susceptible to oxidation, which introduces deep-level defects that trap carriers and hinder charge transport. Therefore, strategies to minimize these oxidations are essential. Surface passivation reduces trap density at the perovskite/transport layer interface, and facilitates smoother charge flow. For extracting charges efficiently in solar cells, the optimization of charge transport layer is essential, and the optimization of these layers involves improving their conductivity, mobility, and interface compatibility with the low-bandgap perovskite.

Reducing Recombination Losses

Optimization of Material Crystallization

First and foremost way to minimize this recombination process is to optimize the material crystallization which is influenced by material composition, solvent used, substrate on which the material grown, fabrication protocol and environmental condition (Luo et al. 2020). In the case of low-bandgap perovskite, researchers have developed mixed cations and mixed halides perovskite to improve the crystallinity and optoelectrical properties. In fact, A-cation mixed perovskite systems of (FA/MA)PbI_3, (MA/Cs)PbI_3,(FA/Cs)PbI_3, (FA/Rb)PbI_3 showed enhanced stability as compared to reference cells comprising simple perovskites ($MAPbI_3$ or $FAPbI_3$). In the similar way, the systems with mixed X-anion halides with A-site constrained to a single-monovalent cation, MAPb(I/Cl), FAPb(I/Cl), MAPb(I/Br), FAPb(I/Br), MAPb(Br/Cl), MAPb(I/Br/Cl), CsPb(I/Br), CsPb(Br/Cl), CsPb(I/Cl) were reported that can suppress the recombination losses. Along with that complex systems of simultaneously mixed-A and mixed-X like (MA/Cs)Pb(I/Br), (FA/MA)Pb(I/Br), (FA/MA)Pb(I/Cl) were also reported with different compositions (Ono et al. 2017). Also with that, solvent engineering found to be one of the most successful approaches to obtain high-quality films instead of using one solvent. The mixed solvent is currently being used as perovskite precursor solution, like N-methyl-2-pyrrolidone (Tu et al. 2018) and acetonitrile (Li et al. 2016b) have been mixed with the most widely used dimethyl formamide. Change in the fabrication process like using non-halide acetate route and using alkali halides for optimization composition are favourable in recombination losses.

Interface Engineering

Interface engineering is one of the best strategies to reduce the recombination losses, since recombination often occurs at the interference between charge transport layer and light absorbing layers. One of the approaches is to use of passivating interlayer to create a barrier and reduce interfacial defects and recombination. As discussed in the previous section, self-assembled monolayer of fullerene and its derivatives and graphene oxide are reported to applied at the perovskite and electron transport layer interface to reduce recombination losses (Li et al. 2017). In the similar way, the choice of HTL and ETL can influence the recombination losses. The appropriate energy alignment of material with perovskite is highly essential like TiO_2, ZnO, SnO_2, SiO_2, Al_2O_3, and ZrO_2 as ETL and NiO_x, spiro-OMeTAD, Cu_2O, CuO, CuI, CuSCN, and MoS_2 as HTL has demonstrated enhanced charge collection efficiency and passivation effects.

Passivation of Grain Boundaries

Grain boundaries in polycrystalline perovskite films are consist of structural defects due to the presence of dangling bonds and vacancies, which eventually form electronic traps and sites for non-radiative recombination (Yin et al. 2014). Several strategies are adopted but use of passivating agents like phenethylammonium iodide (PEAI), thiocyanate (SCN^-), and chloride (Cl^-) ions, have been effectively passivate these sites. These additives help in demote the density of trap states at the grain boundaries. Additionally, solvent engineering methods like anti-solvent dripping methods can reduce the grain boundary defects by improve the crystallinity of perovskite film. For example, passivation in $CsPb_{0.6}Sn_{0.4}I_3$ IHPs by suppressing the oxidation using acetylhydrazine incorporated antioxidative solution system found to be an effective way of solving this crisis. Recently, researchers are focusing more on producing perovskite films with larger grains and fewer grain boundaries to reduce recombination losses.

Employing efficient device architectures like the use of 2D/3D perovskite heterostructure has also been explored as a method to reduce recombination losses. This method creates a passivating barrier of 2D perovskite on top of the 3D perovskite to protect it from moisture, oxygen, and ion migration and to enhance the stability as well as to reduce non-radiative recombination at the surface and grain boundaries of film (Li et al. 2023). Recently, 1D/3D perovskite heterostructure is constructed by Lin and colleagues; they introduced a anion regulation strategy and PSCs with a 1D/3D perovskite heterojunction resulting from tetra-*n*-butylammonium acetate (TBAAc) treatment yielding high PCE and demonstrating excellent stability. This 1D/3D–perovskite interface leads to outstanding defect passivation performance and retains 90% of its initial PCE after 2400 h of storage in the air atmosphere (Fig. 9.5) (Mei et al. 2024).

Multi-junction Arrangements

To overcome the limitations of single-junction devices, particularly their efficiency limits, researchers designed the possibility of multi-junction arrangements in solar cells. The prominent reason for using multi-junction solar cells is to surpass the Shockley–Queisser efficiency limit, which is around 33% for a single-junction solar cell (Yan et al. 2021). This limit arises because single-junction cell is incapable of converting all the portions of the solar spectrum into electrical energy. The bandgap of the material plays the crucial role in absorbing photons were photons

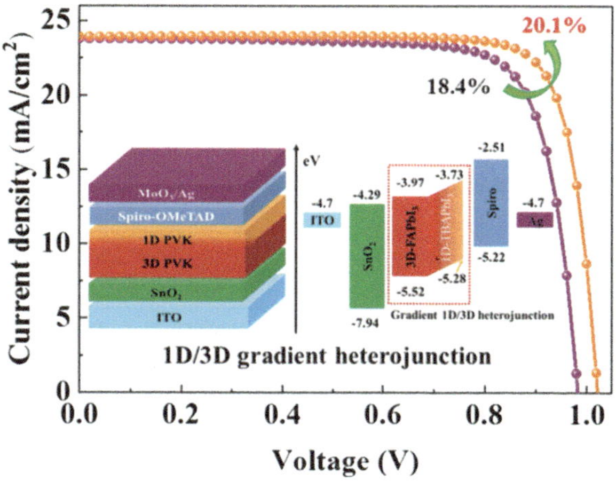

Fig. 9.5 Schematic diagrams of the device structure, energy-level alignment of PSCs with gradient 1D/3D–perovskite heterojunction and *J–V* curves of PSCs with TBAAc (Mei et al. 2024). Copyright 2024. Reproduced with permission from American Chemical Society

with energies below the bandgap pass through the material without being absorbed, and higher-energy photons lose their excess energy as heat. The main motivation behind the multi-junction solar cells is to stack layers with different bandgaps to absorb a wider range of the solar spectrum more efficiently. Here, the high-energy photons are absorbed by the top cell with a wide bandgap and lower-energy photons pass through bottom cell are get absorbed, with low-bandgap materials. By dividing the solar spectrum among different sub-cells of multifunction solar cells, each can operate near its maximum efficiency, resulting in higher overall energy conversion. When we discuss about the power conversion efficiency (PCE), which is the ratio of electrical power generated by the solar cell to the power of the incoming sunlight, multi-junction perovskite solar cells exhibit surprising results compared to single-junction perovskite solar cells. Even though some of the single-junction perovskite cell report efficiencies, around 26% tandem or triple-junction solar cells, which combine different materials such as perovskites with silicon or other wide-bandgap semiconductors, can achieve efficiencies above 30% and, in some experimental settings, exceed 40% (Liu et al. 2024). Another important aspect of the multifunction solar cell is its flexibility in material selection, and each layer can be made from different materials optimized for their specific function enabling researchers to combine the best properties and minimize their weaknesses, leading to overall better device performance. Also it is challenging to balance efficient charge extraction with effective light absorption when dealing with thicker absorber layer in the case of single-junction devices. It can be solved to some extent by multifunction device in which each sub cell can be thinner, allowing for faster and more efficient charge extraction.

Traditional materials like silicon have reached their efficiency limits and perovskite tandem solar cells with silicon can bypass this limit. This hybrid approaches is the current topic of research and perovskite tandem solar cell with different materials and 4T and 2T architecture could drive the future of photovoltaics. Multi-junction arrangements are promising for applications requiring high efficiency and power density, such as space exploration, solar concentrator systems, and next-generation solar power installation.

Conclusion

The notable advancement in the field of low-bandgap perovskite solar cells has marked a history in the development of next-generation photovoltaic technologies. Low-bandgap perovskites, with formamidinium (FA), cesium (Cs), lead–tin (Pb–Sn) mixed compositions, and mixed-anion and cation configurations present a diverse range of tunable properties, especially with unique opportunities for absorbing a broader spectrum of sunlight make them suitable for efficient solar energy harvesting. Compared to higher bandgap materials, these materials offer high power conversion efficiency; however, challenge lies in optimizing their stability and overcoming inherent performance limitations such as defect density and recombination losses. The suitable designing of devices based on low-bandgap perovskites is crucial to maximizing their performance. Sequential deposition techniques play a crucial role as it create uniform and defect-free perovskite films by a controlled and optimized method. This method, coupled with the careful introduction of additives such as small molecules, polymers, and passivating agents, significantly improves film morphology, enhances crystallization and stability, and reduces trap states that limit efficiency. Introduction of tandem configurations, the multi-junction solar cell provide another dimension to solar cell industry. These devices owing to their capability to harness photons of both low and high energy have demonstrated a breaking efficiency beyond those of single-junction cells. The tandem perovskite solar cells are a suitable candidate for commercial applications as it provides stability as well as efficiency. Along with device design, performance optimization is also important for developing commercially viable low-bandgap perovskite solar cells. The primary concern is the defect density which act as non-radiative recombination centres, severely limiting efficiency of solar cell. The reduction in defect density can be achieved through improved synthesis methods and post-deposition treatments. Surface passivation techniques, such as the use of molecular or ionic agents could possibly passivate the grain boundaries and surface defects and enhance PCE. Improving charge extraction is another vital aspect of performance enhancement. The selection of appropriate charge transport layer which in agreement with the bandgap of perovskite layer itself facilitates efficient carrier separation and extraction. To minimize energy losses and ensuring that photogenerated electrons and holes are efficiently transported to the electrodes, optimizing the energy alignment at interfaces is essential which can be achieved by strategies like doping of transport layers, using 2D materials, and employing novel interlayer. Moreover, reducing recombination losses remains a key challenge for perovskite solar cells. Recombination, both at interfaces and within the bulk of the material influence, the overall efficiency and energy-level tuning and the development of better-encapsulated materials are effective strategy to reducing non-radiative recombination pathways. The advancements in low-bandgap perovskites hold enormous promise, not just for increasing the efficiency of photovoltaic systems but also for enabling cost-effective, scalable solutions for energy crisis. The challenges needed to address properly before widespread commercial deployment, especially with degrading nature of perovskite under ambient conditions. Growing emphasis on finding environmentally friendly perovskite material is still under search and researches focusing on the innovations in material science, device engineering, and performance optimization in the solar cells are well-positioned to become a key player in the renewable energy landscape. As the research continues, it is definite that low-bandgap perovskites will be integrated into next-generation solar technologies, providing a path to more efficient, affordable, and sustainable energy solutions.

References

Alam I, Ashraf MA (2020) Effect of different device parameters on tin-based perovskite solar cell coupled with In2S3 electron transport layer and CuSCN and Spiro-OMeTAD alternative hole transport layers for high-efficiency performance. Energy Sources Part A: Recovery Utiliz Environ Effects 46:1–17. https://doi.org/10.1080/15567036.2020.1820628

Al-Ashouri A, Köhnen E, Li B, Magomedov A, Hempel H, Caprioglio P, Márquez JA, Morales Vilches AB, Kasparavicius E, Smith JA, Phung N, Menzel D, Grischek M, Kegelmann L, Skroblin D, Gollwitzer C, Malinauskas T, Jošt M, Matič G, Albrecht S et al (2020) Monolithic perovskite/silicon tandem solar cell with >29% efficiency by enhanced hole extraction. Science 370(6522):1300–1309. https://doi.org/10.1126/science.abd4016

Bailie CD, Christoforo MG, Mailoa JP, Bowring AR, Unger EL, Nguyen WH, Burschka J, Pellet N, Lee JZ, Grätzel M, Noufi R, Buonassisi T, Salleo A, McGehee MD (2015) Semi-transparent perovskite solar cells for tandems with silicon and CIGS. Energy Environ Sci 8(3):956–963. https://doi.org/10.1039/C4EE03322A

Beal RE, Slotcavage DJ, Leijtens T, Bowring AR, Belisle RA, Nguyen WH, Burkhard GF, Hoke ET, McGehee MD (2016) Cesium lead halide perovskites with improved stability for tandem solar cells. J Phys Chem Lett 7(5):746–751. https://doi.org/10.1021/acs.jpclett.6b00002

Castro-Méndez A-F, Hidalgo J, Correa-Baena J-P (2019) The role of grain boundaries in perovskite solar cells. Adv Energy Mater 9(38):1901489. https://doi.org/10.1002/aenm.201901489

Chen G, Li P, Xue T, Su M, Ma J, Zhang Y, Wu T, Han L, Aldamasy M, Li M, Li Z, Ma J, Chen S, Zhao Y, Wang F, Song Y (2021) Design of low bandgap CsPb1−SnI2Br perovskite solar cells with excellent phase stability. Small 17(30):2101380. https://doi.org/10.1002/smll.202101380

Chen P, Xiao Y, Li L, Zhao L, Yu M, Li S, Hu J, Liu B, Yang Y, Luo D, Hou C-H, Guo X, Shyue J-J, Lu Z-H, Gong Q, Snaith HJ, Zhu R (2023) Efficient inverted perovskite solar cells via improved sequential deposition. Adv Mater 35(5):2206345. https://doi.org/10.1002/adma.202206345

Chen Z, Zhang H, Yao F, Tao C, Fang G, Li G (2020) Room temperature formation of semiconductor grade α-FAPbI3 films for efficient perovskite solar cells. Cell Rep Phys Sci 1(9). https://doi.org/10.1016/j.xcrp.2020.100205

Dey K, Roose B, Stranks SD (2021) Optoelectronic properties of low-bandgap halide perovskites for solar cell applications. Adv Mater 33(40):2102300. https://doi.org/10.1002/adma.202102300

Ding J, Han Q, Ge Q-Q, Xue D-J, Ma J-Y, Zhao B-Y, Chen Y-X, Liu J, Mitzi DB, Hu J-S (2019) Fully air-bladed high-efficiency perovskite photovoltaics. Joule 3(2):402–416. https://doi.org/10.1016/j.joule.2018.10.025

Duan L, Uddin A (2022) Defects and stability of perovskite solar cells: a critical analysis. Mater Chem Front 6(4):400–417. https://doi.org/10.1039/D1QM01250A

Dunfield SP, Bliss L, Zhang F, Luther JM, Zhu K, van Hest MFAM, Reese MO, Berry JJ (2020) From defects to degradation: a mechanistic understanding of degradation in perovskite solar cell devices and modules. Adv Energy Mater 10(26):1904054. https://doi.org/10.1002/aenm.201904054

Eperon GE, Leijtens T, Bush KA, Prasanna R, Green T, Wang JT-W, McMeekin DP, Volonakis G, Milot RL, May R, Palmstrom A, Slotcavage DJ, Belisle RA, Patel JB, Parrott ES, Sutton RJ, Ma W, Moghadam F, Conings B, Snaith HJ et al (2016) Perovskite-perovskite tandem photovoltaics with optimized band gaps. Science 354(6314):861–865. https://doi.org/10.1126/science.aaf9717

Forgács D, Gil-Escrig L, Pérez-Del-Rey D, Momblona C, Werner J, Niesen B, Ballif C, Sessolo M, Bolink HJ (2017) Efficient monolithic perovskite/perovskite tandem solar cells. Adv Energy Mater 7(8). https://roderic.uv.es/bitstream/handle/10550/78741/115676.pdf?sequence=1

Gharibzadeh S, Hossain IM, Fassl P, Nejand BA, Abzieher T, Schultes M, Ahlswede E, Jackson P, Powalla M, Schäfer S, Rienäcker M, Wietler T, Peibst R, Lemmer U, Richards BS, Paetzold UW (2020) 2D/3D heterostructure for semitransparent perovskite solar cells with engineered bandgap enables efficiencies exceeding 25% in four-terminal tandems with silicon and CIGS. Adv Func Mater 30(19):1909919. https://doi.org/10.1002/adfm.201909919

Hao F, Stoumpos CC, Cao DH, Chang RPH, Kanatzidis MG (2014) Lead-free solid-state organic–inorganic halide perovskite solar cells. Nat Photonics 8(6):489–494. https://doi.org/10.1038/nphoton.2014.82

Heo JH, Im SH (2016) $CH_3NH_3PbBr_3$-$CH_3NH_3PbI_3$ perovskite-perovskite tandem solar cells with exceeding 2.2 V open circuit voltage. Adv Mater (Deerfield Beach, Fla) 28(25):5121–5125. https://doi.org/10.1002/adma.201501629

Hu M, Wang G, Zhang Q, Gong J, Xing Z, Gao J, Wang J, Zeng P, Zheng S, Liu M, Zhou Y, Yang S (2022) Antioxidative solution processing yields exceptional Sn(II) stability for sub-1.4 eV bandgap inorganic perovskite solar cells. J Energy Chem 72:487–494. https://doi.org/10.1016/j.jechem.2022.05.030

Ibrahim, Shoukat A, Aslam F, Israr Ur Rehman M (2024) Emerging trends in low band gap perovskite solar cells: materials, device architectures, and performance optimization. Mol Phys 122(17):e2316273. https://doi.org/10.1080/00268976.2024.2316273

Ito N, Kamarudin MA, Hirotani D, Zhang Y, Shen Q, Ogomi Y, Iikubo S, Minemoto T, Yoshino K, Hayase S (2018) Mixed Sn–Ge perovskite for enhanced perovskite solar cell performance in air. J Phys Chem Lett 9(7):1682–1688. https://doi.org/10.1021/acs.jpclett.8b00275

Jiang F, Liu T, Luo B, Tong J, Qin F, Xiong S, Li Z, Zhou Y (2016) A two-terminal perovskite/perovskite tandem solar cell. J Mater Chem A 4(4):1208–1213. https://doi.org/10.1039/C5TA08744A

Jiang T, Xu X, Lan Z, Chen Z, Chen X, Liu T, Huang S, Yang Y (Michael) (2022) Efficient MA-free Pb-Sn alloyed low-bandgap perovskite solar cells via surface passivation. Nano Energy 101:107596. https://doi.org/10.1016/j.nanoen.2022.107596

Ke W, Stoumpos CC, Spanopoulos I, Mao L, Chen M, Wasielewski MR, Kanatzidis MG (2017) Efficient lead-free solar cells based on hollow {en}MASnI3 perovskites. J Am Chem Soc 139(41):14800–14806. https://doi.org/10.1021/jacs.7b09018

Kim M, Kim G-H, Lee TK, Choi IW, Choi HW, Jo Y, Yoon YJ, Kim JW, Lee J, Huh D, Lee H, Kwak SK, Kim JY, Kim DS (2019) Methylammonium chloride induces intermediate phase stabilization for efficient perovskite solar cells. Joule 3(9):2179–2192. https://doi.org/10.1016/j.joule.2019.06.014

Koh TM, Fu K, Fang Y, Chen S, Sum TC, Mathews N, Mhaisalkar SG, Boix PP, Baikie T (2014) Formamidinium-containing metal-halide: an alternative material for near-IR absorption perovskite solar cells. J Phys Chem C 118(30):16458–16462. https://doi.org/10.1021/jp411112k

Lee SJ, Shin SS, Kim YC, Kim D, Ahn TK, Noh JH, Seo J, Seok SI (2016a) Fabrication of efficient formamidinium tin iodide perovskite solar cells through SnF2–pyrazine complex. J Am Chem Soc 138(12):3974–3977. https://doi.org/10.1021/jacs.6b00142

Lee M, Park SJ, Hwang YJ, Jun Y, Min BK (2016b) Tandem architecture of perovskite and Cu(In,Ga)(S,Se)2 created by solution processes for solar cells. Adv Opt Mater 4(12):2102–2108. https://doi.org/10.1002/adom.201600373

Lee S, Moon J, Ryu J, Parida B, Yoon S, Lee D-G, Cho JS, Hayase S, Kang D-W (2020) Inorganic narrow bandgap CsPb0.4Sn0.6I2.4Br0.6 perovskite solar cells with exceptional efficiency. Nano Energy 77:105309. https://doi.org/10.1016/j.nanoen.2020.105309

Leijtens T, Prasanna R, Bush KA, EEperon G, Raiford JA, Gold-Parker A, Wolf EJ, Swifter SA, Boyd CC, Wang H-P, Toney MF, Bent SF, McGehee MD (2018) Tin–lead halide perovskites with improved thermal and air stability for efficient all-perovskite tandem solar cells. Sustain Energy Fuels 2(11):2450–2459. https://doi.org/10.1039/C8SE00314A

Li Z, Boix PP, Xing G, Fu K, Kulkarni SA, Batabyal SK, Xu W, Cao A, Sum TC, Mathews N, Wong LH (2016a) Carbon nanotubes as an efficient hole collector for high voltage methylammonium lead bromide perovskite solar cells. Nanoscale 8(12):6352–6360. https://doi.org/10.1039/C5NR06177F

Li H, Tao L, Huang F, Sun Q, Zhao X, Han J, Shen Y, Wang M (2017) Enhancing efficiency of perovskite solar cells via surface passivation with graphene oxide interlayer. ACS Appl Mater Interfaces 9(44):38967–38976. https://doi.org/10.1021/acsami.7b10773

Li Q, Zhao Y, Fu R, Zhou W, Zhao Y, Liu X, Yu D, Zhao Q (2018) Efficient perovskite solar cells fabricated through CsCl-enhanced PbI2 precursor via sequential deposition. Adv Mater 30(40):1803095. https://doi.org/10.1002/adma.201803095

Li C, Song Z, Zhao D, Xiao C, Subedi B, Shrestha N, Junda MM, Wang C, Jiang C-S, Al-Jassim M, Ellingson RJ, Podraza NJ, Zhu K, Yan Y (2019) Reducing saturation-current density to realize high-efficiency low-bandgap mixed tin-lead halide perovskite solar cells. Adv Energy Mater 9(3):1803135. https://doi.org/10.1002/aenm.201803135

Li X, Aftab S, Abbas A, Hussain S, Aslam M, Kabir F, Abd-Rabboh HSM, Hegazy HH, Xu F, Ansari MZ (2023) Advances in mixed 2D and 3D perovskite heterostructure solar cells: a comprehensive review. Nano Energy 118:108979. https://doi.org/10.1016/j.nanoen.2023.108979

Li L, Chen Y, Liu Z, Chen Q, Wang X, Zhou H (2016) The additive coordination effect on hybrids perovskite crystallization and high-performance solar cell. Adv Mater (Deerfield Beach, Fla) 28(44):9862–9868. https://doi.org/10.1002/adma.201603021

Liang P-W, Liao C-Y, Chueh C-C, Zuo F, Williams ST, Xin X-K, Lin J, Jen AK-Y (2014) Additive enhanced crystallization of solution-processed perovskite for highly efficient planar-heterojunction solar cells. Adv Mater 26(22):3748–3754. https://doi.org/10.1002/adma.201400231

Liao W, Zhao D, Yu Y, Shrestha N, Ghimire K, Grice CR, Wang C, Xiao Y, Cimaroli AJ, Ellingson RJ, Podraza NJ, Zhu K, Xiong R-G, Yan Y (2016) Fabrication of efficient low-bandgap perovskite solar cells by combining formamidinium tin iodide with methylammonium lead iodide. J Am Chem Soc 138(38):12360–12363. https://doi.org/10.1021/jacs.6b08337

Liu M, Chen Z, Chen Z, Yip H-L, Cao Y (2019) A cascade-type electron extraction design for efficient low-bandgap perovskite solar cells based on a conventional structure with suppressed open-circuit voltage loss. Mater Chem Front 3(3):496–504. https://doi.org/10.1039/C8QM00620B

Liu S, Guan Y, Sheng Y, Hu Y, Rong Y, Mei A, Han H (2020) A review on additives for halide perovskite solar cells. Adv Energy Mater 10(13):1902492. https://doi.org/10.1002/aenm.201902492

Liu X, Chen L, Yu Y, He D, Shai X, Zhang S, Zhang Z, Feng J, Yi J, Chen J (2024) Advancements of highly efficient perovskite based tandem solar cells. Sci China Mater. https://doi.org/10.1007/s40843-024-3076-3

Luo D, Su R, Zhang W, Gong Q, Zhu R (2020) Minimizing non-radiative recombination losses in perovskite solar cells. Nat Rev Mater 5(1):44–60. https://doi.org/10.1038/s41578-019-0151-y

Mei A, Peng X, Li X, Zhang G, Lin S, Fu N (2024) Anions regulation of 1D perovskite intrusion-behavior for efficient and stable perovskite solar cells. ACS Appl Mater Interfaces 16(24):31209–31217. https://doi.org/10.1021/acsami.4c06087

Nenon DP, Pressler K, Kang J, Koscher BA, Olshansky JH, Osowiecki WT, Koc MA, Wang L-W, Alivisatos AP (2018) Design principles for trap-free CsPbX3 nanocrystals: enumerating and eliminating surface halide vacancies with softer Lewis bases. J Am Chem Soc 140(50):17760–17772. https://doi.org/10.1021/jacs.8b11035

Noel NK, Stranks SD, Abate A, Wehrenfennig C, Guarnera S, Haghighirad A-A, Sadhanala A, Eperon GE, Pathak SK, Johnston MB, Petrozza A, Herz LM, Snaith HJ (2014) Lead-free organic–inorganic tin halide perovskites for photovoltaic applications. Energy Environ Sci 7(9):3061–3068. https://doi.org/10.1039/C4EE01076K

Ogomi Y, Morita A, Tsukamoto S, Saitho T, Fujikawa N, Shen Q, Toyoda T, Yoshino K, Pandey SS, Ma T, Hayase S (2014) $CH_3NH_3Sn_xPb_{(1-x)}I_3$ perovskite solar cells covering up to 1060 nm. J Phys Chem Lett 5(6):1004–1011. https://doi.org/10.1021/jz5002117

Ono LK, Juarez-Perez EJ, Qi Y (2017) Progress on perovskite materials and solar cells with mixed cations and halide anions. ACS Appl Mater Interfaces 9(36):30197–30246. https://doi.org/10.1021/acsami.7b06001

Ono LK, Liu S (Frank), Qi Y (2020) Reducing detrimental defects for high-performance metal halide perovskite solar cells. Angew Chem Int Ed 59(17):6676–6698. https://doi.org/10.1002/anie.201905521

Pereyra C, Xie H, Lira-Cantu M (2021) Additive engineering for stable halide perovskite solar cells. J Energy Chem 60:599–634. https://doi.org/10.1016/j.jechem.2021.01.037

Prasanna R, Leijtens T, Dunfield SP, Raiford JA, Wolf EJ, Swifter SA, Werner J, Eperon GE, de Paula C, Palmstrom AF, Boyd CC, van Hest MFAM, Bent SF, Teeter G, Berry JJ, McGehee MD (2019) Design of low bandgap tin–lead halide perovskite solar cells to achieve thermal, atmospheric and operational stability. Nat Energy 4(11):939–947. https://doi.org/10.1038/s41560-019-0471-6

Prochowicz D, Runjhun R, Tavakoli MM, Yadav P, Saski M, Alanazi AQ, Kubicki DJ, Kaszkur Z, Zakeeruddin SM, Lewiński J, Grätzel M (2019) Engineering of perovskite materials based on formamidinium and cesium hybridization for high-efficiency solar cells. Chem Mater 31(5):1620–1627. https://doi.org/10.1021/acs.chemmater.8b04871

Qin P, Kast H, Nazeeruddin MK, Zakeeruddin SM, Mishra A, Bäuerle P, Grätzel M (2014) Low band gap S,N-heteroacene-based oligothiophenes as hole-transporting and light absorbing materials for efficient perovskite-based solar cells. Energy Environ Sci 7(9):2981–2985. https://doi.org/10.1039/C4EE01220H

Raj A, Kumar M, Anshul A (2021) Recent advancement in inorganic-organic electron transport layers in perovskite solar cell: current status and future outlook. Mater Today Chem 22:100595. https://doi.org/10.1016/j.mtchem.2021.100595

Raj A, Kumar M, Anshul A (2022) Recent progress in cesium-based lead-free halide double perovskite materials for photovoltaic applications. Physica Status Solidi (a) 219(22):2200425. https://doi.org/10.1002/pssa.202200425

Raj A, Kumar M, Anshul A (2023) Topical advances in fabrication technologies of perovskite solar cell heterostructures: performance and future perspective. Mater Lett 340:134171. https://doi.org/10.1016/j.matlet.2023.134171

Rajagopal A, Yang Z, Jo SB, Braly IL, Liang P-W, Hillhouse HW, Jen AK-Y (2017) Highly efficient perovskite-perovskite tandem solar cells reaching 80% of the theoretical limit in photovoltage. Adv Mater 29(34):1702140. https://doi.org/10.1002/adma.201702140

Shen H, Duong T, Peng J, Jacobs D, Wu N, Gong J, Wu Y, Karuturi SK, Fu X, Weber K, Xiao X, White TP, Catchpole K (2018) Mechanically-stacked perovskite/CIGS tandem solar cells with efficiency of 23.9% and reduced oxygen sensitivity. Energy Environ Sci 11(2):394–406. https://doi.org/10.1039/C7EE02627G

Shukla R, Kumar RR, Pandey SK (2021) Theoretical study of charge carrier lifetime and recombination on the performance of eco-friendly perovskite solar cell. IEEE Trans Electron Dev 68(7):3446–3452. https://doi.org/10.1109/TED.2021.3078063

Song Z, Chen C, Li C, Awni RA, Zhao D, Yan Y (2019) Wide-bandgap, low-bandgap, and tandem perovskite solar cells. Semicond Sci Technol 34(9):093001. https://doi.org/10.1088/1361-6641/ab27f7

Sunny F, Kurukkal Balakrishnan S, Kalarikkal N (2024) An overview on lead halide perovskite based composites and heterostructures: synthesis and applications. ChemNanoMat 10(3):e202300484. https://doi.org/10.1002/cnma.202300484

Sutton RJ, Eperon GE, Miranda L, Parrott ES, Kamino BA, Patel JB, Hörantner MT, Johnston MB, Haghighirad AA, Moore DT, Snaith HJ (2016) Bandgap-tunable cesium lead halide perovskites with high thermal stability for efficient solar cells. Adv Energy Mater 6(8):1502458. https://doi.org/10.1002/aenm.201502458

Taheri S, Ahmadkhan Kordbacheh A, Minbashi M, Hajjiah A (2021) Effect of defects on high efficient perovskite solar cells. Opt Mater 111:110601. https://doi.org/10.1016/j.optmat.2020.110601

Tong J, Song Z, Kim DH, Chen X, Chen C, Palmstrom AF, Ndione PF, Reese MO, Dunfield SP, Reid OG, Liu J, Zhang F, Harvey SP, Li Z, Christensen ST, Teeter G, Zhao D, Al-Jassim MM, van Hest MFAM, Zhu K et al (2019) Carrier lifetimes of >1 μs in Sn-Pb perovskites enable efficient all-perovskite tandem solar cells. Science 364(6439):475–479. https://doi.org/10.1126/science.aav7911

Tonui P, Oseni SO, Sharma G, Yan Q, Tessema Mola G (2018) Perovskites photovoltaic solar cells: an overview of current status. Renew Sustain Energy Rev 91:1025–1044. https://doi.org/10.1016/j.rser.2018.04.069

Tu Y, Yang X, Su R, Luo D, Cao Y, Zhao L, Liu T, Yang W, Zhang Y, Xu Z, Liu Q, Wu J, Gong Q, Mo F, Zhu R (2018) Diboron-assisted interfacial defect control strategy for highly efficient planar perovskite solar cells. Adv Mater 30(49):1805085. https://doi.org/10.1002/adma.201805085

Wang H, He J, Xiang H, Ran R, Zhou W, Wang W, Shao Z (2023) Additive engineering for mixed lead-tin narrow-band-gap perovskite solar cells: recent advances and perspectives. Energy Fuels 37(9):6401–6423. https://doi.org/10.1021/acs.energyfuels.3c00462

Wei M, Xiao K, Walters G, Lin R, Zhao Y, Saidaminov MI, Todorović P, Johnston A, Huang Z, Chen H, Li A, Zhu J, Yang Z, Wang Y-K, Proppe AH, Kelley SO, Hou Y, Voznyy O, Tan H, Sargent EH (2020) Combining efficiency and stability in mixed tin lead perovskite solar cells by capping grains with an ultrathin 2D layer. Adv Mater 32(12):1907058. https://doi.org/10.1002/adma.201907058

Werner J, Nogay G, Sahli F, Yang TC-J, Bräuninger M, Christmann G, Walter A, Kamino BA, Fiala P, Löper P, Nicolay S, Jeangros Q, Niesen B, Ballif C (2018) Complex refractive indices of cesium–formamidinium-based mixed-halide perovskites with optical band gaps from 1.5 to 1.8 eV. ACS Energy Lett 3(3):742–747. https://doi.org/10.1021/acsenergylett.8b00089

Yan C, Huang J, Li D, Li G (2021) Recent progress of metal-halide perovskite-based tandem solar cells. Mater Chem Front 5(12):4538–4564. https://doi.org/10.1039/D0QM01085E

Yang WS, Noh JH, Jeon NJ, Kim YC, Ryu S, Seo J, Seok SI (2015) High-performance photovoltaic perovskite layers fabricated through intramolecular exchange. Science 348(6240):1234–1237. https://doi.org/10.1126/science.aaa9272

Yang Z, Rajagopal A, Jen AK-Y (2017) Ideal bandgap organic-inorganic hybrid perovskite solar cells. Adv Mater 29(47):1704418. https://doi.org/10.1002/adma.201704418

Yang G, Qin P, Fang G, Li G (2018) A Lewis base-assisted passivation strategy towards highly efficient and stable perovskite solar cells. Solar RRL 2(8):1800055. https://doi.org/10.1002/solr.201800055

Yang Z, Rajagopal A, Chueh C-C, Jo SB, Liu B, Zhao T, Jen AK-Y (2016) Stable low-bandgap Pb-Sn binary perovskites for tandem solar cells. Adv Mater 28(40):Article DOE-UW-Jen-33. https://doi.org/10.1002/adma.201602696

Yin W-J, Shi T, Yan Y (2014) Unique properties of halide perovskites as possible origins of the superior solar cell performance. Adv Mater (Deerfield Beach, Fla) 26(27):4653–4658. https://doi.org/10.1002/adma.201306281

Zhao D, Yu Y, Wang C, Liao W, Shrestha N, Grice CR, Cimaroli AJ, Guan L, Ellingson RJ, Zhu K, Zhao X, Xiong R-G, Yan Y (2017) Low-bandgap mixed tin–lead iodide perovskite absorbers with long carrier lifetimes for all-perovskite tandem solar cells. Nat Energy 2(4):1–7. https://doi.org/10.1038/nenergy.2017.18

Zhao D, Chen C, Wang C, Junda MM, Song Z, Grice CR, Yu Y, Li C, Subedi B, Podraza NJ, Zhao X, Fang G, Xiong R-G, Zhu K, Yan Y (2018) Efficient two-terminal all-perovskite tandem solar cells enabled by high-quality low-bandgap absorber layers. Nat Energy 3(12):1093–1100. https://doi.org/10.1038/s41560-018-0278-x

Zhumagali S, Isikgor FH, Maity P, Yin J, Ugur E, De Bastiani M, Subbiah AS, Mirabelli AJ, Azmi R, Harrison GT, Troughton J, Aydin E, Liu J, Allen T, Rehman Aur Baran D, Mohammed OF, De Wolf S (2021) Linked nickel oxide/perovskite interface passivation for high-performance textured monolithic tandem solar cells. Adv Energy Mater 11(40):2101662. https://doi.org/10.1002/aenm.202101662

Zong Y, Wang N, Zhang L, Ju M-G, Zeng XC, Sun XW, Zhou Y, Padture NP (2017) Homogenous alloys of formamidinium lead triiodide and cesium tin triiodide for efficient ideal-bandgap perovskite solar cells. Angew Chem Int Ed 56(41):12658–12662. https://doi.org/10.1002/anie.201705965

Chapter 10
Role of 2D Materials in Perovskite Solar Cells

Brijmohan Prajapati and Vijay K. Singh

Abstract Solar photovoltaics, particularly organic/inorganic halide-based perovskite solar cells (PSCs) with high efficiency and cost-effective solution processing, have shown immense potential in harnessing renewable energy to various cause and sustainable economy in recent decades. Though the power conversion efficiency (PCE) of existing PSCs geared up to 26.7% steadily with some fraction of improved stability to earlier ones, there are still some untraceable challenges regarding the upscaling and stability of these promising perovskite-based solar cells. The present chapter highlights the recent developments in PSCs to their peer fabrication, performance, and commercialization. In addition, the chapter also provides in-depth knowledge of various factors affecting solar cell efficiency with possible remedies such as by using two-dimensional (2D) materials as electrodes, additives, and charge transport layers to overcome them. Therefore, this chapter will be helpful to understand the importance of 2D materials, particularly electrode materials and additives, and the existing charge (electron–hole) transport to enhance the power conversion efficiency of perovskite with relatively improved stability of solar cells from laboratory to commercial applications.

Keywords Perovskite solar cell · Flexibility · Efficiency · HTL · ETL · 2D materials

B. Prajapati (✉)
Department of Physics, Deshbandhu College, University of Delhi, Delhi, India
e-mail: brijfizics@gmail.com

V. K. Singh (✉)
Department of Physics, Indian Institute of Technology Jodhpur, Jodhpur, Rajasthan, India
e-mail: vijayks@iitj.ac.in

Department of Physics, Indian Institute of Technology Ropar, Rupnagar, Punjab, India

M. Kumar and A. Raj (eds.), *Perovskite Solar Cells*, Springer Tracts in Electrical and Electronics Engineering, https://doi.org/10.1007/978-981-96-9463-1_10

Introduction

Solar, among various renewable energy sources, such as wind, geothermal, and hydroelectric, can supply exhaustible energy together with facilitating an alternative energy source for mitigating the limited fossil fuel crisis and greenhouse effect. Photovoltaic technology is an efficient way to harvest solar energy. At present, single and polycrystalline silicon-based solar modules are dominating the market; however, in recent decades, other technologies, such as quantum dot solar cells (QDSCc) (Kamat 2013; Singh et al. 2022), dye-sensitized solar cells (DSSCs) (Nazeeruddin et al. 2011; Sharma et al. 2018), organic solar cells (OSCs) (Duan and Uddin 2020; Tress 2014), and perovskite solar cells (PSCs) (Duan and Uddin 2020; Green et al. 2014; Tress 2014), are well explored and have shown immense potential for commercialization. Among these emerging technologies, organic–inorganic halide perovskite-based solar cells have drawn significant interest due to the excellent opto-electronic properties of perovskites (Chouhan et al. 2020; Dong et al. 2023). Its general chemical formula is ABX_3 (Fig. 10.1a), where A refers to organic cations like methylammonium (**A**: $CH_3(NH_2)^{2+}$) or inorganic cations, **B** refers to metal cations like Pb^{2+}, Sn^{2+}, etc., and **X** refers to halogen anions like I^-, Br^-, and Cl^-. The excellent opto-electronic properties of perovskites, such as strong light absorption, small exciton binding energy (B.E.), low effective carrier mass, high carrier mobilities, long carrier lifetimes and diffusion lengths, low trap densities, etc. are ideal for photovoltaics. The high PCE, reaching 26.7%, just above to the single-crystal solar cells (26.1%) (Lamanna et al. 2020), together with solution-processable low fabrication cost, have justified their acceptance for next-generation photovoltaics at commercial scale.

In solar cells, perovskite as an optically active material is generally sandwiched between the hole transport layer (HTL) and electron transport layer (ETL), which scavenges holes and electrons generated in perovskite layer, respectively, as illustrated in Fig. 10.1b–d. HTL and ETL help in minimizing the recombination of charge carriers, thereby enhancing the solar cell efficiency. The two different planar structures, such as regular n-i-p or inverted p-i-n architectures, are illustrated in Fig. 10.1c and d, respectively. The architecture classification is based on ETL/HTL position and independent of light incidence direction. In all the PSC architecture, the light will incident on FTO except for the case of transparent and semitransparent electrodes. Perovskite material can be deposited on a flexible substrate due to their inherent flexibility for a flexible photovoltaic. However, the realization of complete photovoltaic devices in flexible form requires flexible ETL, HTL electrode layers. This can be realized by utilizing the flexible nature of 2D materials owing to their excellent optical, electronic, mechanical, and low-cost easy solution processability.

Thus, the present chapter provides a comprehensive summary of the recent advances in 2D materials-based perovskite solar cells. The role of 2D materials (graphene and its derivatives, transition metal dichalcogenides (TMDs), black phosphorus, MXenes, and metal–organic framework (MOF)) as HTL, ETL, and as an

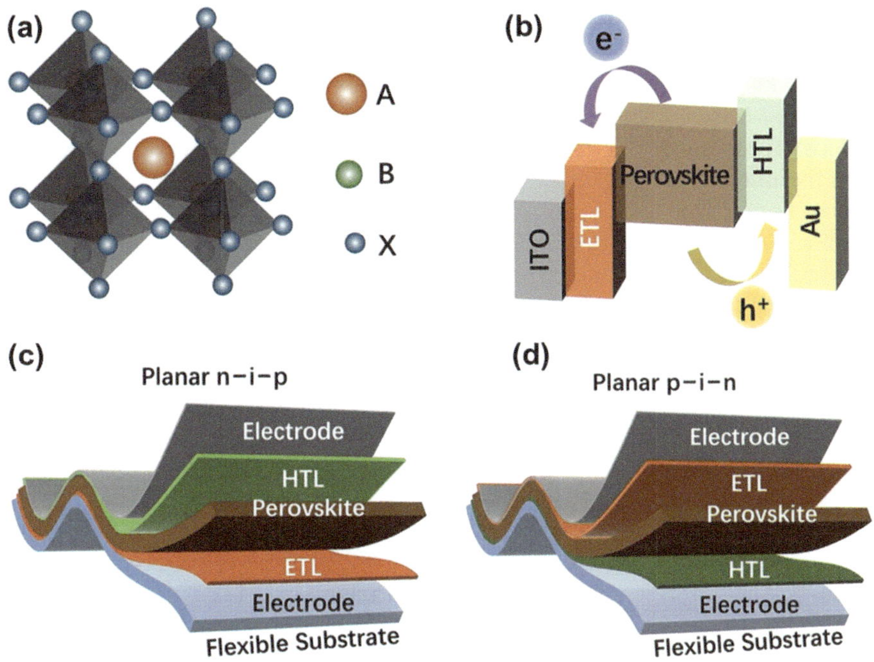

Fig. 10.1 Schematic illustration showing **a** perovskite crystal structure, **b** energy band diagram of photovoltaic device structure, **c** and **d** planar n-i-p and p-i-n device structure, respectively (Li et al. 2023). Copyright 2023. Reproduced with permission from Springer

additive in a photo-absorbing perovskite layer with underlying mechanisms for a better understanding are discussed in this chapter.

Graphene and Its Derivative for Perovskite Solar Cells

In 2004, Geim and Novosolove discovered a true 2D material, graphene, which is a single layer of sp2 hybridized carbon atoms closely packed in a honeycomb lattice (Novoselov et al. 2004). For the discovery of this important 2D material, Geim and Novosolove earned the Nobel Prize for Physics in 2010 (Gerstner 2010). The importance of Graphene can be understood by the exceptional electronic, optical, thermal, and mechanical properties of it, which are enjoyed for various applications. Graphene offers exceptionally high carrier mobility ~ 2×10^5 cm^2 V^{-1} S^{-1}, high thermal conductivity ~ 5×10^3 W m^{-1} K^{-1}, high optical transparency ~ 97.7%, high Young modulus ~ 1 TPa, and very high specific surface area ~ 2600 m^2 g^{-1} (Gong et al. 2021; Geim and Novoselov 2007). Graphene is also known as the mother allotrope, as other graphene derivatives, in principle, can be obtained from graphene, such as its 3D stacking, which provides 3D graphite, rolling provides 1D carbon nanotubes (CNTs), and wrapping in forms of 0D provides fullerene, illustrated in Fig. 10.2a. Graphene oxide (GO) and controllably reduced graphene oxide (RGO) are two well-known graphene derivatives. Where GO is having a mixture of sp2- and sp3-hybridized carbon atoms having covalently bonded oxygen-containing functional groups. These functional groups provide hydrophilicity to GO/RGO and facilitate sites for immobilization of other biomolecules through covalent bonding. In the case of pristine graphene, functional groups are absent, which requires additional chemical modification for creating functional groups over the graphene surface through which immobilization of other molecules can be realized (Singh et al. 2018). Graphene quantum dots (GQDs), which are a quantum-confined structure of double- or few-layer graphene sheets, offer the merits of graphene and quantum dots, including strong photoluminescence, high conductivity, and excellent solubility. Further, the dangling bonds available with GQDs offer strong linkage of GQDs with perovskite materials for solar cell fabrication (Gong et al. 2021). Figure 10.2b–d represents the chemical structures of GO, RGO, and modified GQDs, respectively.

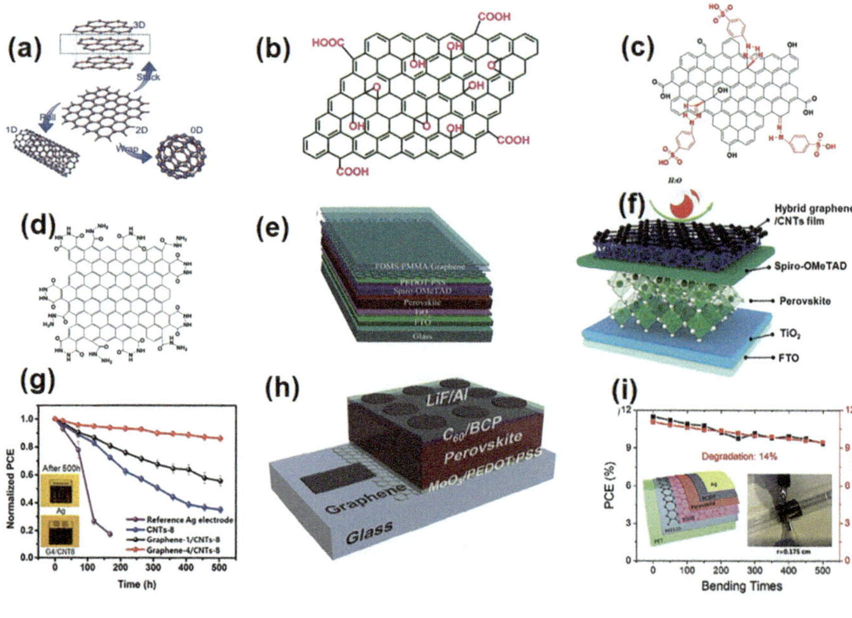

Fig. 10.2 **a** Schematic illustrations of honeycomb lattice of graphene and its derivatives: (i) monolayer graphene stacked to 3D graphite, (ii) graphene rolled to 1D-CNT, (iii) graphene wrapped to 0D fullerenes. **b–d** are chemical structures of GO, RGO, and modified GQDs, respectively (Gong et al. 2021). **e** Device structure of semitransparent PSC with graphene as transparent conducting top electrode (You et al. 2015). **f** PSC architecture with the hybrid graphene/CNT film counter electrode. **g** PCE long-term stability of PSCs under controlled humidity of 50 ± 2%, at 25 °C (Tian et al. 2020), **h** schematic structure of the inverted $MAPbI_3$ perovskite solar cells utilizing graphene as a transparent anode (Sung et al. 2016), and **i** PCE (black) and J_{SC} (red) of a flexible PSC in bending tests with different number of bending cycles, inset showing schematic diagram of flexible PSC and photograph of flexible PSC attached to surface of bending radii (Liu et al. 2016)

As mentioned, due to the excellent properties of graphene and its derivatives, these materials have been enormously used as different components in perovskite solar cells (Gong et al. 2021; Znidi et al. 2024), which are briefly described one by one in the following subsections.

Graphene as Top Electrode

As a transparent conducting electrode in solar cells, ITO- and FTO-coated glass substrates are mostly used. However, ITO deposition over glass is an expensive process that can't fulfill the market demand for a low-cost solar cell. On the other hand, graphene has received universal acceptance as a transparent conducting electrode in solar cells due to its high optical transparency, low sheet resistance, and chemical robustness with low fabrication cost. Graphene deposited over polymer layers also offers flexibility to the system.

Graphene as a top conducting electrode is extensively used for PSCs. For example, in 2015, You et al. (2015), demonstrated a PSCs with multilayer graphene (MLG) as the top electrode, Fig. 10.2e. The developed PSCs offered PCE of 12.02% and 11.65% when illuminated from the bottom FTO side and the top graphene side, respectively. In another work, Tavakoli et al. (2019) reported semi-transparent $CsPbI_3$ QDs photovoltaics using graphene as the top electrode. The optimization process led to the PCE of 4.95% and 5.11% when illuminated from the ITO and graphene side. In 2020, Lee and co-workers (Tian et al. 2020) fabricated printable free-standing PSCs using hybrid graphene/dry-spun carbon nanotube (CNT) film as the top electrode, Fig. 10.2f. This novel device delivered an excellent PCE of 15.36% together with excellent stability, retaining 86% of the initial PCE after storing for 500 h in a high-moisture environment, Fig. 10.2g. These findings suggest that graphene could be the ideal choice as the top transparent conducting electrode for PSCs.

Graphene as the Bottom Electrode

The conventional ITO as the bottom electrode in p-i-n architecture PSCs is not suitable for flexible solar cells. However, graphene, due to its excellent mechanical properties together with high optical transparency and electronic conductivity, can be useful for flexible devices. For example, Sung et al. (2016) first demonstrated transparent conductive oxide-free PSCs having graphene as the bottom electrode, Fig. 10.2h. A thin MoO_3 layer was introduced between HTL and graphene to enable better interface formation and energy alignment between HTL and anode. The optimized device delivered excellent PCE of 17.1%. In another work, Yan and co-workers (Liu et al. 2016) demonstrated ultra-flexible PSCs with graphene as the bottom electrode, Fig. 10.2i. The device delivered PCE of 11.5%, which slightly decreased ~ 14% even after bending for 500 times, Fig. 10.2f. There are several reports suggesting universal acceptance of graphene as a bottom electrode (Lee et al. 2008; Quan et al. 2017).

Graphene and Its Derivatives as ETL

The role of ETL is the efficient extraction of the photogenerated electron in the perovskite layer, thereby limiting the electron–hole recombination. ETL material should offer a low resistance for efficient electron transfer. Graphene has been extensively explored for ETL in PSCs. For example, Wang et al. (2014) demonstrated low-temperature processed PSCs utilizing graphene and TiO_2 nanoparticle composites as ETL. A cross-sectional scanning electron micrograph with a color-enhanced cross section of the full device architecture is shown in Fig. 10.3a. An energy level diagram of different layers used in the device with respect to vacuum level is presented in Fig. 10.3b. In this work, a remarkable PCE of 15.6% was achieved. In 2019, Hahn and co-workers (Mahmoudi et al. 2020) fabricated PSCs using mesoporous Al_2O_3/graphene (mp-AG) as ETL. The schematic of the fabricated device structure with charge separation mechanism and conduction through the mp-AG composite layer are presented in Fig. 10.3c and d. This composite has much higher electron mobility, which facilitates easy electron transport from perovskite to the ETL and hence reduces charge recombination. The optimized device delivered an exceptionally high PCE of 20.58%. Thus, graphene in its pristine and composite form can be utilized as ETL, in PSCs for efficient electron extraction, thus reducing the charge recombination for an efficient solar device.

Graphene and Its Derivatives as HTL

HTL in PSCs is used for the extraction of photo-generated holes from perovskite layers. The efficient extraction of holes minimizes the carrier recombination and hence improves the PCE of PSCs. Graphene, due to its promising electronic and optical properties, has been extensively used in PSCs as HTL. For example, Cao et al. (2015) reported stable and efficient PSCs using functionalized nanographene (perthiolated trisulfur-annulated hexa-peri-hexabenzocoronene, TSHBC) as HTL. The efficiency was further improved to 14.0% from 12.8% with the addition of graphene sheets into TSHBC. In another work, Matsuda and co-workers (Wang et al. 2016) developed highly stable PSCs using single-walled carbon nanotubes (SWNT)/graphene oxide (GO) buffer as a HTL, Fig. 10.3e. The band energy diagram, shown in Fig. 10.3f, clearly depicts that due to the complementary properties of CNTs and GO, the optimized device showed a PCE of 13.3%. A number of similar other reports acceptance of 2D material graphene and its derivatives as HTL in PSCs (Wang et al. 2016; Bati et al. 2020).

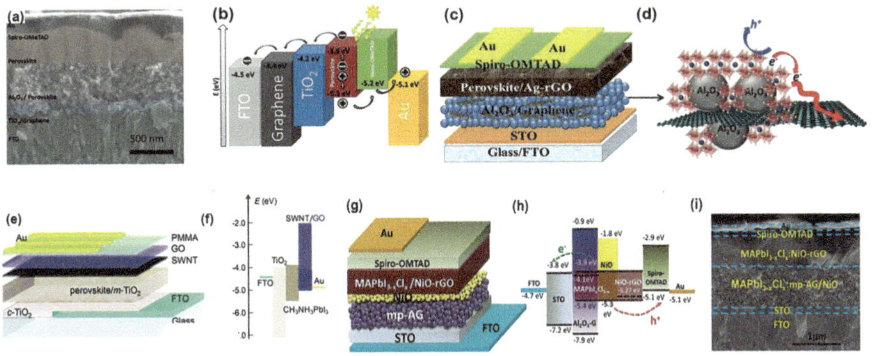

Fig. 10.3 **a** Cross-sectional SEM micrograph with color-enhanced cross-section showing a general schematic of the PSCs structure (Wang et al. 2014). **b** Schematic illustration of energy levels of the materials used for PSC (Wang et al. 2014). **c** and **d** Schematic illustration of FTO/STO/mp-AG/$MAPbI_{3-x}Cl_x$:Ag-rGO/Spiro-OMeTAD/Au and an illustrative mechanism of charge separation and conduction through the mp-AG composite layer, respectively (Mahmoudi et al. 2020). **e** and **f** Schematic configuration of an organo-lead iodide perovskite/SWNT/GO/PMMA PSC and energy band alignment in a perovskite/SWNT/GO solar cell, respectively (Wang et al. 2016). **g–i** Illustration of the best device with FTO/STO/mp-AG/NiO/($MAPbI_{3-x}Cl_x$/NiO-rGO)/Spiro-OMeTAD/Au configuration, energy diagram of the best device NiO-rGO structure and cross-sectional field-emission scanning electron micrograph of the best device, respectively (Mahmoudi et al. 2021)

Graphene in Perovskite Layer

To overcome the device performance degradation in PSCs and instability of the device response, various novel strategies have been explored. Among these, functional nanocomposites of perovskite with graphene as a photoactive layer have been intensively explored. For example, in a recent work, Hahn and co-workers (Mahmoudi et al. 2021) demonstrated highly stable PSCs based on perovskite/NiO-graphene functional nanocomposite as a photo-absorber layer with superb PCE of 20.8%. Figure 10.3g–i shows an illustration of the device configuration, energy diagram, and cross-sectional field emission scanning electron micrograph of the champion device. The optimized device showed long-term thermal, air, and photo-stability retaining its performance ~ 97–100% of initial values even though the device was kept in ambient condition for 310 days. Thus, graphene and its derivatives can also be uniquely employed in the perovskite photo-absorber layer for realizing the highly efficient photovoltaic device.

2D Materials Beyond Graphene for Perovskite Solar Cells

The successful discovery of 2D material graphene gave rise to the birth of plenty of other 2D materials with unique electrical, optical, and mechanical properties (Jariwala et al. 2016; Lei et al. 2022; Rai et al. 2023; Singh et al. 2020). A few 2D materials and their importance for PSCs are briefly discussed in the following sections.

Transition Metal Dichalcogenides

In recent years, transition metal dichalcogenides (TMDCs) have emerged as fascinating 2D materials for PSCs owing to their highly energetic carrier dynamic, optical transparency, layer-dependent tunable bandgap, and chemical inertness, together with an easy solution processability. The general formula of the TMDC family is MX_2, where M stands for transition metal and X for chalcogens (Fig. 10.4a). This family has various members depending on the different combinations of transition metal and chalcogen atoms, which include MoS_2, $MoSe_2$, WS_2, WSe_2, WTe_2, etc. TMDCs generally exhibit two polymorphs: natural semiconducting 2H and metallic 1T phases. The 1T phase offers much higher conductivity, and thus it is centrally important for achieving high performance. Due to the ambipolar nature of TMDCs, it can be used as both ETL and HTL. The use of TMDCs as ETL, HTL, and nanocomposite in the perovskite layer is briefly discussed in this section.

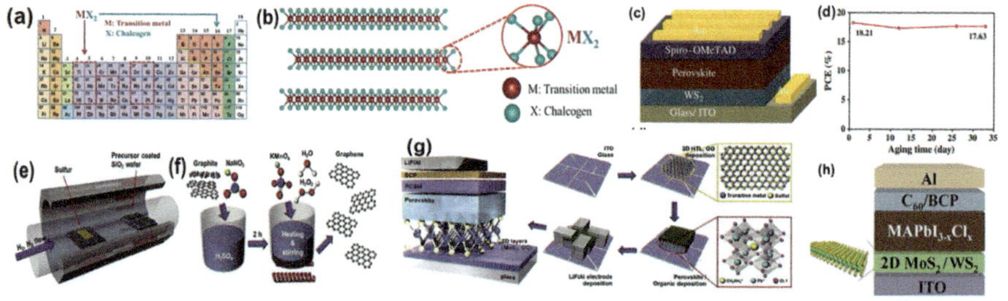

Fig. 10.4 **a, b** TMDCs with the MX_2 structure consisting of M from the 16 transition metals indicated by the red dotted box and X from the three halogen elements indicated by the green dotted box, and layered structure of MX_2, respectively (Hoang Huy et al. 2021). **c** and **d** PSC device architecture with compact WS_2 nanosheets ETL and long-term stability property of the champion device under aging in the N_2 atmosphere (Malek et al. 2020). **e–g** Schematic of CVD synthesis for MoS_2 and WS_2 layers, synthesis of GO, and schematic of PSC fabrication procedure and device structure (Kim et al. 2016). **h** PSC device architecture with TMDs as HTL (Huang et al. 2017)

TMDs as ETL

In recent years, various 2D TMDs such as MoS_2, WS_2, TiS_2, and TaS_2 have been intensively explored as ETL in PSCs. In 2020, Zhan and co-workers (Malek et al. 2020) utilized few-atom-thick 2H-WS_2 as ETL, which drives ultrafast charge transfer from the perovskite layer to the electrode and thus minimizes the charge recombination during the photovoltaic process, Fig. 10.4c. The champion device delivered PCE of 18.21% and retained high stability with 96.87% of its performance when aged in the N_2 atmosphere for 33 days, Fig. 10.4d. Afzali et al. (2020) used double-layered 2H-TaS_2 nanosheet films, synthesized through a robust microwave method as ETL in PSCs. The performance of the device is also compared with the commonly used TiO_2 and SnO_2 as ETL materials. The obtained PCE of 2H-TaS_2, TiO_2, and SnO_2 devices were 15.23%, 16.21%, and 15.82%, respectively. This comparable performance of 2H-TaS_2 clearly indicates that it has potential to replace commonly used ETL. In another work, Yin et al. (2018) used 2D-TiS_2 as ETL in PSCs and obtained a significantly high PCE of 17% as compared to an ETL-free device having a PCE of only 8.71%. These research works clearly indicate that solution processable 2D TMDCs can be an important ETL material for solution processable low-cost PSCs.

TMDs as HTL

In the quest for a better alternative to the commonly used unstable organic HTL PEDOT:PSS, TMDCs are intensively studied in PSCs for efficient extraction of holes and blocking of the electrons. For example, Kim et al. (2016) used MoS_2, WS_2 grown through the chemical vapor deposition (CVD) method, and graphene oxide (GO) grown through modified Hummer's method as HTL in PSCs. Figure 10.4e–g shows an illustration of CVD synthesis for MoS_2 and WS_2 layers, synthesis of GO, and schematic of PSC fabrication procedure and device structure. The PCE of MoS_2, WS_2, and GO-based devices were 9.53%, 8.02%, and 9.62%, respectively, which were comparable to PEDOT:PSS having PCE of 9.93%. This study demonstrates that TMDCs can be used as a replacement of commonly applied unstable organic material PEDOT:PSS as HTL. In another work, Li and co-workers (Huang et al. 2017) replaced commonly used HTL material PEDOT:PSS with water-soluble MoS_2 and WS_2 in p-i-n PSCs, Fig. 10.4h. Interestingly, it is observed that 1T-rich TMDCs showed much higher efficiencies and device stability as compared to PEDOT:PSS. The PSCs having MoS_2 and WS_2 delivered PCE of 14.35% and 15.0%, respectively, whereas devices with PEDOT:PSS showed PCE of 12.44%. These studies suggest that TMDs as HTL significantly boosted the efficiency of PSCs.

TMDs in Perovskite Layer

TMDs are also explored as a photo-absorber material for electron–hole pair generation in the perovskite layer. Despite the attractive properties of perovskite as absorber material, they suffer from defect generation during thin film formation, and these imperfections trap the photogenerated charges, leading to charge recombination and slow charge collections at electrodes. This challenge could be overcome by using TMDCs additives in the perovskite layer, which passivates these trapping sites (Qin et al. 2021). For example, Li and co-workers (2022) demonstrated the use of WS2-QDs in the perovskite layer that helps in reducing the trapping states and leading to the decreased charge recombination. The fabricated device delivered PCE of 16.85% with performance retention of 71% after 50 days of indoors conservation with 30–50% humidity.

MXene

MXene is another very important class of 2D materials from the transition metal carbide/carbonitride family with the general formula $M_{n+1}X_nT_x$, where M stands for an early transition metal, X represents carbon and/or nitrogen, and T_x indicates surface termination groups (–O, –OH, and –F), Fig. 10.5a. MXene possesses a large specific surface area, high electrical conductivity, and tunable electronic properties in their 2D layered forms. These unique properties of MXene endow it as highly useful material as ETL, HTL, and additive in a photo-absorbing perovskite layer for PSCs. The various roles of MXene in PSCs are briefly presented in the following subsections.

MXene as ETL

MXene as ETL and additive in ETL for n-i-p and p-i-n device architectures have been studied and shown quite impressive performance. In a recent example, Wang et al. (2021) studied the performance of $Ti_3C_2T_x$ MXene as an ETL in photovoltaic devices having n-i-p architecture. The effect of oxygen plasma treatment on the MXene ETL layer is also studied in this work. It is observed that plasma treatment generates abundant Ti–O bonds by breaking Ti–C bonds, illustrated in Fig. 10.5b. This surface modification helps in tuning the work function of MXene together with helping in reduced trap states and improved electron transport at the interface of ETL and photo-absorbing layer. The optimized device delivered an impressive PCE of 18.9%. The device response was quite stable, showing ~ 90% PCE even after 750 h exposure to 50% humidity at room temperature (25 °C), Fig. 10.5c. Other similar reports on surface modified MXene through plasma oxidation and normal oxidation revealed that the surface-modified MXene showed better PCE than the oxidized MXene. In another work, Ma and co-workers (Zheng et al. 2021) studied the nanocomposite of SnO_2 and $Ti_3C_2T_x$ MXene as ETL. They observed that changing the wt.% of SnO_2 and $Ti_3C_2T_x$ notably modulates the device response. The optimized device having $Ti_3C_2T_x$(1.4 wt.%)/SnO_2 delivered a significantly high PCE of 20.35%. The device

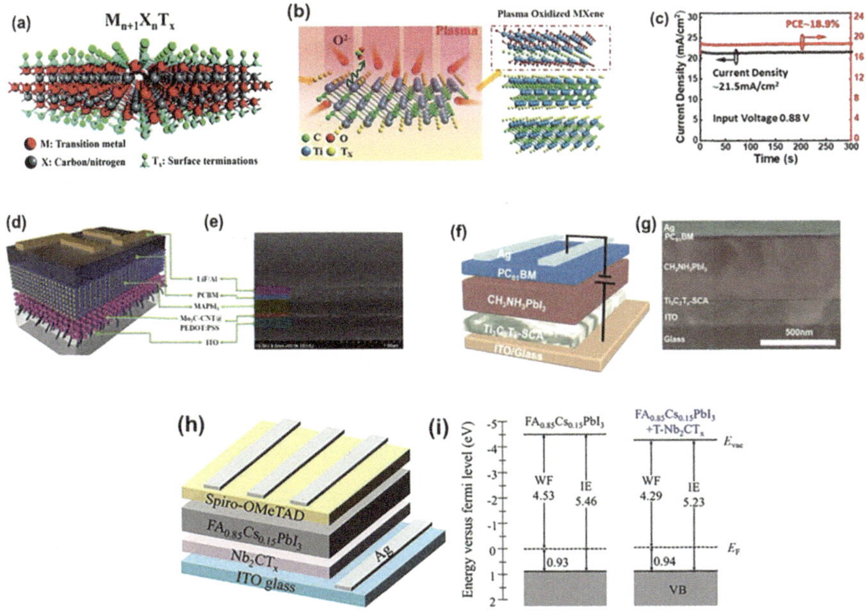

Fig. 10.5 **a** Schematic illustration of $M_{n+1}X_nT_x$ MXene crystal structure (Su et al. 2022). **b** and **c** Schematic illustration of oxygen plasma treatment on pristine MXene and steady-state output of the device showing PCE of ~ 18.9%, respectively (Wang et al. 2021). **d** and **e** Schematic illustration of device architecture and corresponding cross-sectional SEM image, respectively (Hussain et al. 2021). **f** and **g** energy scheme for the pristine and T-Nb_2CT_x MXene-doped perovskite films and schematic of device structure, respectively (Du et al. 2022). **h** and **i** Schematic illustration of device structure and energy diagram showing reduction in work function for the NH_2 terminated T-Nb_2CT_x MXene nanosheet-doped perovskite film, respectively (Zhang et al. 2022)

stability, monitored under environmental conditions, exhibited 74% efficiency retention. These findings and other similar reports attest that MXene can, independently or as a nanocomposite with other materials, be used as ETL in photovoltaic for achieving the significant enhancement in device response.

MXene as HTL

Apart from the use of MXene as ETL, researchers have also explored its use as HTL in perovskite solar cells. For example, Jung and co-workers (Hussain et al. 2021) for the first time used transition metal carbide (Mo_2C) and CNTs nanocomposite as the HTL in PSCs. In this work, the Mo_2C-CNT hybrid was blended with PEDOT:PSS HTL having ITO/HTL/$CH_3NH_3PbI_3$/PCBM/LiF/Al device architecture. A schematic illustration of device architecture is presented in Fig. 10.5d with corresponding cross-sectional scanning electron microscopic image in Fig. 10.5e.

Adopting this strategy, an improved PCE of 11.98% was achieved with Mo_2C-CNT@PEDOT:PSS HTL as compared to pure PEDOT:PSS 9.17%. In a recent work, Yu and co-workers (Zhang et al. 2021b), demonstrated use of Nb_2CT_x Mxene as HTL in PSCs. In this work, MXene work function was modulated through oxygen plasma treatment. The optimized device with Nb_2CT_x as HTL delivered excellent PCE of 20.74% with good stability. Further, Du et al. (2022) fabricated PSCs using $Ti_3C_2T_x$ MXene nanosheets with varying amounts of silane as coupling agents for HTL. The advantage of using a silane coupling agent was observed in terms of flat, pin-hole-free $Ti_3C_2T_x$ HTL film. This facilitates reduced charge transfer resistance at the interface of HTL and perovskite layer, thus provides a superior hole transport path for enhanced hole extraction at the interface. The schematic diagram of the device and its cross-sectional scanning electron microscopy image are shown in Fig. 10.5f and g, respectively. These works clearly indicate that MXene can be used as HTL in PSCs for realizing the device with improved efficiency.

MXene in Perovskite Layer

Several works demonstrated that MXene can also be added to the perovskite photo-absorbing layer to enhance the device response. Addition of MXene in the perovskite layer facilitates easy charge extraction and hence minimizes the losses through charge recombination. Thus, it helps in boosting the device performance with enhanced solar conversion efficiency. For example, Guo et al. (2018) for the first time, incorporated $Ti_3C_2T_x$ MXene into the perovskite photo-absorber layer and observed a significant enhancement in the device PCE. The high electrical conductivity and excellent charge mobility of MXene accelerated the charge transfer, thereby minimizing the recombination rate. Authors achieved 12% enhancement in device PCE from 15.54% to 17.41% with optimized key parameters and 0.03 wt.% of MXene as an additive. In 2020, Sadasivuni and co-workers (Bykkam et al. 2022) used different vol.% of 2D-MXene as an additive in perovskite ($MAPbI_3$) absorber layer. The best device with optimized 5 vol.%. MXene as an additive delivered the PCE of 13.62%. Furthermore, in a recent work, Zhang et al. (2022), incorporated N_2H_4 treated 2D T-Nb_2CT_x MXene nanosheets into the perovskite precursor and also used MXene as ETL layer, device structure shown in Fig. 10.5h. N_2H_4 treatment replaces surface functional groups with NH_2 groups and helps in minimizing the work function from 4.65 to 4.32 eV to match conduction band minimum of perovskite layer for efficient charge transfer. Figure 10.5i shows the reduced work function (WF) and ionization energy (IE) for the NH_2 terminated T-Nb_2CT_x MXene nanosheets doped perovskite film. This helps in electron extraction from the perovskite layer to the ETL. This also helps in retarding the crystallization rate promoting the formation of oriented and high-quality growth of perovskite film. The PSCs with T-Nb_2CT_x MXene as ETL and T-Nb_2CT_x MXene additive in the perovskite layer delivered very high PCE of 21.79%, and the corresponding flexible and large-area devices achieved the highest PCE of 19.15% and 18.31%. The devices maintained 93% of the original PCE after storage for 1500 h.

These works clearly demonstrate that 2D MXenes, a new intriguing family of 2D materials owing to their excellent electrical and optical properties with structure tunability, can be used as an additive in perovskite photo-absorbing layers, along with their potential use as ETL and HTL (described earlier).

Black Phosphorus

Among other 2D materials, another innovative class of natural semiconducting material with some unique opto-electronic properties is black phosphorus (BP), also called phosphorene. It has tunable bandgaps, anisotropic conductivity, and high carrier mobility. 2D phosphorene was initially synthesized from bulk BP (Bridgman produced black phosphorus for the first time in 1914 (Bridgman 1914) and in 2014 by employing a scotch tape-based microcleavage process (Liu et al. 2014). In contrast to carbon atoms in graphene, phosphorene is made up of several phosphorus atoms with puckered honeycomb lattice structure, Fig. 10.6b. At high temperature and pressure, BP is one of the most stable orthorhombic phosphorus allotropes. Prior research revealed that the energy bandgap of monolayer phosphorene is around 2.0 eV, that of bilayer BP was 1.88 ± 0.24 eV, and that of bulk BP was approximately 0.33 ± 0.02 eV (Tran et al. 2014; Woomer et al. 2015). The band structure and crystal structure of black phosphorous are shown in Fig. 10.6a. As shown in Fig. 10.6a, the interlayer spacing in the side view of the BP crystal lattice is approximately 0.53 nm, while the top view of BP along the z direction reveals a hexagonal structure with bond angles of 96.3° and 102.1°. The band structure of bulk BP with a bandgap of approximately 0.3 eV, as determined by angle-resolved photoemission spectroscopy (ARPES). The calculated bands of the bulk crystal are superimposed on top. Bands that are empty or filled are shown by blue solid and red dashed lines, accordingly. Furthermore, Fig. 10.6b provides details on the bandgap determined using various techniques and the energy of the optical absorption peak based on the number of stacking layers in a few-layer BP. BP shows great potential as a material for optoelectronic devices and other future uses due to their bandgap tunability, which is even anticipated to be better than graphene in some circumstances. To improve the stability and efficiency of devices, scientists are now focused on utilizing these materials as additives in perovskite absorber materials. The perovskite material is utilized as an active layer (AL) in the standard PSCs device architecture, which is shown in Fig. 10.6a highlighting their layers and essential components for their operation. The active layer is sandwiched between the HTL and ETL layers. The charge transferring materials such as HTL and ETL have a significant impact on the device's overall performance, which mostly consists of its stability and efficiency. Open-circuit voltage (V_{oc}), fill factor (FF) and short-circuit current (I_{sc}) are the three most often used parameters to characterize a solar cell's performance. For better performance at the commercial scale, layered BP is employed as interfacial layers, specifically the ETL, HTL, and as an additive in perovskite photo-absorber layer in PSCs. Following subsections highlight the roles of BP as ETL, HTL, and additive in perovskite layer.

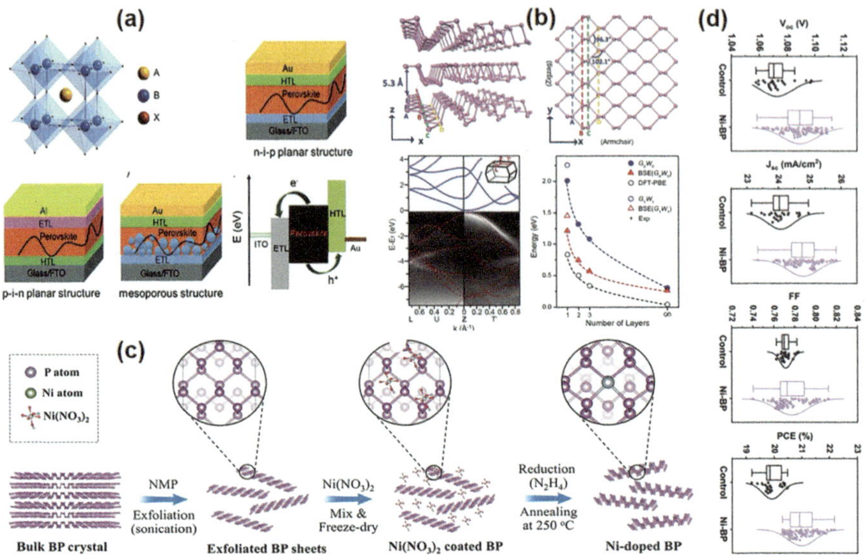

Fig. 10.6 **a** Basic structure of a PSC along with energy level scheme. Reprinted with permission from Pan et al. (2020). **b** Crystal structure of BP (side and top view), band structure of bulk, and energy gap variation with the stacking layers of a few layer BPs. Reprinted with permission from Ling et al. (2015). **c** Diagram showing the steps involved in preparing nickel-doped black phosphorus (Ni-BP). **d** A comparison of PSCs built with and without (control) Ni-BP passivation of the perovskite layer is shown by the statistical distribution of the V_{OC}, J_{SC}, FF, and PCE. Reprinted with permission from Bati et al. (2024)

BP as ETL

The charge-transporting materials, such as electron-transporting material and hole-transporting material, are sandwiched between the perovskite absorber layer in a conventional PSCs structure. Choosing the additive, ETL, or HTL materials that can improve the stability and efficiency of organic or inorganic hybrid PSCs on a commercial scale is, nevertheless, the most difficult task. To achieve higher PCE, TiO_2 has mostly been used as an ETL in solar cells. Its reduced mobility, however, can cause UV illumination to deteriorate the materials used in PSCs. It has been demonstrated in earlier research that PSCs can produce up to 12% power PCE. Due to BP's high carrier mobility, extended charge carrier lifespan, and adjustable bandgaps, the efficiency is further raised up to 20% when added to PSCs (Thirugnanasambandan et al. 2024). Studies have revealed that BPs with unique physical properties, just as other 2D-layered materials, making these as a viable option for raising the efficiency of PSCs (Yin et al. 2023). It has been observed that when BP is incorporated as ETL between TiO_2 and the perovskite layer, the PCE increases to 16.53% (when BP is synthesized by Batmunkh et al.) from 14.32% (when BP is synthesized by traditional sonication-assisted exfoliation, which can take 30 min or up to several hours). The tunable bandgap, high carrier mobility, and layer thickness-dependent properties have all been cited as the prime causes of the increased PCE. When atomically thin Ni-doped BP is added to a PSC, Batmunkh's group (Bati et al. 2024) again anticipated improved power conversion efficiency up to 22% in 2024. Figure 10.6c represents the stepwise procedure for Ni-BP flakes formation. $Ni(NO_3)_2{\cdot}6H_2O$, or nickel(II) nitrate hexahydrate, was agitated over night after the bulk BP was exfoliated in N-methyl-2-pyrrolidone. After centrifuging the solution, the precipitate containing BP and $Ni(NO_3)_2$ was treated with hydrazine monohydrate vapor. The sample was then freeze-dried, and to create the Ni-BP flakes, it was heated for 2 h at 250 °C in a constant argon (Ar) flow. Figure 10.6d shows the detailed photovoltaic (PV) parameters, which include V_{oc}, FF, J_{sc}, and PCE, summarized by their measurements. The best control devices without passivation produced FF of 0.78, V_{oc} of 1.10 V, J_{sc} of 24.0 mA cm^2, and a maximum PCE of 20.5%. On the other hand, the best Ni-BP-treated devices showed a noteworthy FF of 0.81, a V_{oc} of 1.11 V, a J_{sc} of 24.7 mA cm^2, and a significantly improved maximum PCE of 22.2%. They urged that reduced defect densities, minimal hysteresis, longer carrier lifespans, charge transfer, and improved device stability are the primary causes of improved PSCs (Bati et al. 2024). In a different study, efficient solar cells with PCE of up to 18% were fabricated by overlaying MoS_2 with bilayer BP. These cells performed better than MoS_2/Graphene solar cells (Dai and Zeng 2014).

Black Phosphorus as HTL

To increase the PCE and make the device commercially viable, PSCs with high-quality ETL also need high-quality HTL. High hole mobility, an adjustable bandgap, a long charge carrier lifespan, low cost, hydrophobicity (depending on whether the PSCs are n-i-p or p-i-n), solubility, etc. are again the key parameters for an ideal hole transport material. The use of fluorinated BP (F-BP) nanosheets as a hole transfer layer between the perovskite layer and Spiro-OMeTAD layer in PSCs was recently demonstrated by Liu et al. (2022). A standard delamination process was adopted to get the F-BP nanosheets, and it was carried out in a reaction cell using two electrodes. As a result, the PCE of the F-BP-based PSCs eventually reached to 22.06% with exceptional stability (Liu et al. 2022). It was reported by Zhang et al. (2020a) that BP nanosheets may be used as both HTL and ETL at the same time in PSCs, making them a dual functional transport material. The thickness of BP determined the energy levels of the BP nanosheets: a few-layer BP as ETL and a monolayer BP as HTL to generate a cascade band structure. This is advantageous for effective extraction of holes and electrons from the perovskite layer. In a typically PSCs with a mixed cation/halide perovskite absorber, the simultaneous incorporation of BP as HTL and ETL increased the PCE to 19.8%, while devices without BP interlayers only achieved 16.17%. This might be explained by lower interfacial recombination and enhanced energy alignment at the electrode interfaces, which facilitate charge extraction.

Black Phosphorus in the Perovskite Layer

The perovskite material in PSCs functions as an active layer that absorbs light, producing charge carriers. The charge carriers created by light are collected and moved to the appropriate electrodes by means of ETL and HTL (Jung and Park 2015). Perovskite is an appealing photo-absorber material; however, during thin-film development or during traditional synthesis, defects are created that trap photo-generated charges, causing charge recombination and slowing charge collections at electrodes. BP is a unique option for a variety of optical applications, including solar cells, because its bandgap can be regulated from 0.3 eV in bulk form to roughly 2.0 eV in monolayer form. This allows BP's adsorption spectrum to cover a wide range, from infrared light to visible light (Zhang et al. 2021c). Electron–photon interactions, which typically take place at band boundaries, are fundamental to optical adsorption. One or two-layer BP films with a thickness of less than 2 nm have been shown to be able to absorb roughly 10% of the visible light spectrum. For this reason, stacked BP is thought to be a desirable addition to PV systems that already exist, either to increase and expand light harvesting or even to serve as the only light absorber (Zhang et al. 2021c; Aidarkhanov et al. 2022). The incorporation of BP as an interfacial buffer layer to facilitate hole or electron transfer has been widely reported to aid perovskite development (Zhang et al. 2020a; Aidarkhanov et al. 2022; Chen et al. 2017). According to Zhao et al., the addition of black phosphorus (BP), which is well-known for its superior ability to regulate charge recombination and for its ability to heal itself, to the absorber precursor solution of $CH_3NH_3PbI_3$ perovskites ($MAPbI_3$/BP) resulted in a significant improvement in photostability for the associated PSCs, as well as a 20.65% increase in the photovoltaic (PV) performance of PSC (Wang et al. 2019).

Metal–Organic Frameworks (MOFs)

MOFs are a new class of material made up of metal ions or clusters functioning as linkers with organic ligands. These days, many terms such as coordination polymer, metal–organic framework, organic zeolite, hybrid organic–inorganic materials, and metal–organic polymers are used to describe these compounds (Butova et al. 2016). A general schematic for MOF formation has been shown in Fig. 10.7a (Heo et al. 2020). In addition to 3-D and 1-D MOFs, researchers in the photovoltaic community, especially those studying perovskite, are interested in 2-D MOFs, as surface modifiers and additives due to their special ability to improve photovoltaic efficiency (Chang et al. 2015; Li et al. 2018; Ryu et al. 2018; Wu and Lou 2017). These materials' criteria, which include extremely high porosity, extremely high surface area (up to 7000 $m^2 g^{-1}$), and a great deal of possibilities for varied architectures, highlights their importance as possible uses in photovoltaic devices (Khan et al. 2023). Following the exfoliation methods, a variety of MOFs can easily be converted into extremely crystalline, atomic-thin nanosheets that are comparable to the fabrication of conventional 2D materials. To separate MOF nanosheets from their bulk counterparts, exfoliation is the most effective and often utilized technique in this regard. Primary approaches for synthesizing 2D MOFs using bottom-up, top-down, and hybrid techniques are presented in Fig. 10.7b (Khan et al. 2023). It's becoming more and more popular to apply MOFs in PSCs for improving the long-term stability and device efficiency through sensible compositional nanoengineering. MOFs typically have a microporous scaffold to control the development of perovskite layers, improving the quality and crystallinity of perovskites. Furthermore, band-alignment in PSCs and charge transport layers (CTLs) conductivity can also be enhanced by including MOFs into CTLs. MOFs have four purposes: To form a hybrid MOF/perovskite heterojunction layer, they must: (i) blend in with the perovskite; (ii) act as the CTL; (iii) serve as the interface layer between the CTL and the perovskite interface; and (iv) it can minimize Pb^{2+} leakage from degraded devices which is useful to improve the PCE and operational dependability (Ye et al. 2023).

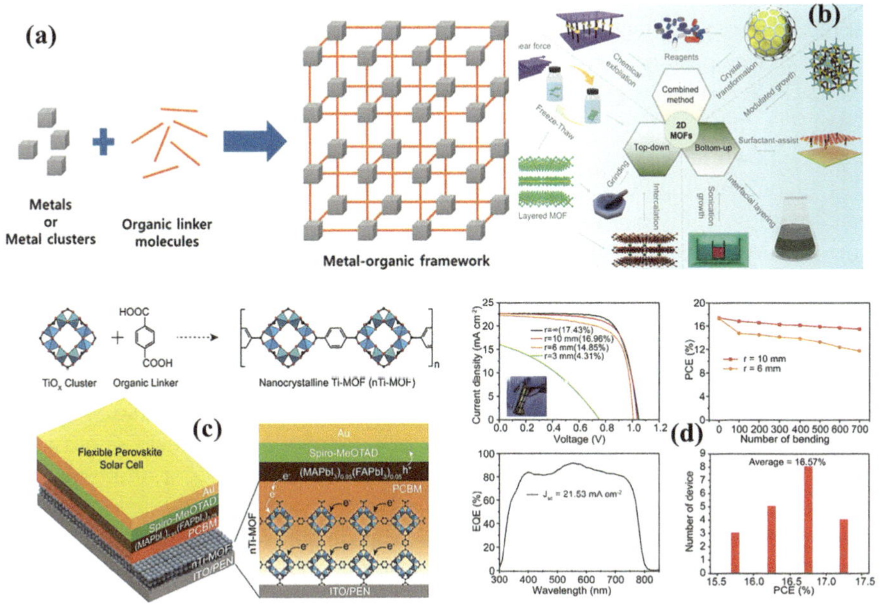

Fig. 10.7 **a** Schematic illustration of MOF formation (Heo et al. 2020). **b** General techniques for synthesizing 2D-MOFs using bottom-up, top-down, and hybrid approaches (Khan et al. 2023). **c** Synthetic protocol for nanocrystalline Ti-MOF (nTi-MOF), where small nanocrystals are produced by the 1,4-benzenedicarboxylate (BDC) molecules' subsequent linking of TiO_2 clusters. A flexible perovskite solar cell with nTi-MOF/PCBM ETL integrated into its device structure. The magnified figure shows that electrons and holes are transferred from the perovskite to interlayers, nTi-MOF/PCBM, and spiro-MeOTAD, respectively. **d** The *J–V* curves of the flexible devices, measured between bending radius $r = \infty$ and $r = 3$ mm, the bending test of the apparatus carried out up to 700 times for $r = 10$ mm and $r = 6$ mm, the external quantum efficiency (EQE) spectrum and a histogram of the PCE for twenty devices (Ryu et al. 2018)

MOFs as ETL

Compared to their roles in the perovskite layer, MOFs play very distinct roles in the ETL. MOFs have an impact on device stability, charge extraction, and UV-region absorption when they are present in the ETL. However, the limited carrier mobility of MOFs as charge-transport layers is a major concern, as most MOFs under study are insulators (Ryu et al. 2018). For example, Wu et al. (2020) demonstrated an easy method for electron-extraction layers by putting 2D integrated MOFs at the perovskite/cathode interface that show n-type behavior together with an appropriate energy level to address this problem. Chung et al. (2018) used an improved technique by applying ZnO, produced from 2D MOFs, as an ETL to boost the extraction and transport of electron. The outcome was an increase in the interface area between PSCs and ZnO produced from MOF, which enhanced the carrier extraction efficiency to 18.1% (Zhang et al. 2020b). To function as a useful ETL in PSCs, Ryu et al. utilized Ti-based nanocrystalline MOF (nTi-MOF) as shown in Fig. 10.7c. [6,6]-phenyl-C61-butyric acid (PCBM) was added to the nTi-MOF ETLs to increase film conductivity and reduce direct contact between the perovskite and substrates. Thus, these offer superior device performance for PSCs topologies that are rigid (PCE = 18.94%) or flexible (PCE = 17.43%) (Ryu et al. 2018). For flexible devices using the nTi-MOF/PCBM ETL, the J–V curves are presented in Fig. 10.7d. The actual device's bending image is shown in the inset. With $V_{oc} = 1.05$ V, $J_{sc} = 22.61$ mA cm^{-2}, and FF = 0.734, an outstanding PCE of 17.43% was achieved prior to bending. When ITO was treated with polyethylenimine-ethoxylated (PEIE), a PCE of 14.14% was achieved in the PCBM device (i.e., absence of nTi-MOFs in the ETL). The external quantum efficiency (EQE) spectrum and durability test for up to 700 bending cycles at bending radii of 10 mm and 6 mm are also shown in Fig. 10.7d. After 700 bending cycles, the initial PCE of 17.37% and 17.28% dropped to 15.43% and 11.73%, respectively. The device's performance dropped linearly at the same bending radius as soon as it was deformed. An integrated J_{sc} of 21.53 mA cm^{-2} was achieved, and high EQE values were noted in the visible range. The histography depicts the PCE of twenty devices as 16.57% on average (varying from 15.68% to 17.43% at minimum and maximum). These findings imply that the nTi-MOF can attribute an exceptional device performance and has a great deal of promise for usage in the fabrication of highly flexible architecture in PSCs. Considering this, it can be said that MOFs can efficiently promote charge transfer across connected surfaces and act as an interface modifiers to facilitate band alignment at the associated interface.

MOFs as HTL

To improve device performance, the hole transport layer in a PSCs device is crucial for facilitating easy hole extraction and transport. By keeping moisture and oxygen from penetrating the perovskite layer beneath to it in a standard forward device construction in photovoltaic cells, the HTL can also serve as an upper covering layer to safeguard the cell, thereby enhancing the device's long-term stability in the outdoors (Huang et al. 2019). As an addition to HTL, Zhou et al. (2023) introduced a 2D graphitic N-rich porous carbon (NPC) generated from MOF and confirmed that the addition of it may effectively improve the characteristics of HTL, facilitating reduced recombination and easy hole extraction and migration. Recently, Zhang et al. (2021a), developed a bi-functional layer by adding anionic 2D MOFs (FJU-17) based on indium $(Me_2NH_2)^+$ encapsulated into hole transport material. While its anionic framework was able to support positively charged oxidized HTM to increase hole mobility, the FJU-17 passivated cation vacancies by discharging the $(Me_2NH_2)^+$ ion. As a result, PSCs performance was improved, with PCE increasing from 18.32% to 20.34%. Huang et al., combined a Spiro-OMeTAD layer with a 2D lead (Pb)-based MOF. When compared to a bare Spiro-OMeTAD layer, the composite film with the haloing effect displayed higher hydrophobicity, smoother surface, and upshifted energy levels. The greater PCE (13.17%) and moisture-resistance feature were exhibited in the cell when the composite layer was used as HTL in PSCs (Huang et al. 2019). Three distinct supramolecular compounds of zinc (II), indium (III), and zirconium (IV) with proton transfer compounds were recently studied in relation to their role in perovskite solutions and how it affected device performance. The addition of Zn metal–organic compound to PSCs improved the perovskite layer, and the PCE increased from 2.9% to 5.6% with the help of the MOF (Seifpanah Sowmehesaraee et al. 2021).

MOFs in Perovskite Layer

It is possible to control the development of perovskite layers, promote charge transport, and minimize non-radiative recombination by creating an appropriate interfacial layer with an appropriate energy level and chemical bonding. Shen et al. (2018), used a thin interlayer of zeolitic imidazolate framework-8 (ZIF-8) to cover the surface of mesoporous TiO_2. They claimed that ZIF-8 has strong chemical resistance and steady temperature. Moreover, ZIF-8 can provide a scaffold to regulate the crystallization of the perovskite layer formed on top by the cross-linking action. The perovskite crystals' grain size and crystallinity are significantly enhanced by the interfacial layer. Using a ZIF-8 coating on mesoporous TiO_2, the device efficiency rises from 14.75% to 16.99%. Chang et al. (2015) found that adding Zr-based 2D porphyrin MOFs (MOF-525) as an additive significantly improved the crystallinity and shape of perovskite. The perovskite crystallization inside the MOF/perovskite composite film was facilitated by the porous MOF525 nanocrystals that were inserted at the bottom of the film.

Carbon Nitride

Research into alternative 2D materials that can overcome graphene's zero bandgap feature, which is a drawback for several applications like switching, light-emitting diodes, and photocatalysis, is being spurred by the system's enormous success in a variety of sectors. We can either modify the graphene structure by cutting into thin nanoribbons or doping (here N, because of the identical relative atomic mass and electronic configuration) to open the bandgap, or we can look for a new structure with an atomic structure like graphene. The former, however, can unavoidably result in some damage to the pure graphene's 2D carbon lattice, which would throw the ratio of C to N out of equilibrium. Thus, to synthesize carbon nitride (which has the structural formula C_xN_y) as a substitute for graphene, one must first use distinct synthesis procedures to synthesize precursors that are rich in C and N. In addition to the extensively studied graphitic carbon nitride (g-C_3N_4), additional extensively studied 2D C_xN_y polymorphs include C_2N, C_3N, C_5N_2, and so on. The crystal structures of these C_xN_y compounds are displayed in Fig. 10.8a. These compounds differ in the configurations of C and N atoms in a basic benzene ring and the matching structural motif in the sp2-hydridized basal plane. Due to this, they display unique textural and bandgap physicochemical characteristics. Because they have already been synthesized in laboratories and have good visible-light adsorption and a sufficient bandgap, g-C_3N_4, C_2N, and C_3N have generated a lot of interest in the field of photocatalysis; the remaining 2D-C_xN_y are still at the theoretical research stage. High-performance planar PSCs are greatly impacted by carrier recombination and charge losses at the perovskite layer interface. Therefore, to increase the PCE, one can use carbon nitride-based 2D materials as additives in the pervoskite-based solar cells.

Carbon Nitride as ETL

The charge transport layer's carrier mobility and energy level alignment are crucial for electron extraction and transport in various derivative of PSCs. The surface imperfections of the perovskite layer operate as interfacial non-radiative recombination sites, causing energy disorder and impeding charge extraction. This leads to a fall in the fill factor, a reduction in the open circuit voltage, and a dramatic decrease in the PCE (Liu et al. 2020). Like other 2D materials, carbon nitride has the potential to be a very promising material that can greatly enhance the performance of PSCs (Xu et al. 2014; Yang et al. 2020; Zhou et al. 2016). For example, Hsieh et al. (2018) found an increase in the PCE from 14.6% to 18.8% in $MAPbI_3$ upon adding urea or thiourea, which is the primary source of g-C_3N_4. It was determined that the improvement was caused by $MAPbI_3$'s larger feature sizes, which may prevent charge recombination at grain boundaries. In another study, Cao et al. (2019) developed perovskite films of superior quality with fewer flaws by incorporating iodine-doped g-C_3N_4 into the films and found PCE up to 18.28%. They have explained this due to a decrease in defect density from 1.43×10^{16} cm^{-3} to 1.07×10^{16} cm^{-3}. According to Y. Guo et al., the g-C_3N_4 alteration improves charge transfer and separation, leading to better PV performance, PCE up to 20%, by raising the conduction band energy (E_c) of TiO_2 and decreasing the energy level difference between TiO_2 and perovskite. (Guo et al. 2024). Moreover, g-C_3N_4 successfully inhibits nonradiative charge recombination

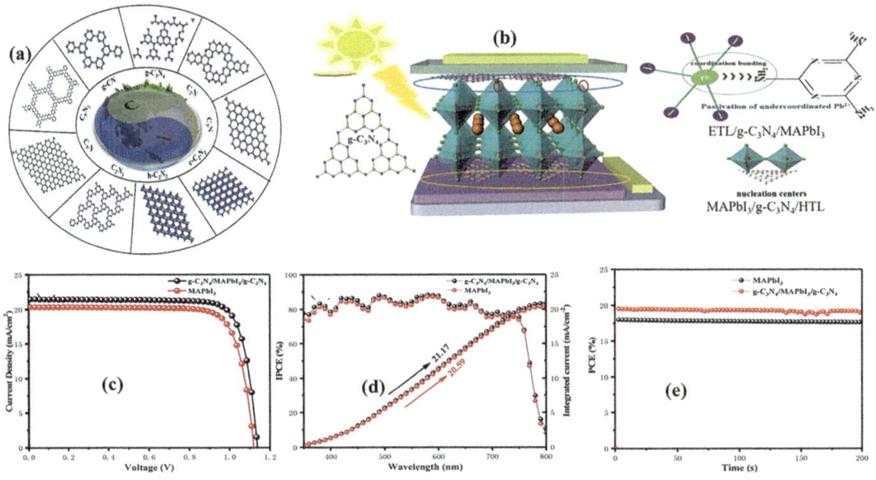

Fig. 10.8 **a** A pictorial representation of various 2D carbon nitride (C_xN_y) structures (Tan et al. 2021). **b** g-C_3N_4 dual-modified PSCs. **c** *J–V* curves of a planar PSCs without any g-C_3N_4 modification and a planar PSC with g-C_3N_4 simultaneously altering the perovskite layer and SnO_2 ETL, **d** IPCE spectra of a device without any modifications and a device with interfaces that have been altered to accommodate g-C_3N_4 (ETL/g-C_3N_4/perovskite/g-C_3N_4/spiro-OMeTAD). **e** PSCs steady-state efficiency, both with and without g-C_3N_4 dual inclusion, as determined by measuring power output at maximum capacity (Liu et al. 2020)

by passivating the interface defects separating the perovskite and ETL (Liu et al. 2023).

Carbon Nitride as HTL

Recent studies have also shown a great deal of interest in the chemical, mechanical, and optoelectronic properties of polymeric 2D materials based on carbon nitride when used as the HTL in PSCs. According to Ahmad's research group, a control device without an interfacial layer recently showed a PCE of 16.69% with an V_{oc} of 1001.89 mV, a J_{sc} of 22.79 mA cm^{-2}, and a FF of 73.07%. The C_3N_5-incorporated device's PCE, on the other hand, rose to 17.17% with a considerably improved FF of 75.95%, a V_{oc} of 1003.48 mV, and a J_{sc} of 22.52 mA cm^{-2}. They found that the C_3N_5 interface layer had a comparatively greater shunt resistance (R_{sh}) and a comparatively reduced series resistance (R_s). Additionally, it is possible to associate the high R_{sh} with the reduced recombination losses and synergistically inhibited charge build-up at the perovskite/NiO_x interface. This allowed them to obtain a good HTL layer and a high FF with the C_3N_5-based PSCs (Ashraf et al. 2023). In a different study by Hemasiri et al. (2023) an interface engineering of inverted PSCs was demonstrated using carbon nitride-based 2D materials, notably g-C_3N_4 and L-C_3N_4. Because of the faster photo-induced charge transfer kinetics and less charge recombination behavior with 2D-carbon nitrides, the PSCs' photovoltaic performances improved by 18.58%. Thus, the distinct qualities of the materials used in 2D-based PSCs have made them attractive options for a variety of photoconversion uses.

Carbon Nitride in the Perovskite Layer

The process of interfacial modification, which involves removing flaws from the perovskite layer's interface, is essential for optimizing perovskite films and preventing reductions in FF, V_{oc}, and PCE. Recent studies have included graphitic carbon nitride (g-C_3N_4), a 2D polymeric material, into the perovskite layer as a workable method to encourage perovskite crystallization and, consequently, their best use in practical applications. Liu et al. (2020), recently reported a straightforward and efficient technique for creating n-type g-C_3N_4 ultrathin films at the interfaces between perovskite and HTL in planar PSCs, resulting in great stability and efficiency over time. They fabricated a standard FTO/SnO_2/$MAPbI_3$/spiro-OMeTAD/Au planar PSCs structure. All other layers were spin coated, whereas the Au layer was deposited through thermal evaporation (Fig. 10.8b). Multilayer g-C_3N_4 was used to alter the perovskite's upper and lower layers. The photovoltaic properties of normal PSCs (0.06 cm^{-2}) and one with a g-C_3N_4-modified $MAPbI_3$ perovskite absorber at the two interfaces under AM 1.5 G irradiation at 100 mW cm^{-2} are shown in Fig. 10.8c–e. The control device had the following measurements: PCE of 18.03%, FF of 0.793, J_{sc} of 20.3 mA cm^{-2}, and V_{oc} of 1.12 V. In comparison to the pristine device, they observed an improved value of FF, V_{oc}, and J_{sc} that 0.807, 1.14 V, and 21.45 mA cm^{-2}, respectively, and this allowed the PCE to rise to 19.67%. The enhanced performance of the planar PSCs after g-C_3N_4 dual-incorporation was mainly caused by two factors: (i) The perovskite layer's absorption of visible light was enhanced by g-C_3N_4 dual-incorporation; (ii) the device exhibiting g-C_3N_4 dual-incorporation showed reduced series resistance (R_s), reduced amplitude, and an increased recombination resistance (Liu et al. 2020).

Current Challenges and Future Aspects

However, the advent of 2D materials has opened a variety of opportunities to scale up solar cell PCE to commercial scale, at the same time many significant challenges remain in this regard. As everyone knows that PSC is always surrounded by the problem of stability and degradation, and controlling its scalability is also a very challenging task, which requires excellent laboratory and scientific temperament. Interface engineering, selection of promising materials, and reproducibility of the same device with enhanced efficiency at every time are again being the big deal to the scientist. But we can say that with the advent of 2D materials and by doing more and more research on them, success can be achieved to some extent in overcoming all the difficulties given above.

Conclusion

This chapter comprehensively outlined the role of 2D materials (graphene and beyond) for perovskite solar cells. 2D materials can easily be processed following the solution preparation method, thereby facilitating a rising hope for the low-cost, solution processable perovskite solar cell. The use of 2D materials as ETL offers work function modulation of ETL for achieving optimal band positioning relative to the perovskite layer, thus facilitating an efficient electron transfer from perovskite layer to electrode. Similarly, 2D materials employed as HTLs minimize the charge transfer resistance at the interface of HTL and perovskite, thereby offering an easy path for hole extraction. Furthermore, 2D materials incorporated as additives in the photo-absorbing perovskite layer helps in enhancing the crystallinity, tunable morphology, together with providing easy charge transfer owing to their high carrier mobility, conductivity, and unique 2D structure. In summary, a number of studies summarized in this chapter demonstrate the broad application of 2D materials as ETL, HTL, and additives in light-absorbing perovskite layers to increase the photo-conversion efficiency and stability of photovoltaic solar cells.

Acknowledgements BMP acknowledges the DBT Star College scheme and i-4 Centre, Deshbandhu College, University of Delhi for research facilities. Vijay K. Singh acknowledges the DST Govt. of India for funding support through the Inspire Faculty Fund (Grant No.: DST/INSPIRE/04/2019/002451), IIT Jodhpur and IIT Ropar for providing the research facility and funding support.

Declaration of Competing Interest
The authors do not have any conflicts of interest to declare.

References

Afzali M, Mostafavi A, Shamspur T (2020) Improved perovskite solar cell with 2H–TaS2 nanosheets as an electron transport layer using microwave irradiation. J Alloy Compd 817:152742. https://doi.org/10.1016/J.JALLCOM.2019.152742

Aidarkhanov D, Surya C, Ng A (2022) The roles of black phosphorus in performance enhancement of halide perovskite solar cells. J Energy Chem 67:672–683. https://doi.org/10.1016/J.JECHEM.2021.11.006

Ashraf M, Hemasiri NH, Kazim S, Ullah N, Khan M, Adewale Ganiyu S, R, Alhooshani K, Tahir MN, Ahmad S (2023) Interface engineering of a hole-transport layer/perovskite with low-band-gap 2D-carbon nitrides for solar cell fabrication. Sustain Energy Fuels 7(3):763–768. https://doi.org/10.1039/D2SE01225A

Bati ASR, Myagmarsereejid P, Fronzi M, Fan K, Liu P, Zhong YL, Burn PL, Gentle IR, Shaw PE, Batmunkh M (2024) Atomically doped 2D black phosphorus for efficient and stable perovskite solar cells. Small Struct 5(2):2300334. https://doi.org/10.1002/SSTR.202300334

Bati ASR, Batmunkh M, Shapter JG (2020) Emerging 2D layered materials for perovskite solar cells. Adv Energy Mater 10(13). https://doi.org/10.1002/aenm.201902253

Bridgman PW (1914) Two new modifications of phosphorus. J Am Chem Soc 36(7):1344–1363. https://doi.org/10.1021/JA02184A002/ASSET/JA02184A002.FP.PNG_V03

Butova VV, Soldatov MA, Guda AA, Lomachenko KA, Lamberti C (2016) Metal-organic frameworks: structure, properties, methods of synthesis and characterization. Russ Chem Rev 85(3):280–307. https://doi.org/10.1070/RCR4554

Bykkam S, Mishra A, Prasad DN, Maurya MR, Cabibihan JJ, Ahmad Z, Sadasivuni KK (2022) 2D-MXene as an additive to improve the power conversion efficiency of monolithic perovskite solar cells. Mater Lett 309:131353. https://doi.org/10.1016/J.MATLET.2021.131353

Cao J, Liu YM, Jing X, Yin J, Li J, Xu B, Tan YZ, Zheng N (2015) Well-defined thiolated nanographene as hole-transporting material for efficient and stable perovskite solar cells. J Am Chem Soc 137(34):10914–10917. https://doi.org/10.1021/JACS.5B06493/ASSET/IMAGES/LARGE/JA-2015-06493Z_0005.JPEG

Cao W, Lin K, Li J, Qiu L, Dong Y, Wang J, Xia D, Fan R, Yang Y (2019) Iodine-doped graphite carbon nitride for enhancing photovoltaic device performance via passivation trap states of triple cation perovskite films. J Mater Chem C 7(40):12717–12724. https://doi.org/10.1039/C9TC04264D

Chang T-H, Kung C-W, Chen H-W, Huang T-Y, Kao S-Y, Lu H-C, Lee M-H, Boopathi KM, Chu C-W, Ho K-C, Chang T-H, Kung C-W, Chen H-W, Huang T-Y, Kao S-Y, Lu H-C, Lee M-HK, Ho C, Boopathi KM, Chu C-W (2015) Planar heterojunction perovskite solar cells incorporating metal-organic framework nanocrystals. Adv Mater 27(44):7229–7235. https://doi.org/10.1002/ADMA.201502537

Chen W, Li K, Wang Y, Feng X, Liao Z, Su Q, Lin X, He Z (2017) Black phosphorus quantum dots for hole extraction of typical planar hybrid perovskite solar cells. J Phys Chem Lett 8(3):591–598. https://doi.org/10.1021/ACS.JPCLETT.6B02843/ASSET/IMAGES/LARGE/JZ-2016-02843Z_0005.JPEG

Chouhan L, Ghimire S, Subrahmanyam C, Miyasaka T, Biju V (2020) Synthesis, optoelectronic properties and applications of halide perovskites. Chem Soc Rev 49(10):2869–2885. https://doi.org/10.1039/C9CS00848A

Chung HY, Lin CH, Prabu S, Wang HW (2018) Perovskite solar cells using TiO_2 layers coated with metal-organic framework material ZIF-8. J Chin Chem Soc 65(12):1476–1481. https://doi.org/10.1002/JCCS.201800173

Dai J, Zeng XC (2014) Bilayer phosphorene: effect of stacking order on bandgap and its potential applications in thin-film solar cells. J Phys Chem Lett 5(7):1289–1293. https://doi.org/10.1021/JZ500409M/SUPPL_FILE/JZ500409M_SI_001.PDF

Dong H, Ran C, Gao W, Li M, Xia Y, Huang W (2023) Metal Halide Perovskite for next-generation optoelectronics: progresses and prospects. ELight 3(1):1–16. https://doi.org/10.1186/S43593-022-00033-Z

Du JB, Yang L, Jin X, Liu CL, Wang HH, Wang XF (2022) Spray deposition of vinyl tris(2-methoxyethoxy) silane-doped $Ti_3C_2T_x$ MXene hole transporting layer for planar perovskite solar cells. J Alloy Compd 900:163372. https://doi.org/10.1016/J.JALLCOM.2021.163372

Duan L, Uddin A (2020) Progress in stability of organic solar cells. Adv Sci 7(11):1903259. https://doi.org/10.1002/ADVS.201903259

Geim AK, Novoselov KS (2007) The rise of graphene. Nat Mater 6(3):183–191. https://doi.org/10.1038/nmat1849

Gerstner E (2010) Nobel Prize 2010: Andre Geim & Konstantin Novoselov. Nat Phys 6(11):836–836. https://doi.org/10.1038/nphys1836

Gong K, Hu J, Cui N, Xue Y, Li L, Long G, Lin S (2021) The roles of graphene and its derivatives in perovskite solar cells: a review. Mater des 211:110170. https://doi.org/10.1016/J.MATDES.2021.110170

Green MA, Ho-Baillie A, Snaith HJ (2014) The emergence of perovskite solar cells. Nat Photonics 8(7):506–514. https://doi.org/10.1038/nphoton.2014.134

Guo Z, Gao L, Xu Z, Teo S, Zhang C, Kamata Y, Hayase S, Ma T, Guo Z, Xu Z, Teo S, Zhang C, Kamata Y, Hayase S, Ma T, Gao L (2018) High electrical conductivity 2D MXene serves as additive of perovskite for efficient solar cells. Small 14(47):1802738. https://doi.org/10.1002/SMLL.201802738

Hemasiri NH, Ashraf M, Kazim S, Graf R, Berger R, Ullah N, Tahir MN, Ahmad S (2023) Interface tweaking of perovskite solar cells with carbon nitride-based 2D materials. Nano Energy 109:108326. https://doi.org/10.1016/J.NANOEN.2023.108326

Heo DY, Do HH, Ahn SH, Kim SY (2020) Metal-organic framework materials for perovskite solar cells. Polymers 12(9):2061. https://doi.org/10.3390/POLYM12092061

Hoang Huy VP, Ahn YN, Hur J (2021) Recent advances in transition metal dichalcogenide cathode materials for aqueous rechargeable multivalent metal-ion batteries. Nanomaterials 11(6):1517. https://doi.org/10.3390/NANO11061517

Hsieh CM, Liao YS, Lin YR, Chen CP, Tsai CM, Wei-Guang Diau E, Chuang SC (2018) Low-temperature, simple and efficient preparation of perovskite solar cells using Lewis bases urea and thiourea as additives: stimulating large grain growth and providing a PCE up to 18.8%. RSC Adv 8(35):19610–19615. https://doi.org/10.1039/C8RA03175D

Huang L, Zhou X, Wu R, Shi C, Xue R, Zou J, Xu C, Zhao J, Zeng W (2019) Oriented haloing metal-organic framework providing high efficiency and high moisture-resistance for perovskite solar cells. J Power Sources 433:226699. https://doi.org/10.1016/J.JPOWSOUR.2019.226699

Huang P, Wang Z, Liu Y, Zhang K, Yuan L, Zhou Y, Song B, Li Y (2017) Water-soluble 2D transition metal dichalcogenides as the hole-transport layer for highly efficient and stable p-i-n perovskite solar cells. ACS Appl Mater Interfaces 9(30):25323–25331. https://doi.org/10.1021/ACSAMI.7B06403/ASSET/IMAGES/LARGE/AM-2017-06403Z_0009.JPEG

Hussain S, Liu H, Vikraman D, Hussain M, Jaffery SHA, Ali A, Kim HS, Kang J, Jung J (2021) Characteristics of Mo2C-CNTs hybrid blended hole transport layer in the perovskite solar cells and X-ray detectors. J Alloy Compd 885:161039. https://doi.org/10.1016/J.JALLCOM.2021.161039

Jariwala D, Marks TJ, Hersam MC (2016) Mixed-dimensional van der Waals heterostructures. Nat Mater 16(2):170–181. https://doi.org/10.1038/nmat4703

Jung HS, Park NG (2015) Perovskite Solar Cells: From Materials to Devices. Small 11(1):10–25. https://doi.org/10.1002/SMLL.201402767

Kamat PV (2013) Quantum dot solar cells. The next big thing in photovoltaics. J Phys Chem Lett 4(6):908–918. https://doi.org/10.1021/JZ400052E/ASSET/IMAGES/LARGE/JZ-2013-00052E_0012.JPEG

Khan U, Nairan A, Gao J, Zhang Q (2023) Current progress in 2D metal-organic frameworks for electrocatalysis. Small Struct 4(6):2200109. https://doi.org/10.1002/SSTR.202200109

Kim YG, Kwon KC, Le QV, Hong K, Jang HW, Kim SY (2016) Atomically thin two-dimensional materials as hole extraction layers in organolead halide perovskite photovoltaic cells. J Power Sources 319:1–8. https://doi.org/10.1016/J.JPOWSOUR.2016.04.032

Lamanna E, Matteocci F, Calabrò E, Serenelli L, Salza E, Martini L, Menchini F, Izzi M, Agresti A, Pescetelli S, Bellani S, Del Río Castillo AE, Bonaccorso F, Tucci M, Di Carlo A (2020) Mechanically stacked, two-terminal graphene-based perovskite/silicon tandem solar cell with efficiency over 26%. Joule 4(4):865–881. https://doi.org/10.1016/J.JOULE.2020.01.015

Lee C, Wei X, Kysar JW, Hone J (2008) Measurement of the elastic properties and intrinsic strength of monolayer graphene. Science (New York, N.Y.) 321(5887):385–388. https://doi.org/10.1126/science.1157996

Lei Y, Zhang T, Lin YC, Granzier-Nakajima T, Bepete G, Kowalczyk DA, Lin Z, Zhou D, Schranghamer TF, Dodda A, Sebastian A, Chen Y, Liu Y, Pourtois G, Kempa TJ, Schuler B, Edmonds MT, Quek SY, Wurstbauer U, Terrones M et al (2022) Graphene and beyond: recent advances in two-dimensional materials synthesis, properties, and devices. ACS Nanosci Au 2(6):450–485. https://doi.org/10.1021/ACSNANOSCIENCEAU.2C00017/ASSET/IMAGES/LARGE/NG2C00017_0019.JPEG

Li M, Xia D, Yang Y, Du X, Dong G, Jiang A, Fan R (2018) Doping of $[In_2(phen)3Cl_6]{\cdot}CH_3CN{\cdot}2H_2O$ indium-based metal-organic framework into hole transport layer for enhancing perovskite solar cell efficiencies. Adv Energy Mater 8(10):1702052. https://doi.org/10.1002/AENM.201702052

Li S, Li Y, Liu K, Chen M, Peng W, Zhang C, Yang Y, Li X (2022) Laser generated WS2 quantum dots for effective charge transport in high-performance carbon-based perovskite solar cells. J Power Sources 518:230766. https://doi.org/10.1016/J.JPOWSOUR.2021.230766

Li X, Yu H, Liu Z, Huang J, Ma X, Liu Y, Sun Q, Dai L, Ahmad S, Shen Y, Wang M (2023) Progress and challenges toward effective flexible perovskite solar cells. Nano-Micro Lett 15(1):1–24. https://doi.org/10.1007/S40820-023-01165-8

Ling X, Wang H, Huang S, Xia F, Dresselhaus MS (2015) The renaissance of black phosphorus. Proc Natl Acad Sci USA 112(15):4523–4530. https://doi.org/10.1073/PNAS.1416581112/ASSET/1866C046-22CC-428B-B1E0-D3A7D9C35B96/ASSETS/GRAPHIC/PNAS.1416581112FIG06.JPEG

Liu H, Neal AT, Zhu Z, Luo Z, Xu X, Tománek D, Ye PD (2014) Phosphorene: an unexplored 2D semiconductor with a high hole mobility. ACS Nano 8(4):4033–4041. https://doi.org/10.1021/NN501226Z/SUPPL_FILE/NN501226Z_SI_001.PDF

Liu T, Liu Y, Chen M, Guo X, Tang S, Zhang R, Xie Z, Wang J, Gu A, Lin S, Wang N, Liu T, Liu Y, Chen M, Guo X, Tang S, Zhang R, Xie Z, Wang J, Wang N et al (2022) Fluorinated black phosphorene nanosheets with robust ambient stability for efficient and stable perovskite solar cells. Adv Funct Mater 32(1):2106779. https://doi.org/10.1002/ADFM.202106779

Liu X, Ma H, Hu C, Keerthi M, Manavalan S, Chen S-M, Guo Y, Zhao D, Yu M, Liu M, Zhang Y, Zheng Z (2023) A simple strategy to obtain graphitic carbon nitride modified TiO_2 layer for efficient perovskite solar cells. Nanotechnology 35(7):075201. https://doi.org/10.1088/1361-6528/AD0D21

Liu Z, Wu S, Yang X, Zhou Y, Jin J, Sun J, Zhao L, Wang S (2020) The dual interfacial modification of 2D g-C3N4 for high-efficiency and stable planar perovskite solar cells. Nanoscale Adv 2(11):5396–5402. https://doi.org/10.1039/D0NA00613K

Liu Z, You P, Xie C, Tang G, Yan F (2016) Ultrathin and flexible perovskite solar cells with graphene transparent electrodes. Nano Energy 28:151–157. https://doi.org/10.1016/J.NANOEN.2016.08.038

Mahmoudi T, Wang Y, Hahn YB (2021) Highly stable perovskite solar cells based on perovskite/NiO-graphene composites and NiO interface with 25.9 mA/cm^2 photocurrent density and 20.8% efficiency. Nano Energy 79:105452. https://doi.org/10.1016/J.NANOEN.2020.105452

Mahmoudi T, Wang Y, Hahn Y-B, Mahmoudi T, Wang Y, Hahn Y-B (2020) $SrTiO_3/Al_2O_3$-graphene electron transport layer for highly stable and efficient composites-based perovskite solar cells with 20.6% efficiency. Adv Energy Mater 10(2):1903369. https://doi.org/10.1002/AENM.201903369

Malek NAA, Alias N, Umar AA, Zhang X, Li X, Saad SKM, Abdullah NA, Zhang H, Weng Z, Shi Z, Li C, Rosli MM, Zhan Y (2020) Enhanced charge transfer in atom-thick 2H–WS2 nanosheets' electron transport layers of perovskite solar cells. Solar RRL 4(10):2000260. https://doi.org/10.1002/SOLR.202000260

Nazeeruddin MK, Baranoff E, Grätzel M (2011) Dye-sensitized solar cells: a brief overview. Sol Energy 85(6):1172–1178. https://doi.org/10.1016/J.SOLENER.2011.01.018

Novoselov KS, Geim AK, Morozov SV, Jiang D, Zhang Y, Dubonos SV, Grigorieva IV, Firsov AA (2004) Electric field in atomically thin carbon films. Science 306(5696):666–669. https://doi.org/10.1126/SCIENCE.1102896/SUPPL_FILE/NOVOSELOV.SOM.PDF

Pan H, Zhao X, Gong X, Li H, Ladi NH, Zhang XL, Huang W, Ahmad S, Ding L, Shen Y, Wang M, Fu Y (2020) Advances in design engineering and merits of electron transporting layers in perovskite solar cells. Mater Horiz 7(9):2276–2291. https://doi.org/10.1039/D0MH00586J

Qin Z, Chen Y, Zhu K, Zhao Y (2021) Two-dimensional materials for perovskite solar cells with enhanced efficiency and stability. ACS Mater Lett 3(9):1402–1416. https://doi.org/10.1021/ACSMATERIALSLETT.1C00327/ASSET/IMAGES/LARGE/TZ1C00327_0009.JPEG

Quan Q, Lin X, Zhang N, Xu Y-J (2017) Graphene and its derivatives as versatile templates for materials synthesis and functional applications. Nanoscale 9(7):2398–2416. https://doi.org/10.1039/c6nr09439b

Rai S, Mishra H, Singh VK, Gupta TK, Srivastava A (2023) 2D materials for flexible electronics. 2D materials for electronics, sensors and devices: synthesis, characterization, fabrication and application, pp 169–206. https://doi.org/10.1016/B978-0-12-821505-0.00005-8

Ryu UJ, Jee S, Park JS, Han IK, Lee JH, Park M, Choi KM (2018) Nanocrystalline titanium metal-organic frameworks for highly efficient and flexible perovskite solar cells. ACS Nano 12(5):4968–4975. https://doi.org/10.1021/ACSNANO.8B02079/ASSET/IMAGES/LARGE/NN-2018-02079E_0005.JPEG

Seifpanah Sowmehesaraee M, Ranjbar M, Abedi M, Mozaffari SA (2021) Fabrication of lead iodide perovskite solar cells by incorporating zirconium, indium and zinc metal-organic frameworks. Solar Energy 214:138–148. https://doi.org/10.1016/J.SOLENER.2020.12.001

Sharma K, Sharma V, Sharma SS (2018) Dye-sensitized solar cells: fundamentals and current status. Nanoscale Res Lett 13(1):1–46. https://doi.org/10.1186/S11671-018-2760-6

Shen D, Pang A, Li Y, Dou J, Wei M (2018) Metal–organic frameworks at interfaces of hybrid perovskite solar cells for enhanced photovoltaic properties. Chem Commun 54(10):1253–1256. https://doi.org/10.1039/C7CC09452C

Singh S, Khan ZH, Khan MB, Kumar P, Kumar P (2022) Quantum dots-sensitized solar cells: a review on strategic developments. Bull Mater Sci 45(2):1–13. https://doi.org/10.1007/S12034-022-02662-Z/FIGURES/8

Singh VK, Kumar S, Pandey SK, Srivastava S, Mishra M, Gupta G, Malhotra BD, Tiwari RS, Srivastava A (2018) Fabrication of sensitive bioelectrode based on atomically thin CVD grown graphene for cancer biomarker detection. Biosens Bioelectron 105:173–181. https://doi.org/10.1016/J.BIOS.2018.01.014

Singh VK, Pendurthi R, Nasr JR, Mamgain H, Tiwari RS, Das S, Srivastava A (2020) Study on the growth parameters and the electrical and optical behaviors of 2D tungsten disulfide. ACS Appl Mater Interfaces 12(14):16576–16583. https://doi.org/10.1021/ACSAMI.9B19820/ASSET/IMAGES/LARGE/AM9B19820_0005.JPEG

Su T, Ma X, Tong J, Ji H, Qin Z, Wu Z (2022) Surface engineering of MXenes for energy and environmental applications. J Mater Chem A 10(19):10265–10296. https://doi.org/10.1039/D2TA01140A

Sung H, Ahn N, Seok Jang M, Lee J-K, Yoon H, Park N-G, Choi M, Sung H, Ahn N, Yoon H, Choi M, Jang MS, Lee J, Park N (2016) Transparent conductive oxide-free graphene-based perovskite solar cells with over 17% efficiency. Adv Energy Mater 6(3):1501873. https://doi.org/10.1002/AENM.201501873

Tan L, Nie C, Ao Z, Sun H, An T, Wang S (2021) Novel two-dimensional crystalline carbon nitrides beyond g-C3N4: structure and applications. J Mater Chem A 9(1):17–33. https://doi.org/10.1039/D0TA07437C

Tavakoli MM, Nasilowski M, Zhao J, Bawendi MG, Kong J (2019) Efficient semitransparent CsPbI3 quantum dots photovoltaics using a graphene electrode. Small Methods 3(12):1900449. https://doi.org/10.1002/SMTD.201900449

Thirugnanasambandan T, Karuppaiah C, Sankar BV, Gopinath SCB (2024) Insights and potentials of two-dimensional black phosphorous-based solar cells. Phys Scr 99(5):052002. https://doi.org/10.1088/1402-4896/AD3A27

Tian M, Woo CY, Choi JW, Seo JY, Kim JM, Kim SH, Song M, Lee HW (2020) Printable free-standing hybrid graphene/dry-spun carbon nanotube films as multifunctional electrodes for highly stable perovskite solar cells. ACS Appl Mater Interfaces 12(49):54806–54814. https://doi.org/10.1021/ACSAMI.0C17141/ASSET/IMAGES/LARGE/AM0C17141_0007.JPEG

Tran V, Soklaski R, Liang Y, Yang L (2014) Layer-controlled band gap and anisotropic excitons in few-layer black phosphorus. Phys Rev B Condens Matter Mater Phys 89(23):235319. https://doi.org/10.1103/PHYSREVB.89.235319/FIGURES/6/MEDIUM

Tress W (2014) Organic solar cells. Springer Ser Mater Sci 208:67–214. https://doi.org/10.1007/978-3-319-10097-5_3

Wang F, Endo M, Mouri S, Miyauchi Y, Ohno Y, Wakamiya A, Murata Y, Matsuda K (2016) Highly stable perovskite solar cells with an all-carbon hole transport layer. Nanoscale 8(23):11882–11888. https://doi.org/10.1039/C6NR01152G

Wang J, Cai Z, Lin D, Chen K, Zhao L, Xie F, Su R, Xie W, Liu P, Zhu R (2021) Plasma oxidized $Ti_3C_2T_x$ MXene as electron transport layer for efficient perovskite solar cells. ACS Appl Mater Interfaces 13(27):32495–32502. https://doi.org/10.1021/ACSAMI.1C07146/ASSET/IMAGES/LARGE/AM1C07146_0005.JPEG

Wang JTW, Ball JM, Barea EM, Abate A, Alexander-Webber JA, Huang J, Saliba M, Mora-Sero I, Bisquert J, Snaith HJ, Nicholas RJ (2014) Low-temperature processed electron collection layers of graphene/TiO_2 nanocomposites in thin film perovskite solar cells. Nano Lett 14(2):724–730. https://doi.org/10.1021/NL403997A/SUPPL_FILE/NL403997A_SI_001.PDF

Wang Y, Zhang H, Zhang T, Shi W, Kan M, Chen J, Zhao Y (2019) Photostability of MAPbI3 perovskite solar cells by incorporating black phosphorus. Solar RRL 3(9):1900197. https://doi.org/10.1002/SOLR.201900197

Woomer AH, Farnsworth TW, Hu J, Wells RA, Donley CL, Warren SC (2015) Phosphorene: synthesis, scale-up, and quantitative optical spectroscopy. ACS Nano 9(9):8869–8884. https://doi.org/10.1021/ACSNANO.5B02599/ASSET/IMAGES/LARGE/NN-2015-02599S_0009.JPEG

Wu HB, Lou XW (2017) Metal-organic frameworks and their derived materials for electrochemical energy storage and conversion: promises and challenges. Sci Adv 3(12). https://doi.org/10.1126/SCIADV.AAP9252/ASSET/A0E618A0-0D1A-49AC-8262-C1FE26C55D5D/ASSETS/GRAPHIC/AAP9252-F7.JPEG

Wu S, Li Z, Li MQ, Diao Y, Lin F, Liu T, Zhang J, Tieu P, Gao W, Qi F, Pan X, Xu Z, Zhu Z, Jen AKY (2020) 2D metal–organic framework for stable perovskite solar cells with minimized lead leakage. Nat Nanotechnol 15(11):934–940. https://doi.org/10.1038/s41565-020-0765-7

Xu J, Brenner TJK, Chabanne L, Neher D, Antonietti M, Shalom M (2014) Liquid-based growth of polymeric carbon nitride layers and their use in a mesostructured polymer solar cell with V_{oc} exceeding 1 V. J Am Chem Soc 136(39):13486–13489. https://doi.org/10.1021/JA508329C/SUPPL_FILE/JA508329C_SI_001.PDF

Yang Q, Yu S, Fu P, Yu W, Liu Y, Liu X, Feng Z, Guo X, Li C, Yang Q, Yu S, Fu P, Yu W, Liu Y, Liu X, Feng Z, Guo X, Li C (2020) Boosting performance of non-fullerene organic solar cells by 2D g-C3N4 doped PEDOT:PSS. Adv Funct Mater 30(15):1910205. https://doi.org/10.1002/ADFM.201910205

Guo Y, Zhao D, Yu M, Liu M, Zhang Y, Zheng Z (2024) A simple strategy to obtain graphitic carbon nitride modified TiO_2 layer for efficient perovskite solar cells. Nanotechnology 35(7):075201. https://doi.org/10.1088/1361-6528/ad0d21

Ye Y, Yin Y, Chen Y, Li S, Li L, Yamauchi Y (2023) Metal-organic framework materials in perovskite solar cells: recent advancements and perspectives. Small 19(25):2208119. https://doi.org/10.1002/SMLL.202208119

Yin G, Zhao H, Feng J, Sun J, Yan J, Liu Z, Lin S, Liu S (2018) Low-temperature and facile solution-processed two-dimensional TiS2 as an effective electron transport layer for UV-stable planar perovskite solar cells. J Mater Chem A 6(19):9132–9138. https://doi.org/10.1039/C8TA01143E

Yin Y, Zhou Y, Rafailovich MH, Nam CY (2023) Recent advances of two-dimensional material additives in hybrid perovskite solar cells. Nanotechnology 34(17):172001. https://doi.org/10.1088/1361-6528/ACB441

You P, Liu Z, Tai Q, Liu S, Yan F, You P, Liu Z, Tai Q, Liu S, Yan F (2015) Efficient semitransparent perovskite solar cells with graphene electrodes. Adv Mater 27(24):3632–3638. https://doi.org/10.1002/ADMA.201501145

Zhang J, Guo S, Zhu M, Li C, Chen J, Liu L, Xiang S, Zhang Z (2021a) Simultaneous defect passivation and hole mobility enhancement of perovskite solar cells by incorporating anionic metal-organic framework into hole transport materials. Chem Eng J 408:127328. https://doi.org/10.1016/J.CEJ.2020.127328

Zhang J, Huang C, Sun Y, Yu H, Zhang J, Huang C, Sun Y, Yu H (2022) Amino-functionalized niobium-carbide MXene serving as electron transport layer and perovskite additive for the preparation of high-performance and stable methylammonium-free perovskite solar cells. Adv Funct Mater 32(24):2113367. https://doi.org/10.1002/ADFM.202113367

Zhang J, Huang C, Yu H (2021b) Modulate the work function of Nb2CTx MXene as the hole transport layer for perovskite solar cells. Appl Phys Lett 119(3):33506. https://doi.org/10.1063/5.0057978/41797

Zhang M, Biesold GM, Lin Z (2021c) A multifunctional 2D black phosphorene-based platform for improved photovoltaics. Chem Soc Rev 50(23):13346–13371. https://doi.org/10.1039/D1CS00847A

Zhang M, Ye M, Wang W, Ma C, Wang S, Liu Q, Lian T, Huang J, Lin Z, Zhang M, Ye M, Lin Z, Wang W, Ma C, Wang S, Liu Q, Lian T, Huang J (2020a) Synergistic cascade carrier extraction via dual interfacial positioning of ambipolar black phosphorene for high-efficiency perovskite solar cells. Adv Mater 32(28):2000999. https://doi.org/10.1002/ADMA.202000999

Zhang YN, Li B, Fu L, Li Q, Yin LW (2020b) MOF-derived ZnO as electron transport layer for improving light harvesting and electron extraction efficiency in perovskite solar cells. Electrochim Acta 330:135280. https://doi.org/10.1016/J.ELECTACTA.2019.135280

Zheng H, Wang Y, Niu B, Ge R, Lei Y, Yan L, Si J, Zhong P, Ma X (2021) Controlling the defect density of perovskite films by MXene/SnO_2 hybrid electron transport layers for efficient and stable photovoltaics. J Phys Chem C 125(28):15210–15222. https://doi.org/10.1021/ACS.JPCC.1C04361/ASSET/IMAGES/LARGE/JP1C04361_0006.JPEG

Zhou G, Liu H, Zhang Y, Zhang K, Liu L, Wang H (2023) 2D graphitic nitrogen-rich porous carbon derived from metal-organic frameworks as efficient hole transport layer for perovskite solar cells. Small 19(22):2206635. https://doi.org/10.1002/smll.202206635

Zhou L, Xu Y, Yu W, Guo X, Yu S, Zhang J, Li C (2016) Ultrathin two-dimensional graphitic carbon nitride as a solution-processed cathode interfacial layer for inverted polymer solar cells. J Mater Chem A 4(21):8000–8004. https://doi.org/10.1039/C6TA01894G

Znidi F, Morsy M, Nizam Uddin M (2024) Recent advances of graphene-based materials in planar perovskite solar cells. Next Nanotechnol 5:100061. https://doi.org/10.1016/J.NXNANO.2024.100061

Chapter 11
Future of Perovskite Solar Cells and Challenges in Large-Scale Energy Production

Mahendra Singh Rathore, Anand Y. Joshi, and Unnati A. Joshi

Abstract In this chapter, future prospects of perovskite solar cells development using solar energy and their opportunity and possible challenges during the large-scale energy production have been discussed. As the demand, the future of solar cells holds immense promise in revolutionizing energy production. The path way of the deployment of solar energy in future to achieve the target of sustainable goal up-to 2035 has been demonstrates. The review of research work done to improve the perovskite solar cell efficiency and future scope of work in milestone to design the solar cell have been summarized. The requirement of advancement in the solar cells as photovoltaic (PV) cells which has significant importance to enhance in efficiency, durability, cost reduction, and scalability at large scale has been discussed. The next-generation PV technology, perovskite-based solar cell, and their system offer to resolve efficiency and manufacturing cost problems against conventional energy sources. The technological challenges deterring the large-scale development of perovskite solar cells including intermittency, grid integration, storage solutions, and material sustainability have been discussed. The supportive policies, market incentive, and demand for the adoption of perovskite solar energy at large scale have been emphasized. The future scope of solar energy and possible solution of the challenges are described in details.

Keywords Perovskite solar cell · Solar energy · Future scope · Challenges · Grid photovoltaic · Efficiency · Materials · Storage

---

M. S. Rathore (✉)
Department of Physics (Applied Science), Parul University, Vadodara, Gujarat, India
e-mail: mahendra.rathore8944@paruluniversity.ac.in

M. S. Rathore · A. Y. Joshi · U. A. Joshi
Micro Nano Research and Development Centre, Parul University, Vadodara, Gujarat, India

A. Y. Joshi
Department of Mechatronics Engineering, Parul University, Vadodara, Gujarat, India

U. A. Joshi
Department of Mechanical Engineering, Parul University, Vadodara, Gujarat, India

M. Kumar and A. Raj (eds.), *Perovskite Solar Cells*, Springer Tracts in Electrical and Electronics Engineering, https://doi.org/10.1007/978-981-96-9463-1_11

Introduction

In this chapter, the various types of solar cell and their mechanism for the production of energy are summarized. The future prospects and technology use in advancement of materials in development of solar cell is described. The review of research work done to improve the perovskite solar cell efficiency and future scope of work in milestone to design the solar cell has been summarized. The requirement of advancement in the solar cells as photovoltaic (PV) cells which has significant importance to enhance in efficiency, durability, cost reduction, and scalability at large scale are discussed. The next-generation perovskite-based PV technology resolves the efficiency issues and manufacturing cost problems against conventional energy sources. The technological challenges deterring the large-scale development of solar cells and the remarkable possible solution are included.

Overview of Solar Cell

As the demand for clean and renewable energy continues to surge, the future of solar cells holds immense promise in revolutionizing energy production. The main objective to achieve the de-carbonization that require accelerating the clean energy deployment. At present, many countries worldwide have set their goal to achieve the neutrality of carbon (Soga 2006). The pathway of the deployment of solar energy in future is to achieve the target of sustainable goal up to 2035. The research work done carried out to improve the solar cell efficiency and milestone to design the solar cell (Ardani et al. 2021). The materials advancement in solar cells as photovoltaic (PV) cells has significant importance to enhance in efficiency, durability, cost reduction, and scalability at large scale (Ananthakumar et al. 2019). The landscape of research for development of solar cells is dynamic and vibrant and needs continuous advancements in future for leading to higher efficiencies. Ongoing efforts in materials science, manufacturing, and integration are crucial for the continued growth and adoption of solar energy worldwide (Al-Ezzi and Ansari 2022). The next-generation PV technology, e.g., tandem cells (Yamaguchi et al. 2018) silicon solar cells, inorganic solar cells, organic solar cells, perovskite-based solar cell and hybrid solar cells, etc. (Soonmin et al. 2023). The mechanism process is described in the system and offers to resolve efficiency and manufacturing cost problems against conventional energy sources (Al-Ezzi and Ansari 2022; Güney 2019).

The generation mechanism of energy through solar cell as shown in Fig. 11.1. The solar cell as photovoltaic (PV) works on the principle of the conversion of light energy into electrical energy through the photovoltaic effect (Ardani et al. 2021; Mamodiya 2019; Soga 2006). The working feature depends on the nature of substance that may capable to absorption of light. The absorbed light offers to generate the charge carrier that may separate due to excitation and responsible for the extraction of current in the external electrical circuit (Tonui et al. 2018).

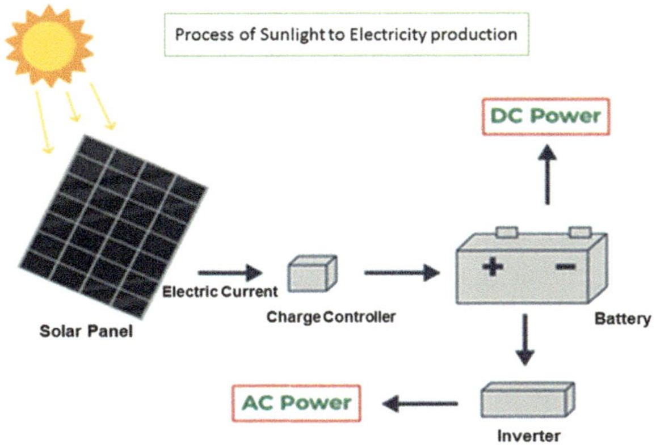

Fig. 11.1 Mechanism of energy generation by solar cell

- Absorption of Light
- Generation of Charge Carriers
- Separation of Charge Carriers
- Extraction of Current

The work toward the efficiency summaries the different types of solar cell in this book. The design and materials utilization in fabrication process depends on the nature of materials. The research directed towards to develop the different type of solar cells such as silicon solar cell, organic solar cell, inorganic solar cell, perovskite solar cell, and thin-film solar cell (Ali et al. 2018; Ananthakumar et al. 2019; Tonui et al. 2018).

Importance of Perovskite Materials for Solar Cell

Perovskite solar cells are capable type of new generation of photovoltaics cell that offers the high efficiency, low production costs, and widespread applications. The unique features are:

- **High Efficiency**: Perovskite-based solar cells rapidly increase the efficiency over a quite short period.
- **Low production cost**: The solution processes are used for the production of perovskite-based solar cells which can be cheaper and less energy-intensive than the production methods for traditional silicon-based and other cells.
- **Tunable Bandgap**: The other benefit of perovskite materials is tunable bandgap that can be modifying by the variation of composition which is the key factor for production of tandem solar cells. A thin layer of perovskite is combined with other materials in the structure of tandem cells, which is capable to capture more solar spectrum and increasing the overall efficiency.

- **Flexibility**: The flexible and lightweight substrates are making the perovskite solar cell is make them attractive and building-integrated photovoltaics.
- **Stability Challenges**: The long-term stability is the drawback and challenges of the perovskite materials. The perovskite materials are sensitive to moisture, heat, and oxygen, which can be degrade the materials and decrease the performance of solar cell. However, stability, scalability, and ecological concerns need to be addressed before they can fully replace or complement traditional solar technologies in the energy market.

Summary and Scope of the Chapter

In summary, this chapter discusses the overview of solar cell. The importance and demand of perovskite solar cell technology as renewable energy source to generate the electrical energy with future goal worldwide is summaries. As the demand for clean, renewable energy continues to surge, the future of solar cells holds immense promise in revolutionizing energy production. The future prospects of solar cells development using solar energy and their opportunity and possible challenges during the large-scale energy production is discuss. The pathway to achieve the de-carbonization which require to accelerate the clean energy deployment and sustainable goal up to 2035 has been studied. The review of research work done to improve the solar cell efficiency and future scope of work in milestone to design the solar cell have been summarized. The future challenges and their possible overcome to resolve the issues in large-scale production of energy by the solar cells as photovoltaic (PV) cells have been discussed. The next-generation PV cell and technological challenges deterring the large-scale development of solar cells including intermittency, grid integration, storage solutions, and material sustainability are also discussing. The supportive policies, market incentive, and demand for the adoption of solar energy at large scale have been emphasized. The future scope of solar energy and possible solution of the challenges are described in detail.

Solar Cell Technology

(a) **Emerging solar cell technology**

Solar cell technology booming because of increase in the demand of energy and electricity in present. It has been reported that, in 2022, increased the generation of the solar PV to achieve a record 270 TWh (up 26%) which was reaching almost 1300 TWh. This generation growth rate matches the level envisaged from 2023 to 2030 in the Net Zero Emissions by 2050 Scenario at worldwide (Ahmadi et al. 2018; Bouich et al. 2023) The economic attractiveness continuous increases the growth of PV, massive development in the supply chain and increasing policy support, especially in China, the USA, the European Union, and India, are expected to further accelerate capacity growth in the coming years (Choubey et al. 2004). The tracking status of solar photovoltaic upgraded in 2023 from "more effort needed" to "on track". Maintaining a generation growth rate aligned with the net zero scenario will require reaching annual capacity additions that are close to three times higher than those of 2022 until 2030. Achieving this will require continuous policy ambition and effort from both public and private stakeholders (Kurokawa n.d.), especially in the areas of grid integration and in addressing policy, regulation, and financing challenges.

The sustainable goal to achieve the energy generation by 2028 using solar cell technology is shown in Fig. 11.2. As comparative other renewable energy production from other sources, such as solar cell, wind power and hydrothermal, bioenergy, the solar energy is most prominent with perovskite materials, which demonstrate that the energy production through the perovskite materials in solar cell is drastically changes continuously due to available source of energy and cost effective than other renewable energy sources and materials (Østergaard et al. 2022). Worldwide it shows the increase in demand. Indian government changes their market policy and promote to development of solar cell use to production of electrical energy at lower level (Güney 2019). The cost comparative chart at worldwide using perovskite solar cell with year wise is shown in Fig. 11.3. It was observed at the initial stage, the cost of energy production through the solar was higher. According to the IEA report, in 2009, approximate cost of energy production through solar cell was 275 USD/MWh that was very high as comparative to others, whereas in 2023, it reduces to 60USD/MWh (IEA 2023a, b). This improvement done because of the continue promoting the research in this direction worldwide (Garg and Ratnesh 2022; Adeyinka et al. 2023; Ardani et al. 2021; Güney 2019; Singh et al. 2021). Research toward the direction to achieve the objective to improve the cost effective so that may develop at large scale with minimal cost with improved efficiency of solar cell. These practices make directive toward the advancements in research which indicate promising future materials for solar energy (Carlé et al. 2017) that may be making it a key component in the global transition to sustainable energy sources. The future trend in solar technology is explain next section.

(b) **Future trends in perovskite solar cell technology**

The future trend of perovskite solar cell looks bright to continue growth to produce the energy with innovation. In innovative point of view, two classes of technology considered in them first are to converting the solar energy into electricity using photovoltaics (PV) and concentrated solar power also known as solar thermal in their current and conceivable future forms (Mamodiya 2019; Sopian et al. 2017). The technologies will dominate in these two classes of solar-powered generation and target to achieve the green energy goal up to 2050 (Gerold and Antrekowitsch 2024; Madon et al. 1991; Wolf et al. 1993). In contrast to some, earlier there are no forecasts for two reasons such as (Gurion n.d.)

1. Expanding the solar industry from its relatively tiny current scale that may produce changes we do not invented to be able to expect today.

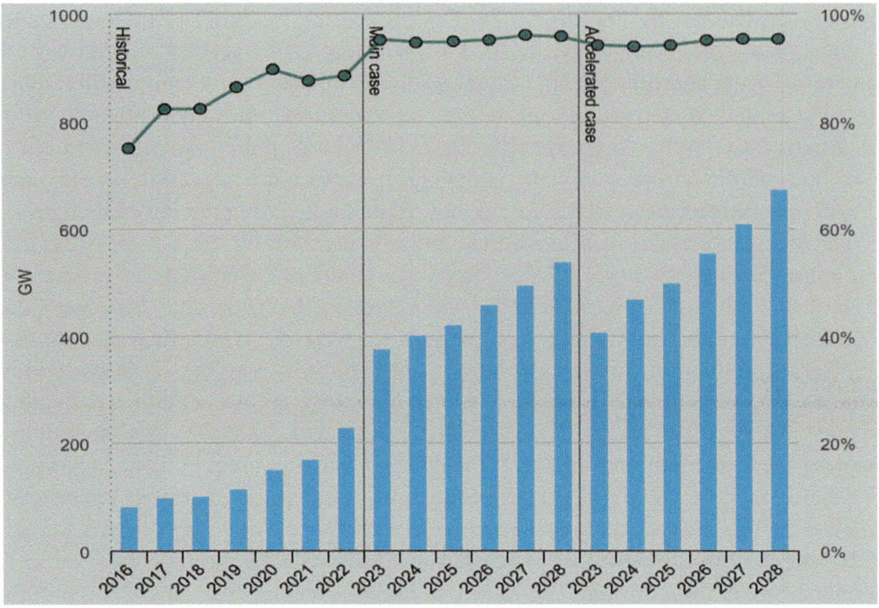

Fig. 11.2 Energy generation trend to achieve the goal by year wise using solar technology (IEA 2023a, b)

Fig. 11.3 Cost comparative in USD/MWh for energy production using perovskite solar cell (IEA 2023a, b)

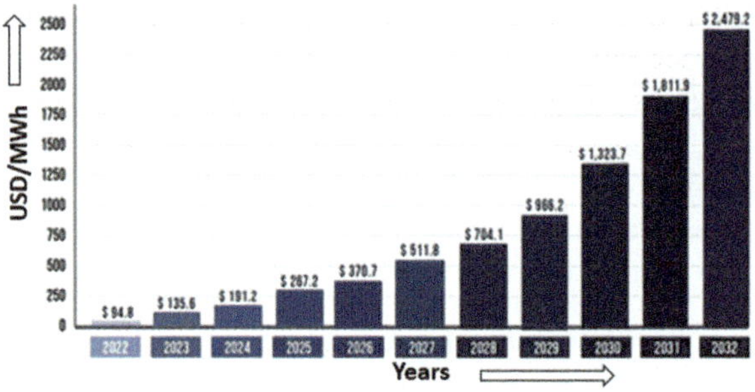

2. Deployment of the future solar energy will depend heavily on uncertain future market conditions and public policies.

The public policies including aimed at mitigating global climate change. The energy supply facilities typically last several decades. The trends and developments of the solar cell that are prospective to shape the future solar energy have been summarized as (VR 2023):

- **Increasing adoption**: To increase the source of electricity as solar power to promote more to individuals and businesses widely. Aware about the environmental benefits and fall the cost of solar technology to widely recognized.

Continue grow rapidly to use of solar energy, setting ambitious targets for the deployment of solar power at large scale.

- **Innovations in technology**: Increase the innovation in solar power production field at solar industries to develop with new technologies. The technology aims to increase the efficiency of solar cell, the use of nanomaterials as most promising way to developments to increase the efficiency of solar cells. The development of transparent solar panels that can be integrated into windows, and monitoring to optimize the performance of solar systems.
- **Growing role in the transportation sector**: The development of the solar-powered vehicles or solar power electric vehicles that may expected to increase the requirement of the use of solar energy in the transportation sector with the use of solar panels.
- **Increased storage capacity**: The main challenges faced as storage of solar energy because it is an intermittent source of electric energy. The production of energy varies with the weather condition and time of the day. There is required to develop the technology for the storage and battery technology, which can overcome this challenge by allowing stored the excess solar energy and used when needed.
- **Policy and regulatory support**: To increase the adoption of solar energy needs to modify the government's policies around the world. These include to providing financial incentives for solar system deployment for set the target of renewable energy and also encourage the use of solar energy ("Global-market-outlook-for-solar-power-2023–2027" 2024).

(c) **Solar Energy Predictions**

The precise predictions about the future of perovskite-based solar energy are difficult, but it plays a crucial role in the global energy production. Future trends suggest that perovskite solar energy will continue to grow in the coming years (Ali et al. 2018). As directed in the innovation toward the PV resulting the cost of solar cell rapidly decreases which makes its cost-competitive form of energy. Additionally, the prediction may be influenced by the technology advancement that made it possible to produce solar panels with higher efficiency, means they are able to generate more electricity from the same amount of sunlight (Mamodiya 2019). At the level of national and international level offers, a number of initiatives and policies programs that are designed to promote and encourage the adoption of solar energy establish the renewable energy targets (Project 2024). However, the growth rate will depend on the factors and is difficult to predict with certainty (Soonmin et al. 2023). At industrial level, solar energy trends are currently shaping the industry by the following advancement at the large-scale production such as

- **Reducing the cost of solar photovoltaic (PV) panels**: Advancement focus to reducing the cost of solar PV panels for economics point of view, which makes solar energy production more cost-competitive with other forms of energy (Kabeyi and Olanrewaju 2023).
- **Increase the efficiency of solar panels**: The efficiency of solar panel needs to advancement for the make solar panels become more efficient to convert the

sunlight into electricity (Ahmadi et al. 2018). This may help to reduce the overall cost of a solar energy system.

- **Growth of utility-scale solar projects**: To generate the more significant amount of electricity, future trend towards the development of large-scale solar projects. Such projects developed by the utility companies and connected to the grid.
- **Increasing adoption of solar energy**: Encourage the adoption of solar energy by the businesses and institutions that are turning to solar energy way to reduce the energy costs and achieve the zero-point carbonization to make eco-friendly environment.
- **Growth of the rooftop solar market**: The rooftop solar market makes aware about the cost of energy, this trend expected to continue for benefits of solar energy cost continues to decline. That has been includes the installation of solar panels on the roofs of homes and other buildings.
- **Rise of solar energy storage**: The major issues is storage problems regarding this future trends to increase the use of solar energy, also growing the efficient and cost-effective ways to store the excess energy generated by solar panels. This can be improving the energy storage and widely available.

Future Prospects and Challenges in Perovskite Solar Cell

Advancements in Materials

The prediction of solar energy marks the pathway of future trends to development of solar energy by the technological advancement that plays a crucial role to achieve the sustainable goal of energy. In the recent years, there have been significant advancements in the field of solar energy. Some of the utmost prominent are includes: More efficient solar panel development requires to enhancement in the efficient solar panel that may capable to convert the solar energy into electricity. The same size of solar panel produces more electricity that may reduce the cost. For improving the efficiency, researchers investigate the new sustainable materials that may able to enhance the conversion of sunlight to electricity. Explore the new materials improve their efficiency and reduce their costs. The research target to develop the technology makes more efficient, affordable, and versatile solar cell (Nozik et al. 2010; Dambhare et al. 2021). Some of the advancements in solar cell to overcoming existing challenges are summarized as:

- **Perovskite Solar Cells**—Perovskite solar cells technology are more attractive due to high absorption coefficient that may capable to rapidly increase in efficiency which reaching over 25% in laboratory settings. The stability and longevity of materials use in perovskite make them use as commercial at large-scale energy production. Further combining the perovskite with silicon solar cells in tandem structures of a single junction cells can able to improve the efficiency up to 30%

(Djurišić et al. 2017; Gerold and Antrekowitsch 2024; Jang et al. 2016; Tonui et al. 2018).

- **Silicon Photovoltaics**—The silicon crystal-based photovoltaics cell such as bifacial solar cells both sides are capable to capturing the sunlight that increasing the probability of energy output using diffusion and reflection from the surface. The heterojunction made by the combining the layers of amorphous silicon with crystalline silicon to improve the efficiency with lower temperature coefficient (Ali et al. 2018; Fraas and Partain 2010).
- **Organic Photovoltaics cell**—The organic photovoltaic cells have the advantage of efficiency gain with flexible, lightweight, and cheaper. The advancement in organic cell is to improve the stability with performance (Hoppe and Sariciftci 2004).
- **Quantum dot-based solar cells**—The quantum dots are efficient to moderate the light absorbance property of materials which improve the efficiency of solar cell. Lead-free solar cells are most prominent in quantum dots to develop the next-generation solar cell (Ananthakumar et al. 2019; Sharma et al. 2021; Singh et al. 2021).
- **Dye-Sensitized perovskite solar cells**—These types of solar cells are relatively lower cost and capable to work in presence of low sunlight that make it suitable to produce energy generation indoor condition. Research going to develop the electrolytes and dye materials for improve the stability and efficiency of dye-synthesized solar cell (Kumavat et al. 2017; Sutthana et al. 2016).
- **Thin-Film or layer-based solar cells**—Some of the materials such as cadmium telluride, copper indium gallium selenide, doped TiO_2 thin films are offered to achieve the high efficiency with simpler manufacturing process (Adeyinka et al. 2023; Chopra et al. 2004; Efaz et al. 2021; Singh et al. 2021).

Generation of Solar Cell Development

The development of PV cell generation is distributed in four groups; the details of different generation is already discussed in previous chapters. The requirement in improving the efficiency makes the significant changes in design and develop or discover such kind of materials which may enhance the use of solar cell at large-scale production (Pastuszak 2022).

First-Generation Solar Cells: The crystalline silicon-based solar cells are introduced in the first generation group of solar cell. In present, solar power markets based on silicon crystalline solar cell are dominates around 90% due to high efficiency. The average efficiency of single-crystalline solar cell is around 25% (Thick single-crystalline concentrator has 27.6%, single-crystalline non-concentrator has 26.1% and multi-crystalline-based solar cell has 23.3% and crystalline thin films have the efficiency around 21.2%) (Sopian et al. 2017; Yamaguchi et al. 2018) although, single-crystalline solar cell has high efficiency but due to high manufacturing cost make its more expansive. Because of this, research attempts develop low-cost fabrication technology of new materials for solar cell (Lu 2023).

Second-Generation Solar Cells: To overcome the disadvantage of first generation of solar cell developed the thin film-based solar cell that is referred as second generation of solar cell. The fabrication cost of second generation solar cell are much lower and cheaper as compared to the manufacturing of first generation. The amorphous silicon, copper indium gallium selenide (CIGS), and cadmium telluride (CdTe) are the most common materials were used in development of second-generation type of solar cell (Fraas and Partain 2010; Palm et al. 2004). The advantage of thin films solar cell is that less mass of materials is required for thin films so that reduce the amount of materials. The higher absorption coefficient of thin films than crystalline silicon makes its superior to gather the large amount of photons and increase the efficiency. It has been reported the efficiencies of CIGS-based solar cell have around 23.4% and CdTe thin films solar cell have around 22.1% (Singh et al. 2021; Tang et al. 2017). However, the second-generation solar cell has the advantage of lower cost but the poor performance as compared to first-generation solar cell, marks its limited in future use. Therefore, research focuses to advancement in materials with high efficiency and lower cost (Palm et al. 2004).

Third Generation solar cell: The objective in research to develop the third generation of solar cell to achieve the cost reduction and high efficiency of solar cell. In this emerging technology, the dye-synthesized solar cell, organic solar cell, and perovskite-based solar cell have been developed. The semiconductor-based technology improved the drawback of first and second generation of solar cell. The novel approach focus to use the semiconductor nanoparticles because at nanoscale semiconductor capable to produce the high current and higher ionization rates to generate more power. These devices are thermodynamic stable and conversion efficiency is high enough and varies between 31 and 40% which termed as Shockley–Queisser efficiency limit that can be overcome from nanoparticles of semiconductor (Ananthakumar et al. 2019; Halim 2013; Mariotti et al. 2020). Many researchers have been reported new noel materials that can be deployed in fabrication of solar cell. In the third generation solar cell, quantum dot sensitized solar cell (Nozik et al. 2010; Halim 2013), dye synthesized solar cell (Halim 2013; Kumavat et al. 2017), hybrid perovskite (Ananthakumar et al. 2019; Tang et al. 2017; Tessler and Vaynzof 2018), and organic solar cell are included (Green 2002). The TiO_2 nanoparticle layer used for sensitized by ruthenium dye in the dye sensitized solar cell (Aboulouard et al. 2020; Hardani et al. 2021; Furugori et al. 2017) and semiconductor nanoparticle are used as quantum dot that utilized in quantum dot sensitized solar cell (Aboulouard et al. 2020). The nanoparticles can enhance the optical absorbance and increase the efficiency of conversion of light to solar power. The reported efficiency high for perovskite solar cell is around 23% as high in compared to organic and DSSC solar cell. However, more research is required to achieve the higher efficiency at large scale. The research is ongoing to develop part of solar cell, high thermal stable electrode, and electrolytes that will lead to high efficiency and lower cost (Al-Ezzi and

Ansari 2022; Ali et al. 2018; Vigil-Galán et al. 2015). The recently research also emerged with develop organic–inorganic perovskite hybrid solar cell that increased efficiency up to 20% which is higher than organic solar cell (14%) (Dubey et al. 2017). In addition, the tandem solar cell and multiband solar cell also emerged in the third-generation solar cell (Green 2002; Yamaguchi et al. 2018).

The fabrication technology groom to develop the new materials that can improve the solar cell efficiency and reduce cost. In future, the advancement in manufacturing process aims to overcome the challenges which leading the efficiency and cost as well as ethically ecofriendly. The fourth-generation solar cell may include all the possible resolution of issues raised during the process of previous generation (Rona et al. 2024; Rehman et al. 2023). In recent research in solar cell, develop the advanced fabrication technology such as bifacial solar cell and tandem solar cell (You et al. 2013; Zhang 2020). In case of bifacial solar cell, can capable to absorb the light from both side of panel and increase the probability of absorbance and enhance the output energy (Tang et al. 2017; Ameri et al. 2009; Wali et al. 2018), whereas the tandem solar cell is the mixture of materials (hybrid) that can capture all kind of wavelength of light and produce the photoelectron to produce the solar power and enhance the efficiency and also can be work in absence or low sunlight (Ameri et al. 2009). Rather than solar cell materials, the advanced materials manufacturing process includes the antireflection coating to reduce the light loss by reflection (Askar et al. 2013), flexible, lightweight and lower production cost of solar panel, also using the nanomaterials for enhance the absorbance behavior at nanostructure which leading to higher efficiency as compared to bulk (Askar et al. 2013; Patidar et al. 2020; Sarkın et al. 2020). Research is also focused on eco-friendly and non-toxic materials to develop the solar cells with sustainable and efficient also may recycle the materials for reuse to develop the materials of solar panel that may also reduce the environmental impact (Nair et al. 2020).

Characteristic Properties of Perovskite Materials

Perovskite materials have emerged as a key focus due to their potential in various energy-related applications, particularly solar cells. The unique crystal structure of perovskites offers a versatile platform for engineering the materials with tailored properties. Therefore, perovskite materials hold tremendous potential applications for solar cells being the primary focus due to their efficiency and ease of fabrication. However, key challenges like stability, lead toxicity, and scalability must be addressed for perovskites to achieve widespread commercial success

- Stability: The long-term operational stability is a major challenge with the other solar cell, perovskite materials has the ability to long life operation stability. Future research must focus on improving resistance to environmental factors like humidity, oxygen, UV light, and thermal stresses on the materials to increase the lifetime of perovskite devices are needed.

- Environmental and Toxicity Concerns: Although lead perovskites are currently the most efficient, they pose environmental hazards. Continued research into lead-free alternatives that maintain or exceed current performance is essential.
- Life cycle analysis and strategies for recycling or safely disposing of lead-based perovskite cells will become increasingly important as the technology matures.
- Scalability: For perovskites to compete with established technologies like silicon, the manufacturing processes must be scalable and cost-effective. This includes developing low-temperature, solution-based deposition techniques that can be applied to large-area devices.
- Interface Engineering: A critical area of research is the optimization of interfaces between the perovskite material and other layers (e.g., charge transport layers). Improving energy level alignment, reducing defect densities, and minimizing carrier recombination will further boost efficiency and stability.
- Lead Substitutes and Alloying: More effort is needed to find viable substitutes for lead that do not compromise the electronic properties of perovskites. Alloys and doped materials that maintain the same high mobility and low recombination rates must be developed.
- Reproducibility and Uniformity: Ensuring consistent performance across large-scale devices requires improvements in fabrication reproducibility. Issues like defect tolerance, grain boundary management, and surface passivation need better solutions to avoid degradation and enhance uniformity across devices and cost efficiency for widespread commercial adoption, perovskites fabrication is at lower cost as compared to other current technologies.

In present, there are many research going on to design and fabricate the perovskite material-based solar cell among some are discussed below.

Types of Solar Cell Based on Perovskite Materials

The solar cell development based on perovskite materials, the several types of solar cell based on their structure, materials used, and fabrication techniques. The efficiency of different perovskite solar cell is tabulated in Table 11.2. The types of PSC's depend on the perovskite layer that incorporated into the solar cell, the most common PSC's as;

1. **Mesoporous Perovskite Solar Cells—The mesoporous** PSCs include a mesoporous scaffold made of titanium dioxide (TiO_2), on which the perovskite material infiltrates. The mesoporous layer provides high surface area, improving the contact between the perovskite and the electron transport layer (ETL). Due to the mesoporous structure, the advantage of the solar cell has the high efficiency and better charge transport property. Major challenge in the mesoporous PSCs is the fabrication and stability issues due to the exposure of perovskite layer to moisture.

2. **Planar Perovskite Solar Cells** These PSCs the perovskite material is directly deposited onto a flat electron transport layer and do not use a mesoporous layer. Planar PSCs can be further divided into two types based on the charge transport materials:

    I. **n-i-p Structure (Normal)**: The perovskite layer is sandwiched between the electron transport layer (ETL) and the hole transport layer (HTL).
    II. **p-i-n Structure (Inverted)**: The positions of the electron and hole transport layers are inverted compared to the n-i-p structure. The perovskite is deposited on the HTL first, followed by the ETL.

The advantage of planar perovskite solar cell is to easier fabrication and higher reproducibility. The stability and less efficient in compared to mesoporous solar cell is the limitation of these planar solar cell.

The market size at the global target to increase the perovskite solar cell uses on the base of structure is listed in Table 11.1.

3. **Flexible Perovskite Solar Cells**: The flexible perovskite solar cells are manufactured on a flexible substrate such as plastic, PET or PEN. The structure of the PSC depending on the application. The advantages of these cell is the lightweight and flexible which makes them suitable and portable electronics, wearable devices, or even curved surfaces like building facades or vehicles. The flexibility of materials and long-term performance on flexible substrates are the major challenges with these solar cells.
4. **Tandem Perovskite Solar Cells**: The Tandem PSCs combine perovskite with another type of solar cell, such as silicon, CIGS (copper indium gallium selenide), or other perovskites. The two different materials absorb different parts of the solar spectrum, which improves overall efficiency. Based on the materials and structure, the tandem solar cell is classified as

    i. **Perovskite-Silicon Tandem**: Combines a perovskite cell with a traditional silicon cell. This is one of the most promising configurations, achieving efficiencies over 30%.
    ii. **Perovskite-Perovskite Tandem**: Uses two different perovskite layers with varied bandgaps to capture a wider range of the solar spectrum.

5. **Dye-Sensitized Perovskite Solar Cells**: In these type of solar cells, the dye-sensitized solar cell combine with the perovskite materials elements. A mesoporous scaffold is infiltrated with both a dye and perovskite, which act together to enhance light absorption and charge transport. The lower cost and flexible application are the key factor and advantages of these kind of solar cell.
6. **Hybrid Perovskite Solar Cells**: The hybrid PSCs involve a combination of organic and inorganic perovskites, which can enhance specific properties such as flexibility, stability, or efficiency. The tenability property of these material is the major advantages that allowing for customization for different applications.

Table 11.1 Global perovskite solar cell market of planar and mesoporous perovskite solar cell (USD million)

PSC's structure	2022	2023	2027	2032
Planar perovskite solar cells	46.0	65.9	250.3	1220.8
Mesoporous perovskite solar cells	48.8	69.6	261.5	1258.4

7. **Low-Dimensional Perovskite Solar Cells**: These types of solar cells incorporate both 2D (layered) and 3D perovskite structures. The 2D perovskite layers typically provide better moisture resistance, while the 3D structure enables efficient charge transport and absorption. Combining the best of both dimensions improves moisture resistance and stability also shows good efficiency. The optimization condition for fabrication is the big challenge for designing the low-dimensional perovskite solar cell (Table 11.2).

Challenges in Energy Production at Large Scale Using Perovskite Solar Cell

The challenges occur in the large energy production using perovskite solar cell are summarized as below. The problems faced while using the PSC-based solar cell are shown in Fig. 11.4.

(a) **Utility-scale solar projects:** Advancement in the large-scale solar energy production projects and development of solar farms that may efficient to generate significant amount of electricity. Advance technology in projects often develops the connectivity to the companies and connected to the grid. Technological advancements for solar cell projects of global shift toward the renewable energy and innovative financing models that are driving the growth of projects significantly contribute to the global energy (Asgher et al. 2018; Wan et al. 2016). The projects focus to develop the solar farms and produce the tracking system that allow to follow the path of sun so panel may capture maximum light throughout the day. For grid projects, the invertor technology improves the efficiency and reliability of converting DC power into AC power for the grid. Smart Grid Technology systems enable better integration of solar power, balancing supply and demand, and improving grid stability the demand response programs match solar power availability with consumption, that may help to optimize the use of energy and reduce strain on the grid (Hossain et al. 2016; International Renewable Energy Agency (IRENA) 2019; Wan et al. 2016; Zhang et al. 2017). At world level some of the notable solar power projects are;

- Noor Complex (Morocco): One of the world's largest CSP installations.
- Bhadla Solar Park (India): Among the largest solar farms, with a capacity of over 2 GW.

Table 11.2 Efficiency of various perovskite solar cells and their structures

Perovskite solar cell	Power conversion efficiency (PCE, %)	Notable features	References
Single-junction perovskite solar cells	23–26	PSC uses materials like methyl ammonium lead halide. Stability and scalability are ongoing challenges	Jiang et al. (2023) and Wu et al. (2021)
Tandem (perovskite-silicon) solar cells	28–32	Combination of silicon and perovskite layers, breaking the silicon efficiency limit	Wali et al. (2018) and Zhang (2020)
Flexible perovskite solar cells	17–21	Used in wearable devices and BIPV applications. Flexible solar cell but has lower efficiency than rigid cells	Jung et al. (2019)
Perovskite-CIGS (copper indium gallium selenide) tandem cells	24–26	Still in research stages. Combines perovskite thin films with CIGS	Shrivastav et al. (2023)
Quantum dot sensitized perovskite solar cells	16–18	Quantum dots enhance light absorption in perovskite layers, improving overall solar efficiency	Pan et al. (2018)
Dye-sensitized perovskite solar cells	12–16	Hybrid configuration of perovskites with dye-sensitized solar cells, low-cost potential	Aboulouard et al. (2020)
Perovskite organic hybrid cells	12–15	Combines perovskite and organic photovoltaic materials, focusing on flexible applications	Jeon et al. (2014)
All-perovskite tandem solar cells	25–27	Tandem cells using multiple perovskite materials for different parts of the solar spectrum	Malii et al. (2024)

- Topaz Solar Farm (USA): Major PV solar farm with a capacity of 550 MW.

Utilization of land that is dual-use as agri-voltaic systems in which solar panels are installed above agricultural land which allows the simultaneous production of crops and electricity. Government policies and incentives also promoting

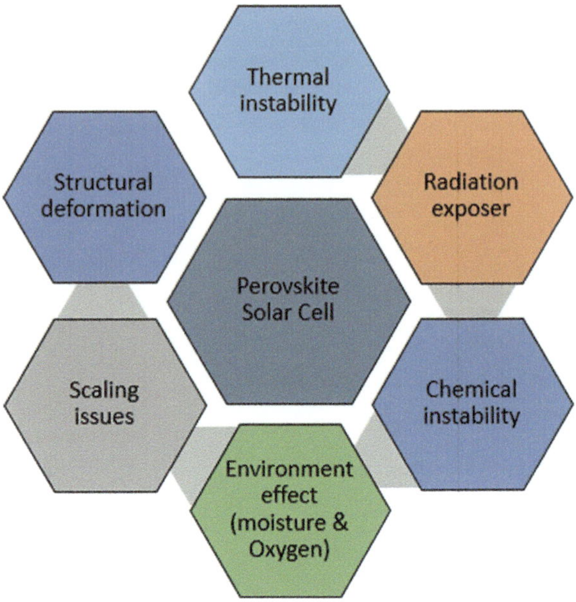

Fig. 11.4 Challenges with the perovskite-based solar cell (Chauhan et al. 2024) (Copyright 2024, Reproduced with permission from Elsevier)

the utility scale solar projects. The financial support and stability make large project sustainable by long term contracts between solar project developers and utility companies. Also encourages the collaborations between government entities and private companies for help in securing the necessary funding for large-scale projects. The large-scale production of solar energy to leading the market to deployment of solar project due to favorable policies and substantial investments. Although the grid capacity upgradation and infrastructure are the significant challenges in future development of solar cell projects (Sharma 2018; Zhang et al. 2017).

(b) **Storage of solar energy**: The storage issue is the big challenge in future that needs to advancement, as increase in the demand of solar energy also increases the efficient and cost-effective storage devices to store the excess solar energy generated by the solar panels. In these trends, researchers will do the practices to improve the storage, develop the solar thermal storage and batteries with new technology. The significant advance is required for the storage of excess of solar energy. The enlargement of research will improve the incorporation of energy storage systems with solar panels. The battery technology is one of the best option in future to store the energy. The advanced battery technologies will be continued grow specially to design of solar energy storage, such as lithium-ion, K-ion and Na-ion, and polymer-gel electrolyte-based batteries (Madhani et al. 2023). The advantage of batteries is offer to higher energy density, longer lifespan, and improved discharging and charging capabilities, allowing for more efficient utilization of stored solar energy. Advanced battery management provides optimize better control of energy storage. Such type of systems permits to maximize the demand of use of solar energy storage and grid conditions, ultimately increased energy efficiency. In this direction, adding large-scale battery storage with solar farms tolerates for better battery management, objective to provide the power during non-sunny periods.

(c) **Land scale problem for large-scale production**: Since the solar power is more dominate and viable method to generate the energy at large scale with lower cost as compared to other renewable energy source. Although production at large scale is still challenges that needs to address in future research. The aim of the present to technological improvement may be offered to develop the resources at large-scale production which also help to promote the solar power at worldwide. In small or low dimensional solar cell devices maximum number of cations are replaced by the large ligands (Gurion n.d.). This exhibits the higher stability over long-term operation against moisture and light soaking, as compared with large perovskite panel devices (Alshahrani et al. 2019; Gurion n.d.; Cai et al. 2017). However, the large ligands hamper charge-transfer transport and sacrifice the efficiency. The conjugated and shorter chain cations potentially increase the efficiency of cell without sacrificing stability by large ligands (Hooda and Rathore 2024; Kurokawa n.d.). Therefore, the manufacturing and fabrication challenge are there for large energy production further challenge as the economical, storage, market, logistical, and technological are also included.

At the large-scale electricity production, for the implement the grand solar plan, two aspects are there as

a. Realize the investment scale that needs to replace the fossil fueled world for energy at large by solar power.
b. Other one is to create advancement in solar technology.

The employment of PV technology to generate electricity about the 10% efficiency is about 2000 kWh/m^2, such type of desert region required and the production of electricity is equivalent to 100B square of solar panel and three to four times larger land area is required. The availability of larger land is likely political complication to projects of grand solar panel at worldwide. In India, the shortage of land for desired weather condition for solar plant has the big challenges. A single political system should be involved to decide the policies and regulation to solar power plant for providing electricity generated through the solar plant to the public sector. But the situation is more complicated to design and develop the power plant (Post and Moore 2008). The far removal of desert land would need to come to agreement. There are several challenges in production of energy at large scale from solar cell are summarized as;

(d) **Efficiency Conversion**: At laboratory scale the theoretical efficiency of energy from solar cell is to be found around 35–40%, while at large-scale energy production efficiency is too difference and obtained around 15–22% only. The energy conversion efficiency makes the limited to use of solar cell at large-scale. The amount of energy is limited that can be collected from the available sunlight. The other most challenging factor is degradation of solar panel that causes the poor efficiency (Kudelas et al. 2019). At large scale, it is difficult to resolve the conversion issues due to degradation, therefore required to further research to resolve the issue of conversion efficiency and degradation of panel (Anusuya et al. 2023; Greijer et al. 2001).

(e) **Environmental impact**: The environmental effects such as weather, temperature, and geographic location are varying that may be also effected on the production of energy from the solar cell. Since the solar energy varies greatly depending on weather and location condition. The cloudy weather or less sunlight condition is less suitable for the production of energy that make the limiting applicability of the solar energy technology. Also the temperature sensitivity causes the lose efficiency of panel for energy production. The presence of sunlight variability or day–night cycle make challenges to continuous generation of energy and power supply (Urbina 2020). Additional, the chemicals involved in manufacturing of solar cell are hazardous, so recycling of solar cell is challenging task that leading the hazards of environment (Celik et al. 2017). Since the production of solar energy is intermittent, that is varying with the weather conditions and sunlight. So that the produced energy needs to be stored in the storage device to long use of energy. But the limited storage capacity makes expansive. In other, the grid is required for the large amount of solar power. The challenges come in handle and maintain the grid for solar energy (Ardani et al. 2021; International Renewable Energy Agency (IRENA) 2019; Wan et al. 2016). The operation of grid deals in transmission system with the fluctuations of frequency and voltage so that the components of the systems are not affected. The grid performed through the subsidiary systems which can generate the active or reactive power into the grid to balance the actual and predicted power. It can maintain the frequency and voltage levels of grid. It is necessary that the demand of electricity from the consumption side is matching the generation and load all the time. The energy generation from the solar cell can be predictable but it cannot be controllable thus have the degree of variability. Although the grid has the low level of variability use in the form of intermittent at low level of penetration because of the integration challenges on the device and local-grid (Gerold and Antrekowitsch 2024). On the other hand, it is quite challenging to manage the level of grid (Fouad et al. 2017), therefore it is needed to overcome the strategies and technology for action of grid level.

(f) **Cost or economic factors**: The installation cost of solar cell is relatively high. At large-scale solar power has high initial cost. So that sufficient investment is required for the solar power establishment that can be create the barrier in deployment of solar plant at large scale. The economic factor influences the development of solar projects at large scale. However, government policies for providing the incentive and subsidies are limited so it needs to change the policy term and resolve the barrier of deployment of solar projects. The cost of raw materials also enhanced the production cost of energy. The energy production at large scale is required 3B square meter area of solar panel to overcome the energy demand. At present for large plant, the cost of per square meter solar panel area is around US $670. At global infrastructure estimated cost is around US $ 2T per year. Thus cost factor is about 17 larger than conventional growth without including the cost of transmission and storage (Rong et al. 2018). The large storage capacity of device needs to require into the account weather forecast and electricity tariffs throughout the day. But it could be utility to use the opportunity of part of device in night has more efficient in base line generation that is possible without large amount of storage capacity (Ahmadi et al. 2018; Zhang et al. 2022). Thus, the total cost may be effected, thus storage needs lower cost, long life, and high efficient to storage. Therefore, the development of such kind of storage device will require more research and innovation to overcome the storage problem throughout the day and year.

- **Market**: The global market of energy production through the solar panel is also influenced the market fluctuation and trade policies. The raw materials demand and supply for solar plants, dynamical create the issues and lead to increase the cost. This reduces the installation demand that affecting the growth of solar industries market.
- **Effect of technology**: The advancement required in materials research, but the limited availability of raw materials could limit the deployment of large-scale production of solar plants. The conversion of light is possible only in the presence of sunlight, PV cells gets lost during the less sunlight or gets reflected into the heat instead of energy conversion during the process. At 30° of typical latitude of desert, the sunlight is available for 10 h and night would be around 14 h throughout the year. Such a day fully overcast, so that it would be necessary to charge the battery in night-time using fossil energy sources. Therefore, 14 GWh storage capacity of battery is required for this function (70% round trip); thus, the technological challenges may effect on the production and transmission of solar plant at large-scale production of electricity. The advancement is required in solar technology.
- **Regulatory and permitting hurdles**: At the large scale, the regulations of industrial solar power plants are varying from region to region and can be created the complication (Oduro et al. 2024). Central and state government norms, rules, and approval procedures are much complicate that develop the challenge the diverse set of policies and timeline that could change the policy for project viability.
- **Transporting and Distribution:** The electricity transportation from the large scale from the power grid to urban areas is difficult. At long distance, the transmission of AC power through the conventional source at high voltage in order to reduce the joule lose. Because of corona discharge to the ground the joule loses may dominate resulting the raise the voltage too far. The Joule loses is reduced to zero by employment of superconducting materials. However, superconductor requires the cryogenic cooling to liquid nitrogen temperature (− 200 °C). So that required to research to development such kind of UHV DC for transmitting the power at large distance. In addition, they lose the energy significantly and time consuming in build the technology for large-scale energy transmission; also the infrastructure is also a big challenge in large-scale production (Alshahrani et al. 2019; Cai et al. 2017).

Thus, the technology required to develop the low cost, high efficient transmission method, and solar plant at large scale. The hurdles are come in the technology development that needs to be addresses. Therefore, in next section discuss resolution of the challenges and overcome in technology advancement at large-scale energy production by solar system.

Overcome the Challenges in Energy Production

In present, ongoing research aims to continually improving the efficiency with resolve and overcome the solution of the challenges in large-scale energy production. The development of technology directed toward the changes in solar technology at the large-scale energy production. The resolution of issues in large-scale energy production including intermittency, grid integration, storage solutions, and material sustainability, policies, market incentive, and demand for the adoption of solar energy at large scale emphasize (Siegler et al. 2022). To overcome the issues, raise in solar cell at large-scale production, there are main three areas are focused:

(a) Advancement in materials for surface modification use for solar panel
(b) Energy storage and energy management
(c) Solar power generation with hybrid system technology

The technological innovation includes the advancement for incredulous the issues such as efficiency at large-scale solar panel, cost, availability of suitable materials and surface modifications and scalability. The possible resolution done by the researcher are summarized as:

Surface texturing and modification—Innovation involved the development of the such kind of anti-reflecting coating and texturing the surface for minimize the loss of light due to reflection from the surface during the PV process of energy production (Fallahazad et al. 2018; Park et al. 2009). Since the nature of surface can restrict the optical properties due to recombination of generated charge carriers and reflection of light (Radfar et al. 2020). The surface texturing reduces the reflection and can allow the maximum absorbance of sunlight (Abdullah et al. 2016) results in increase in the performance and efficiency of energy production. The open circuit voltage and fill factor of solar cell depends on the thickness of layer and current density affects the device performance. The light trapping from the solar cell is also influenced by the surface modification. It has been reported the two methods of modification surface as texturing and rear side polishing (Park et al. 2020). The surface texturing methods adopted according to the types of solar cell such as acid vapor texturing process used for the ZnO-based dye synthesized solar cell (Sutthana et al. 2016), chemical vapor texturing process adopted for the PSS thin film-based organic photovoltaic cell (Ritruksa et al. 2020), saw etching method in acidic solution was used for single crystalline silicon solar cell (Blakers et al. 2013).

Efficiency improvement—The research carried out in improvement of solar panel efficiency to conversion of sunlight into electricity by the advancement in new materials (Schneider et al. 2006). The different generation of solar cell described as earlier in the previous section which indicates the continuous research going on to produce the more efficient materials such as perovskite (Freiburg 2009), tandem cells, and quantum dot-based solar cell drastically boost the efficiency on conversion of light into electricity as compared to first generation (silicon-based) solar cell (Halim 2013; Nair et al. 2020). However, poor stability of single layer, the multijunction solar cells includes the different layer of materials that can capable to enhance the stability and capturing a border range of wavelength of light that can leading the efficiency and capable to capture in presence of less sunlight (Wang et al. 2022).

Reliability and durability of solar cells in harsh environments—The advancement focused on the designed the low-weight PV cell so can improve the mechanical properties of the solar cell. Due to thermal expansion, degradation of back-sheet of solar cell due to UV exposure, and long-term exposure in desert environment are the responsible for the failure of the solar cell (Adothu et al. 2024; Strevel et al. 2013). At large scale of panel back-sheet and encapsulate the surface that are thermally stable and free from the acidic group, the degradation model such as desert module application addressed the environmental safety through the aging tests (Lelièvre et al. 2022; Zhang et al. 1960). It has been reported that the double glass module can perform the better mechanical properties under various stress condition, humidity, thermal cycling, and mechanical loading and degradation (Shaker et al. 2023). The corrosion, moisture and harsh environment may also important factors that reduced the performance of panel at large scale so that technology includes the coating and corrosion resistant materials layer on the surface of solar panel, can protect from the external factors and improve the long life and mechanical properties of solar cell (Guerrero-Lemus et al. 2016). Such the compressive work needs to be done in the direction of reliability and durability of the cell that can improve the performance of solar panel at large scale and increase the efficiency of the cell.

Advanced research—The advanced work aims to resolve the challenges in develop the solar panel that can capable to work less intensity of light such kind of panel works indoor purpose and all weather condition that improve the efficiency. Other hand the design of bifacial solar panel can remove the issue the capturing efficiency of light since the both side of bifacial panel can absorb the sunlight. The characteristic feature of such solar cell is flexible and transparent and also has high mechanical properties to utilize at larger area to develop the long-time solar energy (Uematsu et al. 2003; Vimala et al. 2023). Tandem solar cell module can help to enhance the energy production and improve the efficiency.

Government Policies and Regulatory Framework

The government policies and regulation can promote the industries to establish the solar plant to produce the large-scale electricity. The large electricity production fulfils the requirement and demand of energy and also build up economical background and support to development of nation such opportunities favorable for the sustainability of mankind. Therefore, the government needs to support the new technology for the perspective of long-term development. But the adoption of new technology needs to promotion that is not easy. As new product and materials at initial stage are expansive and making difficult to use at large scale that unable to capture the market (Chen et al. 2023). The promoting of new technology at the market stage, government can reduce the consumer cost of services by the firm. The investment and usage subsidy both can alter the opportunity to use of new solar technology. The subsidy policy and regulation subject to a budget limitation that are depends on the objective function of government. The budget amount announced the interest of investment, since at low budget the investment subsidy and usage subsidy are equally effected. Therefore, the government should increase the financial support. The flexible incentives policies increase the investments in solar technology and their infrastructure (Silveira et al. 2024). The government rules and taxation approaches are play crucial play to promote the financial investments to adoption of new technology. The incentive approach and under one roof taxation policies frequently adjusted subsidy may be best practice for renewable sources (Kaplan Dönmez 2023). The practices to enhance the solar plant development at large scale the following point's needs are to be addressed, the international corporation and market scenario direct impact on the projection of projects at large level.

Government incentives and subsidies for solar energy deployment—The impact of government policies and subsidy in independent innovation at worldwide to achieving the net zero carbon and green energy economy. There are some factors that need to be practiced first, the harmonization is required between the central and local or state-level policies and regulations. Secondly, the efforts in advancement in research continued to develop the new technology that explore the cost-effective and environmental impact and can be recycling and disposal the perovskite PV panels make reuse the panel to make eco-friendly (Alshahrani et al. 2019; Goel 2016; Shukla et al. 2018). The collaborations are also promoting among the policymakers, researcher, stakeholder, and industry; they can facilitate sharing the knowledge and can exchange the technology at large scale in best practices. Such the management of PV panels will be contributing to the effective and sustainable. Government policies create a supportive regulatory framework, drive investments, and establish long-term goals for renewable energy adoption. Some of the government policies that support solar energy deployment such as tax Incentives, mandates for renewable energy production, fixed payments for solar energy producers, financial support for solar deployment, encouraging cleaner energy sources, supporting integration of solar power and public awareness, etc. Regulatory barriers in large-scale energy production, especially for technologies in perovskite solar cells can significantly delay or increase the cost of projects.

Market Projection—The market outlook is highly positive, with the potential for PSCs to capture a significant share of the global solar energy market by 2030. The commercial perovskite cells will achieve efficiencies over 30% in tandem configurations, driving even more interest and investment into the technology.

Conclusion and Future Outlook

In renewable energy sources, the solar energy has demonstrated immense potential application to use as production of electricity at large scale, and their prospects remain promising method for electricity generation with cost effective other than traditional and other renewable sources. The advancement and challenges are key shaping the landscape of solar energy production. The different generation of solar cells have been reported and research focus to improve the efficiency of solar cell. The new technology efficacy by enabling targeted and controlled parameters involve the development of solar cell. The perovskite solar cell with enhanced physical and mechanical properties sustained release capabilities for long term use as solar generation. The strategies for the development and can capable to resolve the issues of solar power plant for produce the large-scale electricity. The materials advancement, environment impact, energy storage devices, transmission of energy and long-time efficacy are the key factors and challenges in large-scale electricity production. The government policies and regulation play a crucial role in rooftop solar power and large solar plant at industrial level can lead to more effective strategies. The taxation and subsidy policies are significantly improving the development of solar plant. The future prospects include the development of new technology in advance materials such as bifacial solar cell and tandem solar cell, and quantum dot dye synthesized solar cell to improve the mechanical properties, long life and improve the efficiency of conversion the light into electricity. The issues or challenges in development of solar plant for large-scale energy production need to resolve. Overcoming the eco-friendly, cost effective, advance research and regulatory for the adoption of solar plant with long-term stability and scalability challenges will be critical for translating these prospects into future electricity production at large scale. Future prospects also involve the design of solar plant to resolution of the land scalability problems with efficiency. Overcoming large-scale production, reproducibility, and cost-effectiveness challenges will be crucial for the widespread adoption of solar energy at industry level for produce electricity.

In conclusion, the prospects for solar power plant for electricity development are promising, with significant potential for advancements in low cost energy production with high efficiency. The challenges such as efficiency, storage, transmission, cost, land scale and policies and regulation at large-scale electricity production using solar cell. However, eco-friendly, scalability, regulatory considerations, and manufacturing challenges need to be addressed for the successful integration of solar technology into practice. Continued research, collaboration between academia and industry, and regulatory frameworks will play vital roles in realizing the full potential to the development of electricity at large scale.

References

Abdullah MF, Alghoul MA, Naser H, Asim N, Ahmadi S, Yatim B, Sopian K (2016) Research and development efforts on texturization to reduce the optical losses at front surface of silicon solar cell. Renew Sustain Energy Rev 66:380–398. https://doi.org/10.1016/j.rser.2016.07.065

Aboulouard A, Gultekin B, Can M, Erol M, Jouaiti A, Elhadadi B, Demic S et al (2020) Dye sensitized solar cells based on titanium dioxide nanoparticles synthesized by flame spray pyrolysis and hydrothermal sol-gel methods: a comparative study on photovoltaic performances. J Market Res 9(2):1569–1577. https://doi.org/10.1016/j.jmrt.2019.11.083

Adeyinka AM, Mbelu OV, Adediji YB, Yahya DI (2023) A review of current trends in thin film solar cell technologies. Int J Energy Power Eng 17(1):1–10

Adothu B, Kumar S, John JJ, Oreski G, Mathiak G, Jäckel B, Alberts V, Jahangir JB, Alam MA, Gottschalg R (2024) Comprehensive review on performance, reliability, and roadmap of c-Si PV modules in desert climates: a proposal for improved testing standard. Appl Res. Retrieved from https://doi.org/10.1002/pip.3827

Ahmadi MH, Ghazvini M, Sadeghzadeh M, Alhuyi Nazari M, Kumar R, Naeimi A, Ming T (2018) Solar power technology for electricity generation: a critical review. Energy Sci Eng 6(5):340–361. https://doi.org/10.1002/ese3.239

Al-Ezzi AS, Ansari MNM (2022) Photovoltaic solar cells: a review. Appl Syst Innov 5(4):1–17. https://doi.org/10.3390/asi5040067

Ali G, Omar M, Khan AK, Faisal Nadeem M (2018) Recent challenges of solar cell technologies; a critical analysis. In: RAEE 2018—international symposium on recent advances in electrical engineering, pp 1–5. https://doi.org/10.1109/RAEE.2018.8706887

Alshahrani A, Omer S, Su Y, Mohamed E, Alotaibi S (2019) The technical challenges facing the integration of small-scale and large-scale PV systems into the grid: a critical review. Electronics (Switzerland) 8(12). https://doi.org/10.3390/electronics8121443

Ameri T, Dennler G, Lungenschmied C, Brabec CJ (2009) Organic tandem solar cells: a review. Energy Environ Sci 2(4):347–363

Ananthakumar S, Kumar JR, Babu SM (2019) Third-generation solar cells: concept, materials and performance—an overview.https://doi.org/10.1007/978-3-030-04474-9_7

Anusuya K, Vijayakumar K, Manikandan S (2023) From efficiency to eternity: a holistic review of photovoltaic panel degradation and end-of-life management. Sol Energy 265:112135. https://doi.org/10.1016/J.SOLENER.2023.112135

Ardani K, Denholm P, Mai T, Margolis R, Silverman T, Zuboy J (2021) Solar futures study. U S Department of Energy, (September), pp 1–279

Asgher U, Rasheed MB, Al-Sumaiti AS, Rahman AU, Ali I, Alzaidi A, Alamri A (2018) Smart energy optimization using heuristic algorithm in smart grid with integration of solar energy sources. Energies 11(12):1–26. https://doi.org/10.3390/en11123494

Askar K, Phillips BM, Fang Y, Choi B, Gozubenli N, Jiang P, Jiang B (2013) Self-assembled self-cleaning broadband anti-reflection coatings. Colloids Surf A 439:84–100. https://doi.org/10.1016/J.COLSURFA.2013.03.004

Blakers A, Zin N, McIntosh KR, Fong K (2013) High efficiency silicon solar cells. Energy Procedia 33:1–10. https://doi.org/10.1016/j.egypro.2013.05.033

Bouich A, Pradas IG, Khan MA, Khattak YH (2023) Opportunities, challenges, and future prospects of the solar cell market. Sustainability (Switzerland) 15(21):1–15. https://doi.org/10.3390/su152115445
Cai L, Liang L, Wu J, Ding B, Gao L, Fan B (2017) Large area perovskite solar cell module. J Semicond 38:014006. Retrieved from https://doi.org/10.1088/1674-4926/38/1/014006
Carlé JE, Helgesen M, Hagemann O, Hösel M, Heckler IM, Bundgaard E, Krebs FC et al (2017) Overcoming the scaling lag for polymer solar cells. Joule 1(2):274–289. https://doi.org/10.1016/j.joule.2017.08.002
Celik I, Phillips AB, Song Z, Yan Y, Ellingson RJ, Heben MJ, Apul D (2017) Environmental analysis of perovskites and other relevant solar cell technologies in a tandem configuration. Energy Environ Sci 10(9):1874–1884. Retrieved from https://doi.org/10.1039/C7EE01650F
Chauhan AK, Kumar P, Sharma SN (2024) Perovskite solar cells: past, present, and future. In: Photovoltaics beyond silicon: innovative materials, sustainable processing technologies, and novel device structures, pp 113–163. Retrieved from https://doi.org/10.1016/B978-0-323-90188-8.00015-4
Chen R, Meng Q, Yu JJ (2023) Optimal government incentives to improve the new technology adoption: subsidizing infrastructure investment or usage? Omega 114:102740. https://doi.org/10.1016/J.OMEGA.2022.102740
Chopra KL, Paulson PD, Dutta V (2004) Thin-film solar cells: an overview. Prog Photovoltaics Res Appl 12(2–3):69–92. https://doi.org/10.1002/pip.541
Choubey PC, Oudhia A, Dewangan R (2004) Solar cell current scenario and future trends. Prog Photovoltaics Res Appl 12(2–3):69–92
Dambhare MV, Butey B, Moharil SV (2021) Solar photovoltaic technology: a review of different types of solar cells and its future trends. J Phys: Conf Ser 1913(1). https://doi.org/10.1088/1742-6596/1913/1/012053
Djurišić AB, Liu FZ, Tam HW, Wong MK, Ng A, Surya C, He ZB et al (2017) Perovskite solar cells—an overview of critical issues. Prog Quantum Electron 53:1–37. https://doi.org/10.1016/j.pquantelec.2017.05.002
Dubey R, Gunasekaran A, Papadopoulos T, Childe SJ, Shibin KT, Wamba SF (2017) Sustainable supply chain management: framework and further research directions. J Clean Prod 142:1119–1130. https://doi.org/10.1016/j.jclepro.2016.03.117
Efaz ET, Rhaman MM, Al Imam S, Bashar KL, Kabir F, Mourtaza ME, Mozahid FA et al (2021) A review of primary technologies of thin-film solar cells. Eng Res Express 3(3). https://doi.org/10.1088/2631-8695/ac2353
Fallahazad P, Naderi N, Eshraghi MJ, Massoudi A (2018) Combination of surface texturing and nanostructure coating for reduction of light reflection in ZnO/Si heterojunction thin film solar cell. J Mater Sci: Mater Electron 29(8):6289–6296. https://doi.org/10.1007/s10854-018-8608-0
Fouad MM, Shihata LA, Morgan ESI (2017) An integrated review of factors influencing the performance of photovoltaic panels. Renew Sustain Energy Rev 80:1499–1511. https://doi.org/10.1016/J.RSER.2017.05.141
Fraas L, Partain L (2010) Solar cells and their applications, 2nd edn. Wiley Series in Microwave and Optical Engineering
Freiburg (2009) World Record: 41.1% efficiency reached for multi-junction solar cells at Fraunhofer ISE. Press Release—Institute Solare Energiiesysteme 49(01):1
Furugori S, Kobayashi A, Watanabe A, Yoshida M, Kato M (2017) Impact of photosensitizing multilayered structure on ruthenium(II)-dye-sensitized TiO_2-nanoparticle photocatalysts. ACS Omega 2(7):3901. Retrieved from https://doi.org/10.1021/acsomega.7b00566
Garg A, Ratnesh RK (2022) Solar cell trends and the future: a review. J Pharmaceut Negat Results 13(6):2051–2060. https://doi.org/10.47750/pnr.2022.13.s06.268
Gerold E, Antrekowitsch H (2024) Advancements and challenges in photovoltaic cell recycling: a comprehensive review. Sustainability (Switzerland) 16(6). https://doi.org/10.3390/su16062542
Global-market-outlook-for-solar-power-2023-2027 (2024)

Goel M (2016) Solar rooftop in India: policies, challenges and outlook. Green Energy Environ 1(2):129–137. https://doi.org/10.1016/j.gee.2016.08.003

Green MA (2002) Third generation photovoltaics: solar cells for 2020 and beyond. Physica E 14(1–2):65–70. https://doi.org/10.1016/S1386-9477(02)00361-2

Greijer H, Karlson L, Lindquist SE, Hagfeldt A (2001) Environmental aspects of electricity generation from a nanocrystalline dye sensitized solar cell system. Renew Energy 23(1):27–39. https://doi.org/10.1016/S0960-1481(00)00111-7

Guerrero-Lemus R, Vega R, Kim T, Kimm A, Shephard LE (2016) Bifacial solar photovoltaics—a technology review. Renew Sustain Energy Rev. https://doi.org/10.1016/j.rser.2016.03.041

Güney T (2019) Renewable energy, non-renewable energy and sustainable development. Int J Sust Dev World 26(5):389–397. https://doi.org/10.1080/13504509.2019.1595214

Gurion B (n.d.) Challenges of large-scale solar cell electricity production

Halim MA (2013) Harnessing sun's energy with quantum dots based next generation solar cell. Nanomaterials 3(1):22–47. https://doi.org/10.3390/nano3010022

Hardani, Idawati S, Mustariani BAA, Dewi YK, Hidayatulloh A, Darmawan MI (2021) Efficient TiO_2 nanoparticle-ruthenium sensitizers with high open-circuit voltage (Voc) for high-performance dye-sensitized solar cells. J Phys: Conf Ser 1816(1). https://doi.org/10.1088/1742-6596/1816/1/012005

Hooda A, Rathore MS (2024) Role of Fe supplantation in tailoring structural, electrical conductivity, and magnetic properties in $La_2Ni_{1-x}Fe_xMnO_6$ double perovskites. Physica B 691:416319. https://doi.org/10.1016/J.PHYSB.2024.416319

Hoppe H, Sariciftci NS (2004) Organic solar cells: an overview. J Mater Res 19(7):1924–1945. https://doi.org/10.1557/JMR.2004.0252

Hossain MS, Madlool NA, Rahim NA, Selvaraj J, Pandey AK, Khan AF (2016) Role of smart grid in renewable energy: an overview. Renew Sustain Energy Rev 60:1168–1184. https://doi.org/10.1016/J.RSER.2015.09.098

IEA (2023) IEA (2020) Projected costs of generating electricity 2020. IEA, Paris. https://www.iea.org/reports/projected-costs-of-generating-electricity-2020. Licence: CC BY 4.0

IEA (2023) Net zero roadmap: a global pathway to keep the 1.5 °C goal in reach. IEA, Paris. https://www.iea.org/reports/net-zero-roadmap-a-global-pathway-to-keep-the-15-0c-goal-in-reach. Licence: CC BY 4.0

International Renewable Energy Agency (IRENA) (2019) Future of solar photovoltaic: deployment, investment, technology, grid integration and socio-economic aspects. Irena. Retrieved from https://www.irena.org/-/media/Files/IRENA/Agency/Publication/2019/Oct/IRENA_Future_of_wind_2019.pdf

Jang YH, Jang YJ, Kim S, Quan LN, Chung K, Kim DH (2016) Plasmonic solar cells: from rational design to mechanism overview. Chem Rev 116(24):14982–15034. https://doi.org/10.1021/acs.chemrev.6b00302

Jeon NJ et al (2014) Solvent engineering for high-performance inorganic–organic hybrid perovskite solar cells. Nat Mater 13:897–903

Jiang Q, Song Z, Bramante RC, Ndione PF, Tirawat R, Berry JJ, Zhu K et al (2023) Highly efficient bifacial single-junction perovskite solar cells. Joule 7(7):1543–1555. https://doi.org/10.1016/J.JOULE.2023.06.001

Jung HS, Han GS, Park N-G, Ko MJ (2019) Flexible perovskite solar cells. Joule 3(1):1850–1880

Kabeyi MJB, Olanrewaju OA (2023) The levelized cost of energy and modifications for use in electricity generation planning. Energy Rep 9:495–534. https://doi.org/10.1016/j.egyr.2023.06.036

Kaplan Dönmez NF (2023) Taxation and incentives in renewable energy investments. Elektronik Sosyal Bilimler Dergisi 22(85):220–245. https://doi.org/10.17755/esosder.1208131

Kudelas D, Taušová M, Tauš P, Gabániová L, Koščo J (2019) Investigation of operating parameters and degradation of photovoltaic panels in a photovoltaic power plant. Energies 12(19). https://doi.org/10.3390/en12193631

Kumavat PP, Sonar P, Dalal DS (2017) An overview on basics of organic and dye sensitized solar cells, their mechanism and recent improvements. Renew Sustain Energy Rev 78:1262–1287. https://doi.org/10.1016/j.rser.2017.05.011

Kurokawa K (n.d.) Energy from the desert; Feasibility of Very Large Scale Photovoltaic Power Generation (VLS-PV) systems. Photovoltaic Power Systems Executive Committee of the International Energy Agency James & James (Science Publishers) Ltd, pp 1–40

Lelièvre J-F, Couderc R, Pinochet N, Sicot L, Munoz D, Kopecek R, Ferrada P, Marzo A, Olivares D, Valencia F, Urrejola E (2022) Desert label development for improved reliability and durability of photovoltaic modules in harsh desert conditions. Sol Energy Mater Sol Cells 236:111508. Retrieved from https://doi.org/10.1016/j.solmat.2021.111508

Lu K (2023) Comparison and evaluation of different types of solar cells. Appl Comput Eng 23(1):263–270. https://doi.org/10.54254/2755-2721/23/20230664

Madhani V, Kumar D, Kanchan DK, Singh Rathore M (2023) Recent advances and prospects of K-ion conducting polymer electrolytes. J Electroanal Chem 935:117334. https://doi.org/10.1016/J.JELECHEM.2023.117334

Madon M, Gillet P, Julien C, Price G (1991) A vibrational study of phase transitions among the GeO2 polymorphs. Phys Chem Miner 18:7–18. https://doi.org/10.1007/BF00199038

Malii SS et al (2024) All-inorganic halide perovskites for air-processed "n–i–p" monolithic perovskite/organic hybrid tandem solar cells exceeding 23% efficiency. Energy Environ Sci 17:1046–1060

Mamodiya UNT (2019) Electricity generation through solar energy. J Gujarat Res Soc 21(5):546–554

Mariotti N, Bonomo M, Fagiolari L, Barbero N, Gerbaldi C, Bella F, Barolo C (2020) Recent advances in eco-friendly and cost-effective materials towards sustainable dye-sensitized solar cells. Green Chem 22(21):7168–7218. https://doi.org/10.1039/d0gc01148g

Nair S, Patel SB, Gohel JV (2020) Recent trends in efficiency-stability improvement in perovskite solar cells. Mater Today Energy 17:100449. https://doi.org/10.1016/j.mtener.2020.100449

Nozik AJ, Beard MC, Luther JM, Law M, Ellingson RJ, Johnson JC (2010) Semiconductor quantum dots and quantum dot arrays and applications of multiple exciton generation to third-generation photovoltaic solar cells. Chem Rev 110:6873–6890. https://doi.org/10.1017/CBO9781139022750.006

Oduro P, Simpa P, Ekechukwu DE (2024) Renewable energy expansion: legal strategies for overcoming regulatory barriers and promoting innovation. World J Adv Eng Technol Sci 12(1):168–186. https://doi.org/10.30574/wjaets.2024.12.1.0208

Østergaard PA, Duic N, Noorollahi Y, Kalogirou S (2022) Renewable energy for sustainable development. Renew Energy. https://doi.org/10.1016/j.renene.2022.09.065

Palm J, Probst V, Karg FH (2004) Second generation CIS solar modules. Sol Energy 77(6):757–765. https://doi.org/10.1016/j.solener.2004.05.011

Pan Z et al (2018) Quantum dot-sensitized solar cells. Chem Soc Rev (20)

Park H, Kwon S, Lee JS, Lim HJ, Yoon S, Kim D (2009) Improvement on surface texturing of single crystalline silicon for solar cells by saw-damage etching using an acidic solution. Sol Energy Mater Sol Cells 93(10):1773–1778. https://doi.org/10.1016/j.solmat.2009.06.012

Park H, Ju M, Khokhar MQ, Cho EC, Kim Y, Cho Y, Yi J (2020) Surface modifications for light trapping in silicon heterojunction solar cells: a brief review. Trans Electr Electron Mater 21(4):349–354. https://doi.org/10.1007/s42341-020-00203-1

Pastuszak J, Węgierek P (2022) Photovoltaic cell generations and current research direction for their development. Materials 15:5542

Patidar R, Burkitt D, Hooper K, Richards D, Watson T (2020) Slot-die coating of perovskite solar cells: an overview. Mater Today Commun 22:100808. https://doi.org/10.1016/j.mtcomm.2019.100808

Post HN, Moore LM (2008) Five years of operating experience at a large, utility-scale photovoltaic generating plant. Prog Photovoltaics: Res Appl 16(1):249–259. https://doi.org/10.1002/pip

Project R (2024) Utility-scale solar photovoltaic power plants. International Finance Corporation World Bank Group. https://doi.org/10.1109/ICIEAM60818.2024.10554002

Radfar B, Es F, Turan R (2020) Effects of different laser modified surface morphologies and post-texturing cleanings on c-Si solar cell performance. Renew Energy 145:2707–2714. https://doi.org/10.1016/j.renene.2019.08.031

Rehman F, Syed IH, Khanam S, Ijaz S, Mehmood H, Zubair M, Mehmood MQ et al (2023) Fourth-generation solar cells: a review. Energy Adv 2(9):1239–1262. https://doi.org/10.1039/d3ya00179b

Ritruksa M, Wongrerkdee S, Lohawet K, Kaewprajak A, Kumnorkaew P, Wongrerkdee S (2020) Surface modification of PEDOT: PSS film by chemical vapor texturing process for enhanced organic photovoltaics. Surf Interfaces 21:100716. https://doi.org/10.1016/j.surfin.2020.100716

Rong Y, Hu Y, Mei A, Tan H, Saidaminov MI, Seok SI, Han H et al (2018) Challenges for commercializing perovskite solar cells. Science 361(6408). https://doi.org/10.1126/science.aat8235

Rono N, Ahia CC, Meyer EL (2024) Recent advances in transition metal dichalcogenides-based materials for fourth-generation perovskite solar cell devices. AIP Adv 14:070702. Retrieved from https://doi.org/10.1063/5.0211610

Sarkın AS, Ekren N, Sağlam Ş (2020) A review of anti-reflection and self-cleaning coatings on photovoltaic panels. Sol Energy 199:63–73. https://doi.org/10.1016/J.SOLENER.2020.01.084

Schneider A, Rubin L, Rubin G (2006) Solar cell efficiency improvement by new metallization techniques-the day4^{TM}electrode concept. In: Conference record of the 2006 IEEE 4th world conference on photovoltaic energy conversion, WCPEC-4, vol 1, pp 1095–1098. https://doi.org/10.1109/WCPEC.2006.279333

Shaker LM, Alamiery A, Isahak WN, Al-Azzawi WK (2023) Corrosion in solar cells: challenges and solutions for enhanced performance and durability. J Opt. Retrieved from https://doi.org/10.1007/s12596-023-01277-9

Sharma M (2018) Development of a 'Green building sustainability model' for Green buildings in India. J Clean Prod 190:538–551. https://doi.org/10.1016/J.JCLEPRO.2018.04.154

Sharma D, Mehra R, Raj B (2021) Comparative analysis of photovoltaic technologies for high efficiency solar cell design. Superlattices Microstruct 153:106861. https://doi.org/10.1016/j.spmi.2021.106861

Shrivastav N, Kashyap S, Madan J, Al-Mousoi AK, Mohammed MK, Hossain MK, Pandey R, Ramanujam J (2023) Perovskite-CIGS monolithic tandem solar cells with 29.7% efficiency: a numerical study. Energy Fuels 37(4)

Shukla AK, Sudhakar K, Baredar P, Mamat R (2018) Solar PV and BIPV system: barrier, challenges and policy recommendation in India. Renew Sustain Energy Rev 82:3314–3322. https://doi.org/10.1016/j.rser.2017.10.013

Siegler TD, Dawson A, Lobaccaro P, Ung D, Beck ME, Nilsen G, Tinker LL (2022) The path to perovskite commercialization: a perspective from the United States Solar Energy Technologies Office. ACS Energy Lett 7(5):1728–1734. https://doi.org/10.1021/acsenergylett.2c00698

Silveira B, Melo dos Santos ME, Ferreira Maia FJ, Basso AP, Singh JN, Costa HK de M (2024) Incentives for photovoltaic energy generation: a comparative analysis of policies in Spain, Germany, and Brazil. Energy Strategy Rev 54:101415. https://doi.org/10.1016/j.esr.2024.101415

Singh BP, Goyal SK, Kumar P (2021) Solar PV cell materials and technologies: analyzing the recent developments. Mater Today: Proc 43:2843–2849. https://doi.org/10.1016/j.matpr.2021.01.003

Soga T (2006) Fundamentals of solar cell. Nanostructured materials for solar energy conversion, pp 3–43. https://doi.org/10.1016/B978-044452844-5/50002-0

Soonmin H, Hardani, Nandi P, Mwankemwa BS, Malevu TD, Malik MI (2023) Overview on different types of solar cells: an update. Appl Sci (Switzerland) 13(4):1–37. https://doi.org/10.3390/app13042051

Sopian K, Cheow SL, Zaidi SH (2017) An overview of crystalline silicon solar cell technology: past, present, and future. AIP Conf Proc 1877. https://doi.org/10.1063/1.4999854

Strevel N, Trippel L, Kotarba C, Khan I (2013) Improvements in CdTe module reliability and long-term degradation through advances in construction and device innovation. Photovoltaics Int 1–8
Sutthana S, Wongratanaphisan D, Gardchareon A, Phadungdhitidhada S, Ruankham P, Choopun S (2016) Enhancement of ZnO dye-sensitized solar cell performance by modifying photoelectrodes using an acid vapor texturing process. Surf Coat Technol 306:30–34. https://doi.org/10.1016/j.surfcoat.2016.04.006
Tang H, He S, Peng C (2017) A short progress report on high-efficiency perovskite solar cells. Nanoscale Res Lett 12. https://doi.org/10.1186/s11671-017-2187-5
Tessler N, Vaynzof Y (2018) Preventing hysteresis in perovskite solar cells by undoped charge blocking layers. ACS Appl Energy Mater 1(2):676–683. https://doi.org/10.1021/acsaem.7b00176
Tonui P, Oseni SO, Sharma G, Yan Q, Tessema Mola G (2018) Perovskites photovoltaic solar cells: an overview of current status. Renew Sustain Energy Rev 91:1025–1044. https://doi.org/10.1016/j.rser.2018.04.069
Uematsu T, Tsutsui K, Yazawa Y, Warabisako T, Araki I, Eguchi Y, Joge T (2003) Development of bifacial PV cells for new applications of flat-plate modules. Sol Energy Mater Sol Cells 75(3–4):557–566. https://doi.org/10.1016/S0927-0248(02)00197-6
Urbina A (2020) The balance between efficiency, stability and environmental impacts in perovskite solar cells: a review. J Phys Energy. https://doi.org/10.1088/2515-7655/ab5eee
Vigil-Galán O, Courel M, Andrade-Arvizu JA, Sánchez Y, Espíndola-Rodríguez M, Saucedo E, Titsworth M et al (2015) Route towards low cost-high efficiency second generation solar cells: current status and perspectives. J Mater Sci: Mater Electron 26(8):5562–5573. https://doi.org/10.1007/s10854-014-2196-4
Vimala M, Ramadas G, Perarasi M, Manokar AM, Sathyamurthy R (2023) A review of different types of solar cell materials employed in bifacial solar photovoltaic panel. Energies 16(8). https://doi.org/10.3390/en16083605
VR A (2023) The future of solar energy; prediction for 2024 and beyond. https://arka360.com/ros/future-of-solar-energy-predictions/
Wali Q, Elumalai NK, Iqbal Y, Uddin A, Jose R (2018) Tandem perovskite solar cells. Renew Sustain Energy Rev 84:89–110. https://doi.org/10.1016/J.RSER.2018.01.005
Wan C, Zhao J, Song Y, Xu Z, Lin J, Hu Z (2016) Photovoltaic and solar power forecasting for smart grid energy management. CSEE J Power Energy Syst 1(4):38–46. https://doi.org/10.17775/cseejpes.2015.00046
Wang Y, Ahmad I, Leung T, Lin J, Chen W, Liu F, Djurišić AB et al (2022). Encapsulation and stability testing of perovskite solar cells for real life applications. ACS Mater Au 2(3):215–236. https://doi.org/10.1021/ACSMATERIALSAU.1C00045
Wolf PJ, Christensen TM, Coit NG, Swinford RW (1993) Thin film properties of germanium oxide synthesized by pulsed laser sputtering in vacuum and oxygen environments. J Vac Sci Technol A 11(1993):2725–2732. https://doi.org/10.1116/1.578633
Wu X, Li B, Zhu Z, Chueh CC, Jen AK (2021) Designs from single junctions, heterojunctions to multijunctions for high-performance perovskite solar cells. Chem Soc Rev 50:13090–13128
Yamaguchi M, Lee KH, Araki K, Kojima N (2018) A review of recent progress in heterogeneous silicon tandem solar cells. J Phys D: Appl Phys 51(13). https://doi.org/10.1088/1361-6463/aaaf08
You J, Dou L, Hong Z, Li G, Yang Y (2013) Recent trends in polymer tandem solar cells research. Prog Polym Sci 38(12):1909–1928. https://doi.org/10.1016/J.PROGPOLYMSCI.2013.04.005
Zhang Y, Chen W, Gao W (2017) A survey on the development status and challenges of smart grids in main driver countries. Renew Sustain Energy Rev 79:137–147. https://doi.org/10.1016/J.RSER.2017.05.032
Zhang J, Chang N, Fagerholm C, Qiu M, Shuai L, Egan R, Yuan C (2022) Techno-economic and environmental sustainability of industrial-scale productions of perovskite solar cells. Renew Sustain Energy Rev 158:112146. https://doi.org/10.1016/J.RSER.2022.112146

Zhang YB, Xu JM, Shu YH, Quan P, Wang Y, Mao J, Gao YY, Fu CG, Feng ZQ, Verlinden P, Yang P, Chu J (1960) High-reliability and long durability double glassmodule with crystalline Silicon Solar cells with fire safety class A certification. Am J Public Health 50:22–24
Zhang HL (2020) Perovskite tandem solar cells: from fundamentals to commercial deployment. Chem Rev 120(18)

Chapter 12
Evaluation of Toxicity of Lead in Perovskite Solar Cells: Current Status and Future Challenges

Anuja Dhingra, O. P. Thakur, and Raghvendra Pandey

Abstract The perovskite solar cells (PSCs) have attracted considerable interest and potential commercial application as lead-based PSCs due to high efficiency and low manufacturing cost. On the other hand, the presence of lead (Pb) in the perovskite structures poses major risk factor in terms of environmental and human health. This chapter focuses and summarises the available acquaintance on Pb toxicity in PSCs, concentrating on physical and chemical properties of Pb in those materials, its environmental hazards, and future perspectives for both solar cell industry and the global ecosystem. The pathways through which Pb could potentially leach out of the perovskite materials throughout the entire life cycle of the product, from manufacturing to operating, how the product is disposed of and the risk of pollution to the environment are also explored. This chapter has also evaluated the existing methodologies for dealing with Pb toxic effects including encapsulation methods and substitutive techniques aiming to develop Pb-free substitutes. In addition, however, as these methods alone will not be sufficient and then outlined what kinds of research will be needed to overcome these issues in the future, in particular, how the safety and sustainability of perovskite solar devices can be improved. This chapter highlights the necessity of cross-disciplinary approaches to allow the full potential of PSCs to be achieved without degrading the environment and human well-being.

Keywords Perovskite solar cell · Lead · Toxicity · Health hazard · Environmental issues

A. Dhingra · O. P. Thakur (✉)
Materials Analysis and Research Lab, Department of Physics, Netaji Subhas University of Technology, New Delhi, India
e-mail: opthakur@nsut.ac.in

A. Dhingra · R. Pandey (✉)
Green Energy Physics and Solid State Ionics Lab, Department of Physics, ARSD College, University of Delhi, New Delhi, India
e-mail: raghvendra.sspl@gmail.com

M. Kumar and A. Raj (eds.), *Perovskite Solar Cells*, Springer Tracts in Electrical and Electronics Engineering, https://doi.org/10.1007/978-981-96-9463-1_12

Introduction

In the last decades, the perovskite solar cells (PSCs) are emerged as the most promising photovoltaic technologies because of their high efficiency, ease of fabrication, and low-cost production. PSCs materials comprise a light-absorbing layer of metal halide perovskites, offer advantages in respect of light absorption, charge transport, and tunable optoelectronic properties. However, Lead (Pb) as a central component and key factor in the rapid development of most of the PSCs. The recent emerging metal halide perovskite with chemical formula $CH_3NH_3PbI_3$ and its derivate has emerged as amazing solar cell formation with power conversion efficiency of 25.5% and is one of the best recorded PSC efficiencies (Jeon et al. 2015; Burschka et al. 2013; Lyu et al. 2017). The efficiency of PSC depends on the several different factors like high absorption coefficient, low exciton binding energy, dominant shallow point defects, high absorption coefficient and long diffusion length of charge carriers (Stranks et al. 1979; Wolf et al. 2014; Yin et al. 2014). Out of these, the economic feasibility due to easily available raw materials, cost-effective solution processing and roll-to-roll technique for thin film synthesis has resulted in outstanding accomplishment of PSCs (Lee et al. 1979; Hwang et al. 2015).

Moreover, the use of Pb in PSCs has raised significant concerns regarding its health and environmental impacts. Pb is a well-known toxic substance, and its release into the environment during the manufacturing, operation, and disposal of PSCs could pose serious risks to both the environment and human health. Pb toxicity can be estimated from the fact that it mimics Ca^{2+}, Zn^{2+}, is carcinogenic and has half-life of nearly 30 years. If lead is ingested, it accumulates inside the bones resulting in the formation of lead phosphate. Moreover, it may result severe health disorders such as anaemia, hypertension, neurological disorder and kidney damage specially to children and pregnant women. Lead concentration > 5 μg/dL is known as lead poising and to treat this, ethylenediamine tetraacetate and other chelating agents are used to bind the Pb and reduce its concentration inside the body (Zhu et al. 2024). PSCs are now-a-days encapsulated to stop Pb from leaking into the environment (Hailegnaw et al. 2015; Snaith 2018). However, they may be exposed if they are destroyed by environmental disasters like fires, landslides, hail, hail, and strong winds, or if waste is not disposed of properly (Jiang et al. 2019; Kwak et al. 2020). Furthermore, rain and snowmelt can cause components from damaged solar panels to seep into the ground (Hailegnaw et al. 2015). For example, rainfall-induced encapsulation degradation has been shown to cause the absorber layer in methylammonium lead iodide solar cells to completely decompose (Hailegnaw et al. 2015). Metal halide PSCs may emit Pb, which could lead to soil and groundwater pollution.

The search to find alternative of Pb in solar cell devices is dependent on the chemical compatibility and neighbour elements in the periodic table. Proposed elements may include Ge, Bi, Ag, Te, Sb, In, and Sn (Zhang et al. 2018). However, the main question arises whether these elements are a superior choice than the conventional Pb-based PSCs and if their replacement is sufficient for the need of over 20 TWp of global installation for cutting of CO_2 emissions by 2050 (Raturi 2019). The maximum efficiency for Pb-free solar cell devices is 9.6% for $Gua_{0.2}FA_{0.8}SnI_3$ (Jokar et al. 2019), with Sn being the most potential alternative due to its similar lone pair of electrons in the s orbital and electronic configuration. $MASnI_3$ is more suitable as compared to $MAPbI_3$ for PV applications due to its reduced bandgap and higher absorption coefficient (Umari et al. 2014). However, high energy $5s^2$ orbital and ease of oxidation from Sn^{2+} to Sn^{4+} in Sn-based perovskites causes damage to their semiconducting properties (Yao et al. 2020). Sn-based perovskites can be slowed with additives or encapsulation to prevent its oxidation, but this can increase manufacturing costs (Ke et al. 2019; Xiao et al. 2019). Other 3D analogues of $CsPbI_3$, $MAPbI_3$, and $FAPbI_3$ can be produced using Ge consisting of small band-gap of 1.6 eV, but their PV performances are worse as instability of +4 oxidation state rises as atomic number decreases (Stoumpos et al. 2015). Other alternatives includes monovalent and trivalent metal co-substitution (Schileo and Grancini 2021). PSC technology is continuously advancing, it is necessary to address these concerns and develop strategies to mitigate Pb toxicity while maintaining the high efficiency.

This chapter aims to review the current state of knowledge regarding the toxicity of Pb in PSCs. The physicochemical properties of lead within the perovskite structure, the pathways through which lead can become hazardous, and the current efforts to mitigate its toxicity are also discussed. The challenges and future research directions needed to make perovskite solar technology more sustainable are also explored in this chapter.

Lead in Perovskite Solar Cells

PSCs are a class of thin-film photovoltaics which utilise materials with a perovskite crystal structure, generally composed of a metal halide, such as lead iodide (PbI_2) enclosed within the electron-transport layer (ETL) and hole-transport layer (HTL) (Fig. 12.1a). The unique structure of perovskite materials offers several advantages for photovoltaic applications, such as long carrier diffusion lengths, high absorption coefficients, and tunable bandgaps. The inclusion of lead in these materials has been a major factor in their rapid rise to prominence, as lead halides enable efficient charge transport and stability.

The most common perovskite structure used in solar cells is a hybrid organic–inorganic material, typically denoted as ABX_3 where A sites are occupied by the methylammonium (MA) cation ($CH_3NH_3^+$), formamidine (FA) cation ($NH_2CH{=}NH_2^+$) or by inorganic alkali metal cation (Cs^+). The X sites consist of halogen anions, such as iodide, bromide, or chloride and impact both stability as well as optoelectronic property of the PSCs. The lead ion, Pb^{2+}, is at the centre of the perovskite lattice at the B site and coordinates with halide anions to form a three-dimensional structure that facilitates the movement of charge carriers (Fig. 12.1b). The prominent characteristics of Pb-based PSCs include (i) a fast-rising direct-bandgap transition within the conduction (Pb-p) and valence states (Pb-s/X-p); (ii) a small exciton binding energy that allows optically generated electrons from holes to disengage

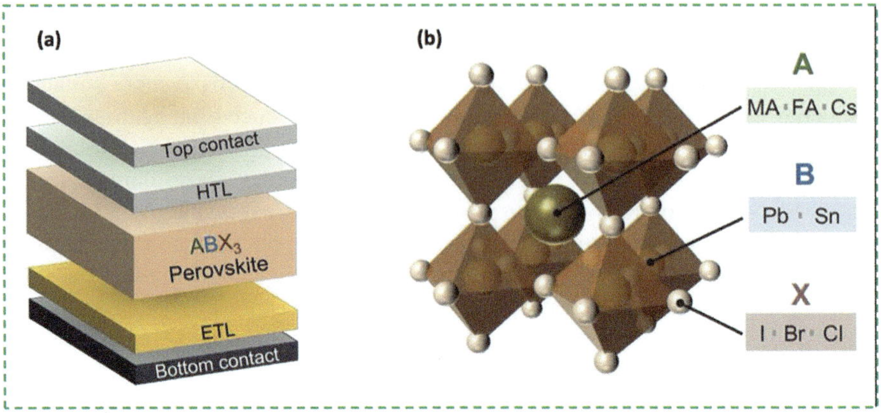

Fig. 12.1 **a** Schematic diagram of PSCs enclosing meta halide layer between the ETL and HTL and **b** Structure of PSCs with chemical formula ABX_3 [**a**, **b** Reprinted with permission from Ref. Zhou et al. (2022)]

quickly; (iii) reduced effective mass of both, electrons as well as holes; (iv) energetically shallow intrinsic defect levels that are advantageous for bipolar conductivity while also reducing scattering and carrier trapping; and (v) cheaper raw material and synthesis route (Stranks et al. 1979; Wolf et al. 2014; Yin et al. 2014).

While the inclusion of Pb has significantly improved the efficiency of PSCs, it has also led to concerns regarding the potential for Pb leakage from the material. The environmental stressors such as temperature fluctuations, moisture, and UV exposure, can result in the degradation of the perovskite material and the release of Pb ions into the surrounding environment. This poses a significant challenge to the widespread adoption of PSCs, as Pb contamination could lead to severe environmental and health consequences.

Toxicity of Lead

The environment and human health are seriously threatened by the poisonous heavy metal lead (Pb). Numerous detrimental health outcomes, such as neurological impairment, kidney failure, cardiovascular disease, and developmental abnormalities in children, have been connected to long-term exposure to lead. Pb is especially harmful because, even at low exposure levels, it can build up in the body over time and produce long-term health issues (Zhu et al. 2024).

Pb exists as Pb^{2+}, naturally occurring in different minerals and extensively utilised from centuries to its widespread availability. The history of Pb usage dates back from ancient civilizations, like in Roman Pb pipes, and is still continued into modern era with products such as leaded gasoline. This widespread usage has contributed significant environmental and human exposure to Pb. Pb enters the bloodstream through inhalation, ingestion, or dermal absorption, typically by the use of lead-based paints (now banned) (Dissanayake et al. 2023).

As Pb enters the body, it mimics essential metals such as Ca Fe and Zn interfering with the function of enzymes and receptors. It essentially effects the central nervous system. By mimicking Ca^{2+}, Pb^{2+} pass through the barrier of blood–brain, accumulating inside the brain tissue. This causes the oxidative damage to genetic material, like Hg, Cd and As. With time, Pb deposits inside the bones as lead phosphate, with a half-life of nearly 20–30 years. Chronic exposure to Pb, even at low level of concentration ~ 5 μg dL^{-1}, has major health consequences. In children, it results in impairment of neurological development, causing reduced intelligence, motor function deficits, and also different behavioural disorders like anxiety, depression, and aggression. Adults specially pregnant women are also vulnerable to these health problems and may also experience non-neurological symptoms, for example reduced fertility, cataract, hypertension, and cardiovascular and renal diseases. Furthermore, Pb also shows genotoxicity in the eukaryotic cells and is moderately genotoxic in test animal, with major damage in the brain, bone marrow, lungs, and liver (Chetyrkina et al. 2023).

In the context of PSCs, Pb toxicity is a concern at multiple stages of the product lifecycle, including manufacturing, installation, and disposal. The primary route of Pb exposure is through the release of Pb ions from the perovskite material. This can occur through various mechanisms, such as the degradation of the perovskite layer due to environmental conditions or the leaching of Pb during recycling or disposal. Pb toxicity can occur both through direct exposure to the substance and through the environmental contamination that results from the improper disposal or degradation of Pb-containing materials. For example, if PSCs are not properly disposed of, Pb can leach into the water or soil, potentially entering the food chain and affecting ecosystems.

The severity of Pb toxicity depends on factors such as the concentration of Pb released, the duration of exposure, and the specific environment in which the exposure occurs. In the case of PSCs, the risk of Pb exposure is heightened by the fact that these devices are often used in large-scale solar arrays that could contribute to significant Pb contamination if proper disposal and recycling protocols are not followed.

Environmental Impact

The environmental impact of Pb in PSCs is a significant concern, particularly in the widespread deployment of solar technologies. While PSCs offer potential benefits in terms of renewable energy generation, their environmental impact must be carefully evaluated to ensure that their adoption does not result in unintended consequences.

The primary environmental concern associated with Pb in PSCs is the potential for Pb contamination during the product's lifecycle. Pb can be released into the environment during the manufacturing process, particularly if proper containment measures are not in place. Additionally, the PSC degradation under environmental stressors such as humidity, temperature fluctuations, and UV radiation can result in the leaching of Pb ions.

Pb-based PSCs also emerged as a concern in the context of environmental contamination. With degradation, Pb-based PSCs release PbI_2, converts into $Pb(OH)_2$ and hydroiodic acid (HI). While $Pb(OH)_2$ is less harmful as compared to organometallic lead compounds like tetra butyl Pb, the acidification by HI amplifies the environmental harm.

Water-Induced Degradation

Several studies have indicated presence of water accelerates the rate of degradation and reduces the chemical stability of PSCs (Li and Liu 2017; Jiang et al. 2018). The degradation of Pb-based PSCs starts by a water molecule and is then driven by the phase shifts of methyl ammonia, which is volatile and soluble in water, and hydrogen iodide which is soluble in water. This pathway give rise to the formulation of yellow solid which is PbI_2 (Fig. 12.2a). The chemical reaction can be given as follows (Frost et al. 2014)

$$(CH_3NH_3)PbI_3 \xrightarrow{H_2O} HI \uparrow + PbI_2 + CH_3NH_2 \uparrow \tag{12.1}$$

Both CH_3NH_2 and HI occurs in gas phase and continuous release of products in gaseous phase favours the forward direction reaction. Therefore, $MAPbI_3$ degrades completely to PbI_2 (Han et al. 2015). Further degradation of HI under ultraviolet (UV) irradiation results in the formation of H_2 and I_2 which can be given as follows (Niu et al. 2014)

$$2HI \xrightarrow{UV} H_2 \uparrow + I_2 \uparrow \tag{12.2}$$

Therefore the overall chemical reaction of $MAPbI_3$ degradation in the presence of H_2O and UV irradiation can be given as follows (Christians et al. 2015)

$$(CH_3NH_3)PbI_3 \xrightarrow{H_2O+UV} \frac{1}{2}H_2 \uparrow + \frac{1}{2}I_2 \uparrow + PbI_2 + CH_3NH_2 \uparrow \tag{12.3}$$

Assuming Pb-based PSCs absorb H_2O to produce hydrated complex, the hydration process can be given as

$$(CH_3NH_3)PbI_3 + H_2O \rightleftharpoons (CH_3NH_3)PbI_3 \cdot H_2O \tag{12.4}$$

$$4(CH_3NH_3)PbI_3 + 8H_2O \rightarrow (CH_3NH_3)_6PbI_6 \cdot 2H_2O + 3Pb(OH)_2 + 6HI \uparrow \tag{12.5}$$

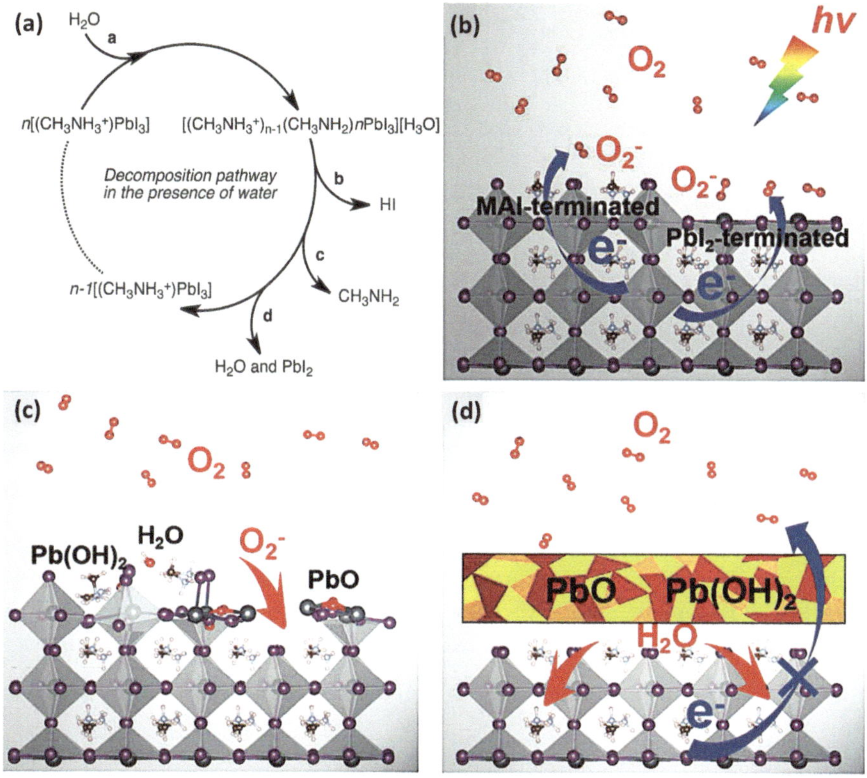

Fig. 12.2 **a** Pathway of decomposition of Pb-based PSCs in presence of water [Reprinted with permission from Ref. Frost et al. (2014)], **b–d** Representation of photo-oxidative degradation process of the $MAPbI_3$ (001) surface [Reprinted with permission from Ref. Ouyang et al. (2019)]

The monohydrated phase obtained in Eq. 12.4 $(CH_3NH_3)PbI_3 \cdot H_2O$ is reversible and can be converted to $MAPbI_3$ using dehydration process (Wang et al. 2017). However, dihydrate $(CH_3NH_3)_6PbI_6 \cdot 2H_2O$ is irreversible in nature. In the hydration process, $[PbI_6]^{4-}$ in three-dimensional network of $MAPbI_3$ decays to zero-dimensional framework of isolated octahedral. The phase $PbI_{2+x}{}^{x-}$ $(0 \leq x < 1)$ is formed under ambient conditions and decomposes to PbO, $Pb(OH)_2$, and $PbCO_3$ using the following chemical reaction.

$$(CH_3NH_3)PbI_6 \cdot H_2O \rightarrow (CH_3NH_3)_xPbI_{2+x} + (4-x)CH_3NH_2 \uparrow + (4-x)HI \uparrow + 2H_2O \quad (12.6)$$

$$(CH_3NH_3)_xPbI_{2+x} + 2CO_2 + O_2 \rightarrow 2PbCO_3 + 2I_2 + 2xCH_3NH_2 \uparrow + 2xHI \uparrow \quad (12.7)$$

$$(CH_3NH_3)_xPbI_{2+x} + H_2O + \frac{1}{2}O_2 \rightarrow Pb(OH)_2 + I_2 + xCH_3NH_2 \uparrow + xHI \uparrow \tag{12.8}$$

$$Pb(OH)_2 \rightarrow PbO + H_2O \tag{12.9}$$

Oxygen-Induced Degradation

Oxygen may also induce Pb-based PSCs degradation as the adsorbed O_2 captures electron forming superoxide ($O_2^{\cdot-}$) leading to the perovskite degradation. O_2 may permeate inside the surface of grain and $MAPbI_3$ which may also reduce to $O_2^{\cdot-}$ via photo-excited electron (Bryant et al. 2016; Zhuang et al. 2023). The $Al_2O_3/MAPbI_3$ system has been observed to produce more $O_2^{\cdot-}$ as compared to the $TiO_2/MAPbI_3$ system. The favourable energy offset at the heterojunction allows the TiO_2 film to accept an electron from the photoexcited $MAPbI_3$. Therefore, $TiO_2/MAPbI_3$ system results in smaller yield of $O_2^{\cdot-}$. Ab initio calculations demonstrate that the $O_2^{\cdot-}$ energetically favours the iodide vacancy sites at the same time (Bryant et al. 2016). The chemical reaction for photo-induced $O_2^{\cdot-}$ formation can be given as follows

$$(CH_3NH_3)PbI_3 \xrightarrow{\text{light}} (CH_3NH_3)PbI_3^* \tag{12.10}$$

$$O_2 \xrightarrow{(CH_3NH_3)PbI_3^*} O_2^{\cdot-} \tag{12.11}$$

$$4(CH_3NH_3)PbI_3^* + O_2^{\cdot-} \rightarrow 4PbI_2 + 2I_2 + 2H_2O + 4CH_3NH_2 \uparrow \tag{12.12}$$

$CH_3NH_3PbI_3^*$ transports photo-induced charge carriers and O_2 captures an electron from $CH_3NH_3PbI_3^*$ to generate $O_2^{\cdot-}$. As large crystallites have reduced surface sites, the $MAPbI_3$ films with comparatively large grains deteriorate more slowly. Sultana et al. (2018) found that under UV irradiation, the PbI_2 might further react with O_2 to form lead oxy-iodide which can be given as follows:

$$PbI_2 \xrightarrow{\text{UV}} e_{cb}^-(PbI_2) + h_{vb}^+(PbI_2) \tag{12.13}$$

$$e_{cb}^-(PbI_2) + O_2 \rightarrow O_2^{\cdot-} + PbI_2 \tag{12.14}$$

$$(x+y)PbI_2 + \frac{y}{2}O_2^{\cdot-} \rightarrow (PbI_2)_x(PbO)_y + yI_2 \tag{12.15}$$

Under UV irradiation, the valence band electron jumps to, where it transfers to the O_2 molecule and produces $O_2^{\cdot-}$. Later, lead oxy-iodide is generated when the $O_2^{\cdot-}$

further reacts with PbI_2. According to studied, when $O_2^{\cdot -}$ approaches the ammonium group, it absorbs an acid proton of the $CH_3NH_3^+$ and generates the hydroperoxyl radical (HO_2) (Godding et al. 2019) which can be given using the following reaction.

$$(CH_3NH_3)^+ + O_2^{\cdot -} \leftrightarrow CH_3NH_2 \uparrow + HO_2^{\cdot} \tag{12.16}$$

The $2HO_2^{\cdot}$ reacts to generate oxygen molecules and hydrogen peroxide (H_2O_2)

$$2HO_2^{\cdot} \leftrightarrow O_2 \uparrow + H_2O_2 \tag{12.17}$$

Finally, H_2O_2 interacts with Pb^0 to produce PbO or $Pb(OH)_2$ which can be represented as follows

$$Pb^0 + H_2O_2 \rightarrow Pb(OH)_2 \tag{12.18}$$

$$Pb^0 + H_2O_2 \rightarrow PbO + H_2O \tag{12.19}$$

The PbO generated possess passivation effect resulting in formation of Pb–Pb dimer which increases open-circuit voltage in PSCs. The $HO_2^{\cdot}$ radicals further dissociate into H_2 and O_2, where the O_2 serves as the reactant and continuous the degradation process

$$2HO_2^{\cdot} \leftrightarrow 2O_2 \uparrow + H_2 \uparrow \tag{12.20}$$

According to a study by Ouyang et al. (2019), $MAPbI_3$ degrades in three steps, as depicted in Fig. 12.2b–d. Initially, O_2 on the $MAPbI_3$ surface absorbs a photo-generated electron from the $MAPbI_3$'s conduction band to create O_2^- (Fig. 12.2b). Thereafter, the Pb–I frameworks shatter and the MAI-terminated surface becomes exposed as a result of O_2^- quickly oxidising the PbI^{2-}terminated surface to form PbO. Then, additional oxidation of the underlying MAI-terminated surface generates H_2O, PbO, or the unstable $Pb(OH)_2$ (Fig. 12.2c). The further oxidation of $MAPbI_3$ can be prevented from further oxidising by the protective layers formed by the oxidation products PbO and $Pb(OH)_2$. The H_2O molecules produced from $Pb(OH)_2$'s surface oxidation and breakdown can gradually hydrate with the inside $MAPbI_3$, ultimately leading to breakdown of $MAPbI_3$ (Fig. 12.2d) (Ouyang et al. 2019). First, O_2 absorbs photogenerated electrons to generate O_2^-, which therefore results in different disintegration pathways. Finally, CH_3NH_2, I_2, PbI_2, and PbO (or $Pb(OH)_2$) are the main degraded products.

Light-Induced Degradation

Light may also cause the degradation of PSCs even if the oxygen and water are not present. Research demonstrated that Pb-based PSCs can be severely degraded by UV light (Li et al. 2016; Leijtens et al. 2013). Li et al. (2017) observed Pb^0 on the $MAPbI_3$ surface after 120 min of UV irradiation with a wavelength of 408 nm. The generalisation of the suggested degradation reactions of $MAPbI_3$ with UV irradiation is given as follows.

$$(CH_3NH_3)PbI_3 \xrightarrow{UV} PbI_2 + HI \uparrow + CH_3NH_2 \uparrow \tag{12.21}$$

$$PbI_2 \xrightarrow{UV} Pb^0 + I_2 \tag{12.22}$$

After 480 min of UV irradiation, Pb^0 ratio is almost constant, indicating that the decomposition of $MAPbI_3$ has already reached saturation.

Furthermore, the potential for lead to enter the food chain through contaminated water and soil presents a long-term risk to human health. This underscores the need for effective disposal and recycling strategies to mitigate the environmental impact of Pb-based PSCs.

In the incident that PSCs are not properly disposed of at the end of their life, Pb contamination could occur through improper recycling or disposal practices. If Pb is released, it can contaminate soil, water, and air, leading to potential harm to local ecosystems. This is of particular concern in regions with high concentrations of solar installations, as the cumulative impact of lead contamination could have significant ecological consequences.

Current Mitigation Strategies

Given the potential risks associated with Pb in PSCs, several mitigation strategies have been proposed and implemented to reduce Pb toxicity and environmental impact. These strategies primarily focus on two main approaches: encapsulation of lead within the PSCs to prevent leaching, and the development of Pb-free alternatives to perovskite materials. To lessen the environmental impact of Pb-based PSCs, appropriate recycling and disposal techniques are also crucial, in addition to encapsulation and lead-free substitutes.

Encapsulation Techniques

Encapsulation is a common strategy used to reduce the environmental impact of Pb in PSCs. By encapsulating the perovskite material in protective layers, it is possible to prevent Pb ions from leaching into the environment during manufacturing, operation, or disposal. Various encapsulation materials have been explored, including polymers, glass, and metal oxides, which act as barriers to moisture, oxygen, and UV radiation. These protective layers help to stabilise the perovskite material and prevent degradation, thereby reducing the risk of Pb release (Dissanayake et al. 2023; Dong et al. 2024; Dipta et al. 2024).

Recent research has focussed on enhancing the stability and durability of encapsulation materials to ensure that they provide long-term protection against lead leaching. Additionally, efforts are underway to develop new encapsulation techniques that are more cost-effective and scalable, which is essential for the widespread commercialization of PSCs. The two main encapsulation techniques are single layer and multilayer encapsulation.

Single Layer Encapsulation

Single layer encapsulation as the name suggest is easy to manufacture and integrate with PSCs as compared to the multilayer encapsulation. The different materials which can be used as a single layer encapsulant materials can be inorganic or organic material.

Inorganic Material

Inorganic materials such as glass, silicon nitrides, and metal oxides such as aluminium oxide, titanium dioxide, tin oxide, etc. are widely used as encapsulants in PSCs (Meyer et al. 2009). Emami et al. (2020, 2018) created a low temperature based glass frit encapsulation technique using commercially available glass frits as an encapsulant material in PSCs. Two distinct glass frits were employed in their study: (i) a blend made up of lead zinc silicate, barium silicate, and silver nanoparticles; and (ii) a composite glass frit made up on vanadate tellurite glass that were sandwiched together. This glass fragment that was sandwiched was used as an encapsulant for the PSCs at a temperature of 120 °C with no damage.

Organic Material

Organic materials such as ethylene–vinyl acetate, epoxy, surlyn, polymethyl methacrylate, thermoplastic polyurethane, fluoro polymers, polyisobutylene are widely as encapsulants. Ethylene–vinyl acetate is one of the most widely used encapsulant material with characteristics such as high transmission of light ~ 91%, strong adhesion strength of 9–12 N/mm (90° peel). Bush et al. (2017) reported $Cs_{0.17}FA_{0.83}Pb(Br_{0.17}I_{0.83})_3$ device enclosed with EVA, glass, and butyl rubber edge seal and had a protective ITO capping layer. The encapsulated Pb-based PSCs had a 10% device efficiency, withstand the 1000-h damp heat test at 85% relative humidity at 85 °C. This encapsulation technique effectively shielded the PVSCs from humidity and temperature changes. Polyethylene glycol has been used by Fumani et al. (2020) as an interlayer between the device and encapsulant serving as an additional barrier layer to limit moisture intrusion. For over 450 days, the polyethylene glycol incorporated PSCs exhibited no decline in the performance, and even after a time of 830 days, 80% of the initial performance was maintained at a relative humidity of 28%.

Multilayer Thin Film Encapsulation

Another well-known encapsulating technique to stop moisture and volatile materials from entering the device is multilayer encapsulation (Zhang et al. 2015). The basic idea behind multilayer thin film encapsulation is stacking of single or multiple-layer film composed of alternating organic and inorganic layers (Fig. 12.3a). The inorganic layers are generally made up of metal oxides acting as a barrier for moisture and vapours, whereas the organic or polymer films are uniform in thickness. The inorganic layer has strong barrier qualities but consist of more flaws and abnormalities that generally allows entry of moisture and oxygen. These flaws are also dependent on the thickness of the film (Klumbies et al. 2015). To solve this problem, organic interlayers are used to avoid moisture penetration and also increase the flexibility without delamination, and with a water vapour transmission rate of less than 10^{-5} g/m^2/day (Yong-Qiang et al. 2014). PSCs are also integrated with numerous single and multilayer thin films, which have been demonstrated to extend the device's lifespan. While the inorganic layers can be produced using atomic layer deposition or chemical vapour deposition, the organic layers are typically deposited using lamination or inkjet printing processes. The multilayer encapsulation technique generates strong barrier properties due to the presence of inorganic layer, whereas the organic layer gives the diffusing species a greater flexibility along with a longer diffusion length (Fig. 12.3b) (Raman et al. 2021).

Lee et al. (2018) used chemical vapour deposition in conjunction with the atomic layer deposition process and created an organic–inorganic multi-layer encapsulant that was immediately placed on the PSC with a minimal loss of only 0.3%. The organic layer was poly (1, 3, 5-trimethyl-1, 3, 5-trivinylcyclotrisiloxane) (Fig. 12.3c), and Al_2O_3 as the inorganic layer. The PSC retained 97% of the efficiency after an exposure time 300 h at a temperature of 50 °C and relative humidity of 50%, confirming the enhanced stability of the PSC against moisture.

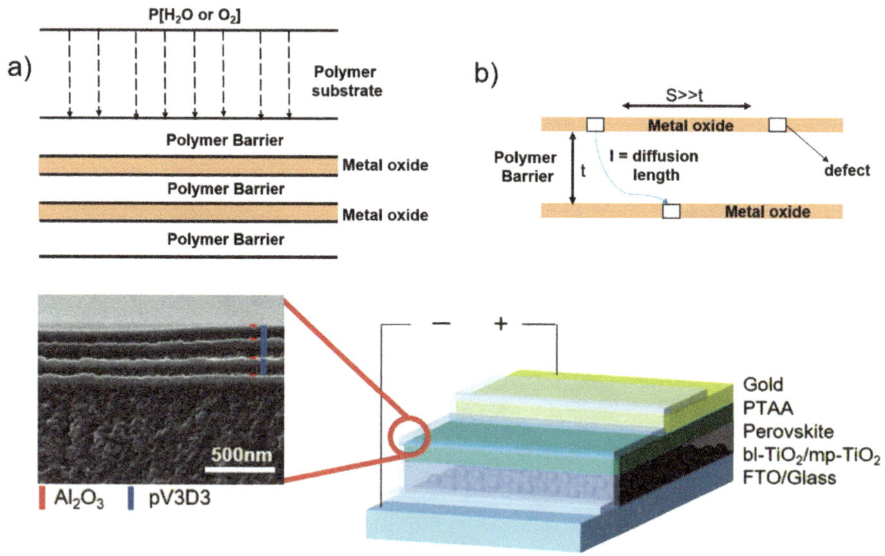

Fig. 12.3 **a** Schematic illustration of multilayer thin film encapsulation, **b** Diffusion pathway of oxygen across multilayer thin film encapsulation (**a**, **b** Reprinted with permission from Ref. Raman et al. (2021)] and **c** multi-layered encapsulated PSC along with the SEM image [Reprinted with permission from Ref. Lee et al. (2018)]

Lead-Free Perovskite Alternatives

Another approach to mitigating the Pb toxicity in PSCs is the development of Pb-free perovskite materials. Researchers have been exploring alternative metal cations, including tin (Sn), germanium (Ge) and bismuth (Bi) to replace Pb in PSCs. Some of the alternatives for Pb-based PSCs are depicted in Fig. 12.4. While these materials show promise in terms of their optoelectronic properties, they generally exhibit lower efficiency and stability in comparison to the Pb-based PSCs. To address these challenges, researchers are focussing on optimising the synthesis of lead-free perovskite materials and improving their stability under environmental conditions. Some of the less toxic Pb alternatives in PSCs can be given as follows:

Homovalent Substitution

Metals with + 2 oxidation state including Sn^{2+}, Ge^{2+} acts as an alternative to Pb^{2+} in Pb-based PSCs. The outer ns^2 orbital possess a significant effect on the opto-electronics properties and Pb(6s) orbital have an inactive characteristic, therefore the alternatives can be environmental friendly. The PSCs produced from different substitutes such as Sn and Ge have shown efficiency of 6.4% for $MASnI_3$ (Noel et al. 2014) and 0.2% for $MAGeI_3$ (Krishnamoorthy et al. 2015). Empirical findings

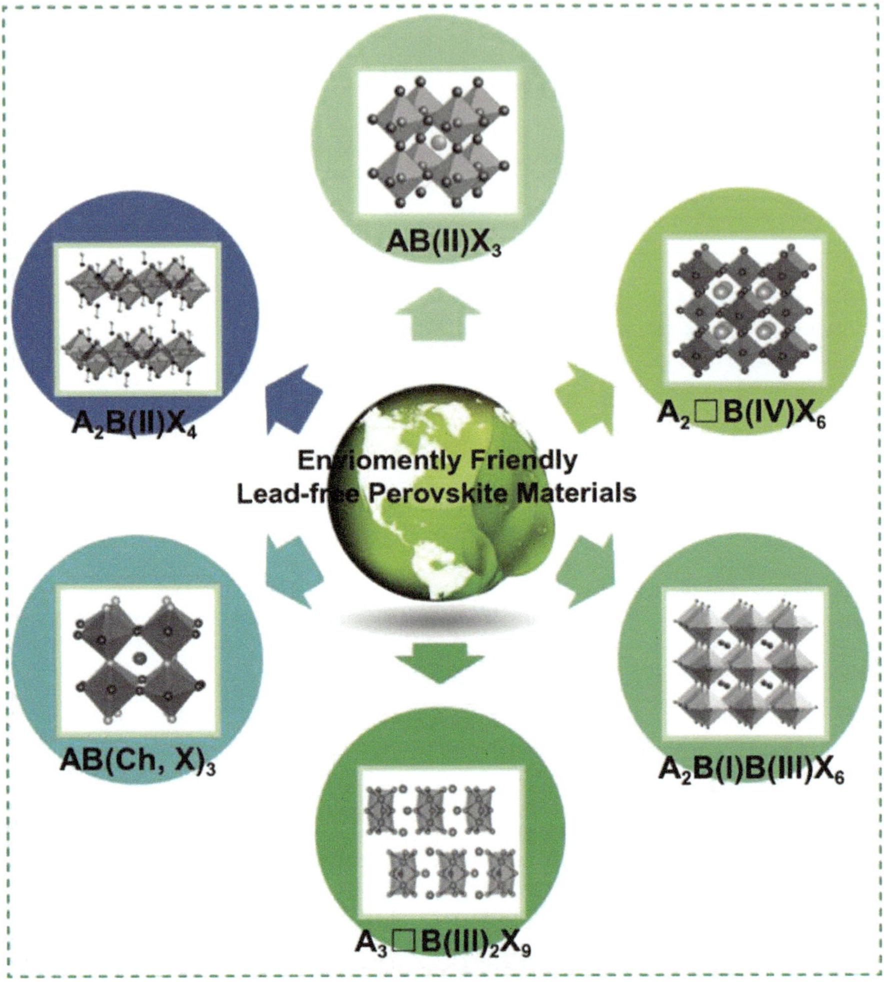

Fig. 12.4 Lead free perovskites for solar cells [Reproduced with permission from Ref. Wang et al. (2021)]

have depicted that substituting polar molecules with flexible orientation at the A site enables the production of a wide range of Pb-free perovskites as well as allows alteration in the phase structure, optoelectronic characteristics, and stability of these perovskites. Using theoretical study, 14 Sn- and Ge-based perovskites exhibited possible higher bulk-material-intrinsic PV performance. There is still much room for research, as only a few perovskites demonstrated some PV capability (Yang et al. 2017).

A new approach to lower the toxicity of Pb-based PSCs is offered by theoretical and experimental research with the partial substitution of Sr^{2+} (Wang et al. 2015), Ca^{2+}, and Cd^{2+}, modifying the electronic structure and carrier density of PSCs (Navas et al. 2015). This can be applicable for homovalent and heterovalent substitution, which involves co-substitution of trivalent and monovalent ions such as In^{3+}, Cu^{+}, and Mn^{+}. Ions such as V^{2+}, Mg^{2+}, Mn^{2+}, Ni^{2+}, Cd^{2+}, Hg^{2+}, Ga^{2+}, In^{2+}, Ge^{2+} and Sn^{2+}, have been proposed by Filip and Giustino (2016), considering the band gaps and stability. Out of the 25 potential compounds, Mg-based halide perovskites, such as $CsMgI_3$, $MAMgI_3$, and $FAMgI_3$, with estimated band gaps of 1.7, 1.5, and 0.9 eV, respectively, were emphasised (Filip and Giustino 2016).

Heterovalent Substitution

Another effective method is heterovalent substitution specifically, the $A_2{}^{I}B^{I}B^{III}X_6$ double PSCs displays ferroelectric, ferromagnetic, and multiferroic characteristics along with a tunable band gap. For photovoltaic applications, these improved optoelectronic characteristics contributes positively (Vasala and Karppinen 2015). Using Cs_2NaBiX_6, the first study on double PSCs was published in the 1970s (Morrs and Robinson 1972). In 2016, the first $A_2{}^{I}B^{I}B^{III}X_6$ double perovskites was created with B^{I} site occupied by the monovalent ions such as Cu, Ag, and Au and B^{III} site occupied by the trivalent ions such as Bi and Sb (Volonakis et al. 2016). This heterovalent replacement preserves the charge neutrality and permits band gap tunability of less than 2.7 eV covering the visible and near-infrared optical spectra. $Cs_2BiAgCl_6$, a double perovskites with an indirect band gap of about 2.2 eV demonstrates this replacement criterion successfully. The B^{I} group element may have empty s and outmost d shell orbital, for example Na^{+}, K^{+} or empty *s* and full outmost *d* shell orbital, for example Cu^{+}, Ag^{+} or full outmost s orbital, for example In^{+} or Tl^{+} (Zhao et al. 2017).

Additionally, new hybrid organic–inorganic materials are being explored to enhance the performance of lead-free PSCs while minimising their toxicity.

Recycling and Disposal Strategies

In addition to encapsulation and Pb-free alternatives, proper recycling and disposal strategies are essential to lessen the environmental impact of Pb-based PSCs. The development of effective recycling processes that can recover valuable materials from used PSCs, including Pb, is crucial to ensuring that perovskite solar technology remains sustainable in the long term. Several methods have been proposed for recycling PSCs, including mechanical, thermal, and chemical processes. These methods aim to recover lead and other materials from degraded or damaged PSCs while minimising the release of toxic substances into the environment. With a recovery efficiency of ~ 98.9%, butylamine assisted closed loop recycling and recrystallisation of PSCs has proved as an effective recycling technique. The performance of a recrystallised solar cell is comparable to that of a new solar cell (Feng et al. 2021).

Widely used techniques, like removing layers of PSCs is also possible as the material collected can be utilised again with a little performance loss. This PbI_2 recycling method has power conversion efficiency of 13.5% and can be recrystallized and reused (Binek et al. 2016). One of the main environmental problem related with PSC recycling is toxicity of the Pb leakage, however when compared to different energy conversion methods such as of coal, the environmental damage from Pb leaks via recycling can be considered as minimal. According to a study (Kadro et al. 2016), the photoanode was fully recovered without any delamination, and the gold was also recovered fully without any residue presence in the recovered materials. It was discovered that the Pb concentration in the solvents was 44 PPM per ml, suggesting material isolation of 99% with no significance drop in the PSC performance.

However, the scalability and cost-effectiveness of these methods remain a significant challenge, and further research is needed to develop efficient recycling technologies.

Future Challenges and Research Directions

Despite the progress made in mitigating the Pb toxicity in PSCs, several challenges remain that must be addressed to ensure the long-term sustainability of this technology. One of the most pressing challenges is improving perovskite material's stability to prevent Pb leaching under real-world operating conditions. Further research is needed to develop more robust encapsulation materials and improve the environmental stability of Pb-based PSCs.

Another challenge is the development of Pb-free alternatives that maintain or exceed the performance of Pb-based PSCs. While significant strides have been made in this area, Pb-based PSCs still face issues related to efficiency, stability, and scalability. Ongoing research into new materials and processing techniques will be critical for overcoming these challenges.

Additionally, the development of effective recycling and disposal strategies is essential to lessen the environmental effect of Pb-based PSCs. The creation of cost-effective, scalable recycling processes that can recover lead and other valuable materials from used PSCs will be a key area of research moving forward.

Finally, interdisciplinary approaches that combine materials science, environmental science, and public health will be essential for addressing the complexities of Pb toxicity in PSCs. Collaborative research efforts will be needed to develop innovative solutions that balance the performance, safety, and sustainability of PSCs.

Conclusion

PSCs becoming a viable next generation photovoltaic technology as they exhibit high conversion efficiencies and low manufacturing costs together with the possibility of mass production. On the other hand, the application of Pb in these devices presents harmful risks to health and the environment that if not resolved will affect the future of the technology. This chapter comprised the existing literature and scientific knowledge on Pb in PSCs, evaluating hazards related to the leaching of Pb, pollution of the environment as well as the risk to human beings. The present measures taken to address these issues including the encapsulation methods, search for alternatives without Pb, and recycling measures etc. Apart from several advantages, a number of gaps still remaining and hence for these additional efforts are needed to enhance the durability of Pb-containing perovskites, efficient recycling methods, and safe and secure Pb-free alternatives as viable solutions. Such measures would position perovskite solar photovoltaic technology to be developed to its full potential without putting health and the environment at risk. Although for attaining a bright future, PSCs require persistence through innovation and cooperation in processes that creates stable and environmentally friendly solutions to impending energy crisis and sustainability.

Acknowledgements Anuja Dhingra would like to acknowledge the support of CSIR-MoE as the form of Junior research fellowship and RP acknowledge the support of SERB for granting the ECRA project (ECR/2016/001152).

References

Binek A, Petrus ML, Huber N, Bristow H, Hu Y, Bein T, Docampo P (2016) Recycling perovskite solar cells to avoid lead waste. ACS Appl Mater Interfaces 8(20):12881–12886

Bryant D, Aristidou N, Pont S, Sanchez-Molina I, Chotchunangatchaval T, Wheeler S, Durrant JR, Haque SA (2016) Light and oxygen induced degradation limits the operational stability of methylammonium lead triiodide perovskite solar cells. Energy Environ Sci 9(5):1655–1660

Burschka J, Pellet N, Moon S-J, Humphry-Baker R, Gao P, Nazeeruddin MK, Grätzel M (2013) Sequential deposition as a route to high-performance perovskite-sensitized solar cells. Nature 499(7458):316–319

Bush KA, Palmstrom AF, Yu ZJ, Boccard M, Cheacharoen R, Mailoa JP, McMeekin DP, Hoye RLZ, Bailie CD, Leijtens T (2017) 23.6%-efficient monolithic perovskite/silicon tandem solar cells with improved stability. Nat Energy 2(4):1–7

Chetyrkina MR, Kameneva L, Mishchenko DV, Klimanova EN, Sashenkova TE, Allayarova UY, Kostyuk SV, Frolova LA, Aldoshin SM, Troshin PA (2023) Lead, tin, bismuth or organics: assessment of potential environmental and human health hazards originating from mature perovskite PV technology. Sol Energy Mater Sol Cells 252:112177

Christians JA, Miranda Herrera PA, Kamat PV (2015) Transformation of the excited state and photovoltaic efficiency of $CH_3NH_3PbI_3$ perovskite upon controlled exposure to humidified air. J Am Chem Soc 137(4):1530–1538

De Wolf S, Holovsky J, Moon S-J, Loper P, Niesen B, Ledinsky M, Haug F-J, Yum J-H, Ballif C (2014) Organometallic halide perovskites: sharp optical absorption edge and its relation to photovoltaic performance. J Phys Chem Lett 5(6):1035–1039

Dipta SS, Rahim MA, Uddin A (2024) Encapsulating perovskite solar cells for long-term stability and prevention of lead toxicity. Appl Phys Rev 11(2)

Dissanayake PD, Yeom KM, Sarkar B, Alessi DS, Hou D, Rinklebe J, Noh JH, Ok YS (2023) Environmental impact of metal halide perovskite solar cells and potential mitigation strategies: a critical review. Environ Res 219:115066

Dong H, Shen G, Fang H, Li Y, Gao X, Song Q, Xu X, Wang Y, Mu C, Xu D (2024) Internal encapsulation strategy using a polymer enables efficient, stable, and lead-safe inverted perovskite solar cells. Adv Funct Mater 2024:2402394

Emami S, Martins J, Madureira R, Hernandez D, Bernardo G, Mendes J, Mendes A (2018) Development of hermetic glass frit encapsulation for perovskite solar cells. J Phys D Appl Phys 52(7):074005

Emami S, Martins J, Ivanou D, Mendes A (2020) Advanced hermetic encapsulation of perovskite solar cells: the route to commercialization. J Mater Chem A Mater 8(5):2654–2662

Feng X, Guo Q, Xiu J, Ying Z, Ng KW, Huang L, Wang S, Pan H, Tang Z, He Z (2021) Close-loop recycling of perovskite solar cells through dissolution-recrystallization of perovskite by butylamine. Cell Rep Phys Sci 2(2)

Filip MR, Giustino F (2016) Computational screening of homovalent lead substitution in organic-inorganic halide perovskites. J Phys Chem C 120(1):166–173

Frost JM, Butler KT, Brivio F, Hendon CH, Van Schilfgaarde M, Walsh A (2014) Atomistic origins of high-performance in hybrid halide perovskite solar cells. Nano Lett 14(5):2584–2590

Mansour Rezaei Fumani N, Arabpour Roghabadi F, Alidaei M, Sadrameli SM, Ahmadi V, Najafi F (2020) Prolonged lifetime of perovskite solar cells using a moisture-blocked and temperature-controlled encapsulation system comprising a phase change material as a cooling agent. ACS Omega 5(13):7106–7114

Godding JSW, Ramadan AJ, Lin Y-H, Schutt K, Snaith HJ, Wenger B (2019) Oxidative passivation of metal halide perovskites. Joule 3(11):2716–2731

Hailegnaw B, Kirmayer S, Edri E, Hodes G, Cahen D (2015) Rain on methylammonium lead iodide based perovskites: possible environmental effects of perovskite solar cells. J Phys Chem Lett 6(9):1543–1547

Han Y, Meyer S, Dkhissi Y, Weber K, Pringle JM, Bach U, Spiccia L, Cheng Y-B (2015) Degradation observations of encapsulated planar $CH_3NH_3PbI_3$ perovskite solar cells at high temperatures and humidity. J Mater Chem A Mater 3(15):8139–8147

Hwang K, Jung Y, Heo Y, Scholes FH, Watkins SE, Subbiah J, Jones DJ, Kim D, Vak D (2015) Toward large scale roll-to-roll production of fully printed perovskite solar cells. Adv Mater 27(7):1241–1247

Jeon NJ, Noh JH, Yang WS, Kim YC, Ryu S, Seo J, Seok S (2015) Compositional engineering of perovskite materials for high-performance solar cells. Nature 517(7535):476–480

Jiang J, Wang Q, Jin Z, Zhang X, Lei J, Bin H, Zhang Z, Li Y, Liu S (2018) Polymer doping for high-efficiency perovskite solar cells with improved moisture stability. Adv Energy Mater 8(3):1701757

Jiang Y, Qiu L, Juarez-Perez EJ, Ono LK, Hu Z, Liu Z, Wu Z, Meng L, Wang Q, Qi Y (2019) Reduction of lead leakage from damaged lead halide perovskite solar modules using self-healing polymer-based encapsulation. Nat Energy 4(7):585–593

Jokar E, Chien C, Tsai C, Fathi A, Diau EW (2019) Robust tin-based perovskite solar cells with hybrid organic cations to attain efficiency approaching 10%. Adv Mater 31(2):1804835

Kadro JM, Pellet N, Giordano F, Ulianov A, Müntener O, Maier J, Grätzel M, Hagfeldt A (2016) Proof-of-concept for facile perovskite solar cell recycling. Energy Environ Sci 9(10):3172–3179

Ke W, Stoumpos CC, Kanatzidis MG (2019) "Unleaded" perovskites: status quo and future prospects of tin-based perovskite solar cells. Adv Mater 31(47):1803230

Klumbies H, Schmidt P, Hähnel M, Singh A, Schroeder U, Richter C, Mikolajick T, Hoßbach C, Albert M, Bartha JW (2015) Thickness dependent barrier performance of permeation barriers made from atomic layer deposited alumina for organic devices. Org Electron 17:138–143
Krishnamoorthy T, Ding H, Yan C, Leong WL, Baikie T, Zhang Z, Sherburne M, Li S, Asta M, Mathews N (2015) Lead-free germanium iodide perovskite materials for photovoltaic applications. J Mater Chem A Mater 3(47):23829–23832
Kwak J, Nam S-H, Kim L, An Y-J (2020) Potential environmental risk of solar cells: current knowledge and future challenges. J Hazard Mater 392:122297
Lee MM, Teuscher J, Miyasaka T, Murakami TN, Snaith HJ (2012) Efficient hybrid solar cells based on meso-superstructured organometal halide perovskites. Science 338(6107):643–647
Lee Y, Jeon NJ, Kim BJ, Shim H, Yang T, Seok S, Seo J, Im SG (2018) A low-temperature thin-film encapsulation for enhanced stability of a highly efficient perovskite solar cell. Adv Energy Mater 8(9):1701928
Leijtens T, Eperon GE, Pathak S, Abate A, Lee MM, Snaith HJ (2013) Overcoming ultraviolet light instability of sensitized TiO_2 with meso-superstructured organometal tri-halide perovskite solar cells. Nat Commun 4(1):2885
Li F, Liu M (2017) Recent efficient strategies for improving the moisture stability of perovskite solar cells. J Mater Chem A Mater 5(30):15447–15459
Li W, Zhang W, Van Reenen S, Sutton RJ, Fan J, Haghighirad AA, Johnston MB, Wang L, Snaith HJ (2016) Enhanced UV-light stability of planar heterojunction perovskite solar cells with caesium bromide interface modification. Energy Environ Sci 9(2):490–498
Li Y, Xu X, Wang C, Ecker B, Yang J, Huang J, Gao Y (2017) Light-induced degradation of $CH_3NH_3PbI_3$ hybrid perovskite thin film. J Phys Chem C 121(7):3904–3910
Lyu M, Yun J, Chen P, Hao M, Wang L (2017) Addressing toxicity of lead: progress and applications of low-toxic metal halide perovskites and their derivatives. Adv Energy Mater 7(15):1602512
Meyer J, Görrn P, Bertram F, Hamwi S, Winkler T, Johannes H-H, Weimann T, Hinze P, Riedl T, Kowalsky W (2009) Al_2O_3/ZrO_2 nanolaminates as ultrahigh gas-diffusion barriers-a strategy for reliable encapsulation of organic electronics. Adv Mater 21(18):1845–1849
Morrs LR, Robinson WR (1972) Crystal structure of $Cs_2NaBiCl_6$. Struct Sci 28(2):653–654
Navas J, Sánchez-Coronilla A, Gallardo JJ, Hernández NC, Piñero JC, Alcántara R, Fernández-Lorenzo C, Desireé M, Aguilar T, Martín-Calleja J (2015) New insights into organic–inorganic hybrid perovskite $CH_3NH_3PbI_3$ nanoparticles. An experimental and theoretical study of doping in Pb^{2+} sites with Sn^{2+}, Sr^{2+}, Cd^{2+} and Ca^{2+}. Nanoscale 7(14):6216–6229
Niu G, Li W, Meng F, Wang L, Dong H, Qiu Y (2014) Study on the stability of $CH_3NH_3PbI_3$ films and the effect of post-modification by aluminum oxide in all-solid-state hybrid solar cells. J Mater Chem A Mater 2(3):705–710
Noel NK, Stranks SD, Abate A, Wehrenfennig C, Guarnera S, Haghighirad A-A, Sadhanala A, Eperon GE, Pathak SK, Johnston MB (2014) Lead-free organic-inorganic tin halide perovskites for photovoltaic applications. Energy Environ Sci 7(9):3061–3068
Ouyang Y, Li Y, Zhu P, Li Q, Gao Y, Tong J, Shi L, Zhou Q, Ling C, Chen Q (2019) Photo-oxidative degradation of methylammonium lead iodide perovskite: mechanism and protection. J Mater Chem A Mater 7(5):2275–2282
Raman RK, Thangavelu SAG, Venkataraj S, Krishnamoorthy A (2021) Materials, methods and strategies for encapsulation of perovskite solar cells: from past to present. Renew Sustain Energy Rev 151:111608
Raturi AK (2019) Renewables 2019 global status report
Schileo G, Grancini G (2021) Lead or no lead? Availability, toxicity, sustainability and environmental impact of lead-free perovskite solar cells. J Mater Chem C Mater 9(1):67–76
Snaith HJ (2018) Present status and future prospects of perovskite photovoltaics. Nat Mater 17(5):372–376
Stoumpos CC, Frazer L, Clark DJ, Kim YS, Rhim SH, Freeman AJ, Ketterson JB, Jang JI, Kanatzidis MG (2015) Hybrid germanium iodide perovskite semiconductors: active lone pairs, structural

distortions, direct and indirect energy gaps, and strong nonlinear optical properties. J Am Chem Soc 137(21):6804–6819

Stranks SD, Eperon GE, Grancini G, Menelaou C, Alcocer MJP, Leijtens T, Herz LM, Petrozza A, Snaith HJ (2013) Electron-hole diffusion lengths exceeding 1 micrometer in an organometal trihalide perovskite absorber. Science 342(6156):341–344

Sultana N, Demarais NJ, Shevchenko D, Derrick PJ (2018) Laser desorption/ionization mass spectrometry of perovskite solar cells: identification of interface interactions and degradation reactions. Solar RRL 2(8):1800022

Umari P, Mosconi E, De Angelis F (2014) Relativistic GW calculations on $CH_3NH_3PbI_3$ and $CH_3NH_3SnI_3$ perovskites for solar cell applications. Sci Rep 4(1):4467

Vasala S, Karppinen M (2015) $A_2B'B''O_6$ perovskites: a review. Prog Solid State Chem 43(1–2):1–36

Volonakis G, Filip MR, Haghighirad AA, Sakai N, Wenger B, Snaith HJ, Giustino F (2016) Lead-free halide double perovskites via heterovalent substitution of noble metals. J Phys Chem Lett 7(7):1254–1259

Wang K, Liang Z, Wang X, Cui X (2015) Lead replacement in $CH_3NH_3PbI_3$ perovskites. Adv Electron Mater 1(10):1500089

Wang Q, Chen B, Liu Y, Deng Y, Bai Y, Dong Q, Huang J (2017) Scaling behavior of moisture-induced grain degradation in polycrystalline hybrid perovskite thin films. Energy Environ Sci 10(2):516–522

Wang M, Wang W, Ma B, Shen W, Liu L, Cao K, Chen S, Huang W (2021) Lead-free perovskite materials for solar cells. Nanomicro Lett 13:1–36

Xiao Z, Song Z, Yan Y (2019) From lead halide perovskites to lead-free metal halide perovskites and perovskite derivatives. Adv Mater 31(47):1803792

Yang D, Lv J, Zhao X, Xu Q, Fu Y, Zhan Y, Zunger A, Zhang L (2017) Functionality-directed screening of Pb-free hybrid organic-inorganic perovskites with desired intrinsic photovoltaic functionalities. Chem Mater 29(2):524–538

Yao H, Zhou F, Li Z, Ci Z, Ding L, Jin Z (2020) Strategies for improving the stability of tin-based perovskite (AS_nX_3) solar cells. Adv Sci 7(10):1903540

Yin W-J, Shi T, Yan Y (2014) Unique properties of halide perovskites as possible origins of the superior solar cell performance. Adv Mater 26(27):4653–4658

Yong-Qiang Y, Yu D, Ya-Hui D, Xiao W, Ping C, Dan Y, Feng-Bo S, Kai-wen X (2014) High barrier properties of transparent thin-film encapsulations for top emission organic light-emitting diodes. Org Electron 15(6):1120–1125

Zhang H, Ding H, Wei M, Li C, Wei B, Zhang J (2015) Thin film encapsulation for organic light-emitting diodes using inorganic/organic hybrid layers by atomic layer deposition. Nanoscale Res Lett 10:1–5

Zhang Q, Hao F, Li J, Zhou Y, Wei Y, Lin H (2018) Perovskite solar cells: Must lead be replaced and can it be done? Sci Technol Adv Mater 19(1):425–442

Zhao X-G, Yang J-H, Fu Y, Yang D, Xu Q, Yu L, Wei S-H, Zhang L (2017) Design of lead-free inorganic halide perovskites for solar cells via cation-transmutation. J Am Chem Soc 139(7):2630–2638

Zhou Y, Herz LM, Jen AKY, Saliba M (2022) Advances and challenges in understanding the microscopic structure–property–performance relationship in perovskite solar cells. Nat Energy 7(9):794–807

Zhu Y, Kang Y, Huang H, Zhuang D, Li M, Ling Z, Peng K, Zhai L, Zou C (2024) Systematic evaluation of the biotoxicity of Pb-based perovskite materials and perovskite solar cells. J Mater Chem A Mater 12(5):2916–2923

Zhuang J, Wang J, Yan F (2023) Review on chemical stability of lead halide perovskite solar cells. Nanomicro Lett 15(1):84

Chapter 13
Perovskite–Perovskite Tandem Solar Cells: Basics to Latest Updates with Challenges

Anne Masih, Vidhi, Anjani Kumar Singh, and O. P. Thakur

Abstract Perovskite–perovskite tandem solar cells exhibit a cutting-edge performance in optoelectronics technology, promising to revolutionize the photovoltaic community with their tunable bandgaps, exceptional efficiency and potential for cost effective production. In this chapter the principles, unique material characteristics and the functioning of perovskite solar cells (PSC) were explored. The design and structure of tandem setups is such that, two perovskite layers with different bandgaps are layered to enhance light absorption and charge collection capabilities over a larger range of the solar spectrum. The chapter also reviews recent accomplishments, stability studies and scalable manufacturing methods for tandem devices and the advancements made in power conversion efficiencies, surpassing 30% recently. This chapter also presents case studies of appreciable effective perovskite–perovskite tandem solar cells and their ongoing research, commercialization and development of this these solar cells for day to day industrial uses. PSCs may become favorable choice for high quality solar cells due to their higher absorption spectrum (~2.23 eV) when compared to their silicon counterparts (~1.48 eV). The flexibility in adjusting the bandgap of perovskite–perovskite materials enables the optimization of tandem cell architectures, further enhancing their performance. Various methods to enhance the carrier diffusion length of narrow bandgap cells and reduce voltage loss in wide bandgap cells is also discussed. Perovskite–perovskite tandem solar cells have different challenges such as stability and degradation problems, interface engineering difficulties, lead toxicity issues, and scalability obstacles. The challenges, current approaches and future research directions to address them are discussed for rapid commercial deployment.

A. Masih · Vidhi · O. P. Thakur (✉)
Materials Analysis and Research Laboratory, Department of Applied Physics, Netaji Subhas University of Technology, Dwarka, New Delhi, India
e-mail: opthakur@nsut.ac.in

A. K. Singh
Department of Physics, Atma Ram Sanatan Dharma College, University of Delhi, Dhaula Kuan, New Delhi, India

M. Kumar and A. Raj (eds.), *Perovskite Solar Cells*, Springer Tracts in Electrical and Electronics Engineering, https://doi.org/10.1007/978-981-96-9463-1_13

Keywords Tandem solar cells · Perovskite–perovskite · Power conversion efficiencies · Interface engineering

Introduction

Over the last few decades the energy consumption have been increased rapidly. This has caused an emergency to explore the scientific areas of carbon-free sources of energy to resolve the spike in utilizing the existing fuel reserves from fossils and carbon energy resources. Photovoltaic solar cells have played an important role in decarbonizing the power sector, since sun is an inexhaustible, reliable, and a pure source of energy (Pietzcker et al. 2014). Perovskite solar cells have become highly attractive since 2009, due to their exceedingly high efficiency as high as 25%. The progress in solar cells can be studied through different generations. First generation solar cells were built around single crystal gallium arsenide and silicon. The second generation solar cells were based on amorphous as well as poly-crystalline thin films. Both of these classifications of solar cell have a higher cost of production. In comparison to these a third generation solar cells were developed which were cost effective as well as had higher efficiency (Noman et al. 2024). Third generation photovoltaic solar cells were made up of different materials like organics (Chamberlain 1983), dye-sensitized (Richhariya et al. 2017), quantum dot (Nozik 2002) as well as halide perovskite (Berhe et al. 2016). Tandem or multijunction cells were proven to amplify the power conversion efficiency of different type of solar cells.

Perovskite materials have shown great promise as prospects in respect to the fabrication of tandem solar cells (TSCs) due of their remarkable optoelectronic characteristics. Some of these include low exciton binding energy, bipolar carrier transport, high carrier mobility, and a high light absorption coefficient. Their easy solution processability and customizable bandgap further add to their desirability. Through the use of compositional engineering, it is possible to continually tune the bandgaps of perovskites between 1.2 and 3.0 eV, which makes it possible to create multijunction solar cells made entirely of perovskite materials (Chen et al. 2015a). Several studies have shown that improving the quality of perovskite crystallization and modifying the interface can lead to considerable increases in power conversion efficiency (PCE). Due to their self-assembling nature, perovskite-based hybrids can be produced in an inexpensive and simple manner. A common cell architecture consists of a hole-transporting material sandwiched between a layer of nanocrystalline metal oxide coated with the UV light-absorbing methylammonium lead halide ($MAPbX_3$) which is porous. Within the field of photovoltaics, these cells are especially appealing because to their high PCE and straightforward construction method (Wang et al. 2019; Ko et al. 2015; Tai et al. 2016; Park et al. 2019; Yang et al. 2024).

It takes skill to create an effective tandem solar cell which is made up of numerous thin-film layers. For maximum and balanced absorption of solar light in each absorber layer, and to permit as nearly perfect photovoltaic diode operation as possible in a multijunctional solar cell, one has to be cautious about the minimization of parasitic absorption and thin-film interference effects (Bush et al. 2018; Husain et al. 2018). The tunable bandgap of $MAPbX_3$-based perovskites, which can be accurately altered through chemical replacement, is one of their most attractive properties. Translucent hues can be produced by altering the bandgap by substituting bromide ions (Br^-) for iodide ions (I^-) (Zhou et al. 2018). The substantial impact of ionic substitution on the electronic structure—specifically, the effective exciton mass—directly relates to this bandgap tuning and governs optical and electrical effects of the material. Due to their tunability, perovskite solar cells can be used in a wide range of architectural applications, allowing for the design of both energy-generating and aesthetically pleasing solar cells. These solar cells could then be installed within windows, roofs, and walls, and countless other structural members (Bing et al. 2022; Lee et al. 2020; Cao et al. 2015). They have two purposes: they quality the architectural specifications and produce solar electricity. Perovskite solar cells are an amazing option for contemporary, environmentally friendly construction solutions because of their exceptional blend of design flexibility and functionality.

Tandem solar cells divide the light spectrum into two regions in order to enhance absorption. Each absorber in a tandem system is set to efficiently absorb photons whose energy matches its specific bandgap in order to maximize energy conversion. The tandem solar cell has a comparatively low parallel resistance (Bett et al. 2020). Photons having energy over band-gap are also absorbed, but energy which is surplus is released as heat through a process known as thermalization. Tandem solar cells use multiple absorbers, each responsible for a different region of the spectrum, to considerably reduce thermalization losses over a single absorber covering the entire spectrum (Bailie and McGehee 2015). Creating perovskite–perovskite tandem devices is an alternate strategy that capitalizes on the readily adjustable bandgap of perovskite absorbers and the utilization of affordable precursor materials. Methylammonium lead iodide ($MAPbI_3$) was used to create two similar subcells in the initial instance of a perovskite–perovskite tandem cell. These subcells were joined by lamination using a doped hole transport layer (HTL) that was 2 μm thick. Even though the two subcells' combined photovoltaic output reached a high open-circuit voltage (V_{oc}) of 2.2 V, PCE was constrained by the same absorbers, which drastically decreased the current density that could be achieved (Heo and Im 2016).

Basic Structure of Perovskite

Perovskite is a family of mineral similar to calcium titanium oxide ($CaTiO_3$). It was discovered in Russia's scientist by Gustav rose and named after L. A. Perovsk at the Ural Mountains. The basic perovskite structure can be represented by the formula AMX_3, in which M is a metal cation, and X is an anion (such as O_2, Cl, Br, I, etc.). Structure features MX_6 octahedra with the M cation in the middle and X anions at the corners as seen in Fig. 13.1. These octahedra establish an extended 3D network, and the A cations fill voids among the octahedra, balancing the network's charge. These materials have unique properties and applications, including colossal magnetoresistance (Tokura and Tomioka 1999), ferroelectricity (Moure and Peña 2015), superconductivity (Rubel et al. 2022), and other transport properties (Souza 2021). Perovskite structures consist of [TiO_6] octahedra with Sr^{2+} ions located in the voids between polyhedral, as depicted in Fig. 13.1. The ABX_3 perovskite structure which is cubic ($a = 3.905$ Å, space group Pm-3m, $Z = 1$), the A cations occupy the Wyckoff configurations 1b (½, ½, ½), B atoms are in 1a (0, 0, 0), and X atoms are in 3d (½, 0, 0; 0, ½, 0; 0, 0, ½). If Sr^{2+} ions (A) are absent, the remaining framework resembles the ReO_3 type (Johnsson and Lemmens 2005).

Researchers have classified perovskite structures based on the radii of their metallic ions. The ABO_3 perovskite structure is adaptable, accommodating a range of cations with varying oxidation states. An important property of perovskites is their potential to undergo multiple cation substitutions (Dimitrovska et al. 2005), forming diverse compounds, including those with different cations at the B position ($AB_xB_{1-x}O_3$), at the A position ($A_xA_{1-x}BO_3$), or substitutions at both cation positions ($A_xA_{1-x}B_xB_{1-x}O_3$). Typically, A and B cations have valences of approximately 2+ and 4+, respectively, though sometimes both may be 3+ if the B cation has sixfold coordination (Kuzmanovski et al. 2007). Perovskite oxide phases are categorized into two main types (Table 13.1):

1. **Ternary Oxide ABO_3 Type**: These can be further classified by oxidation state into $A_1{}^+B_5{}^+O_3$, $A_2{}^+B_4{}^+O_3$, $A_3{}^+B_3{}^+O_3$, along with oxygen and cation-deficient species (Dimitrovska et al. 2005).
2. **Complex Oxide Compounds**: These newer types $(AB'_xB''_y)O_3$ contain two distinct elements at the B position with different oxidation states, where $x + y = 1$ (Bhatti et al. 2016).

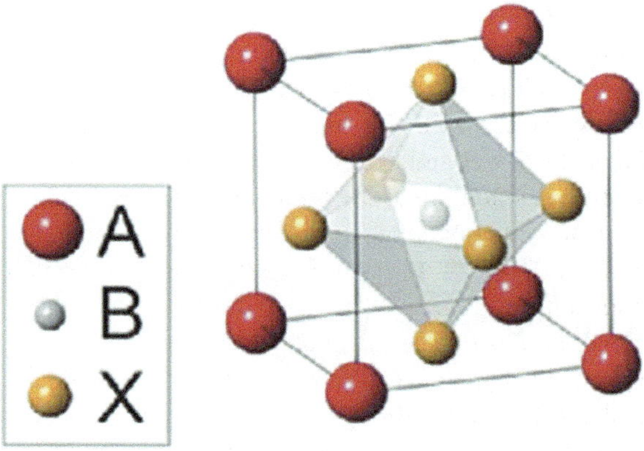

Fig. 13.1 Basic perovskite structure (Dawa and Sajjadi 2024). Copyright 2023 Reproduced with permission from Elsevier

Table 13.1 Examples of perovskite materials corresponding to different properties

Property	Compound
Ferromagnetism	$BaTiO_3$, $PdTiO_3$
Piezoelectricity	$(Pb[Zr_xTi_{1-x}]O_3)$, $(Bi, Na)TiO_3$
Electrical conductivity	ReO_3, $SrFeO_3$, $LaCrO_3$, $LaCoO_3$, $LaNiO_3$
Superconductivity	$La_{0.9}Sr_{0.1}CuO_3$, $YBa_2Cu_3O_7$, $HgBa_2Ca_2Cu_2O_8$
Ion conductivity	$La(Ca)AlO_3$ $BaZrO_3$, $CaTiO_3$, $SrZrO_3$, $BaCeO_3$, $La(Sr)Ga(Mg)O_3$
Magnetism	$LaMnO_3$, $LaFeO_3$, La_2NiMnO_6
Catalytic behavior	$LaCoO_3$, $LaMnO_3$, $BaCuO_3$
Material for electrode	$CH_3NH_3PbI_3$), $La_{0.8}Ca_{0.2}MnO_3$

Distortions in the Structure of Perovskite

Perovskite structures exhibit various distortions, which can be classified into five primary types:

1. **Distortion of the MX_6 Octahedron**: The distortion is often caused by the Jahn–Teller effect in perovskites containing Jahn–Teller ions at the octahedral site, such as Cu^{2+} and Mn^{3+}; it leads to charge ordering, spin ordering, and cooperative distortions. SOJT distortions occur when high-valent d^0 cations (e.g., Ti^{4+}, Nb^{5+}) and/or lone-pair cations (e.g., Pb^{2+}, Bi^{3+}) are displaced from their central positions in the polyhedral (Lufaso and Woodward 2004; Grandhi et al. 2020).
2. **Off-Center Displacement of M Cations**: The off-center displacement, which occurs within the MX_6 octahedron, is the key possibility of ferroelectric behavior. This displacement within the MX_6 octahedron is key to possible ferroelectric behavior (Stokes et al. 2002; King and Woodward 2010).
3. **Tilting of the Octahedral Framework**: Typically it is caused because of an A cation that is too small at the cuboctahedral site, causing the octahedra to tilt (Kim et al. 2021a).
4. **Ordering of Cations or Vacancies**: This involves the arrangement of multiple types of A or M cations, or vacancies, within the structure (Woodwardt 1997).
5. **Ordering of Anions or Vacancies**: Structural distortions occur when anion ordering deviates from ideal stoichiometry, reducing symmetry (Clemens et al. 2016).

The above mentioned distortions significantly impact and change the physical properties of perovskites, influencing electronic, magnetic, and dielectric behaviors essential for various applications. Fine-tuning these properties through controlled cation substitution allows for targeted modifications of perovskite material characteristics for specific application (Hossain et al. 2019).

Tandem Solar Cells (TSCs)

A tandem solar cell consists of two sub-cells with the respective absorbers having complementary absorption ranges and stacked on each other. Each of these sub-cells captures a different segment of the solar spectrum, so that the system can capture a larger portion of the solar radiation. This design greatly enhances the overall efficiency of energy conversion into electrical energy since the whole available sunlight is utilized.

Working Principle of Tandem Solar Cells

A tandem solar cell, comprises of many single junction solar cell with varying absorption ranges are stacked in such a manner that they absorb a wide broad range of photon energy to maximize their conversion efficiency (Chen et al. 2015b). There is a limit up to which efficiency of solar TSC can be exploited. The limit was explored according to the Shockley–Queisser analysis (Wang et al. 2022a). TSC have a tunable band gap, furthermore they have a high structural tolerance. The improvement the opto-electronics in the tandem cells could be increased by reducing the number of active compounds as well as their complexity. The Goldschmidt's tolerance factor is of the parameter (t) which is used to evaluate the geometrical factor is given by the following expression (Fedorovskiy et al. 2020):

$$t = \frac{r_A + r_x}{\sqrt{2}[r_B + r_x]}$$

where $r_{A,}$ $r_{B,}$ r_x are ionic radii of ions occupying the A, B and X sites. The Goldschmidt factor predicts the accuracy of the formation of perovskite. Value of 't' between 0.8 to 1 signifies a 3D structure whereas the if the value of t exceeds 1 then it signifies a 2D structure (Raj et al. 2022).

Perovskite–Perovskite Tandem Solar Cells

Versatility of bandgaps of perovskite–perovskite TSC makes them an excellent alternative in the context to improve the performance of these TSCs as well as reducing the cost of fabricating multijunctional cells. These tandem cells can be classified into two major groups: 4-terminal and 2-terminal devices as shown in Fig. 13.2, which is decided by the degree of electrical conductivity (Zheng et al. 2020).

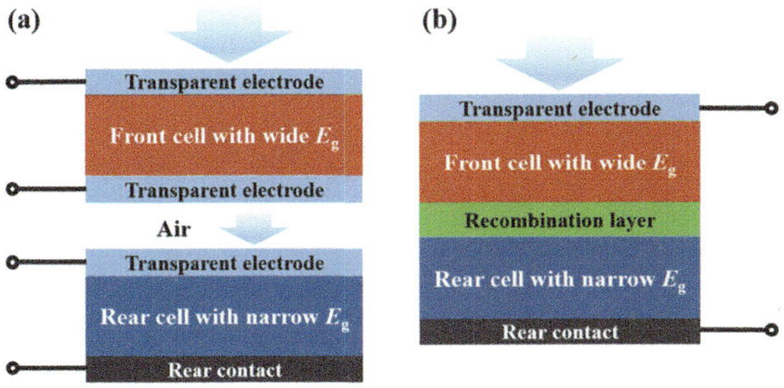

Fig. 13.2 Schematic diagram of **a** 4-terminal and **b** 2-terminal tandem solar cell (Fang et al. 2021). Copyright 2020 reproduced with permission from Elsevier

Fundamentals of Perovskite Tandem Cells

Organic Inorganic Perovskite Tandem Cell

Due to their remarkable optoelectronic capabilities, two-dimensional (2D) perovskites with sheets of Ge(II), Sn(II), and Pb(II) halide have attracted a lot of attention recently (Kim et al. 2021b). These materials naturally form a multi-quantum-well structure, where the {MX6} inorganic semiconductor sheets (Gauthron et al. 2001) behaves as the wells, and the organic layers function as barriers. Because they are inert, the organic layers have a bigger bandgap and a lower dielectric constant than the inorganic components. The generation of stable excitons with high binding energy within the inorganic semiconductor sheets is facilitated by this 2D architecture in conjunction with the dielectric confinement effect. Therefore, these structures display remarkable optical characteristics, such as efficient photoluminescence (Misra and Cohen 2017), heterostructure electroluminescence, and substantial third-order optical nonlinearity (Gippius et al. 1993). The composition of metal halide has an extreme impact on the excitonic absorption and light emission characteristics of 2D perovskites. For example, the substitution of different halogens dramatically alters the absorbance and photoluminescence behavior of $(C_6H_5CH_2NH_3)_2PbX_4$, demonstrating the materials adjustable nature for a variety of optoelectronic applications (Fig. 13.3).

The perovskite structure is incredibly adaptable, enabling a multitude of configurations through changes to the sizes, orientations, and mixes of different inorganic and organic elements. The final structure of perovskites is highly dependent on the organic ligands charge properties, configuration, and structural arrangement.

The Perovskite Top Cell's Structure

The top sub-cells must have a wider bandgap between 1.7 and 1.9 eV, and the bottom sub-cells must have an optical bandgap between 0.9 and 1.3 eV for good tandem solar cells (Eperon et al. 2016).

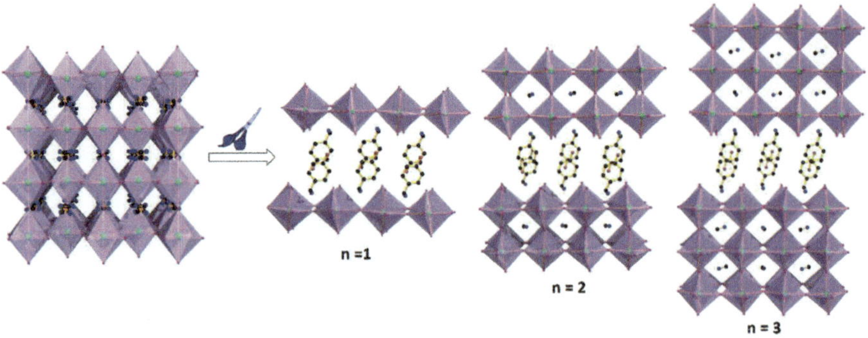

Fig. 13.3 Two dimensional layer of inorganic organic perovskite (ZhenTao et al. 2020). Copyright 2019 Reproduced with permission from Elsevier

Perovskite absorbers with huge bandgaps designed for tandem solar cells were created by Eperon et al. To be more precise, they created $FA_{0.83}Cs_{0.17}Pb(I_{0.63}Br_{0.17})_3$ for the top cell in a four-terminal tandem arrangement, and $FA_{0.83}Cs_{0.17}Pb(I_{0.5}Br_{0.5})_3$ with a 1.8 eV bandgap for monolithic tandems. The sub-cell structures have a p-i-n configuration with indium tin oxide (ITO) serving as the transparent top contact (Nakka et al. 2023). The n-type electron transport layer (ETL) is methyl [6,6]-phenyl-C61-butanoate, and the hole transport layer (HTL) is nickel oxide as shown in Fig. 13.4 (Eperon et al. 2016).

Fig. 13.4 Structure of perovskite top cell

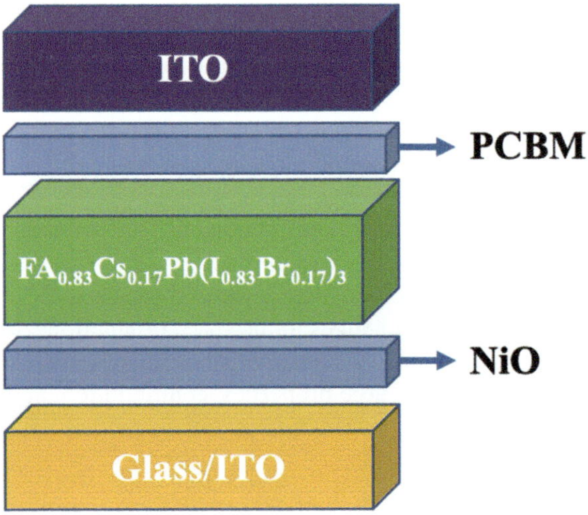

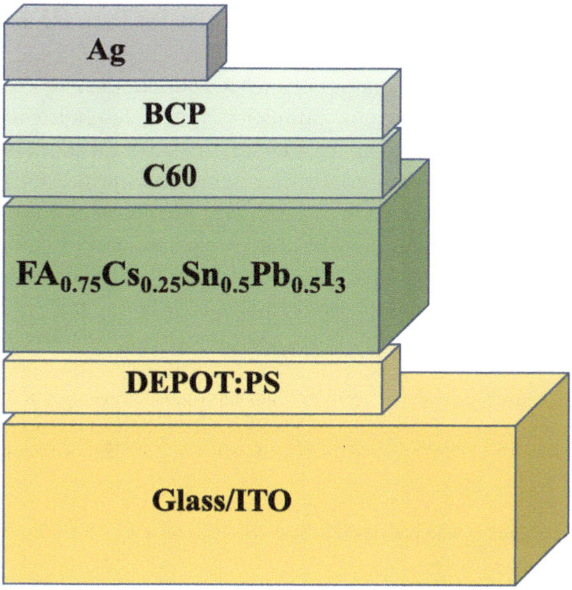

Fig. 13.5 Structure of perovskite bottom cell

Structure of the Perovskite Bottom Cell

The perovskite absorber $FA_{0.75}Cs_{0.25}Sn_{0.5}Pb_{0.5}I_3$ was designed by Eperon et al. with an adjustable bandgap of 1.2 eV, with the purpose of being used as the bottom subcell in tandem photovoltaic setups. Poly(3,4-ethylenedioxythiophene) (styrenesulfonate) (PEDOT) is incorporated into the bottom subcell design, as the hole transport layer (HTL), enabling effective hole extraction and transport. In order to decrease recombination losses and improve efficient electron mobility, fullerene (C_{60}) is utilized as the electron transport layer (ETL). A silver (Ag) electrode acts as the metallic contact, facilitating effective charge extraction and current collection within the device, and bathocuproin functions as the exciton-blocking layer to enhance electron collecting at interface of ETL layer as shown in Fig. 13.5.

2-T Tandem Solar Cells

The recombination layer is important in the formation of an efficient 2-T tandem solar cell. A contrast in work function that exists between the top subcell and the bottom subcell ensures that each of these cells absorbs a greater part of the spectrum of solar radiation, hence increasing efficiency (Figs. 13.6 and 13.7). To avoid any harmful damage to the charge recombination layer, its production should be done under low temperatures (Jiang et al. 2016).

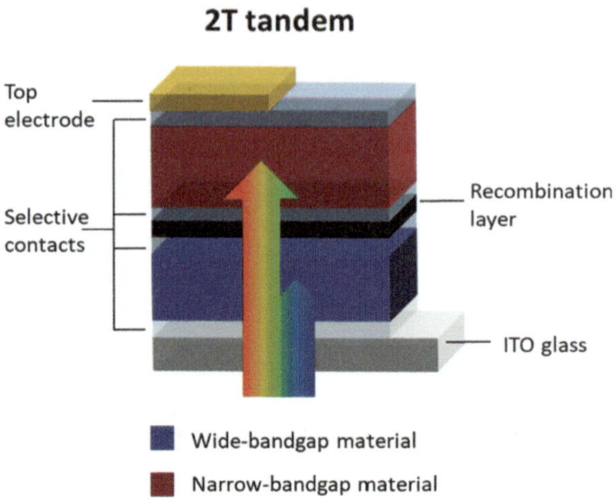

Fig. 13.6 Scheme of 2 terminal (2-T) perovskite solar cell (Zhu et al. 2021). Copyright 2020 reproduced with permission from Elsevier

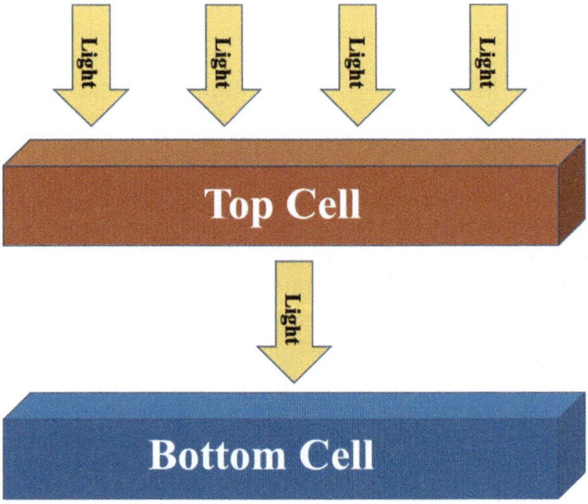

Fig. 13.7 Scheme of 2T tandem solar cell

In 2015, Jiang et al. developed a novel perovskite–perovskite TSC including a novel charge recombination layer, namely 'spiro-OMeTAD/PEDOT:PSS/PEI/PCBM:PEI'. For the effective production of perovskite–perovskite tandem solar cells, the top sub cell should have both wide-bandgap as well as narrow-bandgap perovskite cells. For fabricating a multilayer junction a combination of bottom cell having an bandgap of 1.7–1.9 eV and a bandgap of 0.9–1.2 eV is most effective (Yu et al. 2016).

Four-Terminal Tandem Solar Cell

Two sub-cells are mechanically placed on top of one another in the four-terminal TSC design. The 4-T configuration permits each sub-cell to function independently as shown in Fig. 13.8, in contrast to the two-terminal (2-T) system where the top and bottom subcells are monolithically connected and need to match current (Raza and Ahmad 2022). Being independent allows for somewhat greater theoretical efficiency since it eliminates the need for current coordination among subcells. Perovskite/silicon (Si) (Abbasiyan et al. 2022), perovskite/copper indium gallium diselenide ($Cu(In,Ga)Se_2$) (Chelvanathan et al. 2010), and perovskite/perovskite (Fang et al. 2021) tandem cells are among the effective tandem cells that have been described. Perovskite/perovskite tandem cells show the most promise among them since they can be processed at low temperatures and may require less money to produce than other tandem architectures (Fig. 13.9).

All-perovskite tandem cells have progressed very fast in the last two to three years. There, Hörantner et al. (2017) described a 2-T tandem cell with a power conversion efficiency of 17% and even a 4-T all-perovskite tandem cell with a power conversion efficiency of 20.3%. By merging a vacuum-deposited normal-bandgap perovskite bottom cell with a solution-processed wide-bandgap (E_g) perovskite top cell, Chen et al. (2020) demonstrated a 2 terminal TSC with a champion PCE of 18.1%.

With further effort, Rajagopal et al. (2017) successfully designed a 2-Terminal all-perovskite tandem cell with a PCE as high as 18.5% with the top cell built using a broad band-gap of 1.82 eV and the bottom cell using a narrow bandgap of 1.22 eV. This work increased PCE of 4-T tandem cells with all active layers as perovskite

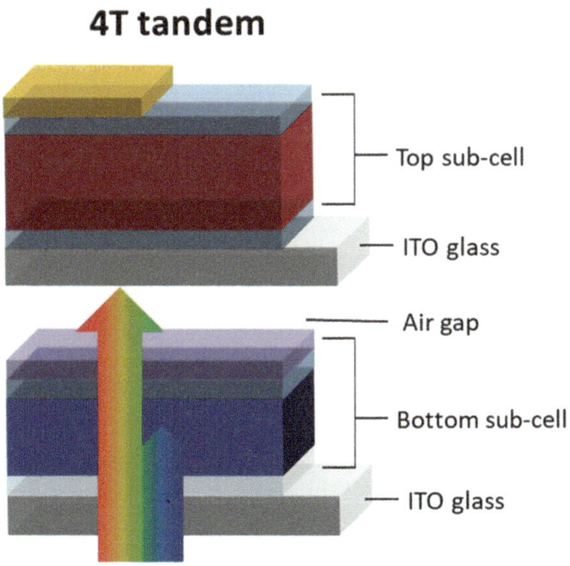

Fig. 13.8 The structure of 4 terminal (4-T) tandem perovskite solar cell (Zhu et al. 2021). Copyright 2020 reproduced with permission from Elsevier

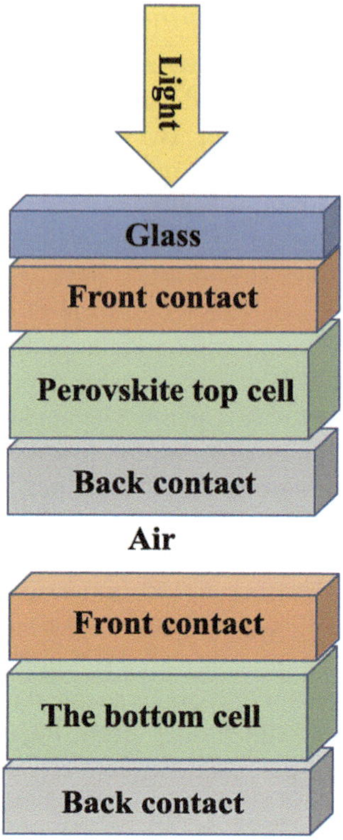

Fig. 13.9 Scheme of 4T tandem solar cell

materials, up to 21% by mechanically stacking on top of a 1.58 eV wide-bandgap semi-transparent perovskite top cell, onto a 1.25 eV low-bandgap perovskite bottom cell. The much higher efficiency of the bottom low-bandgap (1.25 eV) cell is the primary cause for the upgrade. These developments demonstrate the potential for 4-T tandem solar cells to achieve better efficiencies, which is generating interest in tandem architectures based on perovskites that are being further optimized for next-generation solar technologies (Khan et al. 2022).

Fabrication of Tandem Perovskite Solar Cells

PSCs have set high performance standards with their laboratory-scale efficiencies; nonetheless, the available surface area available is often less than 1 cm^2 (Zhou et al. 2024). The transfer of perovskite solar cells from laboratory experiments to industrial-scale uses is hindered by this constraint. It is crucial to build a large surface area, pinhole-free, top grade perovskite films in order to move closer to commercial (Raj et al. 2021). In order to achieve scalability and dependable performance, this criterion is essential. Large-area solar cell film is produced through a variety of

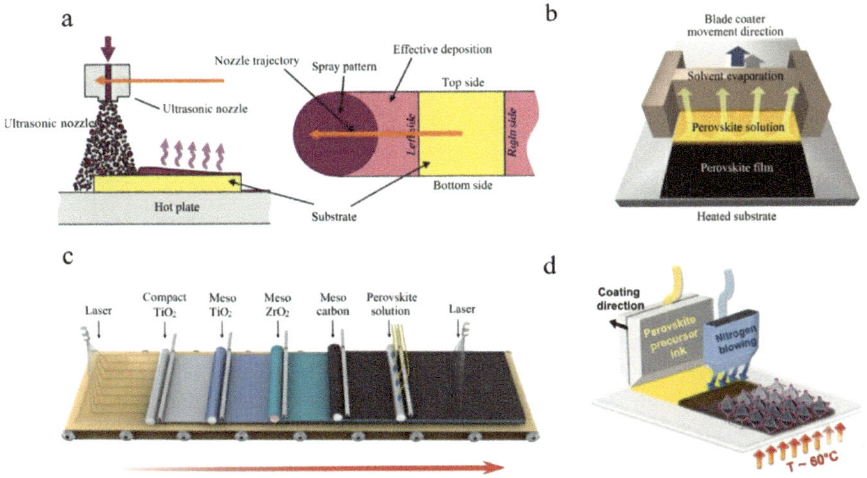

Fig. 13.10 A systematic representation of **a** spray coating, **b** blade coating, **c** inject printing, and **d** die coating fabrication of TSCs (Jiao et al. 2023). Copyright 2023 reproduced with permission from Elsevier

solvent-based deposition techniques, for industrial manufacturing. These techniques concentrate on maximizing film uniformity, reducing flaws, and guaranteeing high crystallinity and quality (Fig. 13.10) (Liu et al. 2024).

Spray Coating

Spray coating is an innovative and adaptable technique that can be used in industrial settings to deposit perovskite films (Alanazi 2023). Using this method, large-area films can be uniformly deposited on the substrate by misting a fine spray of the perovskite precursor solution. The efficiency of resulting perovskite solar cells is greatly impacted by the film's shape, thickness, and crystallization quality, all of which are determined by the solvent used in spray coating. The formation of solar cell can be fine-tuned by carefully choosing solvents with the right stability and solubility for the perovskite precursors, and regulating temperature and solvent evaporation rate in the process of deposition (Hossain et al. 2020). This improvement caused reduced defect levels in the film, increased crystallinity, and higher device performance overall. The regulation of the thickness the film was monitored using spray coating, which is used to customize the tandem cell for optoelectronic devices (Ichwani et al. 2023).

Since its initial application in 2014 to create perovskite films, spray coating has quickly developed into a potential method for industrial production (Alanazi 2023). Three key processes are involved in the spray coating process are as follows: (i) atomization, (ii) droplet falling, and (iii) thin film deposition (Alanazi 2023). Chen et al. (2018) have detailed the principles and parameters regulating each stage of the process, and they contend that each step is an extremely crucial step to the final production of the perovskite film.

Doctor-Blade Coating

One popular method for depositing perovskite films is doctor-blade coating. The efficiency of perovskite solar cells is greatly impacted by the shape, thickness, and crystallization of the final film, which is highly influenced on the solvent used in this process (Peng et al. 2020). Selecting the right solvent is very important since it influences the ink's viscosity, surface tension, and wetting behavior. The deposited film's homogeneity, thickness, and crystal quality are all influenced by these variables (Hong et al. 2023).

The production of the film can be optimized by carefully choosing solvents that guarantee the precursor material's outstanding solubility and stability, and regulating the solvent's evaporation rate and temperature during the coating process (Jiao et al. 2023).

By modifying ink viscosity by solvent mixing, which improves spreading and wetting qualities, solvent engineering can further improve the film forming process. To produce films with a high degree of crystallinity, other methods such as vapor-assisted crystallization or solvent annealing can be utilized (Lee et al. 2021). The study also showed that solid carbonyl hydrazide can successfully prevent void formation at the perovskite–substrate interface (Wang et al. 2023a) during the doctor-blade coating process, enhancing the quality of the film. This replaces dimethyl sulfoxide during the process which gave emphasis on the solvent engineering.

Inject Printing

Due to its accurate control of film thickness, low material consumption, and scale-up capability to large-scale and high-throughput manufacturing, inkjet printing has emerged as a promising method for perovskite film deposition (Okoroafor et al. 2022). The solvents selection and engineering play an important role since they influence the viscosity, surface tension, and rate of evaporation of the ink. Important Features of Inkjet Printing:

Solvent engineering: Stable, homogeneous, and superior perovskite films can be made by carefully selecting and refining the solvent. The form and crystallization of the printed films are directly influenced by the characteristics of the solvent. On the other hand, poor solvent selection might result in issues including agglomeration, nozzle blockage, or uneven coatings.

Process Optimization: To further improve film quality, vacuum-assisted thermal annealing is one technique that can be used in conjunction with inkjet printing. With this method, material scalability is possible without compromising the perovskite layer functionality. Recent Advancements: Li et al. (2020) demonstrated the potential of this technique for scalable solar cell manufacturing by successfully fabricating perovskite films with a larger active area of 4.0 cm^2 through the use of inkjet printing with the use of vacuum assistance during thermal annealing. Precise and versatile, inkjet printing is an advantageous technique for perovskite film deposition, particularly for large-scale, productive solar cell manufacturing (Raj et al. 2023).

Slot Die Coating

A roll-to-roll approach to fabrication under the denomination slot-die coating is envisioned for uniform deposition of large-area perovskite films. This method is a decent choice for producing perovskite solar cells (PSCs) on an industrial scale because of its excellent scalability and reproducibility (Seo et al. 2022). Solvent Role in Film Formation: In the process of slot-die coating, the choice of the solvent determines the quality of perovskite films. Solvents have an impact on the films' shape, thickness, and crystallization, all of which have an impact on PSC performance. Therefore, throughout the coating process, solvent engineering have a crucial contribution to optimizing film quality by regulating variables like temperature and evaporation rate (Remeika et al. 2018).

Process Improvements: By increasing the substrate surface's wettability and spreading behavior, solvent engineering can increase uniformity and thickness control. Further methods to improve the overall quality and crystallinity of the film include vapor-assisted crystallization and solvent annealing (Heo et al. 2017).

Industrial Potential: Roll-to-roll PSC production with slot-die coating has large amount of potential to the industrial sector. But if there isn't any controlled crystallization, the precursor film can dry slowly, which could cause chalcocite to develop and nucleate uncontrollably (Choi et al. 2015). As a result, there is insufficient contact with the substrate and poor crystallization quality. Thus, further methods are needed to encourage improved crystallization.

Comparative Analysis: Slot-die coating offers more control over film formation and scalability than screen printing, making it the preferred method for depositing uniform perovskite films over large-area mesoporous substrates. With ongoing developments in solvent engineering targeted at enhancing film quality, this process is regarded as a promising approaches for producing PSCs in the scientific world for industrial production.

Challenges in Perovskite Tandem Solar Cells

PSCs, or perovskite solar cells, are the subject of intense research in the solar energy sector because of their potential for high efficiency and low cost (Fazal and Rubaiee 2023). However, significant challenges remain, including the need for improved efficiency, enhanced stability, and scalable production methods. This chapter delves into these challenges, providing a detailed analysis of each and reviewing the current strategies aimed at addressing them.

Enhancing Crystal Quality

Enhancing the quality of the crystal structure of perovskite films is critical for boosting the performance of PSCs. The photovoltaic efficiency, strength and durability of perovskite films (Sharma et al. 2022) are strongly influenced by their quality, which is determined by the fabrication methods and materials employed. Enhanced crystallinity can lead to reduced defect density, better charge carrier mobility (Irshad et al. 2024), and overall improved device performance (Tai et al. 2016).

Fabrication of 3-Tandem Solar Cells

All-perovskite triple-junction solar cells hold great potential for achieving high efficiencies, but their experimental exploration has been limited. McMeekin et al. were among the first to demonstrate a proof-of-concept all-perovskite triple-junction solar cell, achieving a (PCE) of just 6.7% (McMeekin et al. 2019). The evolution of efficient all-perovskite triple-junction cells remains challenging due to several factors. Firstly, it is nontrivial to achieve high-performance subcells with near-ideal bandgaps. Secondly, the as-formed perovskite layers are usually degraded in solution processing of a perovskite film. Thirdly, solution processing encounters challenges in achieving thin and conformal charge transport layers such as poly(triaryl amine), PTAA over surface-rough perovskite subcells.

Complex Fabrication Process

Producing high-performance subcells with optimal bandgaps is a complex task. It is critical to identify which subcell limits the current and to determine how much variation in current density between subcells are tolerated irrespective of requiring additional adjustments, like using a cell filter. Current density (J_{sc}) values for the top and bottom cells were 16.10 and 19.41 mA/cm^2, respectively, showing that the top perovskite cell was the current-limiting factor, with the tandem cell's J_{sc} (16.31 mA/cm^2) matching that of the top cell. Therefore, enhancing the J_{sc} of the top cell is essential to boost the overall tandem cell performance. Advancements to attain bandgap engineering of the top cell and optimizing light control at the transparent conductive oxide (TCO) layer (Park et al. 2019).

Damage to Layers

For majority of times, perovskite films must be produced in an inert atmosphere, and unassembled devices cannot withstand prolonged exposure to air, which hinders the actual manufacturing and use of perovskite solar cells.

The underlying layers of perovskites that have already developed are often damaged during the solution processing of perovskite films. Because amine salts are hygroscopic, perovskite–type hybrids are known to break down quickly in the presence of moisture, which restricts the usefulness of halide compounds (Noh et al. 2013).

Preheating substrates before depositing PbI_2 precursors has been shown to improve the morphology of PbI_2 films, enhancing the quality of the resulting $MAPbI_3$ films (Ko et al. 2015). Using Lead(II) thiocyanate ($Pb(SCN)_2$) has been reported to produce enhanced perovskite films in ambient air with relative humidity exceeding 70% (Tai et al. 2016). Rapid solvent removal, induced by antisolvent exposure, significantly reduces the impact of humidity on the crystallization process. The demonstrated Hot Injection Fabrication (HIF) method for producing perovskite solar cells (PSCs) under ambient conditions could enable large-scale, cost-effective fabrication of high-efficiency PSCs (Wang et al. 2019).

Complex Transport Layers

While increasing conversion efficiency in solar cells remains a focus, developing reliable methods to measure it is a greater challenge for advancing tandem solar cells, due to the spectral mismatch under simulated sunlight. This spectral mismatch can be mitigated for single-junction solar cells by adjusting light intensity using a reference cell, but this solution doesn't work for series-connected tandem cells. Different light wavelengths are absorbed by subcells in tandem cells, which causes inconsistent calibration and current mismatch. Using the external quantum efficiency (EQE) curve to measure current density is the most practical method of resolving this mismatch. It can be challenging to calculate each subcell's EQE precisely, though. As such, care must be used when evaluating the performance of tandem solar cells. Thus far, no research has been conducted on accurate measurements for monolithic tandem solar cells established on perovskites.

New Recombination Layer Materials Without Rare Earth Metals

In 2-terminal (2T) perovskite-based TSCs, transparent conducting oxides (TCOs) containing rare earth metals are currently frequently employed as recombination layers. However, the use of rare earth elements drives up production costs and hinders industrialization on a wide scale. Investigating substitute recombination materials that can function as effective interconnecting layers (ICLs) in tandem devices is crucial in order to get over these restrictions. While several materials have been suggested for use as transparent electrodes, they have not yet been successfully incorporated into tandem cells based on perovskites. Before being used for ICL applications, these materials also need to solve other technical problems such surface morphology, chemical stability, and fabrication compatibility.

Metal nanowires (Zhang et al. 2016), nanoparticles (Lu et al. 2015; Zhang et al. 2011), and grids have emerged as promising candidates for transparent electrodes in optoelectronic devices. For instance, copper and silver nanowires have good conductivity and transmittance, which makes them suitable options for ICLs. With their enhanced electrical conductivity, which facilitates effective carrier recombination, copper and gold in particular provide viable solutions to decrease metal-perovskite reactions. Moreover, their superior optical transparency reduces parasitic absorption. Nevertheless, the rough surface morphology of metal nanowires may not significantly affect the top transparent electrodes when used as recombination layers, but it may result in short circuits in the back sub-cell. Graphene is another potential material for transparent electrodes in perovskite solar cells (PSCs) due to its exceptional properties (Zhang et al. 2011). In the solar spectrum, single-layer or thin multilayer graphene exhibits optical transparency higher than 90% and has low sheet resistance and high electron mobility (>150,000 $cm^2\ V^{-1}\ s^{-1}$), all of which are necessary for effective electron and hole recombination. Because graphene is made entirely of carbon atoms, it also improves the chemical stability of perovskite materials, which increases the overall durability of the device. The integration of graphene onto perovskite films, however, presents the biggest obstacle and calls for the development of appropriate manufacturing methods.

Charge Recombination Layers

A major obstacle to the development of perovskite tandem solar cells is the creation of charge recombination layers (CRLs) that minimize ohmic losses while connecting the two subcells. To guarantee optimum performance and integration, deposition procedures must also be established that are completely compatible with the underlying solar cells (Mailoa et al. 2015).

Methods for Increasing Perovskite Solar Cell Performance

Major goals for the fabrication of perovskite solar cells is to enhance their stability and performance. Their stability and efficiency can be enhanced by various methods:

Methods to Boost Perovskite Performance

Superior perovskite films reduce the possibility of short circuits between the layers of the solar cell by guaranteeing consistent surface coverage free of pinholes. A number of techniques have been investigated to improve the perovskite thin-film quality (Hossain et al. 2020). This results in films that are homogenous, continuous, and void-free. High crystallinity, improved light absorption, more effective electron transport, and quick electron extraction at the nearby electron transport layer are the characteristics of these film's grains. These films were used in solar cells, which demonstrated efficiencies of up to 14.17% (Uyanga et al. 2020; Jiao et al. 2023). Moreover, it has been demonstrated that methylammonium chloride (CH_3NH_3Cl) added to the precursor solution during a one-step spin-coating procedure enhances the grade of the perovskite thin film (Huang et al. 2021). The performance of perovskite films was further improved by the addition of 1% 1,8-diiodooctane. Furthermore, the standard of the film is greatly impacted by the stoichiometric ratio of the precursors, lead iodide (PbI_2) and methylammonium iodide (MAI) (Zhang and Park 2024). An efficiency gain of up to 19% and a continuous film were obtained with a 1:3 precursor ratio (Qin et al. 2022; Yang et al. 2015).

Methods to Increase the Stability of Perovskites

Perovskite materials are generally brittle by nature, soluble in water, and prone to quick, irreversible deterioration (Agresti et al. 2024; Ibn-Mohammed et al. 2017). Temperature, exposure to ultraviolet radiation, and moisture infiltration are important environmental factors that affect stability (Tang et al. 2024). Many approaches have been developed to improve the robustness of devices based on perovskites. One method makes use of solar cell's ability to self-heal when exposed to no light. After a minute of darkness, the cells return to their initial efficiency. Charge carriers may become caught in lattice imperfections during the day, which would interrupt their flow. The cell's functionality is restored by darkness, which permits the release of these trapped charges (Yang et al. 2024). Optimizing the crystalline quality of perovskite to reduce grain boundary defects—which can serve as locations for charge recombination and lower efficiency—is another method for improving stability. Suitable crystallization conditions, such as heating substrates to 190 °C, encourage the production of bigger grains and reduce recombination (Cheng et al. 2019; Khatoon et al. 2023).

It has also been demonstrated that increasing the titanium dioxide (TiO_2) content of the electron transport layer in perovskite solar cells increases stability (Banoth et al. 2023; Fatima et al. 2024). Since SnO_2 has a wider bandgap, less UV absorption, and more electron mobility than TiO_2, its cells showed 700 h of stability compared to 500 h for TiO_2 cells (Shilpa et al. 2023). Alternative compounds, like zinc stannate (Zn_2SnO_4) (Mali et al. 2016) and antimony sulfide (Sb_2S_3) (Farhana et al. 2023), have been studied to further address TiO_2's UV sensitivity. These materials have shown improved UV resistance as well as the overall device stability (Wang et al. 2023b). Another essential method for shielding perovskite solar cells from environmental deterioration is encapsulation. Much work has gone into making perovskite solar cells more stable as a result of technological developments. Growing interest is being shown in encapsulating techniques that can tolerate harsh environments, such as extended outdoor exposure or wet heat tests, as more devices show they can withstand these kinds of tests. To ensure performance under these rigorous testing conditions, compatibility between the device architecture intended for increased stability and efficient encapsulation techniques is essential (Belich et al. 2023). Because the encapsulating materials are hydrophobic and UV-resistant, encapsulated cells exhibit much greater stability than non-encapsulated ones (Belich et al. 2023; Cao et al. 2023). Encapsulation combined with a UV filter can further extend the life of the gadget for even more protection.

Future Applications

Flexible Tandem Devices

A cutting-edge field of study is flexible electronics, particularly in light of the growing market for wearable technology. The widespread adoption of these technologies has attracted noteworthy attention, mostly due to the ongoing advancements in internet technology. Because of its small form factors, asymmetrical shapes, and high energy consumption, wearable technology demands flexible and effective power sources. Renowned for their high power conversion efficiency (PCE), flexible tandem solar cells present a promising way to power these gadgets (Yang et al. 2019).

Perovskite films exceed traditional silicon-based solar cells in terms of flexibility, with thicknesses in the range of several hundred nanometres, making them perfect for flexible applications. Perovskite films have been successfully used in single-junction flexible devices, as shown by a number of papers that emphasize its potential. The creation of a suitable flexible interconnecting layer (ICL) is a significant obstacle for flexible perovskite-based tandem devices, though. Since they are usually more than 100 nm thick, the widely utilized sputtered transparent conductive oxides (TCOs) are brittle by nature and cannot be used in flexible arrangements. This difficulty highlights the necessity for cutting-edge fabrication techniques and innovative materials designed especially for flexible device applications in wearable technologies.

Low Cost and Large Area Fabrication Process

The industrialization and adoption of solar cell (SC) modules are heavily influenced by cost factors, which include power conversion efficiency (PCE), raw material expenses, and fabrication costs. Two-terminal (2T) perovskite-based tandem devices, with their higher PCE, have the potential to lower the cost per unit of power generated. However, the overall cost of tandem devices remains high, primarily due to complex fabrication processes. For instance, in the process of sputtering of transparent conductive oxides (TCOs) in the interconnecting layer (ICL), high-energy ions can cause damage. To mitigate this, a buffer layer of metal oxides is often introduced using atomic layer deposition (ALD). This method is both tedious and expensive due to high value of the equipment. Additionally, vacuum-based processes such as sputtering and thermal evaporation present compatibility issues with large-scale device manufacturing. Alternative methods like roll-to-roll printing and doctor blading offer promising solutions for fabricating large-area perovskite devices with high performance. Future research should focus on developing ICL fabrication processes that are compatible with large-scale device production, enabling more cost-effective manufacturing of perovskite tandem solar cells.

Low-Dimensional Perovskite in LEDs

Because perovskite-based LEDs could only operate in freezing temperatures and had low performance and limited stability, early research on the technology received little attention. Using phenylethylammonium lead iodide ($(C_6H_5C_2H_4NH_3)_2PbI_4$), Era et al. (1994) investigated 2D-layered perovskite quantum-well structures and created electroluminescent (EL) devices that functioned at low temperatures. But as perovskite-based solar cells attained a PCE of more than 22%, attention turned back to perovskite LEDs, specifically investigating the application of 3D perovskites to create quantum-confined 2D-layered structures.

Methylammonium lead iodide ($MAPbI_3$)-based perovskites perform poorly in LEDs (Benmessaoud et al. 2016) because of low exciton binding energies (between 9 and 60 meV), which caused free charge carriers and mostly nonradiative recombination. Li et al. (2017) have effectively shown a room-temperature 2D perovskite-based green LED utilizing phenylmethanamine lead iodide ($(C_6H_5CH_2)_2PbI_4$). Over the course of five days, they observed no deterioration in the X-ray diffraction (XRD) pattern at 50% relative humidity, indicating enhanced stability.

Low-Dimensional Halide Perovskite in Photodetectors

Li et al. (2017) investigated high-performance photodetectors with methylammonium lead iodide (2D) nanosheets ($MAPbI_3$) (Wang et al. 2022b). These nanosheets, had thickness of around 20 nm, were created as planar devices by chemical vapor deposition (CVD) technique on PbI_2 crystals that had already been deposited on top of SiO_2 substrates. Vertical devices could be created with the substrate made of indium tin oxide (ITO). The substrate was prepared by sequentially photoetching ITO, etching with O_2 plasma, and removing the photoresist. A two-step process was then used to install the perovskite layer. At an excitation power of 50.82 mW, the vertical device had a higher photoresponse than the planar structure, reaching a maximum responsivity of 36 mA/W. The short carrier diffusion length and low trap density found in 2D perovskite structures are responsible for this improved performance (Shrestha et al. 2022). A counter electrode was then evaporated to finish the photodetector manufacturing process. Because of the low trap density and short carrier diffusion distance, the 2D $CH_3NH_3PbI_3$ nanosheets used in the vertical-type photodetector demonstrated the benefits of strong photoresponsivity, low-voltage operation, and large response (Kim et al. 2022).

Conclusion and Future Outlook

Metal-halide perovskites having special properties of being thin film semiconductors by solution process with higher band gap and lesser energy loss as well as free of toxicity, if proper deployment strategies are devised, could alter the landscape of the photovoltaic industry. When implemented for tandem solar cells, the efficiency limits are expected to have surpassed those of the single-junction Shockley–Queisser limit. Conversely, for perovskite/silicon tandems whose performance is extensively modeled, the expectations are entirely within 36–37% efficiency which is far beyond what single junction silicon cell can accomplish. The capability is however yet to be commercialized because of the existing challenges of efficiency enhancement and stability issues because of the inherent disadvantages of degradation of 'inherently moisture susceptible' perovskite materials. Newer reports have demonstrated that both two-terminal (2T) and four-terminal (4T) tandem devices developed, including the larger graded design 1 cm^2 versions, are capable of efficiency (13.8%) and 16.0%, further modifications would push the 4T tandem beyond 20.3 in excess of the grade design limits. These metrics are already above the levels of other low-cost semiconductor competitors, such as organic small molecules, amorphous and microcrystalline silicon Additionally, there are expectations in tandem cells that to the best single junction commercial perovskite cells at least 4–5% will be gained which means, tandem approaches will make the technology surpass the 30% efficiency barrier further single junction efficiencies improving.

References

Abbasiyan A, Noori M, Baghban H (2022) A highly efficient 4-terminal perovskite/silicon tandem solar cells using QIBC and IBC configurations in the top and bottom cells, respectively. Mater Today Energy 28:101055. https://doi.org/10.1016/J.MTENER.2022.101055

Agresti A, Di Giacomo F, Pescetelli S, Di Carlo A (2024) Scalable deposition techniques for large-area perovskite photovoltaic technology: a multi-perspective review. Nano Energy 122:109317. https://doi.org/10.1016/J.NANOEN.2024.109317

Alanazi TI (2023) Current spray-coating approaches to manufacture perovskite solar cells. Results Phys 44:106144. https://doi.org/10.1016/J.RINP.2022.106144

Bailie CD, McGehee MD (2015) High-efficiency tandem perovskite solar cells. MRS Bull 40:681–685. https://doi.org/10.1557/mrs.2015.167

Banoth R, Raja Shekar PV, Ramana CV, Kumari K (2023) Surface engineering of mesoporous-TiO_2 electron transport layer for improved performance of organic–inorganic perovskite solar cells via suppressing interface defects, enhancing charge extraction and boosting carrier transport. Colloids Surf A Physicochem Eng Asp 676:132075. https://doi.org/10.1016/J.COLSURFA.2023.132075

Belich NA, Petrov AA, Ivlev PA et al (2023) How to stabilize standard perovskite solar cells to withstand operating conditions under an ambient environment for more than 1000 hours using simple and universal encapsulation. J Energy Chem 78:246–252. https://doi.org/10.1016/J.JECHEM.2022.12.010

Benmessaoud IR, Mahul-Mellier AL, Horváth E et al (2016) Health hazards of methylammonium lead iodide based perovskites: cytotoxicity studies. Toxicol Res (Camb) 5:407–419. https://doi.org/10.1039/c5tx00303b

Berhe TA, Su WN, Chen CH et al (2016) Organometal halide perovskite solar cells: degradation and stability. Energy Environ Sci 9:323–356

Bett AJ, Schulze PSC, Winkler KM et al (2020) Two-terminal perovskite silicon tandem solar cells with a high-bandgap perovskite absorber enabling voltages over 1.8 V. Prog Photovoltaics Res Appl 28:99–110. https://doi.org/10.1002/pip.3208

Bhatti HS, Hussain ST, Khan FA, Hussain S (2016) Synthesis and induced multiferroicity of perovskite $PbTiO_3$; a review. Appl Surf Sci 367:291–306. https://doi.org/10.1016/J.APSUSC.2016.01.164

Bing J, Caro LG, Talathi HP et al (2022) Perovskite solar cells for building integrated photovoltaics—glazing applications. Joule 6:1446–1474. https://doi.org/10.1016/J.JOULE.2022.06.003

Bush KA, Frohna K, Prasanna R et al (2018) Compositional engineering for efficient wide band gap perovskites with improved stability to photoinduced phase segregation. ACS Energy Lett 3:428–435. https://doi.org/10.1021/acsenergylett.7b01255

Cao DH, Stoumpos CC, Farha OK et al (2015) 2D homologous perovskites as light-absorbing materials for solar cell applications. J Am Chem Soc 137:7843–7850. https://doi.org/10.1021/JACS.5B03796

Cao M, Ji W, Chao C et al (2023) Recent advances in UV-cured encapsulation for stable and durable perovskite solar cell devices. Polymers (Basel). https://doi.org/10.3390/POLYM15193911

Chamberlain GA (1983) Organic solar cells: a review. Solar Cells 8:47–83. https://doi.org/10.1016/0379-6787(83)90039-X

Chelvanathan P, Hossain MI, Amin N (2010) Performance analysis of copper–indium–gallium–diselenide (CIGS) solar cells with various buffer layers by SCAPS. Curr Appl Phys 10:S387–S391. https://doi.org/10.1016/J.CAP.2010.02.018

Chen CC, Bae SH, Chang WH et al (2015a) Perovskite/polymer monolithic hybrid tandem solar cells utilizing a low-temperature, full solution process. Mater Horiz 2:203–211. https://doi.org/10.1039/c4mh00237g

Chen Q, De Marco N, Yang Y et al (2015b) Under the spotlight: the organic–inorganic hybrid halide perovskite for optoelectronic applications. Nano Today 10:355–396

Chen H, Ding X, Pan X et al (2018) Comprehensive studies of air-brush spray deposition used in fabricating high-efficiency $CH_3NH_3PbI_3$ perovskite solar cells: combining theories with practices. J Power Sources 402:82–90. https://doi.org/10.1016/J.JPOWSOUR.2018.07.097

Chen X, Jia Z, Chen Z et al (2020) Efficient and reproducible monolithic perovskite/organic tandem solar cells with low-loss interconnecting layers. Joule 4:1594–1606. https://doi.org/10.1016/J.JOULE.2020.06.006

Cheng N, Li W, Zhang M et al (2019) Enhance the performance and stability of methylammonium lead iodide perovskite solar cells with guanidinium thiocyanate additive. Curr Appl Phys 19:25–30. https://doi.org/10.1016/J.CAP.2018.10.020

Choi KJ, Lee JY, Park J, Seo YS (2015) Multilayer slot-die coating of large-area organic light-emitting diodes. Org Electron 26:66–74. https://doi.org/10.1016/j.orgel.2015.07.025

Clemens O, Reitz C, Witte R et al (2016) Anion ordering, magnetic structure and properties of the vacancy ordered perovskite $Ba_3Fe_3O_7F$. J Solid State Chem 243:31–37. https://doi.org/10.1016/J.JSSC.2016.07.033

Dawa T, Sajjadi B (2024) Exploring the potential of perovskite structures for chemical looping technology: a state-of-the-art review. Fuel Process Technol 253:108022

De Souza RA (2021) Transport properties of dislocations in $SrTiO_3$ and other perovskites. Curr Opin Solid State Mater Sci 25:100923. https://doi.org/10.1016/J.COSSMS.2021.100923

Dimitrovska S, Aleksovska S, Kuzmanovski I (2005) Prediction of the unit cell edge length of cubic $A^{2+}{}_2BB'O_6$ perovskites by multiple linear regression and artificial neural networks. Open Chem 3:198–215

Eperon GE, Leijtens T, Bush KA et al (2016) Perovskite-perovskite tandem photovoltaics with optimized band gaps. Science 354:861–865. https://doi.org/10.1126/science.aaf9717

Era M et al (1994) Organic–inorganic heterostructure electroluminescent device using a layered perovskite semiconductor $(C_6H_5C_2H_4NH_3)_2PbI_4$. Appl Phys Lett 65:676–678

Fang Z, Zeng Q, Zuo C et al (2021) Perovskite-based tandem solar cells. Sci Bull 66:621–636. https://doi.org/10.1016/J.SCIB.2020.11.006

Farhana MA, Manjceevan A, Bandara J (2023) Recent advances and new research trends in Sb_2S_3 thin film based solar cells. J Sci Adv Mater Dev 8:100533. https://doi.org/10.1016/J.JSAMD.2023.100533

Fatima Q, Haidry AA, Zhang H et al (2024) A critical review on advancement and challenges in using TiO_2 as electron transport layer for perovskite solar cell. Mater Today Sustain 27:100857. https://doi.org/10.1016/J.MTSUST.2024.100857

Fazal MA, Rubaiee S (2023) Progress of PV cell technology: feasibility of building materials, cost, performance, and stability. Sol Energy 258:203–219. https://doi.org/10.1016/j.solener.2023.04.066

Fedorovskiy AE, Drigo NA, Nazeeruddin MK (2020) The role of Goldschmidt's tolerance factor in the formation of A_2BX_6 double halide perovskites and its optimal range. Small Methods. https://doi.org/10.1002/smtd.201900426

Gauthron K, Lauret J, Doyennette L et al (2001) Organic–inorganic electronics. IBM J Res Dev 45(1):29–45

Gippius NA, Ishihara T, Keldysh LV et al (1993) Dielectrically confined excitons and polaritons in natural superlattices—perovskite lead iodide semiconductors. J Phys IV JP 3:437–440. https://doi.org/10.1051/jp4:1993594

Grandhi GK, Likhit HCM, Ong SP, Bin IW (2020) Jahn-Teller distortion-driven robust blue-light-emitting perovskite nanoplatelets. Appl Mater Today 20:100668. https://doi.org/10.1016/J.APMT.2020.100668

Heo JH, Im SH (2016) $CH_3NH_3PbBr_3$–$CH_3NH_3PbI_3$ perovskite–perovskite tandem solar cells with exceeding 2.2 V open circuit voltage. Adv Mater 28:5121–5125. https://doi.org/10.1002/adma.201501629

Heo YJ, Kim JE, Weerasinghe H et al (2017) Printing-friendly sequential deposition via intra-additive approach for roll-to-roll process of perovskite solar cells. Nano Energy 41:443–451. https://doi.org/10.1016/j.nanoen.2017.09.051

Hong S, Park B, Balamurugan C et al (2023) Impact of solvents on doctor blade coatings and bathocuproine cathode interlayer for large-area organic solar cell modules. Heliyon 9:e18209. https://doi.org/10.1016/J.HELIYON.2023.E18209
Hörantner MT, Leijtens T, Ziffer ME et al (2017) The potential of multijunction perovskite solar cells. ACS Energy Lett 2:2506–2513. https://doi.org/10.1021/acsenergylett.7b00647
Hossain A, Roy S, Sakthipandi K (2019) The external and internal influences on the tuning of the properties of perovskites: an overview. Ceram Int 45:4152–4166. https://doi.org/10.1016/J.CERAMINT.2018.11.102
Hossain MA, Khoo KT, Cui X et al (2020) Atomic layer deposition enabling higher efficiency solar cells: a review. Nano Mater Sci 2:204–226. https://doi.org/10.1016/J.NANOMS.2019.10.001
Huang HH, Liu QH, Tsai H et al (2021) A simple one-step method with wide processing window for high-quality perovskite mini-module fabrication. Joule 5:958–974. https://doi.org/10.1016/J.JOULE.2021.02.012
Husain AAF, Hasan WZW, Shafie S et al (2018) A review of transparent solar photovoltaic technologies. Renew Sustain Energy Rev 94:779–791
Ibn-Mohammed T, Koh SCL, Reaney IM et al (2017) Perovskite solar cells: an integrated hybrid lifecycle assessment and review in comparison with other photovoltaic technologies. Renew Sustain Energy Rev 80:1321–1344. https://doi.org/10.1016/J.RSER.2017.05.095
Ichwani R, Price S, Oyewole OK et al (2023) Using machine learning for prediction of spray coated perovskite solar cells efficiency: from experimental to theoretical models. Mater Des 233:112161. https://doi.org/10.1016/J.MATDES.2023.112161
Irshad Z, Lee W, Adnan M et al (2024) Elucidating charge carrier dynamics in perovskite-based tandem solar cells. Small Methods. https://doi.org/10.1002/SMTD.202300238
Jiang F, Liu T, Luo B et al (2016) A two-terminal perovskite/perovskite tandem solar cell. J Mater Chem A Mater 4:1208–1213. https://doi.org/10.1039/c5ta08744a
Jiao J, Yang C, Wang Z et al (2023) Solvent engineering for the formation of high-quality perovskite films: a review. Results Eng 18:101158. https://doi.org/10.1016/J.RINENG.2023.101158
Johnsson M, Lemmens P (2005) Crystallography and chemistry of perovskites
Khan F, Rezgui BD, Khan MT, Al-Sulaiman F (2022) Perovskite-based tandem solar cells: device architecture, stability, and economic perspectives. Renew Sustain Energy Rev 165:112553. https://doi.org/10.1016/J.RSER.2022.112553
Khatoon S, Kumar Yadav S, Chakravorty V et al (2023) Perovskite solar cell's efficiency, stability and scalability: a review. Mater Sci Energy Technol 6:437–459. https://doi.org/10.1016/J.MSET.2023.04.007
Kim S, Nam H, Fujii I et al (2021a) A-site cation off-centering contribution on ferroelectricity and piezoelectricity in pseudo-cubic perovskite structure of Bi-based lead-free piezoelectrics. Scr Mater 205:114176. https://doi.org/10.1016/J.SCRIPTAMAT.2021.114176
Kim EB, Akhtar MS, Shin HS et al (2021b) A review on two-dimensional (2D) and 2D–3D multidimensional perovskite solar cells: Perovskites structures, stability, and photovoltaic performances. J Photochem Photobiol C 48:100405. https://doi.org/10.1016/J.JPHOTOCHEMREV.2021.100405
Kim EB, Akhtar MS, Ameen S et al (2022) Improving the performance of 2D perovskite solar cells by carrier trappings and minifying the grain boundaries. Nano Energy 102:107673. https://doi.org/10.1016/J.NANOEN.2022.107673
King G, Woodward PM (2010) Cation ordering in perovskites. J Mater Chem 20:5785–5796. https://doi.org/10.1039/b926757c
Ko HS, Lee JW, Park NG (2015) 15.76% efficiency perovskite solar cells prepared under high relative humidity: importance of PbI_2 morphology in two-step deposition of $CH_3NH_3PbI_3$. J Mater Chem A Mater 3:8808–8815. https://doi.org/10.1039/c5ta00658a
Kuzmanovski I, Dimitrovska-Lazova S, Aleksovska S (2007) Classification of perovskites with supervised self-organizing maps. Anal Chim Acta 595:182–189. https://doi.org/10.1016/J.ACA.2007.04.062

Lee B, Lahann L, Li Y, Forrest SR (2020) Cost estimates of production scale semitransparent organic photovoltaic modules for building integrated photovoltaics. Sustain Energy Fuels 4:5765–5772. https://doi.org/10.1039/D0SE00910E

Lee KM, Chan SH, Hou MY et al (2021) Enhanced efficiency and stability of quasi-2D/3D perovskite solar cells by thermal assisted blade coating method. Chem Eng J 405:126992. https://doi.org/10.1016/J.CEJ.2020.126992

Li P, Shivananju BN, Zhang Y et al (2017) High performance photodetector based on 2D $CH_3NH_3PbI_3$ perovskite nanosheets. J Phys D Appl Phys. https://doi.org/10.1088/1361-6463/aa5623

Li Z, Li P, Chen G et al (2020) Ink engineering of inkjet printing perovskite. ACS Appl Mater Interfaces 12:39082–39091. https://doi.org/10.1021/acsami.0c09485

Liu J, Zheng D, Wang K et al (2024) Evolutionary manufacturing approaches for advancing flexible perovskite solar cells. Joule 8:944–969. https://doi.org/10.1016/J.JOULE.2024.02.025

Lu H, Zhang D, Cheng J et al (2015) Locally welded silver nano-network transparent electrodes with high operational stability by a simple alcohol-based chemical approach. Adv Funct Mater 25:4211–4218. https://doi.org/10.1002/adfm.201501004

Lufaso MW, Woodward PM (2004) Jahn-Teller distortions, cation ordering and octahedral tilting in perovskites. Acta Crystallogr B 60:10–20. https://doi.org/10.1107/S0108768103026661

Mailoa JP, Bailie CD, Johlin EC et al (2015) A 2-terminal perovskite/silicon multijunction solar cell enabled by a silicon tunnel junction. Appl Phys Lett. https://doi.org/10.1063/1.4914179

Mali SS, Shim CS, Kim H, Hong CK (2016) Reduced graphene oxide (rGO) grafted zinc stannate (Zn_2SnO_4) nanofiber scaffolds for highly efficient mixed-halide perovskite solar cells. J Mater Chem A Mater 4:12158–12169. https://doi.org/10.1039/c6ta04726b

McMeekin DP, Mahesh S, Noel NK et al (2019) Solution-processed all-perovskite multi-junction solar cells. Joule 3:387–401. https://doi.org/10.1016/j.joule.2019.01.007

Misra RK, Cohen BA-E, Etgar L (2017) Low-dimensional organic–inorganic halide perovskite: structure, properties, and applications. ChemSusChem 10(19):3712–3721. https://doi.org/10.1002/cssc.v10.19

Moure C, Peña O (2015) Recent advances in perovskites: processing and properties. Prog Solid State Chem 43:123–148. https://doi.org/10.1016/J.PROGSOLIDSTCHEM.2015.09.001

Nakka L, Shen G, Aberle AG, Lin F (2023) Mitigation of potential-induced degradation in glass-encapsulated perovskite solar cells using a NiO_x barrier layer. Solar RRL. https://doi.org/10.1002/SOLR.202300582

Noh JH, Im SH, Heo JH et al (2013) Chemical management for colorful, efficient, and stable inorganic–organic hybrid nanostructured solar cells. Nano Lett 13:1764–1769. https://doi.org/10.1021/nl400349b

Noman M, Khan Z, Jan ST (2024) A comprehensive review on the advancements and challenges in perovskite solar cell technology. RSC Adv 14:5085–5131

Nozik AJ (2002) Quantum dot solar cells. Phys E Low Dimens Syst Nanostruct 14:115–120. https://doi.org/10.1016/S1386-9477(02)00374-0

Okoroafor T, Maalouf A, Oez S et al (2022) Life cycle assessment of inkjet printed perovskite solar cells. J Clean Prod 373:133665. https://doi.org/10.1016/J.JCLEPRO.2022.133665

Park IJ, Park JH, Ji SG et al (2019) A three-terminal monolithic perovskite/Si tandem solar cell characterization platform. Joule 3:807–818. https://doi.org/10.1016/j.joule.2018.11.017

Peng Y, Zeng F, Cheng Y et al (2020) Fully doctor-bladed efficient perovskite solar cells in ambient condition via composition engineering. Org Electron 83:105736. https://doi.org/10.1016/J.ORGEL.2020.105736

Pietzcker RC, Stetter D, Manger S, Luderer G (2014) Using the sun to decarbonize the power sector: the economic potential of photovoltaics and concentrating solar power. Appl Energy 135:704–720. https://doi.org/10.1016/j.apenergy.2014.08.011

Qin X, Xu B, Lin J et al (2022) Above 10% efficient electrodeposited $Cu_2ZnSn(S,Se)_4$ solar cell achieved by modifying precursor. Sol Energy Mater Sol Cells 242:111781. https://doi.org/10.1016/J.SOLMAT.2022.111781

Raj A, Kumar M, Anshul A (2021) Recent advancement in inorganic–organic electron transport layers in perovskite solar cell: current status and future outlook. Mater Today Chem 22:100595. https://doi.org/10.1016/J.MTCHEM.2021.100595

Raj A, Kumar M, Anshul A (2022) Recent progress in cesium-based lead-free halide double perovskite materials for photovoltaic applications. Phys Status Solidi A 219:2200425. https://doi.org/10.1002/pssa.202200425

Raj A, Kumar M, Anshul A (2023) Topical advances in fabrication technologies of perovskite solar cell heterostructures: Performance and future perspective. Mater Lett 340:134171. https://doi.org/10.1016/J.MATLET.2023.134171

Rajagopal A, Yang Z, Jo SB et al (2017) Highly efficient perovskite-perovskite tandem solar cells reaching 80% of the theoretical limit in photovoltage. Adv Mater. https://doi.org/10.1002/adma.201702140

Raza E, Ahmad Z (2022) Review on two-terminal and four-terminal crystalline-silicon/perovskite tandem solar cells; progress, challenges, and future perspectives. Energy Rep 8:5820–5851. https://doi.org/10.1016/J.EGYR.2022.04.028

Remeika M, Ono LK, Maeda M et al (2018) High-throughput surface preparation for flexible slot die coated perovskite solar cells. Org Electron 54:72–79. https://doi.org/10.1016/j.orgel.2017.12.027

Richhariya G, Kumar A, Tekasakul P, Gupta B (2017) Natural dyes for dye sensitized solar cell: a review. Renew Sustain Energy Rev 69:705–718. https://doi.org/10.1016/J.RSER.2016.11.198

Rubel MHK, Mitro SK, Hossain MK et al (2022) First-principles calculations to investigate physical properties of single-cubic $(Ba_{0.82}K_{0.18})(Bi_{0.53}Pb_{0.47})O_3$ novel perovskite superconductor. Mater Today Commun 33:104302. https://doi.org/10.1016/J.MTCOMM.2022.104302

Seo YH, Cho SP, Lee HJ et al (2022) Temperature-controlled slot-die coating for efficient and stable perovskite solar cells. J Power Sources 539:231621. https://doi.org/10.1016/J.JPOWSOUR.2022.231621

Sharma R, Sharma A, Agarwal S, Dhaka MS (2022) Stability and efficiency issues, solutions and advancements in perovskite solar cells: a review. Sol Energy 244:516–535. https://doi.org/10.1016/j.solener.2022.08.001

Shilpa G, Kumar PM, Kumar DK et al (2023) Recent advances in the development of high efficiency quantum dot sensitized solar cells (QDSSCs): a review. Mater Sci Energy Technol 6:533–546. https://doi.org/10.1016/J.MSET.2023.05.001

Shrestha S, Li X, Tsai H et al (2022) Long carrier diffusion length in two-dimensional lead halide perovskite single crystals. Chemistry 8:1107–1120. https://doi.org/10.1016/J.CHEMPR.2022.01.008

Stokes HT, Kisi EH, Hatch DM, Howard CJ (2002) Group-theoretical analysis of octahedral tilting in ferroelectric perovskites. Struct Sci 58:934–938

Tai Q, You P, Sang H et al (2016) Efficient and stable perovskite solar cells prepared in ambient air irrespective of the humidity. Nat Commun. https://doi.org/10.1038/ncomms11105

Tang X, Yang T, Yu D et al (2024) Current insights and future perspectives of ultraviolet radiation (UV) exposure: friends and foes to the skin and beyond the skin. Environ Int 185:108535. https://doi.org/10.1016/J.ENVINT.2024.108535

Tokura Y, Tomioka Y (1999) Colossal magnetoresistive manganites. J Magn Magn Mater 200:1–23. https://doi.org/10.1016/S0304-8853(99)00352-2

Uyanga KA, Ezike SC, Onyedika AT et al (2020) Effect of acetic acid concentration on optical properties of lead acetate based methylammonium lead iodide perovskite thin film. Opt Mater (Amst) 109:110456. https://doi.org/10.1016/J.OPTMAT.2020.110456

Wang F, Ye Z, Sarvari H et al (2019) Humidity-insensitive fabrication of efficient perovskite solar cells in ambient air. J Power Sources 412:359–365. https://doi.org/10.1016/j.jpowsour.2018.11.013

Wang K, Zheng L, Hou Y et al (2022a) Overcoming Shockley-Queisser limit using halide perovskite platform? Joule 6:756–771

Wang H, Sun Y, Chen J et al (2022b) A review of perovskite-based photodetectors and their applications. Nanomaterials 12
Wang K, Tong Y, Cao L et al (2023a) Progress of inverted inorganic cesium lead halide perovskite solar cells. Cell Rep Phys Sci 4:101726. https://doi.org/10.1016/J.XCRP.2023.101726
Wang R, Han M, Wang Y et al (2023b) Recent progress on efficient perovskite/organic tandem solar cells. J Energy Chem 83:158–172. https://doi.org/10.1016/j.jechem.2023.04.036
Woodwardt PM (1997) Octahedral tilting in perovskites. I. Geometrical considerations. Struct Sci 53(1):32–43
Yang Z, Cai B, Zhou B et al (2015) An up-scalable approach to $CH_3NH_3PbI_3$ compact films for high-performance perovskite solar cells. Nano Energy 15:670–678. https://doi.org/10.1016/J.NANOEN.2015.05.027
Yang C, Hu W, Liu J et al (2024) Achievements, challenges, and future prospects for industrialization of perovskite solar cells. Light Sci Appl 13:227
Yang D, Yang R, Priya S, Liu S (Frank) (2019) Recent advances in flexible perovskite solar cells: fabrication and applications. Angew Chem Int Ed 58:4466–4483. https://doi.org/10.1002/ANIE.201809781
Yu Z, Leilaeioun M, Holman Z (2016) Selecting tandem partners for silicon solar cells using spectral efficiency. Nat Energy 1(11):1–4
Zhang C, Park N-G (2024) Materials and methods for cost-effective fabrication of perovskite photovoltaic devices. Commun Mater 5:194. https://doi.org/10.1038/s43246-024-00636-8
Zhang D, Choy WCH, Wang CCD et al (2011) Polymer solar cells with gold nanoclusters decorated multi-layer graphene as transparent electrode. Appl Phys Lett. https://doi.org/10.1063/1.3664120
Zhang H, Cheng J, Lin F et al (2016) Pinhole-free and surface-nanostructured NiO_x film by room—temperature solution process for high-performance flexible perovskite solar cells with good stability and reproducibility. ACS Nano 10:1503–1511. https://doi.org/10.1021/acsnano.5b07043
Zhang T, Dar MI, Li G et al (2017) Bication lead iodide 2D perovskite component to stabilize inorganic a-$CsPbI_3$ perovskite phase for high-efficiency solar cells
Zheng X, Alsalloum AY, Hou Y et al (2020) All-perovskite tandem solar cells: a roadmap to uniting high efficiency with high stability. Acc Mater Res 1:63–76. https://doi.org/10.1021/accountsmr.0c00017
ZhenTao J, Yanhuan H, Hui W et al (2020) Two bilayer organic–inorganic hybrid perovskite compounds exhibiting reversible phase transition and dielectric anomalies. J Solid State Chem. https://doi.org/10.1016/j.jssc.2019.121104
Zhou C, Ou Q, Chen W et al (2018) Illumination-induced halide segregation in gradient bandgap mixed-halide perovskite nanoplatelets. Adv Opt Mater. https://doi.org/10.1002/adom.201801107
Zhou J, Tan L, Liu Y et al (2024) Highly efficient and stable perovskite solar cells via a multifunctional hole transporting material. Joule 8:1691–1706. https://doi.org/10.1016/J.JOULE.2024.02.019
Zhu Z, Mao K, Xu J (2021) Perovskite tandem solar cells with improved efficiency and stability. J Energy Chem 58:219–232

Chapter 14
Perovskite-Silicon Tandem Solar Cells: Basics to Latest Updates with Challenges

Vidhi, Anne Masih, Anjani Kumar Singh, and O. P. Thakur

Abstract The perovskite-silicon tandem solar cells (PSTSCs) have gained a lot of attention since they have the potential to revolutionize entire photovoltaic industry. They utilize the proven performance and stability of silicon solar cell along with high efficiency of the perovskite materials. It is anticipated that this combination will outperform single-junction silicon solar cells in terms of efficiency and increase the accessibility and competitiveness of solar energy. Single-junction perovskite solar cells have achieved the power conversion efficiency (PCE) above 26% while silicon solar cells have over 25% but it is only in the tandem arrangement that the Shockley–Queisser efficiency limit was surpassed recording PCEs of 33%. Yet, concerns toward stability of perovskite materials are still hindering the possibilities of high efficiency and durable PSTSCs. This chapter aims to analyze the trends and developments in field of PSTSC, covering last ten years and presenting in detail the state of the art, persisting issues, and manufacturing processes of tandem solar cells. Moreover, in this context, the chapter presents existing and new strategies to tackle these issues, with particular regard to material stability, scalability, and market penetration. Finally, this chapter sets forth a vision for the future of efficient, scalable, and profitable PSTSCs.

Keywords Tandem solar cells · Perovskites · Silicon · Shockley-Queisser efficiency · Power conversion efficiency (PCE)

Vidhi · A. Masih · O. P. Thakur (✉)
Materials Analysis and Research Laboratory, Department of Physics, Netaji Subhas University of Technology, Dwarka, New Delhi, India
e-mail: opthakur@nsut.ac.in

A. K. Singh
Department of Physics, Atma Ram Sanatan Dharma College, University of Delhi, Dhaula Kuan, New Delhi, India

M. Kumar and A. Raj (eds.), *Perovskite Solar Cells*, Springer Tracts in Electrical and Electronics Engineering, https://doi.org/10.1007/978-981-96-9463-1_14

Introduction

In the past decade, solar photovoltaic (PV) technology has established itself as one of the most trustworthy, affordable, and sustainable sources of energy for addressing energy needs. Accordingly, it provides a clean replacement for fossil fuels. PV cells (in particular silicon PV cells) have become more cost effective over the past decade due to advances in manufacturing and installation (Victoria et al. 2021; Dada and Popoola 2023; Bouich et al. 2023; Raj et al. 2022). In 1954, Shockley and his team reported the first silicon solar cell with an efficiency of 16%. The efficiency of a solar cell is the ratio of the energy produced to that of incoming solar energy. It is mainly based on material properties, temperature, and the process of fabrication. These parameters have a major impact on a PV cell's efficiency. Not all of the sunlight that strikes on the solar panels is transformed into usable energy. Some of it gets wasted due to thermalization and absorption losses. The amount of sunlight which is effectively converted into electrical energy is referred to as a solar panel's efficiency. When sunlight falls on the solar cell, a current is produced. Subsequently, it is transmitted through wires and busbars. Alternating current (AC) energy is created from this current and distributed for usage in residences (Oni et al. 2024; Zanatta 2022).

In recent years, a lot of focus has been placed on improving efficiency and reducing costs. Despite such advancements, the power conversion efficiency (PCE) was limited. Single junction silicon cells remain as the dominant player in the solar market and have reached the PCE to a maximum of 26.7% (Lin et al. 2023). Silicon-based photovoltaic cells are one of the most important PV technologies because of their abundance of material availability, reliability, and well-established manufacturing processes. The leader of the pack is crystalline silicon, with stunning efficiency yields-again, largely monocrystalline varieties. As is usual with efficient crystalline silicon, they are not inexpensive, and they have a limiting principle—the Shockley–Queisser (SQ) limit—state that means for single-junction cells, the maximum efficiency is theoretically capped at about 33%. In practice, thermalization losses and the inability to absorb low-energy photons that fall below silicon's bandgap reduce this efficiency further to a practical upper limit of around 29.4%, accounting for Auger recombination (Ho-Baillie et al. 2021).

There are limitations that prevent us from surpassing a certain threshold. Even if we create a solar cell with exceptional efficiency or a material that absorbs sunlight extremely well, enhancing both charge transfer and recombination to achieve optimal efficiency is not possible. Various applications require higher output power, and single-junction solar cells, providing inadequate efficiencies. The semiconductor absorber layer's band characteristics determine the maximum energy that can be extracted per photon and influence both the maximum voltage (V_M) and the open-circuit voltage (V_{OC}). A solar cell's efficiency is mainly restricted by the proportion of solar photons absorbed by the cell, and the amount of electrical energy generated per photon. The maximum power output of the device is the result of multiplying the average electrical energy produced per photon by the photon absorption rate. The

PCE of solar cells can be elevated due to stacking solar cells called tandem solar cells. Tandem cells stack materials with dissimilar bandgaps to absorb a wider range of sunlight; thereby, considerably enhancing power conversion efficiency. Such concept has been found to operate better than single-junction solar cells which is limited by the Shockley–Queisser (SQ) limit.

The configuration of tandem solar cells consists of stacking two or more materials and each of them is optimized to absorb different parts of the solar spectrum. Theoretical power conversion efficiencies (PCEs) of multi-junction cells can exceed 40%, making them a focal point for increasing solar energy output (Oni et al. 2024). Tandem cells PCEs can be further optimized further using various optical engineering techniques such as anti-reflective coatings, light trapping structures, and textured interfaces. All of these techniques help mitigate reflection while enhancing light absorption. Different device configurations such as two-terminal and four-terminal structure are employed, where the goal is to mitigate losses in either configuration. These configurations can help us ensure optimal current matching between the sub-cells which further enhances the device's overall performance. The tandem solar cells can be of different types, such as perovskite-crystalline silicon or perovskite-perovskite combinations, or they can be identical.

Recently, researchers are eager to investigate perovskite-crystalline silicon tandem solar cells because of the favorable optoelectronic properties of perovskite solar cells (PSCs). PSCs have predicted PCEs exceeding 29% by 2040 (Yamaguchi et al. 2018). In PSTSCs, the perovskite layer, which has a large bandgap of about ~ 1.7 eV, is the top cell that absorbs high energy photons (primarily blue and green light). Owing to their unique optoelectronic properties, perovskites have become a promising material for top cells in silicon tandem solar cells. Furthermore, perovskite cells' low cost and facile processing, along with their ability to overcome conventional lattice mismatch restrictions, have made them a popular choice for multi-junction cell designs. Silicon serves as the lower cell, possessing a narrower bandgap (approx. 1.1 eV) and is capable of capturing lower-energy photons, such as red and infrared light; however, the use of a tandem setup guarantees that visible and infrared light is harvested more effectively than traditional silicon-only cells. This broad absorption range translates into higher photocurrent and, ultimately, greater efficiency. On the other hand, silicon solar cells have a long history because of their wide range of applications in both industrial and research applications, their abundant supply of materials, and their bandgap that extends beyond visible light (about 500 nm to below 1200 nm). Stacked solar cells, which are made up of a combination of silicon and perovskite, can therefore reach greater solar efficiencies, with a theoretical PCE of over 40% (Oni et al. 2024). One important factor of this breakthrough is that it allows solar technology to be more efficient and cost-effective.

In this chapter, we have focused on the perovskite-silicon tandem solar cells (PSTSCs). We begin this chapter with some fundamental concepts. Then we dwell into tandem solar cells and their various configurations like 2 T, 3 T, and 4 T. Next, we discuss the PSTSCs in detail with their advancements over the years. The second half of the chapter elucidates various fabrication methods employed in making PSTSCs. In the end, we have discussed the challenges and future scope of the PSTSCs.

Tandem Solar Cells

Tandem solar cells are an exciting and innovative solution to improving PV efficiency by stacking different types of solar cells. Compared to conventional single-junction cells, tandem cell configuration enables the capture of a broader region of the solar spectrum. Thus, it can greatly increase PCE of a solar cell. Tandem solar cells are categorized into many varieties according on the materials utilized or the manner in which the sub-cells are joined.

(i). Classification according to material: Tandem cells can be organic, inorganic, or hybrid. Organic tandem cells are typically more cost-effective but have lower efficiency due to the nature of organic materials. Currently, the only commercial tandem cells that are available in the market are the inorganic tandem cells, which are derived from materials belonging to the III–V group. They can only be used in specialist applications and niche markets like satellites due to their high cost. Hybrid tandem cells make use of organic and inorganic materials. They aim at balancing the cost and efficiency by leveraging the strengths of both. In a multi-junction design, each layer or junction is tailored to absorb a specific part of the solar spectrum at different voltages, minimizing thermalization losses. This has led to record-breaking PCEs, with III–V multi-junction solar cells achieving up to 38.8% efficiency under standard sunlight conditions and up to 46% under concentrated sunlight (France et al. 2015; Chiu et al. 2014).

(ii). Classification based on physical integration and connections: Tandem cells can also be categorized based on the connections between the sub-cells. Tandem solar cells can be categorized as mechanically stacked or monolithic from a physical perspective. Sub-cells in monolithic tandem cells are integrated onto the same substrate and electrically coupled. These cells frequently encounter difficulties because of their expensive production costs and the intricate technical integration of various layers. Conversely, mechanically stacked tandem cells feature separate, functional sub-cells layered on top of each other, but because of the extra electrodes and substrates, they are more expensive to make and have more optical losses. As an alternative, tandem solar cells (TSCs) can be categorized according to the number of electrical contacts they have—two to four. While four-terminal (4 T) tandems are mechanically stacked, two-terminal (2 T) and three-terminal (3 T) tandems are often monolithic (Fig. 14.1).

2 T Tandem Solar Cell

The two-terminal (2 T) TSC arrangement has been the subject of the great interest among the researchers. This is due to its similarity in exterior electrical connections with single-junction solar cells (SJSCs). This similarity exists because it can simply replace SJSCs in ordinary photovoltaic modules. But manufacturing it is not easy, especially when the sub-cells are integrated in a monolithic fashion. To maximize

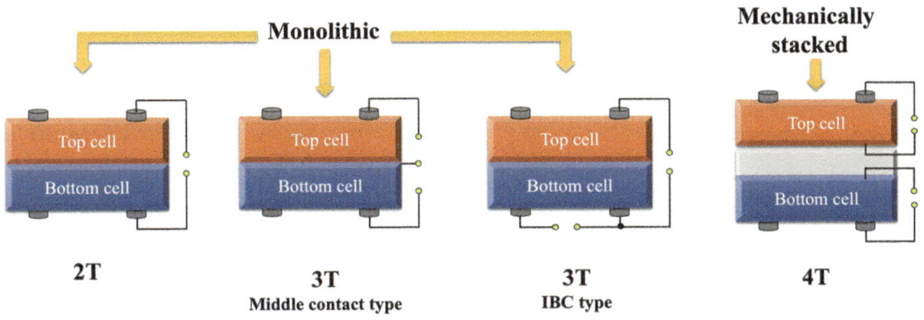

Fig. 14.1 Classification of tandem solar cells based on physical integration and connections

performance, the sub-cells in a 2 T TSC connected in series also need to have their photocurrents precisely matched. An interconnection layer connects the sub-cells in a 2 T TSC in series (Fig. 14.1), giving the device the characteristics of a single-junction cell but with a larger power output. Generally speaking, 2 T TSCs can also be measured using the same standard techniques as are used for SJSCs. However, there are a number of significant obstacles to precisely characterizing 2 T TSCs, such as complicated electrical and optical interactions between the sub-cells, spectrum mismatch, and difficulty in sub-cell characterization.

3 T Tandem Solar Cell

Unlike the two-terminal TSC, where sub-cells are connected in series, a three-terminal (3 T) TSC design allows for greater flexibility due to the presence of an additional electrode and source measure unit (SMU). Two distinct power outputs can be collected with this setup. But because of the need for a new module arrangement, this design might be more expensive to commercialize. Due to the relatively new concept, only a small number of devices, such as organic/organic, III–V/III–V, III–V/silicon (Si), Si/Si and perovskite/silicon (PVSK/Si) TSCs, have been reported up to this point (Sista et al. 2010; Geisz et al. 2021; Steiner et al. 2009; Rienäcker et al. 2019; Schnabel et al. 2020; Tockhorn et al. 2020). 3 T TSCs can be classified as middle-contact or interdigitated back contact (IBC) depending on the kind of bottom cell (Fig. 14.1). Since a top contact is not feasible, the only feasible method of integrating highly efficient IBC bottom cells is through the placement of two electrical connections at the back and one at the front of a conventional monolithic IBC TSC. In contrast, middle-contact 3 T TSCs have an extra electrical contact between the sub-cells for parallel connections or to help with characterization. Here, the internal contact is exposed by reducing the active area of one sub-cell. A 3 T TSC's sub-cells may be linked in parallel (bipolar junction transistor type) or series (tunnel junction type). In any configuration, the two sub-cells of a 3 T TSC operate independently, each with its own source measure unit. As a result, a 3 T TSC generates two smaller power outputs instead of a single high-power output. Because of this design, unlike 2 T TSCs, 3 T TSCs are not limited by the requirement for exact current matching between sub-cells (Gota et al. 2020).

4 T Tandem Solar Cell

A 4 T TSC can be constructed by stacking two solar cells as long as the top cell's back is transparent (Fig. 14.1). Direct connections between the sub-cells are not necessary. The ease of integration of 4 T TSCs makes them attractive, especially for solar cells with fabrication limits, despite the possibility of optical losses and increased module prices due to the additional transparent electrode and insulation between the sub-cells. Similar to 3 T TSCs, 4 T TSCs require unique module designs. For example, modules with independent top (transparent) and bottom cells are required. As a result, many different 4 T TSC configurations have been developed, such as PVSK/PVSK, PVSK/Si, PVSK/CIGS, and PVSK/CZTS (Moghadamzadeh et al. 2021; Essig et al. 2017; Jung et al. 2021; Huang et al. 2022; Kim et al. 2019a; Wang et al. 2022a; Whitehead et al. 2021). The two electrical connections in each of the four sub-cells in a 4 T TSC can be connected in different ways to resemble 2 T or 3 T arrangements. The cells share many benefits and limits with 2 T and 3 T TSCs since they can be operated with one or two source measure units (SMUs) and coupled in series or parallel. Although there are some common inaccuracies in the literature when characterizing 4 T TSCs, one typical mistake is to measure the sub-cells independently, which ignores the electrical and luminous connection between them. Both sub-cells should be monitored concurrently, one lit and the other biassed to its maximum power point voltage, in order to replicate real-world operation. Furthermore, it is crucial to accurately identify the cell areas because differences in cell area might have a substantial impact on reported conversion efficiencies. The total 4 T TSC efficiency is calculated by summing the efficiency of each sub-cell (Dewi et al. 2020).

Perovskite-Silicon Tandem Solar Cells (PSTSCs)

Top Cell-Perovskite

Halide perovskites are named after the Russian mineralogist L.A. Perovski. The mineral perovskite ($CaTiO_3$) was named in his honor. Structurally, halide perovskites are characterized by their unique crystal formation, adhering to the general formula ABX_3. In this formula, A represents a larger cation, B a smaller cation, and X an anion, often from the halogen family such as chlorine or bromine. This structure is illustrated in the perovskites' depiction in scientific diagram shown in Fig. 14.2. Perovskite

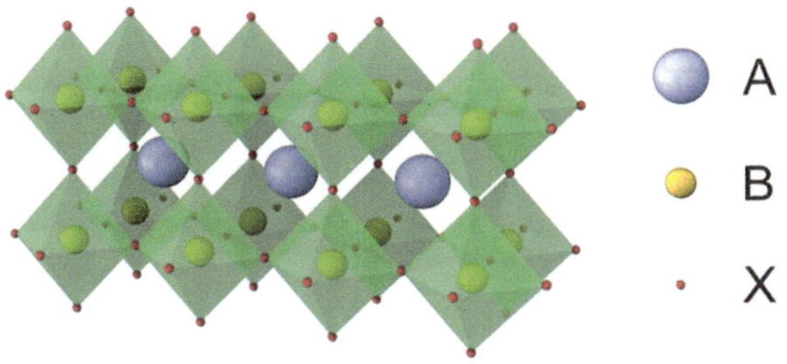

Fig. 14.2 ABX_3 structure of perovskite (Anaya et al. 2017). Copyright 2017 Elsevier. Reproduced with permission from Elsevier

structure and its detailed description is given in Chap. 13 already; therefore, we have discussed it only in brief here.

The top cell PSTSCs are made up of PSCs, which are classified into two architectures based on the order of the electron transport layer (ETL) and hole transport layer (HTL): n-i-p and p-i-n (Fig. 14.3). In n-i-p configuration, arrangement is as follows: transparent conductive oxide (TCO), ETL, perovskite, hole transport layer HTL, and metal electrode. On the other hand, for p-i-n configuration, they are in order as follows: TCO, HTL, perovskite, ETL, and metal electrode. In PSTSCs, both structures have shown excellent performance (Köhnen et al. 2021). But for the top cell, the p-i-n arrangement is becoming more and more well-liked because of its easier low-temperature fabrication method, improved stability and decreased hysteresis. Moreover, the p-i-n structure's thinner ETL contributes to a decrease in parasitic absorption and a rise in J_{SC} (Wu et al. 2022; Xu et al. 2020). According to theoretical calculations, the best tandem structure is produced by pairing a silicon bottom cell with a 1.1 eV bandgap with a perovskite top cell with a 1.73 eV bandgap (Leijtens et al. 2018; Futscher and Ehrler 2016). Common perovskites, with bandgaps of 1.60 and 1.5 eV, respectively, like $MAPbI_3$ and $FAPbI_3$, fall short of this ideal (Jesper Jacobsson et al. 2016). Researchers frequently alter the composition of perovskites, usually by changing the X-site anion or A-site cation, in order to maximize the bandgap. Through modifications to the lattice structure, the size of the A-site cation indirectly influences the bandgap, hence modifying the electrical characteristics and B-X-B bond angles. For example, smaller cations such as caesium and rubidium (Cs^+ or Rb^+) have a higher bandgap, which makes them appropriate for use in tandem cells. The bandgap and stability of perovskites are adjusted for tandem applications by adjusting the composition of the perovskite using cationic additions, such as Cs^+ or Rb^+ (Amat et al. 2014; Prasanna et al. 2017).

Single-Cation Perovskites

The initial PSTSCs were based on single-cation perovskites, $MAPbI_3$. But, their poor bandgap caused them to function poorly (Wu et al. 2017; Werner et al. 2016; Albrecht et al. 2016). Tandem applications are more appealing for inorganic perovskites, which are mainly Cs-based and have greater stability and longer lifetimes due to their more

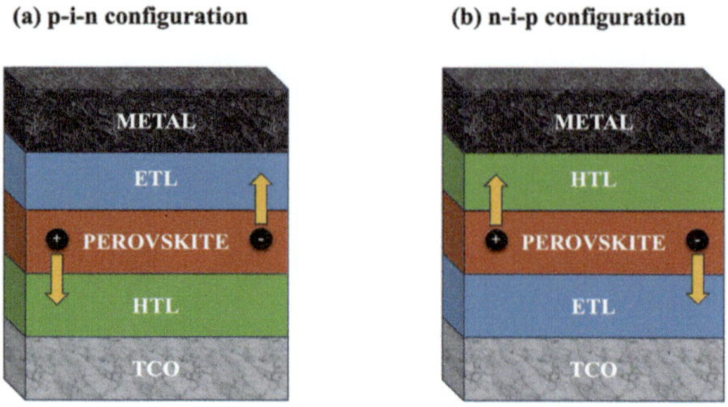

Fig. 14.3 p-i-n and n-i-p configurations of perovskite solar cells

appropriate bandgaps (e.g., $CsPbI_3$ at 1.72 eV) (Wang et al. 2024). Improved PCE and V_{OC} were achieved in $CsPbI_xBr_{3-x}$ films by employing inorganic nickel iodide as a passivation agent for halogen vacancies. With a PCE of 22.95% and a high V_{OC} of 2.04 V, this method produced the first inorganic perovskite/Si tandem cell that retained 95.7% of its initial efficiency after 300 h of continuous illumination. The efficiency of inverted inorganic tandem devices was further increased to 25.31% by the use of surface passivation materials such as 2-amino-5-bromobenzamide (ABA), demonstrating good thermal and optical stability (Wang et al. 2022b, 2023).

Double-Cation Perovskites

Double-cation perovskites have drawn interest because they are more optically and thermally stable than single-cation systems (Li et al. 2016). Under extreme environmental circumstances, these double-cation complexes, have proven to be remarkably stable. For example, K. Bush et al. prepared $Cs_{0.17}FA_{0.83}Pb(Br_{0.17}I_{0.83})_3$. Even after extended exposure to high temperatures and humidity, this perovskites with silicon heterojunctions (SHJ) have obtained PCEs as high as 23.6% (Bush et al. 2017). Further, with similar composition, efficiency levels as high as 26% have been attained by B. Chen and his team. They achieved it with additional tuning, which includes changing the ratio of halide anions and adding textured silicon surfaces (Chen et al. 2020).

Triple-Cation Perovskites

High-efficiency PSTSCs frequently use triple-cation perovskites because they provide even more stability and compactness than double-cation systems. Under operating conditions, films derived from triple-cation perovskites are usually more stable and free of pinholes (Al-Ashouri et al. 2020; Saliba et al. 2016). For example, M. Saliba and et al. fabricated PSTSCs with the composition $Cs_{0.05}(FA_{0.77}MA_{0.23})_{0.95}Pb(I_{0.77}Br_{0.23})_3$. They achieved efficiencies of up to 29.15% and, after 300 h of operation without encapsulation, have retained 95% of their initial efficiency (Saliba et al. 2016). Further, P. Tockhorn and his team worked with a similar composition or perovskite and have reached to an efficiency of 29.8%. They attained this result by making some improvements, such as the insertion of dielectric buffer layers (Tockhorn et al. 2022).

Bottom Cell-Silicon

With almost 90% of the market for PV cells, single-junction crystalline silicon (c-Si) solar cells are reaching the limits of efficiency (Wawer et al. 2011; Tang et al. 2017). PVK-Si tandem solar cells present a possible alternative. Even though c-Si cells have set a record efficiency of 26.7%, intrinsic losses like thermalization and transparency make further advancements challenging as it is near the theoretical Auger limit of 29.4% (Yoshikawa et al. 2017). Since, it is not practical to replace the well-established Si PV infrastructure, researchers have concentrated on combining Si with other semiconductors to create tandem cells, such as hybrid organic–inorganic halide-based perovskites. In addition to potentially exceeding the Auger limit for silicon and the Shockley–Queisser limit of 33% for single-junction devices, this design reduces the energy losses linked with single-junction cells (Futscher and Ehrler 2016). Depending on the doping element utilized, the silicon bottom cell in PSTSCs can be categorized as *p*-type or *n*-type. Due to its easier production method and lower cost, *p*-type silicon has historically been widely popular. *p*-type silicon is commonly doped with boron (Qiang et al. 2022). On the other hand, phosphorus-doped *n*-type silicon is currently receiving more recognition because of its better qualities, which include a longer carrier lifespan, a lower temperature coefficient, and a higher conversion efficiency. For commercial PV applications, *n*-type silicon technologies such as SHJ, tunnel oxide passivated contact (TOPCon), and interdigitated back contact (IBC) are proving to be more effective and promising (Mazzarella et al. 2019). PSTSCs are becoming more and more efficient as a result of the integration of these various Si bottom cells with perovskite top layers that have been improved, which is also hastening the process of commercialization (Wu et al. 2022, 2017).

Aluminum Back Surface Field (Al-BSF)

One of the conventional methods used in industrial screen-printed *p*-type crystalline silicon (C-Si) batteries is the Al-BSF cell. In this arrangement, an aluminum film is deposited on the back side to generate a p^+ layer. This film functions as a metal electrode and a way to lessen carrier accumulation. J. Mailoa and et al. reported the first PSTSC with an Al-BSF bottom cell, attaining a moderate 13.7% efficiency (Mailoa et al. 2015). Four years later, C. Kim and his team reported that the efficiency of Al-BSF and p-i-n structured PSCs was increased to 21.1% with subsequent advances. The PCE was nevertheless below commercial standards, mostly because of recombination losses (Kim et al. 2019b).

Passivated Emitter Rear Cell (PERC)

Compared to Al-BSF, the PERC cell has more sophisticated technology and has an emitter, front contacts, and an anti-reflective coating (ARC) on its front side. Open-circuit voltage is improved, optical coupling is strengthened, and surface recombination is greatly reduced by the passivation layer on the back side (Lee et al. 2015; Xiao and Xu 2014). With a PERC cell used as the bottom layer in PSTSCs, Y. Wu and his team reported PCE of 22.5%, which is significantly higher than tandem cells based on Al-BSF. Unfortunately, the comparatively high electrical resistance of PERC cells resulted in a reduced fill factor (FF), which restricted the tandem device's overall performance (Wu et al. 2017).

Tunnel Oxide Passivated Contact (TOPCon)

With a structure that combines poly-silicon (poly-Si) for rear contact creation and a thin silicon oxide (SiO_2) layer for passivation, TOPCon technology offers a novel method for *n*-type Si bottom cells. This combination improves the V_{OC} and overall cell performance by reducing recombination losses brought on by dangling bonds at the surface and greatly enhancing charge transfer (Padi et al. 2021). A notable PCE of 28.2% was attained for perovskite/TOPCon tandem device by Z. Ying and et al. by employing surface nano-structuring to enhance perovskite development (Ying et al. 2022).

Silicon Heterojunction (SHJ)

High-performance *n*-type silicon technology is also demonstrated by SHJ cells, which are characterized by hydrogenated amorphous silicon (a-Si: H) passivation layers that substitute the SiO_2/poly-Si combination found in TOPCon cells. Because of their greater V_{OC} (about 750 mV) and reduced surface recombination, SHJ devices are excellent choices for incorporation into perovskite tandem devices (Liu et al. 2020; Panigrahi and Komarala 2021; Chang et al. 2023). The utilization of SHJ cells as the bottom cell in PSTSCs is advantageous due to its capacity to efficiently handle optical losses and harvest near-infrared photons (Boccard and Ballif 2020). S. Mariotti and et al. attained an even greater efficiency of 32.5% by minimizing non-radiative recombination losses with interfacial changes, while P. Tockhorn and et al. reported a PCE of 29.8% for perovskite/SHJ TSCs by adding periodic nanostructures (Tockhorn et al. 2022; Mariotti et al. 2023).

Alliance of Silicon with Perovskites

At first, III–V semiconductors were employed in conjunction with Si, but additional development was impeded by their high manufacturing costs. In recent times, perovskite materials have surfaced as a financially viable substitute, with independent efficiency levels of 25.6%. Perovskites and silicon have been combined to provide notable improvements in tandem cell efficiency. P. Löper and et al. reported PVK/Si cell as early as 2014 (Löper et al. 2015). They proposed a perovskite/Si 4 T tandem structure with a 13.4% efficiency. They reveiled that the 4 T tandem device might exceed 30% PCE by reducing optical losses and increasing the transparency of the connections. One of the first 2 T perovskite/Si tandem cell was created by J. Mailoa and et al. in the beginning of 2015. The fabricated 2 T perovskite/Si tandem cell achieved a stable efficiency of 13.7%. The top and bottom cells of this device were connected via a narrow inter-band tunnel connection made of silicon. The efficiency of the 2 T device was found to be restricted by the perovskite top cell. The researchers proposed that substituting the transparent hole-transporting material for the spiro-OMeTAD layer could enhance the device's efficiency (Mailoa et al. 2015). Then in 2016, D. McMeekin and et al. utilized composition engineering to optimize a 1.74 eV wide bandgap perovskite. This resulted in a 4 T tandem device, which combined a silicon bottom cell and a 1.74 eV perovskite top cell, achieving a total efficiency of 25.2% and demonstrating the potential for tandem structures to achieve higher efficiency (McMeekin et al. 2016). The majority of efforts prior to 2017 were directed toward the development of 4 T perovskite/Si tandem cells. This was due to ease of fabrication of a semi-transparent perovskite cell that was placed on top of a silicon bottom cell. However, later, the focus switched to 2 T tandem designs with an emphasis on transparent top electrodes, improved perovskite composition, and connectivity layers. H. Zhou and et al., for instance, synthesized perovskite compositions with different bandgaps for 2 T tandem cells and found that a silicon bottom cell matched well with a 1.69 eV perovskite top cell, resulting in over 20% PCE for the 2 T tandem cell (Fan et al. 2017). Extensive research was carried out on perovskite optimization, interface engineering, bottom cell enhancements, tunnel junctions, and transparent top electrodes in the ensuing years (Kim et al. 2019b; Shen et al. 2024; Dewi et al. 2019; Amri et al. 2021).

In 2023, E. Aydin and his team fabricated monolithic perovskite/silicon tandem solar cell. They obtained a certified PCE of 32.5% (Aydin et al. 2023). Recently, in 2024, PVK-Si cells achieved 33.89%, breaking the silicon record of 27% as well as single-junction Shockley–Queisser efficiency limit (33.7%). Notable contributors to these advancements include researchers from LONGi Green Energy Technology Co. Ltd, Shanghai, China (Liu et al. 2024a). J. Liu, Y. He, L. Ding, and et al. from LONGi published this work. They reported that the two-terminal monolithic PSTSCs have significantly improved, overcoming issues with charge transport and interfacial recombination. They created a bilayer passivation technique that improved electron extraction and inhibited non-radiative recombination by first applying a nanoscale LiF ultrathin layer and then depositing diammonium diiodide molecules. With a strongly textured rear surface and a minimally textured front, Czochralski (CZ)-based silicon heterojunction cells served as the foundation for the tandem cells, which were designed to maximize photocurrent generation while preserving rear surface passivation. J. Liu and his team reported PCE of 33.89%, an open-circuit voltage of almost 1.97 V, a fill factor (FF) of 83.0%, and the first documented two-junction tandem cell to exceed the 33.7% single-junction Shockley–Queisser efficiency limit (Liu et al. 2024a) (Table 14.1).

Deposition/Fabrication Techniques or Designing PVK-Si Tandem Solar Cells

The performance of PSTSCs is influenced not only by the structure of the two sub-cells but also by the surface morphology of the silicon bottom cell on the light side. Planar or textured silicon surfaces are both possible, and the choice of surface affects the device's light-capturing efficiency and the quality of the top perovskite layer deposit (Al-Ashouri et al. 2020; Chen et al. 2019). For solution-based perovskite deposition techniques like spin-coating, which is well-liked in lab-scale fabrication because of its simplicity, planar silicon structures are frequently employed. Nevertheless, due to interfacial reflection, this method may result in optical losses that reduce photocurrent and overall efficiency. Thus far, the maximum spin-coating efficiency in a tandem device is 32.5%; nonetheless, this still falls short of the realistic bounds (Mariotti et al. 2023; Kim et al. 2020). Given below are some of the fabrication techniques employed in making of perovskite/Si tandem solar cells.

Spin-Coating

Spin-coating makes use of centrifugal force and gravity to evenly distribute the solution across a fast-spinning substrate (Shi and Zhang 2023; Raj et al. 2023). It is employed in order to help build a smooth and compact perovskite film. There are two methods of doing spin coating. (i) One-Step Method for Spin Coating: Lead halide salts like PbI_3 and $PbBr_3$ are mixed with organic halide salts like methylammonium iodide (MAI) and formamidinium iodide (FAI), and the mixture is then dissolved in solvents like dimethyl sulfoxide (DMSO) or N, N-dimethylformamide (DMF). After annealing, the resultant precursor solution is spin-coated onto the conductive or electron/hole transport layer. Perovskite films are formed as a result of this process, as Fig. 14.4a shows. (ii) Two-Step Process for Spin Coating: This method involves the separate dissolution of lead halide and organic halide salts in isopropanol and DMF, respectively. The electron/hole transport layer (ETL/HTL) is first spin-coated with the lead halide solution, and it is then annealed. The lead halide layer is then spin-coated with the organic halide solution (such as MAI), and a second annealing

Table 14.1 Advancements in PSTSCs over the years

Year	Tandem structure	Perovskite composition	Perovskite bandgap (eV)	Top electrode	Tandem PCE (%)	References
2014	4 T	$MAPbI_3$	1.58	Sputter ITO	13.4	Löper et al. 2015)
2015	2 T	$MAPbI_3$	1.58	Ag nanowire	13.7	Mailoa et al. 2015)
2016	4 T	$FA_{0.83}Cs_{0.17}Pb(I_{0.6}Br_{0.4})3$	1.74	Sputter ITO	25.2	McMeekin et al. 2016)
2017	2 T	$MA_{0.37}FA_{0.48}Cs_{0.15}PbI_{2.01}Br_{0.99}$	1.69	Sputter ITO	20.57	Fan et al. 2017)
2019	4 T	$Cs_{0.05}(MA_{0.17}FA_{0.83})_{0.95}Pb(I_{0.83}Br_{0.17})3$	1.58	Sputter ITO	25.5	Dewi et al. 2019)
2021	2 T	$Cs_2AgBi_{0.75}Sb_{0.25}Br_6$	1.8	Sputter IZO	24.4	Amri et al. 2021)
2023	4 T	$Cs_{0.05}FA_{0.8}MA_{0.15}Pb(I_{0.755}Br_{0.255})_3$	1.68	Sputter IZO	32.5	Aydin et al. 2023)
2024	2 T	$FA_{0.8}MA_{0.15}Cs_{0.05}Pb(I_{0.76}Br_{0.24})_3$	1.69	Sputter IZO	33.89	Liu et al. 2024a)

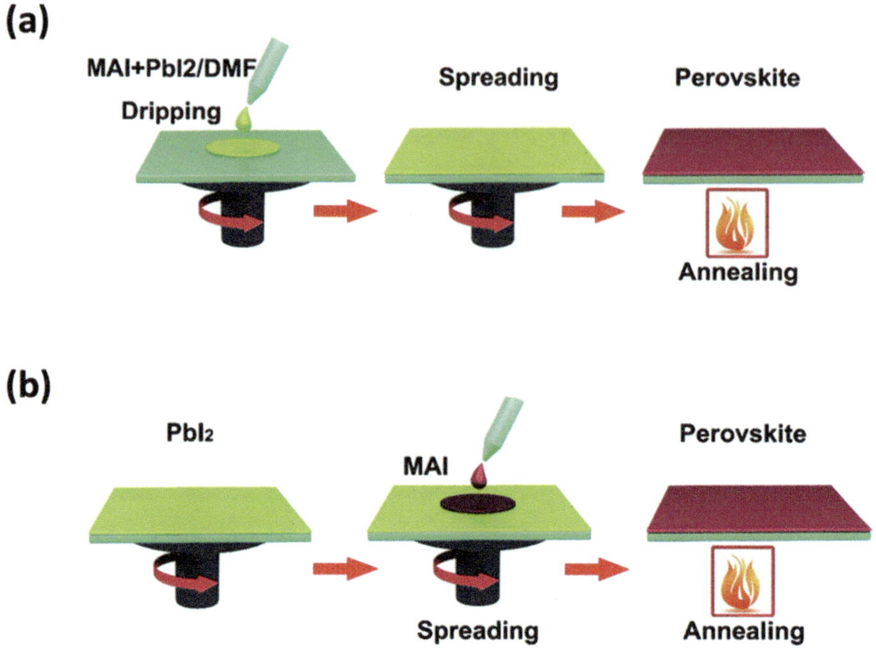

Fig. 14.4 **a** One-step method for spin coating and **b** two-step method for spin coating (Chen et al. 2018). Copyright 2018 royal society of chemistry. (Open access)

step is then performed to promote interdiffusion and the development of perovskite films, as shown in Fig. 14.4b.

In PSTSCs, spin-coating has produced efficiency greater than 30% (Mariotti et al. 2023). Nevertheless, this is still below the achievable efficiency limit. This restriction can be attributed, in part, to optical losses brought on by flat silicon surfaces' interfacial reflections, which lower photocurrent density (Chen et al. 2022). Textured silicon surfaces, with their tiny pyramidal structures, are better at lowering reflection losses, which raises the tandem device's PCE and photocurrent density (J_{SC}) (Wang et al. 2021). But large perovskite layers are difficult to deposit on textured silicon using traditional spin-coating procedures, therefore new methods for more uniform coatings have to be developed like, chemical vapor deposition (CVD) and two-step hybrid deposition, which blends solution processing with thermal evaporation.

Two-Step Hybrid Deposition

Two-stage hybrid technique introduces the organic precursor via a solution-based procedure after an inorganic lead halide layer is thermally evaporated onto the textured silicon in the first step. This combination creates a more homogeneous

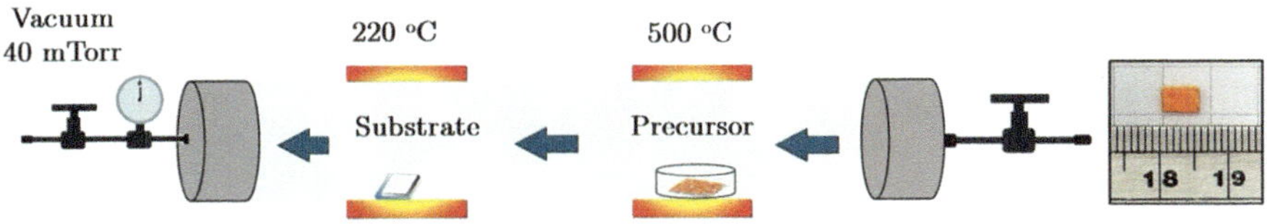

Fig. 14.5 Schematic illustration of CVD setup (Sosa-Acosta et al. 2024). Copyright 2024 Elsevier. Reproduced with permission from Elsevier

layer by enabling the perovskite material to completely penetrate the spaces in the textured silicon (Luo et al. 2023). For instance, this technique was utilized by J. Liu and et al. to create a TSC having J_{SC} of 20.01 mA/cm^2 by depositing a perovskite layer on a crystalline silicon bottom cell (Liu et al. 2024b). Although efficient, this process's scalability is restricted by the second step's dependency on spin-coating. Nevertheless, it might be modified to employ more practically applicable techniques.

Chemical Vapor Deposition (CVD)

A horizontal furnace with a cylindrical quartz tube and two independently regulated heating zones is used in the CVD technique for perovskite film formation (Fig. 14.5). In order to facilitate homogeneous film deposition, the substrate, such as glass or silicon, is positioned downstream in a colder zone maintained at 220 °C, while the precursor material is positioned in the central heating zone at 500 °C. To ensure controlled vapor transfer, the system runs at a 40 mTorr vacuum pressure. The furnace is shut off and the system is left to cool naturally to room temperature after the deposition is finished. For the large-scale fabrication of films, CVD is a proven process. This technique is widely employed to synthesize metal and metal oxide films, provide exact control over the composition of perovskite films, and is highly suitable for industrial uses, especially in the creation of PSTSCs (Gao et al. 2021; Tao et al. 2015; Leyden et al. 2016). Using a mix of sputtering and CVD, D. Kim and colleagues successfully achieved perovskite film deposition on a 100 cm^2 textured silicon substrate (Hwang et al. 2020). Nevertheless, the expensive setup and hardware development costs prevent these approaches from being widely adopted for PSTSC manufacturing, even if they have the ability to produce enormous areas.

Blade-Coating

Blade-coating is a simple, effective process that may be used in large-scale manufacturing to produce high-quality perovskite films. It is also called knife-over-edge coating or doctor-blading. Figure 14.6 shows, entails moving a blade over a surface

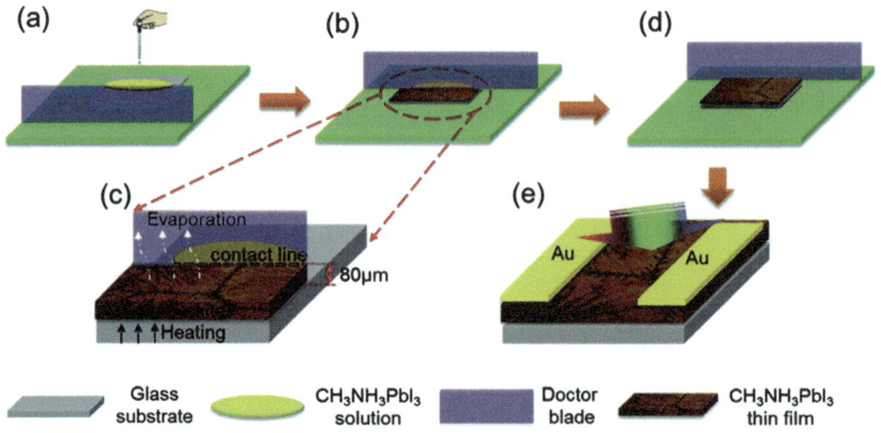

Fig. 14.6 Blade-coating technique (Tong et al. 2017). Copyright 2017 Elsevier. Reproduced with permission from Elsevier

(or the surface across the blade in roll-to-roll processes) in order to disperse pre-dispensed ink into a thin liquid layer. After that, the liquid film is dried to create a solid thin layer. This method is frequently used to create large-area perovskite films. According to studies, perovskite films formed by blade-coating frequently have longer carrier diffusion lengths than films made by other techniques (Yang et al. 2022; Xiao et al. 2021). This process offers a viable, economical way to create PSTSCs with high PCE. Ink dispersal still presents some difficulties, though, and this would require improvement for high-throughput and large-scale production (Chen et al. 2020).

Slot-Die Coating

Similar but more industrialized, slot-die coating is meant to solve the problems with ink distribution that arise with blade-coating. A typical slot-die coating procedure includes dispensing perovskite ink through a small aperture in the slot-die head onto a moving substrate, as shown in Fig. 14.7. Two menisci occur, one upstream and one downstream of the coated bead, when the liquid contacts the substrate. The downstream meniscus then directs the wet film as it extends along the substrate in the coating's direction. Because of its cost-effectiveness and scalability, this approach has gained popularity in the PV industry. It is employed to deposit perovskite films on large area (Xiao et al. 2021). S. De Wolf and et al. constructed a 1.68 eV bandgap perovskite on a 1 cm^2 SHJ cell with a PCE of more than 23% via slot-die coating. It was discovered that the perovskite layer was excessively thick, which prevented effective carrier extraction and transmission. Slot-die coating offers great promise

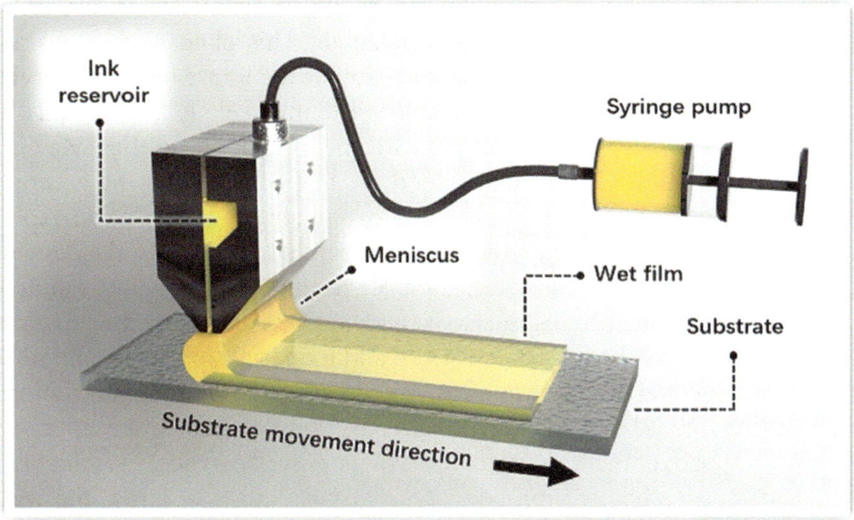

Fig. 14.7 Schematic illustration of slot-die coating setup. Copyright 2023 Elsevier. Reproduced with permission from Elsevier

for furthering R&D and commercialization in PSTSC manufacturing as long as ink formulations and precision tooling remain superior (Subbiah et al. 2020).

Challenges to PSTSCs

PSTSCs have a great deal of promise to boost PCEs, but there are a few important issues that need to be resolved. Improving stability, scaling up production, optimizing transparent electrodes, reducing costs, and ensuring long-term reliability are key areas requiring further research and development.

Transparent Electrode Optimization

Creating an effective transparent electrode is one of the main issues with both 2 T and 4 T tandem devices. A careful balance between outstanding conductivity and exceptional transparency is needed for this component. Transparent conducting oxides (TCOs), like indium tin oxide (ITO), graphene, carbon nanotubes, thin metal layers, and silver nanowires have all been investigated as potential transparent electrodes for perovskite solar cells (Yoon et al. 2021a; Shi et al. 2020; Mujahid et al. 2021; Rahmany and Etgar 2020). ITO has shown to be especially successful among them in extending the life of semi-transparent perovskite cells (Cheng et al. 2021a). Perovskite films, however, are unable to tolerate the high temperatures (>300 °C) required to produce extremely transparent and conductive ITO (Adil Afroz et al. 2020; Balasundaraprabhu et al. 2009). In order to preserve perovskite integrity, low-temperature or room-temperature deposition techniques for ITO are being developed. ITO has been effectively deposited at room temperature directly onto the spiro-OMeTAD layer in n-i-p shaped cells without affecting performance (Bett et al. 2019). However, in order to shield organic layers from deterioration during deposition, p-i-n shaped cells frequently need a protective metal oxide buffer layer, such as ALD SnO_2. Unfortunately, optical losses may result from these buffer layers, decreasing the device's overall efficiency (Cheacharoen et al. 2018). Additionally, a large portion of parasitic absorption losses are caused by ITO layers thicker than 500 nm, which are generally required to attain optimum conductivity. Creating ITO layers that are thinner (about 100 nm) in addition to using metal grid fingers to increase conductivity and transparency is one way to address issue (Cheng et al. 2020; Bush et al. 2016; Jacobs et al. 2019). But metal grids come with their own set of drawbacks, namely optical loss from shadowing. Effectively balancing transparency and conductivity requires careful grid pattern optimization.

Stability of PSTSCs

The stability of PSTSCs is one of the main obstacles. Under environmental stressors including as heat, humidity, light, and electrical potential, perovskite top cells degrade more quickly than commercial silicon cells, which can retain over 80% of their initial efficiency for 25 years. This instability can be attributed to flaws in perovskite films. Device stability can be increased by decreasing interface flaws and raising the quality of the perovskite film. When utilized as transparent electrodes, transparent conducting oxides (TCOs) can also increase stability by shielding perovskite layers from moisture and the volatilization of organic components (Dong et al. 2016). Certain encapsulation techniques, such as the use of glass sheets and butyl rubber, have demonstrated some efficacy in enhancing stability by protecting the cells from outside environmental influences. To pass more demanding tests, such as heat cycling, outdoor testing, and potential-induced degradation (PID), further advancements are necessary. More harsh temperatures can still shorten the lifespan of encapsulated devices, even though some have done well in accelerated tests (maintaining 95% of initial function under 85 °C and 85% relative humidity for over 1,000 h). Maintaining long-term durability is still the main obstacle (Xu et al. 2020; Liu et al. 2022).

Scaling up Device Area

Although the majority of high-efficiency PSTSCS are constructed on tiny surfaces (typically 1 cm^2 or less), scaling these cells up continues to be a difficult task. Although transparent electrodes fabricated from TCOs work well on tiny cells, their reduced lateral conductivity causes problems in bigger devices. Performance declines as this problem becomes more noticeable in the active region (Al-Ashouri et al. 2020; Chen et al. 2019; Hou et al. 2020; Zheng et al. 2018a). B. Niesen and et al. for example, solved this problem by adding silver grid fingers to a tandem cell of 12.96 cm^2 (Sahli et al. 2018). Later, A. Ho-Baillie and et al. enhanced this by creating a new metal grid for a tandem cell measuring 16 cm^3, which produced an astounding FF of 78% (Zheng et al. 2018b). Still, another problem with using silver grids is that the perovskite can be severely degraded by silver atoms diffusing through the ITO layer, which makes the device much less stable.

Perovskite-Silicon Integration and Stability

The fabrication of effective TSCs depends on proper integration of silicon (c-Si) and perovskite cells. The perovskite layer should ideally be put straight onto the c-Si cells' rough surface in order to maximize light trapping (Shi et al. 2015). Nevertheless, the deposition process is complicated by the significant size disparity between the much thinner perovskite layer and the massive surface features of the c-Si pyramids. Polishing the c-Si surface is a popular method of resolving this issue, although this can result in optical losses that need to be made up for with extra precautions such anti-reflective coatings (Swatowska et al. 2011; Raut et al. 2011). Spin-coating has been used to create the majority of PSTSCs that have been reported to far, however this method is not scalable for industrial manufacturing. For conventional 6-inch c-Si wafers, new fabrication techniques that can manage the necessary throughput must be developed in order to fulfill the demands of large-scale manufacturing (Fu et al. 2022). To maximize total power output, the cells must be optimized for various voltages and currents. In addition, the material stack needs to be able to endure thermal expansion without compromising its mechanical integrity in order to guarantee the tandem device's long-term performance and dependability (Akhil et al. 2021).

Manufacturing Scalability

PSTSCs have broken records for efficiency in small-scale laboratory settings, but increasing production volume is still a challenge. Large-scale perovskite film deposition is still a challenge. Additionally, front metal grids need to be adjusted to reduce electrical leakage (Khan et al. 2023; Saki et al. 2021). Larger-scale film production is possible with solution techniques like slot-die coating and blade coating. Blade-coated perovskite films on microtextured silicon substrates, for instance, have demonstrated encouraging efficiency outcomes. To guarantee dependable large-area production, these procedures must be further improved (Duarte and Andrade 2024; Vijayan et al. 2020; Cheng et al. 2021b).

Cost-Effectiveness

Reaching a competitive levelized cost of energy (LCOE) is essential for PSTSCs to be commercially successful. Nowadays, *n*-type silicon bottom cells in PSTSCs function more efficiently than their *p*-type counterparts; nonetheless, *p*-type silicon still dominates because of its lower cost of production. It is imperative to strike a balance between cost savings and efficiency gains (Aydin et al. 2024; Nogay et al. 2019). Perovskite-based cells' comparatively short lifespan (15 years as opposed to 25 years) in comparison with conventional silicon modules also affects how competitively priced they are. Thus, maintaining a low LCOE and making these technologies feasible requires extending the life of PSTSCs and lowering their rate of degradation (Li et al. 2018; Duan et al. 2023).

Future Prospects

PSTSCs have demonstrated an impressive 33.7% efficiency (Liu et al. 2024a). These cells' production process is compatible with currently used commercial silicon fabrication techniques, opening the door to commercialization without requiring a sizable increase in capital expenditure for manufacturing facilities. TSCs that combine perovskite and silicon offer a viable way to push PV technology's efficiency boundaries. An intelligent approach to the commercialization of tandem configurations is to combine the perovskite top cell with an established silicon (Si) bottom cell from the industry. Nonetheless, intrinsic degradation mechanisms persist in impeding the attainment of the targeted benchmarks for PSTSCs. Optimizing the perovskite material composition, reducing strain, and enhancing the top cell's light-trapping mechanisms should improve the technology's long-term stability and appeal to the PV market. Because of its resilience to strain and defect tolerance, single-crystal quantum dot perovskites have showed promise in lowering phase segregation (Bastiani et al. 2021). Additionally, materials with a perfect bandgap and low volatility, such as $CsPbI_3$ perovskites, improve the stability of the top cell in tandem setups (Yoon et al. 2021b).

Moreover, investigating the effects of albedo and performing experiments in multiple climates could potentially improve the stability and energy yield of tandem devices. Encapsulation issues are another source of obstacles since perovskite compositions selected for their efficacy often lack the stability required for long-term use. Promising options for extending the lifespan of devices include thermoplastic polyurethane and polyolefin elastomers, two new encapsulation materials (Lu et al. 2021; Li et al. 2021; Kaltenbrunner et al. 2015; Salhi et al. 2018). Although less common, research into alternate tandem designs like as four-terminal and three-terminal cells may yield fresh insights into stability. For example, 4 T designs provide a possible route to long-term use by enabling the replacement of the degraded perovskite top cell. Economic study has shown that a PCE of between 24 and 27% with a lifetime of 15–21 years is required to establish parity with single-junction silicon cells for PSTSCs to be competitive. The thermal mismatch between the Si and perovskite layers may be addressed by the all perovskite TSC arrangement. Efficiency gains of up to 28% for small-area devices and 21.7% for mini-modules have previously been shown using this tandem architecture (Wang et al. 2022c; Lin et al. 2022). Long-term stability in this structure is still quite difficult to achieve, though, particularly when using high Br content to widen the bandgap (Wen et al. 2022). Despite these difficulties, PSTSCs have made some great leaps in the past decade. PSTSCs have achieved PCEs of over 33% efficiency. Over the next five to ten years, more research will probably overcome stability issues and enable PSTSCs become the leading solar technology. The PSTSCs promise great advancement in the PV industry in the future. In particular developments of the future will provide processes of scalable manufacturing aiming to reduce costs, processes for optical designs which will optimize efficiency, stability augmentation in view of new materials introduction. Tandem cell architecture developments together with interdisciplinary collaborations will most probably yield significant advancements. TSCs will be incorporated to be more environmentally friendly and resource efficient by integrating the principle of a circular economy into their, manufacturing philosophy emphasizing clean technologies, sustainable materials, and recycling methods. These developments place PSTSCs in a good prospect for rapid commercial development and provide an effective method of solar energy which is produced economically and is sustainable.

Conclusion and Outlook

The development of PSTSCs is one of the most promising areas in solar science (Raj et al. 2021). Perovskite and silicon single-junction cell efficiencies are getting close to their Shockley–Queisser limits. PSTSCs have advanced quickly, surpassing commercial silicon solar cells with a record efficiency of 33.9% after starting out at 13.7% Laboratory NR (n.d.). The main reason for this progress has been the attempts to reduce optical and electrical losses while enhancing the device stability. These cells are now more stable and on the verge of commercial viability because to material advancements, especially in the perovskite absorber layer. Calculations by academics like M. Futscher and et al. indicate that the realistic efficiency limit of

PSTSCs is anticipated to approach 41%, which is much higher than present values (Futscher and Ehrler 2016). PSTSCs still need to clear a few obstacles before they can be extensively sold. Among these are the integration with silicon bottom cells and the stability of perovskite top cells, both of which need to be strengthened for large-scale production. Long-term operating stability must be guaranteed, particularly because commercial silicon cells have an average lifespan of more than 25 years. In the future, stability and efficiency need to take precedence. Although transparent contact layers are necessary for both photon harvesting and device stability, minimization of optical losses is an ongoing top priority (Jiang et al. 2023). Thereafter, a decrease in the open-circuit voltage losses would require higher efficiency without compromising on multi-decadal stability. Scaling up manufacturing processes and removing lead poisoning are but two of the challenges that will require a significant collaboration between industrial development and basic science. The potential uses for PSTSCs are growing beyond solar energy as advancements continue. In the near future, the commercial adoption of high-efficiency PSTSCs could be further accelerated by new applications in water splitting and energy storage.

References

Adil Afroz M, Ghimire N, Reza KM et al (2020) Thermal stability and performance enhancement of perovskite solar cells through oxalic acid-induced perovskite formation. ACS Appl Energy Mater 3:2432–2439. https://doi.org/10.1021/acsaem.9b02111

Akhil S, Akash S, Pasha A et al (2021) Review on perovskite silicon tandem solar cells: status and prospects 2T, 3T and 4T for real world conditions. Mater Des 211:110138. https://doi.org/10.1016/j.matdes.2021.110138

Al-Ashouri A, Köhnen E, Li B et al (2020) Monolithic perovskite/silicon tandem solar cell with> 29% efficiency by enhanced hole extraction. Science 370:1300–1309

Albrecht S, Saliba M, Correa Baena JP et al (2016) Monolithic perovskite/silicon-heterojunction tandem solar cells processed at low temperature. Energy Environ Sci 9:81–88. https://doi.org/10.1039/C5EE02965A

Amat A, Mosconi E, Ronca E et al (2014) Cation-induced band-gap tuning in organohalide perovskites: interplay of spin-orbit coupling and octahedra tilting. Nano Lett 14:3608–3616. https://doi.org/10.1021/nl5012992

Amri K, Belghouthi R, Aillerie M, Gharbi R (2021) Device optimization of a lead-free perovskite/silicon tandem solar cell with 24.4% power conversion efficiency. Energies (Basel). https://doi.org/10.3390/en14123383

Anaya M, Lozano G, Calvo ME, Míguez H (2017) ABX3 perovskites for tandem solar cells. Joule 1:769–793. https://doi.org/10.1016/j.joule.2017.09.017

Aydin E, Ugur E, Yildirim BK et al (2023) Enhanced optoelectronic coupling for perovskite/silicon tandem solar cells. Nature 623:732–738. https://doi.org/10.1038/s41586-023-06667-4

Aydin E, Allen TG, De Bastiani M et al (2024) Pathways toward commercial perovskite/silicon tandem photovoltaics. Science 383:3849. https://doi.org/10.1126/science.adh3849

Balasundaraprabhu R, Monakhov EV, Muthukumarasamy N et al (2009) Effect of heat treatment on ITO film properties and ITO/p-Si interface. Mater Chem Phys 114:425–429. https://doi.org/10.1016/j.matchemphys.2008.09.053

Bett AJ, Winkler KM, Bivour M et al (2019) Semi-transparent perovskite solar cells with ITO directly sputtered on spiro-OMeTAD for tandem applications. ACS Appl Mater Interfaces 11:45796–45804. https://doi.org/10.1021/acsami.9b17241
Boccard M, Ballif C (2020) Influence of the subcell properties on the fill factor of two-terminal perovskite-silicon tandem solar cells. ACS Energy Lett 5:1077–1082. https://doi.org/10.1021/acsenergylett.0c00156
Bouich A, Pradas IG, Khan MA, Khattak YH (2023) Opportunities, challenges, and future prospects of the solar cell market. Sustainability. https://doi.org/10.3390/su152115445
Bush KA, Bailie CD, Chen Y et al (2016) Thermal and environmental stability of semi-transparent perovskite solar cells for tandems enabled by a solution-processed nanoparticle buffer layer and sputtered ITO electrode. Adv Mater 28:3937–3943. https://doi.org/10.1002/adma.201505279
Bush KA, Palmstrom AF, Yu ZJ et al (2017) 23.6%-efficient monolithic perovskite/silicon tandem solar cells with improved stability. Nat Energy 2:17009. https://doi.org/10.1038/nenergy.2017.9
Chang NL, Poduval GK, Sang B et al (2023) Techno-economic analysis of the use of atomic layer deposited transition metal oxides in silicon heterojunction solar cells. Prog Photovoltaics Res Appl 31:414–428
Cheacharoen R, Rolston N, Harwood D et al (2018) Design and understanding of encapsulated perovskite solar cells to withstand temperature cycling. Energy Environ Sci 11:144–150. https://doi.org/10.1039/C7EE02564E
Chen Y, Zhang L, Zhang Y et al (2018) Large-area perovskite solar cells—A review of recent progress and issues. RSC Adv 8:10489–10508. https://doi.org/10.1039/C8RA00384J
Chen B, Yu Z, Liu K et al (2019) Grain engineering for perovskite/silicon monolithic tandem solar cells with efficiency of 25.4%. Joule 3:177–190. https://doi.org/10.1016/j.joule.2018.10.003
Chen B, Yu ZJ, Manzoor S et al (2020) Blade-coated perovskites on textured silicon for 26%-efficient monolithic perovskite/silicon tandem solar cells. Joule 4:850–864. https://doi.org/10.1016/j.joule.2020.01.008
Chen B, Ren N, Li Y et al (2022) Insights into the development of monolithic perovskite/silicon tandem solar cells. Adv Energy Mater 12:2003628. https://doi.org/10.1002/aenm.202003628
Cheng Y, Xie C, Liu X et al (2020) High-power bifacial perovskite solar cells with shelf life of over 2000 h. Sci Bull Beijing 65:607–610. https://doi.org/10.1016/j.scib.2020.02.017
Cheng Y, Liu X, Guan Z et al (2021) Revealing the degradation and self-healing mechanisms in perovskite solar cells by sub-bandgap external quantum efficiency spectroscopy. Adv Mater 33:2006170. https://doi.org/10.1002/adma.202006170
Cheng J, Liu F, Tang Z, Li Y (2021b) Scalable blade coating: a technique accelerating the commercialization of perovskite-based photovoltaics. Energ Technol. https://doi.org/10.1002/ente.202100204
Chiu PT, Law DC, Woo RL et al (2014) 35.8% space and 38.8% terrestrial 5J direct bonded cells. In: 2014 IEEE 40th Photovoltaic specialist conference (PVSC) pp 11–13
Dada M, Popoola P (2023) Recent advances in solar photovoltaic materials and systems for energy storage applications: a review. Beni Suef Univ J Basic Appl Sci 12:66. https://doi.org/10.1186/s43088-023-00405-5
De Bastiani M, Mirabelli AJ, Hou Y et al (2021) Efficient bifacial monolithic perovskite/silicon tandem solar cells via bandgap engineering. Nat Energy 6:167–175. https://doi.org/10.1038/s41560-020-00756-8
Dewi HA, Wang H, Li J et al (2019) Highly efficient semitransparent perovskite solar cells for four terminal perovskite-silicon tandems. ACS Appl Mater Interfaces 11:34178–34187. https://doi.org/10.1021/acsami.9b13145
Dewi HA, Wang H, Li J et al (2020) Four-terminal perovskite on silicon tandem solar cells optimal measurement schemes. Energ Technol 8:1901267
Dong Q, Liu F, Wong MK et al (2016) Encapsulation of perovskite solar cells for high humidity conditions. ChemSusChem 9:2597–2603. https://doi.org/10.1002/cssc.201600868

Duan L, Walter D, Chang N et al (2023) Stability challenges for the commercialization of perovskite–silicon tandem solar cells. Nat Rev Mater 8:261–281. https://doi.org/10.1038/s41578-022-00521-1

Duarte VCM, Andrade L (2024) Recent advancements on slot-die coating of perovskite solar cells: the lab-to-fab optimisation process. Energies Basel 17(16):3896

Essig S, Allebé C, Remo T et al (2017) Raising the one-sun conversion efficiency of III–V/Si solar cells to 32.8% for two junctions and 35.9% for three junctions. Nat Energy 2:1–9

Fan R, Zhou N, Zhang L et al (2017) Toward full solution processed perovskite/Si monolithic tandem solar device with PCE exceeding 20%. Solar RRL 1:1700149. https://doi.org/10.1002/solr.201700149

France RM, Geisz JF, García I et al (2015) Design flexibility of ultra-high efficiency 4-junction inverted metamorphic solar cells. In: 2015 IEEE 42nd Photovoltaic specialist conference (PVSC), pp 1–3

Fu F, Li J, Yang TC-J et al (2022) Monolithic perovskite-silicon tandem solar cells: from the lab to fab? Adv Mater 34:2106540. https://doi.org/10.1002/adma.202106540

Futscher MH, Ehrler B (2016) Efficiency limit of perovskite/Si tandem solar cells. ACS Energy Lett 1:863–868. https://doi.org/10.1021/acsenergylett.6b00405

Gao B, Hu J, Tang S et al (2021) Organic-inorganic perovskite films and efficient planar heterojunction solar cells by magnetron sputtering. Adv Sci 8:2102081. https://doi.org/10.1002/advs.202102081

Geisz JF, McMahon WE, Buencuerpo J et al (2021) Characterization of multiterminal tandem photovoltaic devices and their subcell coupling. Cell Rep Phys Sci 2:100677. https://doi.org/10.1016/j.xcrp.2021.100677

Gota F, Langenhorst M, Schmager R et al (2020) Energy yield advantages of three-terminal perovskite-silicon tandem photovoltaics. Joule 4:2387–2403

Ho-Baillie AWY, Zheng J, Mahmud MA et al (2021) Recent progress and future prospects of perovskite tandem solar cells. Appl Phys Rev 8:41307. https://doi.org/10.1063/5.0061483

Hou Y, Aydin E, De Bastiani M et al (2020) Efficient tandem solar cells with solution-processed perovskite on textured crystalline silicon. Science 367:1135–1140. https://doi.org/10.1126/science.aaz3691

Huang T, Wang R, Nuryyeva S et al (2022) Wide-gap perovskite via synergetic surface passivation and its application toward efficient stacked tandem photovoltaics. Small 18:2103887

Hwang J-K, Lee S-W, Lee W et al (2020) Conformal perovskite films on 100 cm^2 textured silicon surface using two-step vacuum process. Thin Solid Films 693:137694. https://doi.org/10.1016/j.tsf.2019.137694

Jacobs DA, Langenhorst M, Sahli F et al (2019) Light management: a key concept in high-efficiency perovskite/silicon tandem photovoltaics. J Phys Chem Lett 10:3159–3170. https://doi.org/10.1021/acs.jpclett.8b03721

Jesper Jacobsson T, Correa-Baena J-P, Pazoki M et al (2016) Exploration of the compositional space for mixed lead halogen perovskites for high efficiency solar cells. Energy Environ Sci 9:1706–1724. https://doi.org/10.1039/C6EE00030D

Jiang Q, Tirawat R, Kerner RA et al (2023) Towards linking lab and field lifetimes of perovskite solar cells. Nature 623:313–318. https://doi.org/10.1038/s41586-023-06610-7

Jung H, Kim G, Jang GS et al (2021) Transparent electrodes with enhanced infrared transmittance for semitransparent and four-terminal tandem perovskite solar cells. ACS Appl Mater Interfaces 13:30497–30503

Kaltenbrunner M, Adam G, Głowacki ED et al (2015) Flexible high power-per-weight perovskite solar cells with chromium oxide–metal contacts for improved stability in air. Nat Mater 14:1032–1039. https://doi.org/10.1038/nmat4388

Khan D, Qu G, Imran M et al (2023) Overcoming two key challenges in monolithic perovskite-silicon tandem solar cell development: wide bandgap and textured substrate—a comprehensive review. Adv Energy Mater. https://doi.org/10.1002/aenm.202302124

Kim DH, Muzzillo CP, Tong J et al (2019a) Bimolecular additives improve wide-band-gap perovskites for efficient tandem solar cells with CIGS. Joule 3:1734–1745

Kim CU, Yu JC, Jung ED et al (2019) Optimization of device design for low cost and high efficiency planar monolithic perovskite/silicon tandem solar cells. Nano Energy 60:213–221. https://doi.org/10.1016/j.nanoen.2019.03.056

Kim D, Jung HJ, Park IJ et al (2020) Efficient, stable silicon tandem cells enabled by anion-engineered wide-bandgap perovskites. Science 368:155–160. https://doi.org/10.1126/science.aba3433

Köhnen E, Wagner P, Lang F et al (2021) 27.9% Efficient monolithic perovskite/silicon tandem solar cells on industry compatible bottom cells. Solar RRL 5:2100244. https://doi.org/10.1002/solr.202100244

Laboratory NR (n.d.) National renewable energy laboratory best research-cell efficiency chart. https://www.nrel.gov/pv/cell-efficiency.html. Accessed on Sep 11 2024

Lee Y, Park C, Balaji N et al (2015) High-efficiency silicon solar cells: a review. Isr J Chem 55:1050–1063. https://doi.org/10.1002/ijch.201400210

Leijtens T, Bush KA, Prasanna R, McGehee MD (2018) Opportunities and challenges for tandem solar cells using metal halide perovskite semiconductors. Nat Energy 3:828–838. https://doi.org/10.1038/s41560-018-0190-4

Leyden MR, Jiang Y, Qi Y (2016) Chemical vapor deposition grown formamidinium perovskite solar modules with high steady state power and thermal stability. J Mater Chem A 4:13125–13132. https://doi.org/10.1039/C6TA04267H

Li Z, Yang M, Park J-S et al (2016) Stabilizing perovskite structures by tuning tolerance factor: formation of formamidinium and cesium lead iodide solid-state alloys. Chem Mater 28:284–292. https://doi.org/10.1021/acs.chemmater.5b04107

Li Z, Zhao Y, Wang X et al (2018) Cost analysis of perovskite tandem photovoltaics. Joule 2:1559–1572. https://doi.org/10.1016/j.joule.2018.05.001

Li J, Xia R, Qi W et al (2021) Encapsulation of perovskite solar cells for enhanced stability: structures, materials and characterization. J Power Sources 485:229313. https://doi.org/10.1016/j.jpowsour.2020.229313

Lin R, Xu J, Wei M et al (2022) All-perovskite tandem solar cells with improved grain surface passivation. Nature 603:73–78. https://doi.org/10.1038/s41586-021-04372-8

Lin H, Yang M, Ru X et al (2023) 26.7% efficiency silicon heterojunction solar cells achieved by electrically optimized nanocrystalline-silicon hole contact layers. Nat Energy 8:789–99

Liu Y, Li Y, Wu Y et al (2020) High-efficiency silicon heterojunction solar cells: materials, devices and applications. Mater Sci Eng R Rep 142:100579. https://doi.org/10.1016/j.mser.2020.100579

Liu J, De Bastiani M, Aydin E et al (2022) Efficient and stable perovskite-silicon tandem solar cells through contact displacement by MgFx. Science 377:302–306. https://doi.org/10.1126/science.abn8910

Liu J, He Y, Ding L et al (2024a) Perovskite-silicon tandem solar cells with bilayer interface passivation. Nature. https://doi.org/10.1038/s41586-024-07997-7

Liu J, Shi B, Xu Q et al (2024b) Textured perovskite/silicon tandem solar cells achieving over 30% efficiency promoted by 4-fluorobenzylamine hydroiodide. Nanomicro Lett 16:189. https://doi.org/10.1007/s40820-024-01406-4

Löper P, Moon S-J, de Nicolas S et al (2015) Organic–inorganic halide perovskite/crystalline silicon four-terminal tandem solar cells. Phys Chem Chem Phys 17:1619–1629. https://doi.org/10.1039/C4CP03788J

Lu Q, Yang Z, Meng X et al (2021) A review on encapsulation technology from organic light emitting diodes to organic and perovskite solar cells. Adv Funct Mater 31:2100151. https://doi.org/10.1002/adfm.202100151

Luo X, Luo H, Li H et al (2023) Efficient perovskite/silicon tandem solar cells on industrially compatible textured silicon. Adv Mater 35:2207883. https://doi.org/10.1002/adma.202207883

Mailoa JP, Bailie CD, Johlin EC et al (2015) A 2-terminal perovskite/silicon multijunction solar cell enabled by a silicon tunnel junction. Appl Phys Lett 106:121105. https://doi.org/10.1063/1.4914179

Mariotti S, Köhnen E, Scheler F et al (2023) Interface engineering for high-performance, triple-halide perovskite–silicon tandem solar cells. Science 381:63–69. https://doi.org/10.1126/science.adf5872

Mazzarella L, Lin Y-H, Kirner S et al (2019) Infrared light management using a nanocrystalline silicon oxide interlayer in monolithic perovskite/silicon heterojunction tandem solar cells with efficiency above 25%. Adv Energy Mater 9:1803241. https://doi.org/10.1002/aenm.201803241

McMeekin DP, Sadoughi G, Rehman W et al (2016) A mixed-cation lead mixed-halide perovskite absorber for tandem solar cells. Science 351:151–155. https://doi.org/10.1126/science.aad5845

Moghadamzadeh S, Hossain IM, Loy M et al (2021) In2O3: H-based hole-transport-layer-free tin/lead perovskite solar cells for efficient four-terminal all-perovskite tandem solar cells. ACS Appl Mater Interfaces 13:46488–46498

Mujahid M, Chen C, Zhang J et al (2021) Recent advances in semitransparent perovskite solar cells. InfoMat 3:101–124. https://doi.org/10.1002/inf2.12154

Nogay G, Sahli F, Werner J et al (2019) 25.1%-efficient monolithic perovskite/silicon tandem solar cell based on a *p*-type monocrystalline textured silicon wafer and high-temperature passivating contacts. ACS Energy Lett 4:844–845. https://doi.org/10.1021/acsenergylett.9b00377

Oni AM, Mohsin ASM, Rahman MM, Hossain Bhuian MB (2024) A comprehensive evaluation of solar cell technologies, associated loss mechanisms, and efficiency enhancement strategies for photovoltaic cells. Energy Rep 11:3345–3366. https://doi.org/10.1016/j.egyr.2024.03.007

Padi SP, Khokhar MQ, Chowdhury S et al (2021) Nanoscale SiOx tunnel oxide deposition techniques and their influence on cell parameters of TOPCon solar cells. Trans Electr Electron Mater 22:557–566. https://doi.org/10.1007/s42341-021-00356-7

Panigrahi J, Komarala VK (2021) Progress on the intrinsic a-Si: H films for interface passivation of silicon heterojunction solar cells: a review. J Non Cryst Solids 574:121166. https://doi.org/10.1016/j.jnoncrysol.2021.121166

Prasanna R, Gold-Parker A, Leijtens T et al (2017) Band gap tuning via lattice contraction and octahedral tilting in perovskite materials for photovoltaics. J Am Chem Soc 139:11117–11124. https://doi.org/10.1021/jacs.7b04981

Qiang Z, Wang C, Gao X et al (2022) Challenges of scalable development for perovskite/silicon tandem solar cells. ACS Appl Energy Mater 5:6499–6515. https://doi.org/10.1021/acsaem.2c00354

Rahmany S, Etgar L (2020) Semitransparent perovskite solar cells. ACS Energy Lett 5:1519–1531. https://doi.org/10.1021/acsenergylett.0c00417

Raj A, Kumar M, Anshul A (2021) Recent advancement in inorganic-organic electron transport layers in perovskite solar cell: current status and future outlook. Mater Today Chem 22:100595. https://doi.org/10.1016/j.mtchem.2021.100595

Raj A, Kumar M, Anshul A (2022) Recent progress in cesium-based lead-free halide double perovskite materials for photovoltaic applications. Phys Status Solidi (a) 219:2200425. https://doi.org/10.1002/pssa.202200425

Raj A, Kumar M, Anshul A (2023) Topical advances in fabrication technologies of perovskite solar cell heterostructures: performance and future perspective. Mater Lett 340:134171. https://doi.org/10.1016/j.matlet.2023.134171

Raut HK, Ganesh VA, Nair AS, Ramakrishna S (2011) Anti-reflective coatings: a critical{,} in-depth review. Energy Environ Sci 4:3779–3804. https://doi.org/10.1039/C1EE01297E

Rienäcker M, Warren EL, Schnabel M et al (2019) Back-contacted bottom cells with three terminals: maximizing power extraction from current-mismatched tandem cells. Prog Photovoltaics Res Appl 27:410–423

Sahli F, Kamino BA, Werner J et al (2018) Improved optics in monolithic perovskite/silicon tandem solar cells with a nanocrystalline silicon recombination junction. Adv Energy Mater 8:1701609. https://doi.org/10.1002/aenm.201701609

Saki Z, Byranvand MM, Taghavinia N et al (2021) Solution-processed perovskite thin-films: the journey from lab- to large-scale solar cells. Energy Environ Sci 14:5690–5722. https://doi.org/10.1039/D1EE02018H

Salhi B, Wudil YS, Hossain MK et al (2018) Review of recent developments and persistent challenges in stability of perovskite solar cells. Renew Sustain Energy Rev 90:210–222. https://doi.org/10.1016/j.rser.2018.03.058

Saliba M, Matsui T, Seo J-Y et al (2016) Cesium-containing triple cation perovskite solar cells: improved stability{,} reproducibility and high efficiency. Energy Environ Sci 9:1989–1997. https://doi.org/10.1039/C5EE03874J

Schnabel M, Schulte-Huxel H, Rienäcker M et al (2020) Three-terminal III–V/Si tandem solar cells enabled by a transparent conductive adhesive. Sustain Energy Fuels 4:549–558

Shen H, Omelchenko ST, Jacobs DA et al (2024) In situ recombination junction between p-Si and TiO2 enables high-efficiency monolithic perovskite/Si tandem cells. Sci Adv 4:eaau9711. https://doi.org/10.1126/sciadv.aau9711

Shi Y, Zhang F (2023) Advances in encapsulations for perovskite solar cells: from materials to applications. Solar RRL 7:2201123. https://doi.org/10.1002/solr.202201123

Shi D, Zeng Y, Shen W (2015) Perovskite/c-Si tandem solar cell with inverted nanopyramids: realizing high efficiency by controllable light trapping. Sci Rep 5:16504. https://doi.org/10.1038/srep16504

Shi B, Duan L, Zhao Y et al (2020) Semitransparent perovskite solar cells: from materials and devices to applications. Adv Mater 32:1806474. https://doi.org/10.1002/adma.201806474

Sista S, Hong Z, Park M-H et al (2010) High-efficiency polymer tandem solar cells with three-terminal structure. Adv Mater 22:E77

Sosa-Acosta JR, Fernández-Izquierdo L, del Río R et al (2024) CsPbBr 3-based photoanode prepared by single-step chemical vapor deposition of tunable thickness perovskite films. Appl Surf Sci 678:161049. https://doi.org/10.1016/j.apsusc.2024.161049

Steiner MA, Wanlass MW, Carapella JJ et al (2009) A monolithic three-terminal GaInAsP/GaInAs tandem solar cell. Prog Photovoltaics Res Appl 17:587–593

Subbiah AS, Isikgor FH, Howells CT et al (2020) High-performance perovskite single-junction and textured perovskite/silicon tandem solar cells via slot-die-coating. ACS Energy Lett 5:3034–3040. https://doi.org/10.1021/acsenergylett.0c01297

Swatowska B, Stapinski T, Drabczyk K, Panek P (2011) The role of antireflection coatings in silicon solar cells—the influence on their electrical parameters. Opt Appl XLI 41(2):487–92

Tang H, He S, Peng C (2017) A short progress report on high-efficiency perovskite solar cells. Nanoscale Res Lett 12:410. https://doi.org/10.1186/s11671-017-2187-5

Tao J, Chai J, Lu X et al (2015) Growth of wafer-scale MoS2 monolayer by magnetron sputtering. Nanoscale 7:2497–2503. https://doi.org/10.1039/C4NR06411A

Tockhorn P, Wagner P, Kegelmann L et al (2020) Three-terminal perovskite/silicon tandem solar cells with top and interdigitated rear contacts. ACS Appl Energy Mater 3:1381–1392

Tockhorn P, Sutter J, Cruz A et al (2022) Nano-optical designs for high-efficiency monolithic perovskite–silicon tandem solar cells. Nat Nanotechnol 17:1214–1221. https://doi.org/10.1038/s41565-022-01228-8

Tong S, Wu H, Zhang C et al (2017) Large-area and high-performance CH3NH3PbI3 perovskite photodetectors fabricated via doctor blading in ambient condition. Org Electron 49:347–354. https://doi.org/10.1016/j.orgel.2017.07.011

Victoria M, Haegel N, Peters IM et al (2021) Solar photovoltaics is ready to power a sustainable future. Joule 5:1041–1056. https://doi.org/10.1016/j.joule.2021.03.005

Vijayan A, Johansson MB, Svanström S et al (2020) Simple method for efficient slot-die coating of MAPbI3 perovskite thin films in ambient air conditions. ACS Appl Energy Mater 3:4331–4337. https://doi.org/10.1021/acsaem.0c00039

Wang J, Zhong F, Liu H et al (2021) Influence of the textured pyramid size on the performance of silicon heterojunction solar cell. Solar Energy 221:114–119. https://doi.org/10.1016/j.solener.2021.04.021

Wang D, Guo H, Wu X et al (2022a) Interfacial engineering of wide-bandgap perovskites for efficient perovskite/CZTSSe tandem solar cells. Adv Funct Mater 32:2107359
Wang S, Wang P, Chen B et al (2022) Suppressed recombination for monolithic inorganic perovskite/silicon tandem solar cells with an approximate efficiency of 23%. eScience 2:339–346. https://doi.org/10.1016/j.esci.2022.04.001
Wang F, Han Y, Duan D et al (2022) Recent progress of scalable perovskite solar cells and modules. Energy Rev 1:100010. https://doi.org/10.1016/j.enrev.2022.100010
Wang S, Wang P, Shi B et al (2023) Inorganic perovskite surface reconfiguration for stable inverted solar cells with 20.38% efficiency and its application in tandem devices. Adv Mater 35:2300581. https://doi.org/10.1002/adma.202300581
Wang Y, Wang Y, Gao F, Yang D (2024) Efficient monolithic perovskite/silicon tandem photovoltaics. Energy Environ Mater 7:e12639. https://doi.org/10.1002/eem2.12639
Wawer P, Müller J, Fischer M et al (2011) Latest trends in development and manufacturing of industrial, crystalline silicon solar-cells. Energy Proced 8:2–8. https://doi.org/10.1016/j.egypro.2011.06.093
Wen J, Zhao Y, Liu Z et al (2022) Steric engineering enables efficient and photostable wide-bandgap perovskites for all-perovskite tandem solar cells. Adv Mater 34:2110356. https://doi.org/10.1002/adma.202110356
Werner J, Weng C-H, Walter A et al (2016) Efficient monolithic perovskite/silicon tandem solar cell with cell area >1 cm^2. J Phys Chem Lett 7:161–166. https://doi.org/10.1021/acs.jpclett.5b02686
Whitehead RC, Vansant KT, Warren EL et al (2021) Optimization of four terminal rear heterojunction GaAs on Si interdigitated back contact tandem solar cells. Appl Phys Lett 3(18):118
Wu Y, Yan D, Peng J et al (2017) Monolithic perovskite/silicon-homojunction tandem solar cell with over 22% efficiency. Energy Environ Sci 10:2472–2479. https://doi.org/10.1039/C7EE02288C
Wu Y, Zheng P, Peng J et al (2022) 27.6% Perovskite/c-Si tandem solar cells using industrial fabricated TOPCon device. Adv Energy Mater 12:2200821. https://doi.org/10.1002/aenm.202200821
Xiao S, Xu S (2014) High-efficiency silicon solar cells—materials and devices physics. Crit Rev Solid State Mater Sci 39:277–317. https://doi.org/10.1080/10408436.2013.834245
Xiao Y, Zuo C, Zhong J-X et al (2021) Large-area blade-coated solar cells: advances and perspectives. Adv Energy Mater 11:2100378. https://doi.org/10.1002/aenm.202100378
Xu J, Boyd CC, Yu ZJ et al (2020) Triple-halide wide-band gap perovskites with suppressed phase segregation for efficient tandems. Science 367:1097–1104. https://doi.org/10.1126/science.aaz5074
Yamaguchi M, Lee K-H, Araki K, Kojima N (2018) A review of recent progress in heterogeneous silicon tandem solar cells. J Phys D Appl Phys 51:133002. https://doi.org/10.1088/1361-6463/aaaf08
Yang J, Lim EL, Tan L, Wei Z (2022) Ink engineering in blade-coating large-area perovskite solar cells. Adv Energy Mater 12:2200975. https://doi.org/10.1002/aenm.202200975
Ying Z, Yang Z, Zheng J et al (2022) Monolithic perovskite/black-silicon tandems based on tunnel oxide passivated contacts. Joule 6:2644–2661. https://doi.org/10.1016/j.joule.2022.09.006
Yoon J, Kim U, Yoo Y et al (2021) Foldable perovskite solar cells using carbon nanotube-embedded ultrathin polyimide conductor. Adv Sci 8:2004092. https://doi.org/10.1002/advs.202004092
Yoon SM, Min H, Kim JB et al (2021) Surface engineering of ambient-air-processed cesium lead triiodide layers for efficient solar cells. Joule 5:183–196. https://doi.org/10.1016/j.joule.2020.11.020
Yoshikawa K, Kawasaki H, Yoshida W et al (2017) Silicon heterojunction solar cell with interdigitated back contacts for a photo conversion efficiency over 26%. Nat Energy 2:17032. https://doi.org/10.1038/nenergy.2017.32

Zanatta AR (2022) The Shockley-Queisser limit and the conversion efficiency of silicon-based solar cells. Res Opt 9:100320. https://doi.org/10.1016/j.rio.2022.100320

Zheng J, Lau CFJ, Mehrvarz H et al (2018a) Large area efficient interface layer free monolithic perovskite/homo-junction-silicon tandem solar cell with over 20% efficiency. Energy Environ Sci 11:2432–2443. https://doi.org/10.1039/C8EE00689J

Zheng J, Mehrvarz H, Ma F-J et al (2018b) 21.8% efficient monolithic perovskite/homo-junction-silicon tandem solar cell on 16 cm^2. ACS Energy Lett 3:2299–2300. https://doi.org/10.1021/acsenergylett.8b01382

Chapter 15
Solar-Driven Halide Perovskite Photocatalysts for Water Splitting and CO_2 Reduction

Esther Dimngaihvungi, Andrew Okram, Balaram Pani, Nasima Arshad, Ashish Kumar Singh, and Manjeet Singh

Abstract To fulfill the global demand of fuel, the generation of fuel must be from renewable sources of energy such as solar radiation. Halide-based perovskite (HLP) is one of the materials which utilizes maximum of irradiant solar radiation and can drive water splitting and CO_2 reduction processes in an efficient way. The HLPs are generally used in the form of perovskite solar cells or as such to enable these reactions. In this chapter, we provide the basic understanding for the utilization of HLP for water splitting and CO_2 reduction to generate valuable chemical energy using HLP-based solar cell devices. The different strategies such as photocatalysis using HLP particles, photoelectrode using HLP in photoelectrocatalysis and photovoltaic cell driven electrochemical process for generation of chemical energy are discussed in detail. Further modifications in the strategies of utilization of HLP itself and perovskite solar cells for chemical fuel generation are presented. In this chapter, we present the discussion on the utilization of HLP in its usual form and solar cell devices fabricated using HLP for chemical energy generation.

Keywords Chemical fuel · Photocatalyst · Halide-based perovskite solar cells · H_2O splitting · CO_2 reduction

---

E. Dimngaihvungi · A. Okram · M. Singh (✉)
Department of Chemistry, School of Physical Sciences, Mizoram University, Aizawl, Mizoram, India
e-mail: msitbhu@gmail.com

B. Pani
Department of Chemistry, Bhaskaracharya College of Applied Sciences, University of Delhi, Delhi, India

N. Arshad
Department of Chemistry, Allama Iqbal Open University, Islamabad, Pakistan

A. K. Singh (✉)
Department of Chemistry, Netaji Subhas University of Technology, New Delhi, India
e-mail: ashish.singh.rs.apc@itbhu.ac.in

Department of Chemistry, Hansraj College, University of Delhi, New Delhi, India

M. Kumar and A. Raj (eds.), *Perovskite Solar Cells*, Springer Tracts in Electrical and Electronics Engineering, https://doi.org/10.1007/978-981-96-9463-1_15

Introduction

The requirements of the fuel generated from renewable energy need to be settled as standard method for fulfilling the fuel demand by the society. However, the electricity generated form the renewable sources of energy have been continuously increasing but still not sufficient to fulfill the increasing global demand (Luna et al. 2019; Yang and Moon 2019; Shen et al. 2018; Ding et al. 2020; Hu et al. 2020). More and more simple and cost-effective way of production of chemical energy and fuel are required at industrial scale to achieve the sustainable development goal. The two major processes that are solar energy driven CO_2 reduction and H_2O splitting are crucial and might fulfill the global demand of fuel energy. Water splitting produces H_2 gas as fuel while CO_2 reduction generates many other fuels such as CH_4, CO, CH_3OH (Tofanello et al. 2020; Suzuki et al. 2020; Creel et al. 2019; Hisatomi and Domen 2019; Zhang et al. 2020). The light absorbing photocatalysts drive these reactions. Photocatalysts are the materials which absorbs solar light and use it for driving some chemical reactions and the process is known as photocatalysis. (Nahar et al. 2017; Tahir 2018; Ahmad et al. 2015; Maeda and Domen 2010; Bagheri et al. 2017; Deng et al. 2017; König 2017). The photocatalysis process occurs in different steps as (i) The materials used as photocatalyst absorb the light photon and generation of electron and hole pair takes place. Thus, generated electron and hole pairs, which are termed as charge carriers may recombine because of their very less life time on the surface and release of the absorbed energy takes place. (ii) The charge carriers which are intact from recombination reached to the surface as reactive sites. (iii) Charge carriers cause the reduction and oxidation reactions (Fig. 15.1a). Thus, it is clear that there is a race between the generation of the charges and their consumption into the redox reaction for photocatalytic process. There are various materials which can be used for the purpose of photocatalysis either for fuel generation through water splitting and CO_2 reduction processes or in photocatalytic degradation of the pollutants. Some of the excellent photocatalysts are titanium dioxide (TiO_2), zinc oxide (ZnO), cadmium sulfide (CdS), tungsten (VI) oxide (WO_3), graphitic carbon nitride (g-C_3N_4), etc. (Fajrina and Tahir 2019; Cheng et al. 2018; Zhang et al. 2019a; Liu et al. 2019a; Dong et al. 2017; Khalafi et al. 2019; Jaramillo-Páez et al. 2018; Wen et al. 2017; Fu et al. 2018). These materials have their own advantages and disadvantages. The major disadvantages are faster charge carriers' recombination and poor band alignment for the specific photocatalytic reactions. Thus, it is required to exploit new materials best suited for photocatalytic applications.

In recent years, perovskite materials have been found having extraordinary properties such as very high absorption coefficient, longer diffusion length, less trapping sites, controllable bandgap, and lower binding of electron and hole generated after light absorption especially the halide perovskites (Lin et al. 2016; Lee et al. 2012; Xing et al. 2013). Some perovskite such ashalide-based perovskite (e.g., $MAPbCl_3$, $MAPbBr_3$, etc.), chalcogenide perovskites (e.g., $BaZrS_3$, $Ba_3Zr_2S_7$, etc.), and ferroelectric oxide perovskites, are found suitable for photocatalysis applications. Although, non-halide perovskite has higher stability, but large bandgap, lower

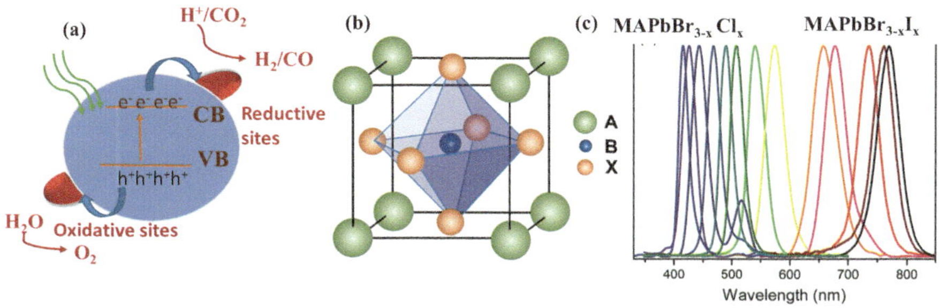

Fig. 15.1 **a** Schematic diagram of photochemical process at the surface of photocatalyst and **b** crystal structure of the halide-based perovskite (HLP). Kim et al. (2014), Jang et al. (2015) Reproduced with permission from ref. (). Copyright 2014, American Chemical Society. **c** PL spectra of the halide perovskite at different halide compositions. Reproduced with permission from ref. (). Copyright 2015, American Chemical Society

conductivity of carriers has limited the photocatalytic efficiency (Xu et al. 2018). As a consequence of large bandgap, poor band alignment in non-halide perovskite could not utilized complete solar radiation and thereby exhibited lower energy conversion efficiency (Teh et al. 2020). Research is going on to modify the structure and composition to increase the solar conversion efficiency of the non-halide perovskite.

Contrary to non-halide perovskite, halide perovskite has shown excellent power conversation efficiency when utilized as absorber layer in solar cell device. The highest efficiency achieved by a perovskite solar cells in comparison with others solar cells is due to its unique properties such as significant absorption in the visible region, low charge carrier recombination, charge carriers low-binding energy (Lin et al. 2016; Lee et al. 2012; Xing et al. 2013). Halide perovskite (HLP) is an ionic crystal having formula ABX_3, where A represents any organic group like $CH_3NH_2^+$, Cs^+, B represents the Pb^{2+}, Sn^{2+} or Cu^{2+}, and X represents halide ions. The crystal structure of the HLP is presented in Fig. 15.1b (Kim et al. 2014).Owing to the significant absorption in the visible and near-IR region, HLP shows strong PL spectra which can be regulated by composition change of halide ions as presented in Fig. 15.1c (Jang et al. 2015), which confirms its unique optical properties as result of that HLP has wide applications in the field of optoelectronics devices in addition to the solar cells (Cao et al. 2018; Lin et al. 2018; Zhang et al. 2019b; Zhou and Zhao 2019). It is very important to note that the standard potential with respect to standard hydrogen electrode required for all the photocatalytic fuel generation reactions such as water splitting, CO_2 reduction fall in between the band edge value of the different halide perovskite (Singh and Sinha 2020). This enables scientist to utilize the halide perovskite for the photochemical solar-to-chemical fuel generation as well as in the organic pollutant degradation. However, limited stability of the perovskite makes it difficult to full utilize for processes which includes moisture, heat, long duration light irradiations. The different arrangements for solar-to-fuel generation using HLP materials are described in the following sections.

HLP-Based Photocatalytic Systems for Water Splitting and CO_2 Reduction

Halide Perovskite Photovoltaic Cell (HLP-Based PVs) Incorporated with Electrocatalytic Cell

HLP-based photovoltaic (PVs) are widely employed in fuel generation from solar energy, in which cathode and anode are acting as the sites of the reaction. When the HLP-based PVs are used in combination with the electrocatalytic system, two modes of arrangement are possible—one way is where the electrodes are photoactive, while in the other way, electrocatalytic reaction was powered by HLP-based PVs as an external sources. The HLP-based PVs can generate voltage as high as 1–1.4 eV. When two photovoltaic devices are connected in tandem, they can split the water much more than the other cell made with Si and CIGS. Recently, some reports demonstrated such type of solar-to-fuel generation system.

In an HLP-based PVs integrated photoelectrocatalysis, the PVs are connected with photocathode and photoanode. Liang et al. demonstrated a structure in which two series-connected HLP-based PV device and CoP-based catalyst electrodes were immersed in water. Such device demonstrated H_2 generation efficiency of 6.7% (Fig. 15.2) (Liang et al. 2020). Baek et al. fabricated a layered photoanode with structure having $BiVO_4/WO_3/SnO_2$ together with an HLP-based PVs for H_2O splitting reaction. When the PV is covered with such triplet layer, they achieved H_2 efficiency of 3.5% (Baek et al. 2017). Fe_2O_3(Hematite) can also be employed as a photoanode in a HLP-based PV-integrated photoelectrocatalysis as they possess a large range of absorption band for the light. Gurudayal et al. integrated HLP-based solar cell with a photoanode. They have produced an unbiased H_2 with a generation efficiency of 2.4% (Gurudayal and Boix 2017). Liu et al. demonstrated a PV-integrated with a TiO_2nanorod coated with CdS as a photoanode which exhibited an H_2 generation efficiency of 1.54% (Liu et al. 2019b). Like the photoanode, photocathode can also be integrated with the HLP-based PVs for chemical fuel generation. Dias et al. prepare a tandem devices from a Cu_2O, photocathode with $FA_xMA_{1-x}PbI_3$-based PV along with anode, which delivered a H_2 generation with efficiency of 2.5% to split H_2O in an unbiased manner. The bottom absorber was the PV and top absorber was Cu_2O (Dias et al. 2015). Luo et al. coupled small bandgap CIGS as photocathode with high bandgap Br-based HLP for tandem device as photocathode. When the perovskite was coated above $CuIn_xGa_{1-x}Se_2$ photocathode, the low energy photons was absorbed and H_2 was produce at efficiency of 6% (Luo et al. 2015).

In a PV-cell integrated with electrocatalysis in which external power was received using PV, the reactions are being driven by the PVs, which provide an external power supply. Thermodynamically, water splitting requires 1.23 V as external power (Luo et al. 2014). However, the actual voltage required a voltage ranges between 1.8 and 2.0 V. Even though, the traditionally used materials for solar cell like Si, copper

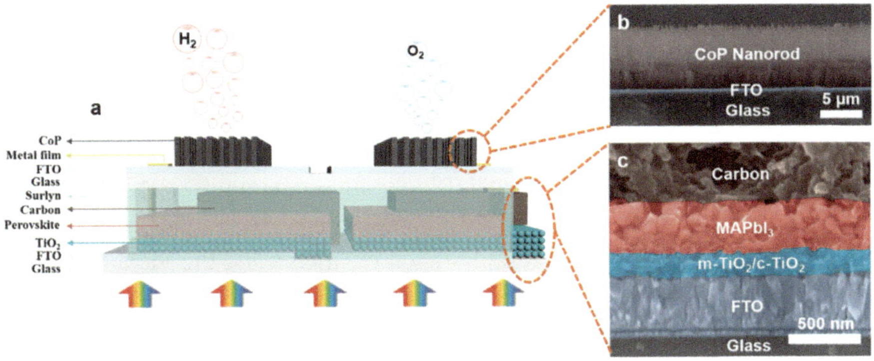

Fig. 15.2 Presentation of the device used. **a** Schematic structure of the two series-connected HLP-based device. Cross-sectional SEM images of **b** glass/FTO/CoP electrode **c** PSC with the structure of glass/FTO/TiO_2/perovskite/carbon. Liang et al. (2020) Reproduced with permission from ref. (). Copyright 2020, American Chemical Society

indium selenide (CIGS) and cadmium telluride have high efficiency their low open-circuit voltage had limited their application for water splitting reaction. The H_2 generation could be achieved by making tandem devices from a traditional solar cells, which can increase the voltage required. The tandem devices made from the traditional solar cell material exhibits a multi-junction solar cells for splitting water to generate chemical fuel (Tan et al. 2019; Jacobsson et al. 2013). However, the PVs-based solar cell exhibits an open-circuit voltage that ranges between 0.9 and 1.5 V with 23% efficiency (El-Ballouli et al. 2019). The main advantage of these PVs is that they can split the water even when only two cells are connected. Gratzel et al. (Luo et al. 2014) demonstrated that when two HLP-based PVs were connected and delivered a Vocof 2.0 V and poser conversion efficiency of 15.7%. The H_2 generation efficiency of 12.3% was observed. The PV-integrated with electrocatalysis types of structures require an electrocatalyst to be present in the same solution for both reduction and oxidation of water. But this is not always a possible process, since there is a limited number of such types of electrocatalysts. To overcome this issue, a bipolar membrane was prepared by Luo et al. (Luo et al. 2016), to separate the anode and cathode. Cobalt phosphide serves as a catalyst for H_2 generation and a H_2 generation efficiency of 12.7% was achieved with this PV-integrated electrocatalysis.

Park et al. had integrated a perovskite solar cell with NiFe(oxy)hydroxide and Ni_4Mo electrocatalysts for H_2 energy generation from solar energy. The fabricated tandem solar cells have exhibited H_2 generation efficiency exceeding 17% under simulated solar light (Park et al. 2019).

Schreier et al. demonstrated electrocatalytic cell were IrO_2 acted as anode while Au acted as cathode. This electrocatalytic cell integrated with HLP-based tandem solar cell achieved CO generation efficiency of more than 6.5%. This result serves as an important benchmark in photoelectrocatalysis for CO_2–CO generation (Schreier et al. 2015). Hydrocarbon production from CO_2 reduction using PV-integrated electrocatalytic cell system were demonstrated by Huan et al. and achieved H_2 generation efficiency of 2.3% (Huan et al. 2019).

Esiner et al. demonstrated an electrocatalytic cell in which cathode and anode was made separately was Au-nanowire and RuO_2 deposited on the Ti-substrate, respectively, for CO_2 reduction. This triple-cation $FA_{0.66}MA_{0.34}PbI_{2.85}Br_{0.15}$ absorber-based PV delivered power conversion efficiency of 17.4%. The perovskite tandem solar cell devices serve as external power source and delivered CO generation efficiency of 8.9% (Esiner et al. 2020).

Halide Perovskite (HLP)-Based Photoelectrode in a Photoelectrochemical Cell

In this type of photoelectrochemical cell, a photoelectrode is linked with the reference electrode and counter electrode. The photoelectrode has a dual function either as a photocathode or photoanode. The electrons produced in the photoelectrode drives the fuel generation reaction while oxidation reaction is driven by the holes created in the counter electrode. The photoelectrode's activity is dependent on the semiconductor. A photoelectrode typically consists of a conductive layer for efficient charge movement deposited on the perovskite. The issues with the stability of the perovskite in presence of moisture or air are resolved by the deposition of a stable layer on the top of perovskite. This type of coating has been greatly employed for photoelectrolysis in preparation of photoelectrode (Da et al. 2015; Zhang et al. 2018; Poli et al. 2019). The HLP-based photoelectrode for the generation of O_2, H_2, and the reduction of CO_2 demonstrated recently are described below for basic understanding of the process.

Da et al. demonstrated a photo anode coated with Ni, which acted as a photoanode. This tandem solar devices achieved poser conversion efficiency of 10%. The Ni coated layer not only function as a protective layer for HLP-based solar cell and interface but also assist in hole transmission for water oxidation. The schematic diagram of the HLP-based photoelectrode to construct photoelectrochemical cell is presented in Fig. 15.3a and b along with energy level diagram for efficient charge carrier transport (Da et al. 2015). Wang et al. fabricated a photoanode by treating HLP-based cell with tetra-ethyl ammonium (TEA) and an ultrathin Ni passivation layer (Wang et al. 2017).

Nam et al. modified a photoelectrode by putting a thin layer of Ni protecting layer with perovskite solar cell. The tandem PCs devices exhibited efficiency of 15.6%. The photocurrent, which was obtained as 13 mA/cm^2 remained stable for 6 h. The reversible hydrogen electrode in the KOH electrolyte acted as a reference electrode (Nam et al. 2018).

HLP-based photoanodes have also reportedly been used by coating by carbon materials. Tao et al. demonstrated a cation substituted HLP as photoanode coated with carbon and silver conductive paint. The HLP-based photoanode shows higher

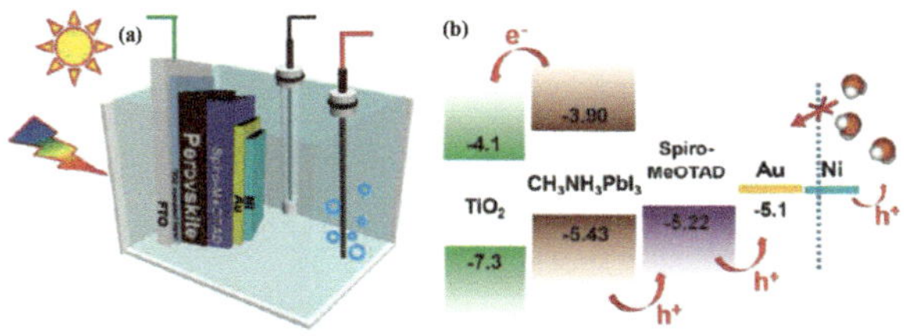

Fig. 15.3 **a** Schematic presentation of the photoelectrochemical cell with three electrodes configuration, **b** corresponding energy level diagram. Da et al. (2015) Reproduced with permission from ref. (). Copyright 2015, American Chemical Society

stability due to the thick layer of carbon over coated to it (Tao et al. 2019). For O_2 evolution in water, Poli et al. used $CsPbBr_3$ perovskite with graphite sheet and mesoporous carbon as photoanode. The fabricated $TiO_2/CsPbBr_3$/m-C/GS tandem devices achieved sufficient photocurrent and was stable for 30 h. By application of water oxidation catalyst (WOC) sufficient amount H_2 was generated (Poli et al. 2019).

Crespo-Quesada et al. demonstrated a stable Field's metal (fusible InBiSn alloy) coated $MAPbI_3$-based photocathode for splitting water. This multilayer device utilizing $MAPbI_3$ as absorber allows transfer of electrons and drive the H_2 production reaction. The function of the photocathode was examined in a three-electrode configuration with the perovskite as working, an Ag/AgCl reference electrode and Pt mesh counter electrode separated by a Nafion membrane. Approximately, 80% of the photocurrent could be sustained under continuous illumination. The Field's metal protects the perovskite from water while at the same time Pt nanoparticles coated over PCs aids in photocatalysis for the generation of H_2. The photocurrent could be for more than 1.5 h under constant irradiation. Due to the outcome and improved stability of the various components in the reaction system, their discovery had a significant contribution in this field (Crespo-Quesada et al. 2016). Andrei et al. fabricated a tandem devices photoelectrochemica lPVs that comprises of CsFAMA, three cation photocathode with a NiOx hole transport layer and photoanode, $BiVO_4$|TiCo. The device integrated with photoanode delivered higher performance without photoanode. The outstanding replicability and versatility of the fabrication processes offered the same efficiency with tandem devices even with a 10 cm^2 area (Andrei et al. 2018).

A sandwich structure photocathode [Pt-Ti/$CH_3NH_3PbI_3$], in which the PVs are protected by the Ti-foil pre-coated with Pt nanoparticles were demonstrated by Zhang et al. This photocathode maintains a photocurrent density of 18 mA/cm^2 at 0 V versus RHE for 12 h (Zhang et al. 2018). Similarly different setup of the photoelectrochemical device was fabricated by other research group and achieved the outstanding results (Edwardes Moore et al. 2019; Chen et al. 2019; Andrei et al. 2020; Jang et al. 2016).

Water Splitting and CO_2 Reduction Generation Using HLP Particulate System

In a semiconductor photocatalysis, the incoming photons are being absorbed and converted into chemical energy. It comprises of four different stages of energy transmission, which is already explained in the introduction section, the electrons and holes generated on the active sites of the semiconductor photocatalyst serve as reduction and oxidation sites. Thermodynamically, the electronic band structure of a semiconductor and the reaction redox potential thus need to be in balance appropriately (Zhou et al. 2012; Takata and Domen 2019).

The wavelength of radiation required for the splitting of water and reduction of CO_2 lies between UV–visible light and the near IR region via. 1008 nm and 954 nm are sufficient for chemical fuel generation. For a reduction reaction to occur, the electron should have a higher negative potential value while higher positive potential value for oxidation. The efficiency of the photocatalysis depends on whether the excited electrons bind to the active sites of the photocatalyst or charge carriers are recombine as the recombination can takes place on the interface of semiconductor (Qian et al. 2019).

Due to instability of HLP photocatalyst when they experience thermal strain, air exposure, phase transitions, their optoelectronic properties get changed. When polar additives come in contact with the charged HLP crystals, the ionic interactions that hold them together make them susceptible to structural changes. Kwon et al. studied the quantitative and systematic mechanism of degradation of HLP-based PVs using an ultrafast transient absorption spectroscopy (Kwon et al. 2014). When HLP materials are exposed to polar solvents, HLP forms coordination compound with it. The solution-processed perovskite thin films possess more defects when fabricated through polar coordinating solvents in comparison with non-coordinating polar solvents (Jeon et al. 2014; Yang et al. 2015; Jung et al. 2019). The stability is the major concern as the HLP dissociates under the operational conditions.

Several researches conducted by researcher to tackle the instability of HLP which include the structural or compositional modifications. Halo acids is one of the substitutes for water (Kulbak et al. 2016; Lee et al. 2015). The replacement of organic group in HLP by Cs^+ is also one of the technique to increase stability of HLP (Zheng and Rubel 2017). This substitution has also increased the stability of the HLP ($CsPbBr_3$) to an organic solvents that are applicable for different photocatalytic processes (Kim et al. 2015).

Ju et al. reported an eco-friendly $MACH_3{}^+SnI_3$ photo-catalytic reaction taking place in an aqueous medium. The H_2 evolution rate of 0.64 μmol h^{-1} was achieved from water splitting which remain stable for 5 h (Ju et al. 2018). Similar work was performed by Pisano et al. in which $MA_{1-x}DMA_xSnBr_3$ in DI water was used as photocatalysis (Pisanu et al. 2019).

Li et al. demonstrated environmentally stable photocatalyst of structure $CsPbBr_3$/ TiO_2 in which the $CsPbBr_3$ are being coated by TiO_2 layer (Li et al. 2018). Zhang et al. demonstrated a $CsPbBr_3$ nanocrystals encapsulated by silica that last in an abrasive acidic media for 50 days. More stable $CsPbBr_3$ based photocatalyst enclosed by silicone/silica of structure $CsPbBr_3$/Polystyrene@Silica, thinly-wrapped $CsPbX_3$ @silicone, $CsPbBr_3$@SiO_2 are also being reported (Zhang et al. 2020; Yang et al. 2019; Zhao et al. 2019a; Zhong et al. 2018).

Park et al. reported a photocatalytic hydriodic acid (HI) splitting using $MAPbI_3$ powder. When this system was exposed to visible light for 9 h, 25.7 μmol of H_2 was produced, and 27 μmol of I_3^- accumulated during the photocatalysis. The reaction required for H_2 evolution is dependent on the electrons while the I_3^- has been generated from I^- by a photocatalytically generated holes. The Pt coating on HLP stabilizes the perovskites in an aqueous medium that are applicable in photocatalysis (Park et al. 2016).

Wang et al. coated a TiO_2 above $MAPbI_3$ and Pt beneath it to further enhances the photocatalytic activity (Wang et al. 2018). The schematic of the H_2 production system using Pt/TiO_2-$MAPbI_3$ nanocrystal powder is presented in Fig. 15.4 (Wang et al. 2018). Wu et al. reported a PCs integrated with rGO. An H_2 production rate of 939 μmol/g. h was achieved which remain stable for 200 h. The electrons produced from the PCs are delivered to R-GO where H_2 evolution results from HI splitting (Wu et al. 2018a). Zhao et al. demonstrated a highly stable photocatalyst by covering $MAPbI_3$ with Ni_3C which achieved a high H_2 generation. The $MAPbI_3$/Ni_3C hybrid structure could sustain the photocurrent for 90 h cycles and continued to function just as well after being stored for a month (Zhao et al. 2019b).

When $MAPbI_3$ are coupled with MoS_2 as a co-catalyst, the composite structure exhibits a 206 μmol/g. h H_2 production rate (Wang et al. 2020). The conducting polymer can also be coated with HLP to enhance the stability for photochemical reactions (Wu et al. 2018b). Wu et al. and Guan et al. developed a $MAPbBr_{3-x}I_x$ and all-inorganic $CsPbBr_{3-x}I_x$ by replacing Br^- with I^- which produces a PCs with tuned bandgaps. The difference in the concentration of I^- in range from bulk to surface, facilitates movement of charge carriers (Guan et al. 2019).

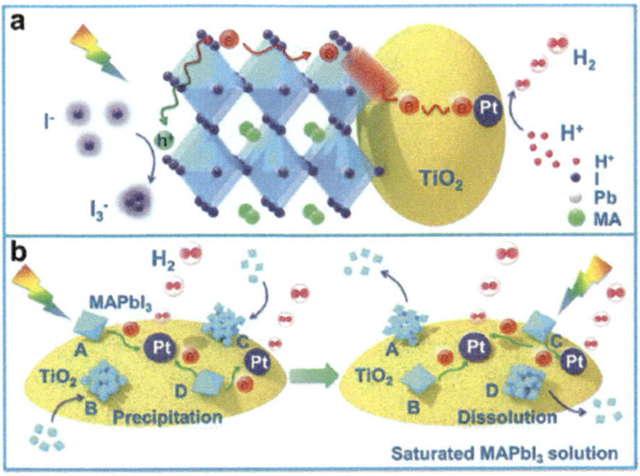

Fig. 15.4 Schematic presentation of H_2 production from HI splitting **a** electron transport from perovskite to TiO_2 for photocatalysis, **b** dissolution and reformation of $MAPbI_3$ in solution of HI. Wang et al. (2018) Reproduced with permission from ref. (). Copyright 2018, American Chemical Society

Conclusion and Outlook

As summarized above, the HLP has been proved as an extraordinary material for chemical fuel generation owing to excellent optical and electronic properties. Due to stability issues of HLP in the presence of moisture, sun light, air, the deliberate strategies are required for utilization of HLP for fuel generation reactions such as H_2O splitting and CO_2 reduction. The three main strategies for this purpose are using HLP perovskite as particles, using HLP films as photoelectrode in photoelectrochemical system and electrochemical cells driven by HLP-based photovoltaic. The solar-to-fuel conversion rate over 10% achieved using HLP demonstrates it capability for future application in the fuel generation process. Although, due to stability issue, special arrangements required for HLP to use as photocatalysis for CO_2 reduction and H_2O splitting, make it difficult to employ it as an efficient and easy to use photocatalysis. Some of the important points are required to be incorporated to improve photocatalytic solar-to-fuel efficiency of HLP. These points are pointed out below.

(i). One of the most important issues to address is the stability of the HLP in the aqueous medium. Most of the solar-to-fuel conversion reactions occur in the aqueous medium, in which HLP get degraded. This problem can be solved by making HLP and other conductive materials as a core–shell structure. The high conductivity of the shell may provide much better charge carrier transfer and HLP as core will remain stable under the longer operational duration.

(ii). It is important to note that nanostructured HLP have not been thoroughly explored in the literature for solar-to-fuel conversion. Nanostructure of HLP will provide higher surface area for the highly efficient photocatalytic reactions.

(iii). Theoretical efficiency of the HLP photovoltaic device has been reported as high as 30%. However, solar-to-fuel conversion efficiency has been reported as close to 10%. This means that the modifications in the processing in the system in which HLP-based photovoltaic integrated system so that the fuel production rate can be enhanced.

(iv). Not only HLP materials, the modification in each parameters of the system employed for photocatalytic solar-to-fuel generation is required such as solvent, photoelectrode, configuration of system, incident light intensity, and sacrificial materials.

(v). Another important point to consider in the case of HLP is to use Pb-free HLP. As we can see in the discussion section of this chapter most of the HLP employed for fuel generation are Pb-based HLP; however, toxicity of the Pb is topic of concern in view of health and environmental safety. The toxicity of the waste product of Pb-based HLP materials has been accessed recently and found to contaminate our ecosystem by releasing Pb as a result of raining on it, when it comes in contact with ground water. Therefore, it is important to replace the Pb from HLP and use of other metal is desirable. The best alternative which is recently discovered is the Sn-based HLP materials, which are supposed to be highly utilized in the future for solar-to-fuel conversion.

Moreover, Sb and Bi-based HLP have been utilized but similar stability issues were face in these materials also. For solar-to-fuel conversion Sn-based HLP or mixed HLP which are having equivalent properties as Pb-based HLP must be explored in the future.

References

Ahmad H, Kamarudin SK, Minggu LJ et al (2015) Hydrogen from photo-catalytic water splitting process: a review. Renew Sustain Energy Rev 43:599–610. https://doi.org/10.1016/j.rser.2014.10.101

Andrei V, Hoye RL, Crespo-Quesada M et al (2018) Scalable triple cation mixed halide perovskite–bivo4 tandems for bias-free water splitting. Adv Energy Mater 8(25):1801403. https://doi.org/10.1002/aenm.201801403

Andrei V, Reuillard B, Reisner E (2020) Bias-free solar syngas production by integrating a molecular cobalt catalyst with perovskite–$BiVO_4$ tandems. Nat Mater 19(2):189–194. https://doi.org/10.1038/s41563-019-0501-6

Baek JH, Kim BJ, Han GS et al (2017) $BiVO_4/WO_3/SnO_2$ double-heterojunction photoanode with enhanced charge separation and visible-transparency for bias-free solar water-splitting with a perovskite solar cell. ACS Appl Mater Interfaces 9(2):1479–1487. https://doi.org/10.1021/acsami.6b12782

Bagheri S, TermehYousefi A, Do TO (2017) Photocatalytic pathway toward degradation of environmental pharmaceutical pollutants: structure, kinetics and mechanism approach. Catal Sci Technol 7(20):4548–4569. https://doi.org/10.1039/C7CY00468K

Cao Y, Wang N, Tian H et al (2018) Perovskite light-emitting diodes based on spontaneously formed submicrometre-scale structures. Nature 562(7726):249–253. https://doi.org/10.1038/s41586-018-0576-2

Chen J, Yin J, Zheng X et al (2019) Compositionally screened eutectic catalytic coatings on halide perovskite photocathodes for photoassisted selective CO_2 reduction. ACS Energy Lett 4(6):1279–1286. https://doi.org/10.1021/acsenergylett.9b00751

Cheng L, Xiang Q, Liao Y, Zhang H (2018) CdS-based photocatalysts. Energy Environ Sci 11(6):1362–1391. https://doi.org/10.1039/C7EE03640J

Creel EB, Corson ER, Eichhorn J et al (2019) Directing selectivity of electrochemical carbon dioxide reduction using plasmonics. ACS Energy Lett 4(5):1098–1105. https://doi.org/10.1021/acsenergylett.9b00515

Crespo-Quesada M, Pazos-Outón LM, Warnan J et al (2016) Metal-encapsulated organolead halide perovskite photocathode for solar-driven hydrogen evolution in water. Nat Commun 7(1):12555. https://doi.org/10.1038/ncomms12555

Da P, Cha M, Sun L et al (2015) High-performance perovskite photoanode enabled by Ni passivation and catalysis. Nano Lett 15(5):3452–3457. https://doi.org/10.1021/acs.nanolett.5b00788

De Luna P, Hahn C, Higgins D et al (2019) What would it take for renewably powered electrosynthesis to displace petrochemical processes? Science 364(6438):eaav3506. https://doi.org/10.1126/science.aav3506

Deng W, Zhao H, Pan F et al (2017) Visible-light-driven photocatalytic degradation of organic water pollutants promoted by sulfite addition. Environ Sci Technol 51(22):13372–13379. https://doi.org/10.1021/acs.est.7b04206

Dias P, Schreier M, Tilley SD et al (2015) Transparent cuprous oxide photocathode enabling a stacked tandem cell for unbiased water splitting. Adv Energy Mater 5(24):1501537. https://doi.org/10.1002/aenm.201501537

Ding P, Jiang T, Han N, Li Y (2020) Photocathode engineering for efficient photoelectrochemical CO_2 reduction. Mater Today Nano 10:100077. https://doi.org/10.1016/j.mtnano.2020.100077

Dong P, Hou G, Xi X et al (2017) WO_3-based photocatalysts: morphology control, activity enhancement and multifunctional applications. Environ Sci Nano 4(3):539–557. https://doi.org/10.1039/C6EN00478D

Edwardes Moore E, Andrei V, Zacarias S et al (2019) Integration of a hydrogenase in a lead halide perovskite photoelectrode for tandem solar water splitting. ACS Energy Lett 5(1):232–237. https://doi.org/10.1021/acsenergylett.9b02437

El-Ballouli AAO, Bakr OM, Mohammed OF (2019) Compositional, processing, and interfacial engineering of nanocrystal-and quantum-dot-based perovskite solar cells. Chem Mater 31(17):6387–6411. https://doi.org/10.1021/acs.chemmater.9b01268

Esiner S, Wang J, Janssen RA (2020) Light-driven electrochemical carbon dioxide reduction to carbon monoxide and methane using perovskite photovoltaics. Cell Rep Phys Sci 1(5):1. https://doi.org/10.1016/j.xcrp.2020.100058

Fajrina N, Tahir M (2019) A critical review in strategies to improve photocatalytic water splitting towards hydrogen production. Int J Hydrogen Energy 44(2):540–577. https://doi.org/10.1016/j.ijhydene.2018.10.200

Fu J, Yu J, Jiang C, Cheng B (2018) G-C3N4-Based heterostructured photocatalysts. Adv Energy Mater 8(3):1701503. https://doi.org/10.1002/aenm.201701503

Guan Z, Wu Y, Wang P et al (2019) Perovskite photocatalyst $CsPbBr_{3-x}I_x$ with a bandgap funnel structure for H_2 evolution under visible light. Appl Catal B 245:522–527. https://doi.org/10.1016/j.apcatb.2019.01.019

Gurudayal John RA, Boix PP et al (2017) Atomically altered hematite for highly efficient perovskite tandem water-splitting devices. ChemSusChem 10(11):2449–2456. https://doi.org/10.1002/cssc.201700159

Hisatomi T, Domen K (2019) Reaction systems for solar hydrogen production via water splitting with particulate semiconductor photocatalysts. Nat Catal 2(5):387–399. https://doi.org/10.1038/s41929-019-0242-6

Hu Y, Zhan F, Wang Q et al (2020) Tracking mechanistic pathway of photocatalytic CO_2 reaction at Ni sites using operando, time-resolved spectroscopy. J Am Chem Soc 142(12):5618–5626. https://doi.org/10.1021/jacs.9b12443

Huan TN, Dalla Corte DA, Lamaison S et al (2019) Low-cost high-efficiency system for solar-driven conversion of CO_2 to hydrocarbons. Proc Natl Acad Sci 116(20):9735–9740. https://doi.org/10.1073/pnas.1815412116

Jacobsson TJ, Fjällström V, Sahlberg M et al (2013) A monolithic device for solar water splitting based on series interconnected thin film absorbers reaching over 10% solar-to-hydrogen efficiency. Energy Environ Sci 6(12):3676–3683. https://doi.org/10.1039/C3EE42519C

Jang DM, Park K, Kim DH et al (2015) Reversible halide exchange reaction of organometal trihalide perovskite colloidal nanocrystals for full-range band gap tuning. Nano Lett 15(8):5191–5199. https://doi.org/10.1021/acs.nanolett.5b01430

Jang YJ, Jeong I, Lee J et al (2016) Unbiased sunlight-driven artificial photosynthesis of carbon monoxide from CO_2 using a ZnTe-based photocathode and a perovskite solar cell in tandem. ACS Nano 10(7):6980–6987. https://doi.org/10.1021/acsnano.6b02965

Jaramillo-Páez CA, Navío JA, Hidalgo MC et al (2018) ZnO and Pt-ZnO photocatalysts: Characterization and photocatalytic activity assessing by means of three substrates. Catal Today 313:12–19. https://doi.org/10.1016/j.cattod.2017.12.009

Jeon NJ, Noh JH, Kim YC et al (2014) Solvent engineering for high-performance inorganic–organic hybrid perovskite solar cells. Nat Mater 13(9):897–903. https://doi.org/10.1038/nmat4014

Ju D, Zheng X, Liu J et al (2018) Reversible band gap narrowing of Sn-based hybrid perovskite single crystal with excellent phase stability. Angew Chem Int Ed 57(45):14868–14872. https://doi.org/10.1002/anie.201810481

Jung M, Ji SG, Kim G et al (2019) Perovskite precursor solution chemistry: from fundamentals to photovoltaic applications. Chem Soc Rev 48(7):2011–2038. https://doi.org/10.1039/C8CS00656C

Khalafi T, Buazar F, Ghanemi K (2019) Phycosynthesis and enhanced photocatalytic activity of zinc oxide nanoparticles toward organosulfur pollutants. Sci Rep 9(1):6866. https://doi.org/10.1038/s41598-019-43368-3

Kim HS, Im SH, Park NG (2014) Organolead halide perovskite: new horizons in solar cell research. J Phys Chem C 118(11):5615–5625. https://doi.org/10.1021/jp409025w

Kim Y, Yassitepe E, Voznyy O et al (2015) Efficient luminescence from perovskite quantum dot solids. ACS Appl Mater Interfaces 7(45):25007–25013. https://doi.org/10.1021/acsami.5b09084

König B (2017) Photocatalysis in organic synthesis–past, present, and future. Eur J Org Chem 2017(15):1979–1981. https://doi.org/10.1002/ejoc.201700420

Kulbak M, Gupta S, Kedem N et al (2016) Cesium enhances long-term stability of lead bromide perovskite-based solar cells. J Phys Chem Lett 7(1):167–172. https://doi.org/10.1021/acs.jpclett.5b02597

Kwon YS, Lim J, Yun HJ et al (2014) A diketopyrrolopyrrole-containing hole transporting conjugated polymer for use in efficient stable organic–inorganic hybrid solar cells based on a perovskite. Energy Environ Sci 7(4):1454–1460. https://doi.org/10.1039/C3EE44174A

Lee MM, Teuscher J, Miyasaka T et al (2012) Efficient hybrid solar cells based on meso-superstructured organometal halide perovskites. Science 338(6107):643–647. https://doi.org/10.1126/science.1228604

Lee JW, Kim DH, Kim HS et al (2015) Formamidinium and cesium hybridization for photo-and moisture-stable perovskite solar cell. Adv Energy Mater 5(20):1501310. https://doi.org/10.1002/aenm.201501310

Li ZJ, Hofman E, Li J, Davis AH et al (2018) Photoelectrochemically active and environmentally stable $CsPbBr_3/TiO_2$ core/shell nanocrystals. Adv Funct Mater 28(1):1704288. https://doi.org/10.1002/adfm.201704288

Liang J, Han X, Qiu Y et al (2020) A low-cost and high-efficiency integrated device toward solar-driven water splitting. ACS Nano 14(5):5426–5434. https://doi.org/10.1021/acsnano.9b09053

Lin Q, Armin A, Burn PL, Meredith P (2016) Organohalide perovskites for solar energy conversion. Acc Chem Res 49(3):545–553. https://doi.org/10.1021/acs.accounts.5b00483

Lin K, Xing J, Quan LN et al (2018) Perovskite light-emitting diodes with external quantum efficiency exceeding 20%. Nature 562(7726):245–248. https://doi.org/10.1038/s41586-018-0575-3

Liu X, Zhai H, Wang P et al (2019a) Synthesis of a WO_3 photocatalyst with high photocatalytic activity and stability using synergetic internal Fe^{3+} doping and superficial Pt loading for ethylene degradation under visible-light irradiation. Catal Sci Technol 9(3):652–658. https://doi.org/10.1039/C8CY02375A

Liu X, Wang Y, Cui X et al (2019b) Enabling highly efficient photocatalytic hydrogen generation and organics degradation via a perovskite solar cell-assisted semiconducting nanocomposite photoanode. J Mater Chem A 7(1):165–171. https://doi.org/10.1039/C8TA08998A

Luo J, Im JH, Mayer MT et al (2014) Water photolysis at 12.3% efficiency via perovskite photovoltaics and earth-abundant catalysts. Science 345(6204):1593–1596. https://doi.org/10.1126/science.1258307

Luo J, Li Z, Nishiwaki S et al (2015) Targeting ideal dual-absorber tandem water splitting using perovskite photovoltaics and CuInxGa1-xSe2 photocathodes. Adv Energy Mater 5(24):1501520. https://doi.org/10.1002/aenm.201501520

Luo J, Vermaas DA, Bi D et al (2016) Bipolar membrane-assisted solar water splitting in optimal pH. Adv Energy Mater 6(13):1. https://doi.org/10.1002/aenm.201600100

Maeda K, Domen K (2010) Photocatalytic water splitting: recent progress and future challenges. J Phys Chem Lett 1(18):2655–2661. https://doi.org/10.1021/jz1007966

Nahar S, Zain MFM, Kadhum AAH et al (2017) Advances in photocatalytic CO_2 reduction with water: a review. Materials 10(6):629. https://doi.org/10.3390/ma10060629

Nam S, Mai CTK, Oh I (2018) Ultrastable photoelectrodes for solar water splitting based on organic metal halide perovskite fabricated by lift-off process. ACS Appl Mater Interfaces 10(17):14659–14664. https://doi.org/10.1021/acsami.8b00686

Park S, Chang WJ, Lee CW et al (2016) Photocatalytic hydrogen generation from hydriodic acid using methylammonium lead iodide in dynamic equilibrium with aqueous solution. Nat Energy 2(1):1–8. https://doi.org/10.1038/nenergy.2016.185

Park H, Park IJ, Lee MG et al (2019) Water splitting exceeding 17% solar-to-hydrogen conversion efficiency using solution-processed Ni-based electrocatalysts and perovskite/Si tandem solar cell. ACS Appl Mater Interfaces 11(37):33835–33843. https://doi.org/10.1021/acsami.9b09344

Pisanu A, Speltini A, Quadrelli P et al (2019) Enhanced air-stability of Sn-based hybrid perovskites induced by dimethylammonium (DMA): synthesis, characterization, aging and hydrogen photogeneration of the $MA_{1-x}DMA_xSnBr_3$ system. J Mater Chem C 7(23):7020–7026. https://doi.org/10.1039/C9TC01743G

Poli I, Hintermair U, Regue M et al (2019) Graphite-protected $CsPbBr_3$ perovskite photoanodes functionalised with water oxidation catalyst for oxygen evolution in water. Nat Commun 10(1):2097. https://doi.org/10.1038/s41467-019-10124-0

Qian R, Zong H, Schneider J et al (2019) Charge carrier trapping, recombination and transfer during TiO_2 photocatalysis: an overview. Catal Today 335:78–90. https://doi.org/10.1016/j.cattod.2018.10.053

Schreier M, Curvat L, Giordano F et al (2015) Efficient photosynthesis of carbon monoxide from CO_2 using perovskite photovoltaics. Nat Commun 6(1):7326. https://doi.org/10.1038/ncomms8326

Shen S, Chen J, Wang M et al (2018) Titanium dioxide nanostructures for photoelectrochemical applications. Prog Mater Sci 98:299–385. https://doi.org/10.1016/j.pmatsci.2018.07.006

Singh M, Sinha I (2020) Halide perovskite-based photocatalysis systems for solar-driven fuel generation. Sol Energy 208:296–311. https://doi.org/10.1016/j.solener.2020.08.007

Suzuki K, Li X, Wang Y et al (2020) Active intermediates in plasmon-induced water oxidation at Au nanodimer structures on a single crystal of TiO_2. ACS Energy Lett 5(4):1252–1259. https://doi.org/10.1021/acsenergylett.0c00478

Tahir M (2018) Photocatalytic carbon dioxide reduction to fuels in continuous flow monolith photoreactor using montmorillonite dispersed Fe/TiO_2 nanocatalyst. J Clean Prod 170:242–250. https://doi.org/10.1016/j.jclepro.2017.09.118

Takata T, Domen K (2019) Particulate photocatalysts for water splitting: recent advances and future prospects. ACS Energy Lett 4(2):542–549. https://doi.org/10.1021/acsenergylett.8b02209

Tan CS, Kemp KW, Braun MR et al (2019) > 10% solar-to-hydrogen efficiency unassisted water splitting on ALD-protected silicon heterojunction solar cells. Sustain Energy Fuels 3(6):1490–1500. https://doi.org/10.1039/C9SE00110G

Tao R, Sun Z, Li F et al (2019) Achieving organic metal halide perovskite into a conventional photoelectrode: outstanding stability in aqueous solution and high-efficient photoelectrochemical water splitting. ACS Applied Energy Materials 2(3):1969–1976. https://doi.org/10.1021/acsaem.8b02072

Teh YW, Chee MK, Kong XY et al (2020) An insight into perovskite-based photocatalysts for artificial photosynthesis. Sustain Energy Fuels 4(3):973–984. https://doi.org/10.1039/C9SE00526A

Tofanello A, Freitas AL, Carvalho WM Jr et al (2020) Hematite surface modification toward efficient sunlight-driven water splitting activity: the role of gold nanoparticle addition. J Phys Chem C 124(11):6171–6179. https://doi.org/10.1021/acs.jpcc.9b11966

Wang C, Yang S, Chen X et al (2017) Surface-functionalized perovskite films for stable photoelectrochemical water splitting. J Mater Chem A 5(3):910–913. https://doi.org/10.1039/C6TA08812K

Wang X, Wang H, Zhang H et al (2018) Dynamic interaction between methylammonium lead iodide and TiO_2 nanocrystals leads to enhanced photocatalytic H_2 evolution from HI splitting. ACS Energy Lett 3(5):1159–1164. https://doi.org/10.1021/acsenergylett.8b00488

Wang F, Liu X, Zhang Z et al (2020) A noble-metal-free MoS_2 nanosheet-coupled $MAPbI_3$ photocatalyst for efficient and stable visible-light-driven hydrogen evolution. Chem Commun 56(22):3281–3284. https://doi.org/10.1039/D0CC00095G

Wen J, Xie J, Chen X, Li X (2017) A review on g-C3N4-based photocatalysts. Appl Surf Sci 391:72–123. https://doi.org/10.1016/j.apsusc.2016.07.030

Wu Y, Wang P, Zhu X et al (2018a) Composite of CH3NH3PbI3 with reduced graphene oxide as a highly efficient and stable visible-light photocatalyst for hydrogen evolution in aqueous HI solution. Adv Mater 30(7):1704342. https://doi.org/10.1002/adma.201704342

Wu Y, Wang P, Guan Z et al (2018b) Enhancing the photocatalytic hydrogen evolution activity of mixed-halide perovskite $CH3NH3PbBr_{3-x}I_x$ achieved by bandgap funneling of charge carriers. ACS Catal 8(11):10349–10357. https://doi.org/10.1021/acscatal.8b02374

Xing G, Mathews N, Sun S et al (2013) Long-range balanced electron-and hole-transport lengths in organic-inorganic $CH_3NH_3PbI_3$. Science 342(6156):344–347. https://doi.org/10.1126/science.1243167

Xu Q, Yang D, Lv J, Sun YY, Zhang L (2018) Perovskite solar absorbers: materials by design. Small Methods 2(5):1700316. https://doi.org/10.1002/smtd.201700316

Yang W, Moon J (2019) Recent advances in earth-abundant photocathodes for photoelectrochemical water splitting. Chemsuschem 12(9):1889–1899. https://doi.org/10.1002/cssc.201801554

Yang WS, Noh JH, Jeon NJ et al (2015) High-performance photovoltaic perovskite layers fabricated through intramolecular exchange. Science 348(6240):1234–1237. https://doi.org/10.1126/science.aaa9272

Yang W, Gao F, Qiu Y et al (2019) $CsPbBr_3$-quantum-dots/polystyrene@ silica hybrid microsphere structures with significantly improved stability for white LEDs. Adv Opt Mater 7(13):1900546. https://doi.org/10.1002/adom.201900546

Zhang H, Yang Z, Yu W et al (2018) A sandwich-like organolead halide perovskite photocathode for efficient and durable photoelectrochemical hydrogen evolution in water. Adv Energy Mater 8(22):1800795. https://doi.org/10.1002/aenm.201800795

Zhang F, Zhuang HQ, Zhang W et al (2019a) Noble-metal-free CuS/CdS photocatalyst for efficient visible-light-driven photocatalytic H_2 production from water. Catal Today 330:203–208. https://doi.org/10.1016/j.cattod.2018.03.060

Zhang Y, Sun R, Ou X et al (2019b) Metal halide perovskite nanosheet for X-ray high-resolution scintillation imaging screens. ACS Nano 13(2):2520–2525. https://doi.org/10.1021/acsnano.8b09484

Zhang Q, Wang B, Zheng W et al (2020) Ceramic-like stable $CsPbBr_3$ nanocrystals encapsulated in silica derived from molecular sieve templates. Nat Commun 11(1):31. https://doi.org/10.1038/s41467-019-13881-0

Zhao H, Wei L, Zeng P et al (2019a) Formation of highly uniform thinly-wrapped $CsPbX_3$@ silicone nanocrystals via self-hydrolysis: suppressed anion exchange and superior stability in polar solvents. J Mater Chem C 7(32):9813–9819. https://doi.org/10.1039/C9TC01216H

Zhao Z, Wu J, Zheng YZ et al (2019b) Ni_3C-decorated $MAPbI_3$ as visible-light photocatalyst for H_2 evolution from HI splitting. ACS Catal 9(9):8144–8152. https://doi.org/10.1021/acscatal.9b01605

Zheng C, Rubel O (2017) Ionization energy as a stability criterion for halide perovskites. J Phys Chem C 121(22):11977–11984. https://doi.org/10.1021/acs.jpcc.7b00333

Zhong Q, Cao M, Hu H et al (2018) One-pot synthesis of highly stable $CsPbBr_3$@ SiO_2 core–shell nanoparticles. ACS Nano 12(8):8579–8587. https://doi.org/10.1021/acsnano.8b04209

Zhou Y, Zhao Y (2019) Chemical stability and instability of inorganic halide perovskites. Energy Environ Sci 12(5):1495–1511. https://doi.org/10.1039/C8EE03559H

Zhou H, Qu Y, Zeid T et al (2012) Towards highly efficient photocatalysts using semiconductor nanoarchitectures. Energy Environ Sci 5(5):6732–6743. https://doi.org/10.1039/C2EE03447F

Chapter 16
Deposition Methods of Perovskite Absorber

Ashish K. Ranjan, Manish Kumar, Vishal Singh, Vani Pawar, and Prabhakar Singh

Abstract In recent years, perovskite absorbers have emerged as a promising class of materials for next-generation photovoltaic devices due to their exceptional light-harvesting properties and tunable bandgaps. The important feature of the perovskite absorbers is that they can be easily modified to improve stability and achieve optimum performance. Thus, with ongoing research and development, we can expect high-performing perovskite solar cells in the coming years that are efficient and affordable. For the development of high-performing and cost-effective perovskite absorbers, the deposition methods play a crucial role in advancing perovskite solar cell technology. This chapter offers a detailed look at the different methods used to create perovskite layers, from traditional solution-based techniques to newer methods like vapor deposition and hybrid approaches. It explains each technique's basics, benefits, and drawbacks and discusses how the conditions during deposition can affect the perovskite film's structure, appearance, and optoelectronic properties. Additionally, the chapter also examines the challenges and future directions in optimizing deposition processes to achieve high-efficiency and stable perovskite solar cells, providing a valuable reference for researchers and engineers working in this rapidly evolving field.

Keywords Perovskite absorber · Deposition methods · Perovskite solar cells · Physical vapor deposition · Chemical vapor deposition · Pulsed laser deposition

---

A. K. Ranjan · V. Singh · P. Singh (✉)
Department of Physics, Indian Institute of Technology (Banaras Hindu University), Varanasi, India
e-mail: psingh.app@iitbhu.ac.in

A. K. Ranjan
e-mail: ashish.kumarranjan.rs.phy18@itbhu.ac.in

M. Kumar
Department of Physics, School of Sciences, IFTM University, Moradabad, Uttar Pradesh, India

V. Pawar
Centre for Nano Science and Engineering, Indian Institute of Science, Bengaluru, India

M. Kumar and A. Raj (eds.), *Perovskite Solar Cells*, Springer Tracts in Electrical and Electronics Engineering, https://doi.org/10.1007/978-981-96-9463-1_16

Introduction

Overview of Perovskite Solar Cells

Solar energy is one of the most plentiful, natural, and environmentally friendly renewable energy sources, which can reduce the environmental impacts of non-renewable fossil fuels and, at the same time, support the growing world's electricity demand (Kannan and Vakeesan 2016). However, modern photovoltaic technologies have several challenges: expensive costs, inefficiency, inconsistent performance, and environmental concerns (Hamakawa 1994). Consequently, there must be innovations in materials and techniques to overcome these limitations so that solar cells can be widely used. A layer of perovskite material makes up PSCs where perovskites are composed of organic–inorganic hybrid compounds having ABX_3 general formula with X being a halide anion (like chlorine, bromide, and iodide), A being monovalent cation (like formamidinium, or cesium, and methylammonium) and B is a divalent metal (tin or lead) (Qiu et al. 2020). Perovskite solar cells, or PSCs, have emerged as a ground-breaking photovoltaic technology throughout the last ten years (Shi and Jayatissa 2018). The mineral perovskite, or calcium titanate, was discovered by Gustav Rose in 1839, and the term "perovskite" refers to a class of material with a crystalline structure (Ortega-San-Martin 2020). However, the use of perovskite absorber materials in solar cells has recently increased. The journey of perovskite solar cells began when a Japanese research group in 2009 led by Tsutomu Miyasaka first reported the use of an organometallic halide perovskite ($CH_3NH_3PbX_3$, containing X halide ions) as a solar cell (DSSC) (Kojima et al. 2009). These initial devices had a deficient capacity of about 3.8%, and the perovskite material was unstable in the liquid electrolyte. However, in 2012, significant discoveries were made, and researchers affirmed that organic–inorganic perovskites can be employed as light-harvesting layers material for solid-state photovoltaics (Boix et al. 2014). This breakthrough leads to a swift enhancement in efficiency, reaching 10% in a year. This breakthrough was achieved when researchers used simple and scalable fabrication methods, such as the solution processing deposition method, to produce highly efficient and stable PSCs (Roy et al. 2020). Since then, the PSCs' power conversion efficiency (PCE) has significantly increased by more than 25% in just a few years compared to traditional solar cells (Li et al. 2024).

Advantages of Perovskite Materials

Perovskite solar cells are among the most studied and emerging technologies that have seen rapid growth and gained researcher's significant interest in photovoltaics due to the following possibilities.

(a) ***Higher Efficiency:*** A rapid increase in PCE from ~ 3.9% in 2009 to more than 25% at present represents the plausible potential of PSCs, and that can be achieved on account of several properties of perovskite materials such as longer carrier diffusion length, high absorption coefficient, and tunability of bandgap (Kumar and Naidu 2021).

(b) ***Low Cost:*** Perovskite absorbers can be processed at low temperatures using inexpensive and scalable deposition methods like spin coating and printing, unlike typical silicon (Si) solar cells, which require high temperature and energy-intensive processes, making them less economical than PSCs (Brenner et al. 2016).

(c) ***Versatile:*** Deposition of perovskite absorbers can be made on flexible substrates, marking the possibility of making lightweight and flexible solar panels. This will create pathways for the incorporation of solar cells in several applications such as building integrated photovoltaics (BIPV), and wearable electronics (Jung and Park 2015).

(d) ***High Absorption Coefficient:*** Perovskite absorber materials are known for their high absorption coefficient, which allows efficient absorption of solar light even by thin layers. This property reduces the amount of material needed and increases the solar cell's overall efficiency (Park 2015).

(e) ***Tandem Solar Cells:*** Perovskite materials have been combined with other types of solar cells (silicon, CIGS, etc.) to form tandem cells where different layers absorb different parts of the sun's spectrum; this technique may take single-junction solar cells near their ultimate limits of efficiency (Lal et al. 2017).

(f) ***Environmental Impact:*** Compared to traditional photovoltaics, PSCs have a lower temperature coefficient and lesser material consumption, making them less hazardous to the environment. However, issues such as the toxicity of lead in perovskites and the stability of the devices are still becoming concerns, although researchers are working hard to counter them (Wang et al. 2015).

Perovskite solar cells are among the most promising developments in new-generation renewable energy sources. Thanks to its incredibly fast growth, high efficiency, low cost, and flexibility, it is an essential element in the further evolution of solar energy systems. Overall, many researchers and industry participants continue to address the remaining challenges for stability, scalability, and eventually the environmental footprint that will define the ultimate success and mainstream adoption of this exhilarating technology.

Challenges of Perovskite Absorber Materials

(a) ***Long-Term Stability:*** One of the biggest issues with perovskite solar cells is their long-term stability. Perovskite absorber materials are sentient to environmental conditions like moisture, oxygen, light, and heat and degrade over time (Fu et al. 2018). Research is being done to improve these materials.

(b) ***Lead Toxicity:*** Most of the high-efficiency PSCs are made of lead-based materials, which are environmentally and health-related concerns. Lead toxicity is a hurdle for the commercialization of PSCs, and research is being done on non-lead-based cells (Ren et al. 2022).

(c) ***Current–Voltage (I–V) Hysteresis:*** Many PSCs show hysteresis in current–voltage measurement, making it difficult to measure their efficiency. This is often related to ion migration in the absorber layer and needs to be researched (Chen et al. 2016).

(d) ***Scalability of Production:*** The fabrication process of PSCs is cost-effective. However, scaling up the industry while maintaining efficiency and stability is quite challenging. Issues of consistency, donor management, and inclusion of large areas need to be addressed (Li et al. 2018).

(e) ***Interface Engineering:*** The interface between the perovskite absorber layer and charge carrier layers such as perovskite–hole transport layer (HTL), and perovskite–electron transport layer (ETL) interfaces are crucial for device performance. Poor interface engineering can cause recombination losses, which reduce overall efficiency. Optimization of these interfaces is an ongoing challenge (Zhou et al. 2014).

(f) ***Reproducibility:*** Perovskite absorber materials are highly sensitive to processing conditions, and it is challenging to attain persistence in performance and reproducibility between different batches of PSCs. Differences in deposition processes can make a big difference in device performance (Angmo et al. 2021).

(g) ***Environmental Impact:*** Careful consideration of the environmental impacts on the high-scale production of PSCs is necessary, including materials extraction, processing, and disposal. For sustainable commercialization, developing environment-friendly materials and recycling strategies is essential (Vidal et al. 2021).

The above points summarize the present state of perovskite absorber materials and their future in the photovoltaic (PV) industry. Research and development must go on, to understand thoroughly the perovskite devices and overcome the challenges.

Importance of Deposition Methods

The deposition method of perovskite absorber is critical in determining the overall performance, efficiency, and stability of perovskite solar cells (PSC). The deposition technique essentially influences the crystal structure and uniformity of the absorber layer, which are pivotal for efficient charge transport. Precisely controlling film thickness and morphology of the material during the deposition process ensures optimal light absorption and reduces defects that can act as charge recombination centers.

Furthermore, deposition techniques impact the material compatibility with the underlying layer, affecting the device's overall stability and performance. Here are some of the main reasons why deposition techniques are so important:

(i) **Stability and Durability**

(a) ***Thermal Stability:*** The thermal stability of the perovskite layer is vital for long-term perovskite solar cell performance and can be influenced by the deposition method (Ava et al. 2019).

(b) ***Moisture and Oxygen Sensitivity:*** The deposition method affects the stability of the perovskite layer, especially its resistance to moisture and oxygen. Some methods make the perovskites better encapsulation or passivation, thus extending the longevity of solar cells (Bryant et al. 2016).

(ii) **Film Quality and Morphology**

(a) ***Crystallinity:*** Deposition methods can affect the crystallization process of perovskite materials. High-quality crystalline films with large grain sizes show better lower recombination rates and charge transport properties (Tsai et al. 2017).

(b) ***Surface Coverage:*** A defect-free surface coverage is critical for light absorption and prevents direct contact between the hole and electron transport layers, resulting in a short circuit (Haruyama et al. 2016).

(c) ***Uniformity:*** The deposition method affects the uniformity of the perovskite layer. A uniform film reduces defects such as grain boundaries and pinholes (Zhang et al. 2020).

(iii) **Cost and Process Complexity**

(a) ***Process Simplicity:*** Less complicated deposition techniques could save processing expenses and time, improving the efficiency of the production process (Teixeira et al. 2022).

(b) ***Material Utilization:*** Certain deposition methods can make more efficient use of the material, reducing waste and material costs (Jung and Park 2015).

(iv) **Control of Composition and Stoichiometry**

(a) ***Phase Purity:*** Preparing proper perovskite materials is essential to achieve high device performance. Some deposition techniques provide better control over phase and stoichiometry (Yang et al. 2017).

(b) ***Elemental Distribution:*** The distribution of elements in the perovskite film can be affected by the deposition method. Therefore, for optimal optoelectronic properties, a homogeneous composition is essential (Stuckelberger et al. 2016).

Deposition Methods

16. 3. 1. Spin Coating

Spin coating is the easiest and most extensively utilized method in the fabrication of perovskite films in laboratories. It is a highly adaptable technique that can be employed and allows for straightforward control over the deposition thickness. The coating is applied by dispensing the ink onto the substrate and rapidly spinning. During spinning, the centrifugal force forms a uniform wet layer (see Fig. 16.1). The thickness of the deposited layer can be determined as well as optimized by the rotation speed, the viscosity, and the concentration of the ink, while the duration of rotation mainly affects the substrate's drying process (Hu et al. 2024). Although different models can describe the process, the rapid dynamics of ink spreading and evaporation often necessitate a calibration curve where the thickness is optimized by experimenting with various spinning speeds and measuring the dry thickness. For a given rotation speed, the dry thickness is linearly related to the ink concentration, provided that the viscosity does not significantly change. The turbulence of the rotating sample helps in the quick drying of the film, making it able to apply washing solvents (or for the perovskite to apply antisolvents) since most of the deposited material is expelled due to centrifugal forces. Spin coating can be categorized into one-step and two-step methods.

Deposition Steps in the Spin Coating Method

The spin-coating process involves four key steps:

(i) ***Substrate Preparation:*** Substrate cleaning to ensure a smooth and contaminant-free surface.
(ii) ***Film Coating:*** Deposition of a small quantity of the solution onto the center of the substrate.
(iii) ***Spinning Process:*** During deposition, spin the substrate at high speeds (typically between 1,000–10,000 RPM) to ensure the solution is evenly distributed across the surface.
(iv) ***Drying and Annealing:*** After film deposition, the substrate is dried to vaporize the solvent and make a thin film, followed by annealing to obtain better crystallinity.

By adapting the above deposition steps, thin films can be fabricated with less required time. The spin speed can be optimized to control film thickness and uniformity. However, this deposition method suffers from several challenges, such as the formation of bubbles, swirling patterns, and pinhole defects.

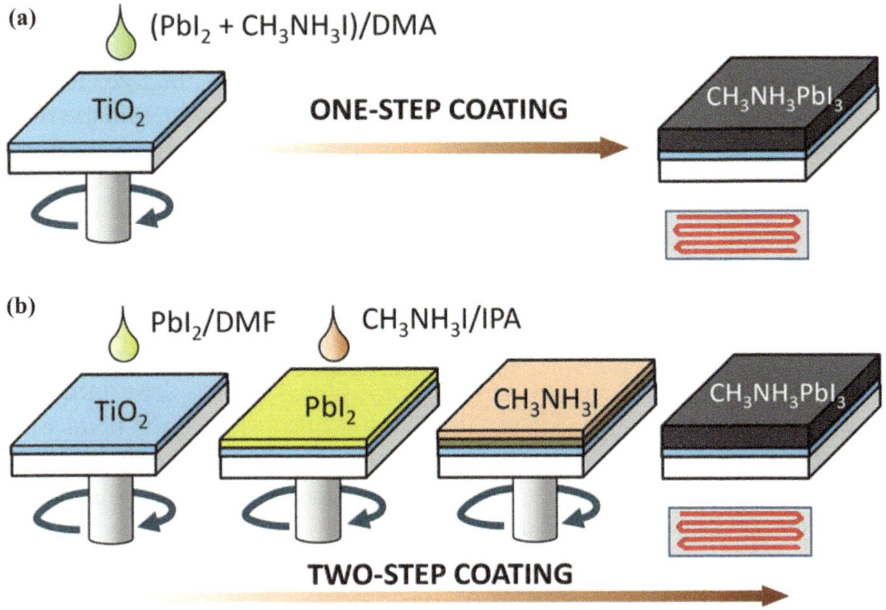

Fig. 16.1 One—and two-step deposition techniques to prepare perovskite thin films. **a** One-step deposition techniques use a pre-mixed stoichiometric solution of the alkyl ammonium salt and lead halide in a polar aprotic solvent like DMF. **b** Two-step deposition techniques generally proceed via the deposition of a lead halide film, followed by conversion of the lead halide to the inorganic–organic lead halide perovskite by exposure to the alkylammonium salt either as a vapor, aerosol or liquid (Im et al. 2014). Copyright 2014. Reproduced with permission from American Institute of Physics Publishing

Solution-Based Methods: One-Step Versus Two-Step Coatings

One-Step Deposition: In single-step deposition methods, the precursors of the final perovskite, specifically PbX_2 (where X can be I, Br, or Cl) and ANH_2X (with A usually being CH_3), are mixed in a solvent that can dissolve both substances (Jamal et al. 2018). This solution is then applied to the substrate, and the solvent evaporates, forming a perovskite film. The key factor to manage in this process is the precipitation of perovskite from the solution; rapid and uncontrolled crystallization can result in low-quality films that negatively impact device performance (Bing et al. 2020). Controlling the morphology is crucial for optimizing device efficiency. However, this method allows precise control over the film's elemental composition and crystallization conditions (Calisi et al. 2022). Two-step deposition techniques have been developed to address these challenges.

Two-Step Deposition: The perovskite layer can be created using a two-step method: first, by depositing lead salts (sometimes with a small amount of cations) and then converting them to perovskite through exposure to organic cations (Han et al. 2022).

These steps can be performed using various solution processing techniques or vacuum deposition, offering significant flexibility in process design (Jamal et al. 2018). In the two-step approach, films are created by individually synthesizing two separate precursors. This technique allows for more straightforward control over the process compared to the one-step method (Wang et al. 2018a).

Figure 16.1 schematically illustrates the one-step and two-step deposition techniques. For coating $CH_3NH_3PbI_3$ in the one-step method, a dimethylformamide (DMA) solution having equal molar amounts of CH_3NH_3I and PbI_2 is deposited onto the mesoporous TiO_2 layer using the spin-coating method. In the two-step technique, PbI_2 is first deposited, subsequently CH_3NH_3I solution is deposited using the spin-coating. The two-step spin-coating method is more precise and controlled, compared to the two-step dipping process (Burschka et al. 2013), which is due to their quantitatively managed steps. The quantity of CH_3NH_3I and the spin-coating parameters must be carefully optimized based on the amount of PbI_2 deposited (Im et al. 2014).

Generally, the fabricated films are annealed after deposition to create a highly crystalline perovskite layer, as annealing promotes strong bonding and adhesion between metal cations and halide anions (Karavioti et al. 2021). The thickness and quality of the film can be optimized by varying the rotation speed, acceleration, and coating duration (Schulze et al. 2020). (Matteocci et al. 2014) produced a larger surface area device (3.36 cm^2) using the spin coating method, demonstrating a 6.3% efficiency in converting electrical. Additionally, a perovskite solar panel with a 16.8 cm^2 active area exhibited a 5.1% increase in PCE. In 2014, a study demonstrated the fabrication of a 60 cm^2 perovskite solar module with a comparable design using a spin coating method comprising a series configuration of interconnected cells (Seo et al. 2014). A 10 cm indium tin oxide-coated perovskite solar panel demonstrated an 8.7% power conversion efficiency. The associated photovoltaic parameters for this module were as follows: fill factor (FF) = 57%, open-circuit voltage (V_{OC}) = 8.1 V, and short-circuit current (I_{SC}) = 1.9 mA-cm^{-2} (Matteocci et al. 2014). The spin coating method has been successfully applied to fabricate perovskite devices with large surface areas. However, the efficiency of these devices may be reduced due to uneven film thickness, especially in the central and edge regions during the spin coating process. Increased recombination events are likely to occur in thicker regions, while thinner areas may have lower light absorption and poor morphology, often characterized by pinholes. These phenomena were evident in the films that increased the active surface area through spin coating. As a result, optimizing the spin coating method for surface area preparation posed significant challenges. The spin coating technique is applicable for fabricating both inverted and regular PSC structures. This method can yield high efficiencies with excellent reproducibility and morphology control for smaller modules (Mbumba et al. 2021). However, it struggles to produce uniform films over larger areas. Additionally, due to its slow processing speed and significant material wastage, spin coating is not ideal for large-scale PSC manufacturing.

Pulsed Laser Deposition Method

Pulsed laser deposition (PLD) is a type of physical vapor deposition (PVD) method frequently employed for the deposition of thin films onto the substrate (Ranjan et al. 2024). This technique is beneficial in fabricating uniform and crystalline films without causing damage to the underlying layers, making it a highly desirable deposition method in the fabrication of perovskite absorber layers as well as complete solar cell devices. Unlike usual solution-based methods that involve dissolving halide salt precursors in organic solvents such as dimethylformamide (DMF) or dimethyl sulfoxide (DMSO), PLD uses a solid target composed of the desired material. This target can be synthesized using either dry mechanochemical synthesis (MCS) (Hong et al. 2019) or wet chemical processes (Leupold and Panzer 2021). However, achieving precise control over the stoichiometry is a pivotal challenge in fabricating perovskite films via PLD.

Key Deposition Steps in PLD Techniques

The PLD technique comprises several elementary steps, as shown in Fig. 16.2, which are discussed as follows:

(i) Laser Ablation

A pulsed laser with high energy, such as Nd: YAG or excimer, is aimed at a solid perovskite target within a vacuum or a controlled gaseous environment. The laser energy vaporizes the target material, creating a plasma plume composed of atoms, ions, and molecules.

(ii) Film Deposition

The plasma plume extends and deposits material onto a substrate positioned opposite of the target. The deposition process is affected by parameters including

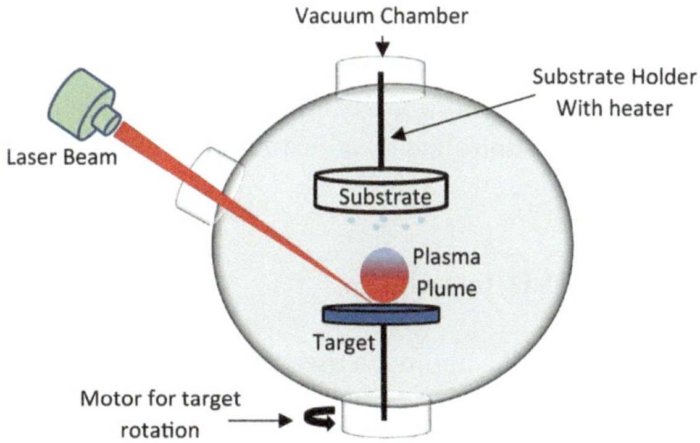

Fig. 16.2 Schematic diagram of pulsed laser deposition technique (Haider et al. 2022). Copyright 2022. Reproduced with permission from Springer Nature

laser energy, repetition rate, distance between the target and substrate, and chamber pressure.

(iii) Post-Fabrication Processing of the Film

After fabrication, the films may be subjected to heat treatment or annealing to improve the crystallinity and phase purity.

Physical Vapor Deposition Method

Physical vapor deposition (PVD) is a thin film deposition or coating technology that transfers material at the atomic level through vaporization to the substrate. PVD is commonly utilized to create films that are only a few nanometers to several thousand nanometers thick, and it also can generate multilayered coatings, graded deposits, thicker layers, and even self-supporting structures.

Principles of PVD

The PVD process typically consists of three major steps:

(i) **Vaporization of Target Materials:** The target material (also known as the source) undergoes exposure to an energy source such as heat, ion bombardment, or a laser, which results in atoms or molecules from the solid material transitioning into the gas phase. Some standard techniques for the vaporization of materials are thermal evaporation, sputtering, pulsed laser deposition, and electron beam evaporation.
(ii) **Transfer of Vaporized Materials:** The vaporized atoms and molecules travel from the target to the substrate through the vacuum and low-pressure region. The vacuum creates conditions where the vapor can move without interacting with other gas molecules, enabling it to move unrestricted from the source to the substrate.
(iii) **Condensation and Film Formation:** In the final process, the vaporized atoms and molecules of the target materials are condensed and accumulated to form a thin film. Additionally, the thickness and uniformity of the deposited thin films can be controlled by adjusting the deposition time, deposition rate, energy, and substrate.

Types of PVD Techniques

According to the approach used to obtain the evaporation of desired material, some of the prevalent PVD techniques can be classified as follows: (a) sputtering (Silva Filho et al. 1563; Calisi et al. 2023), (b) electron beam physical vapor deposition

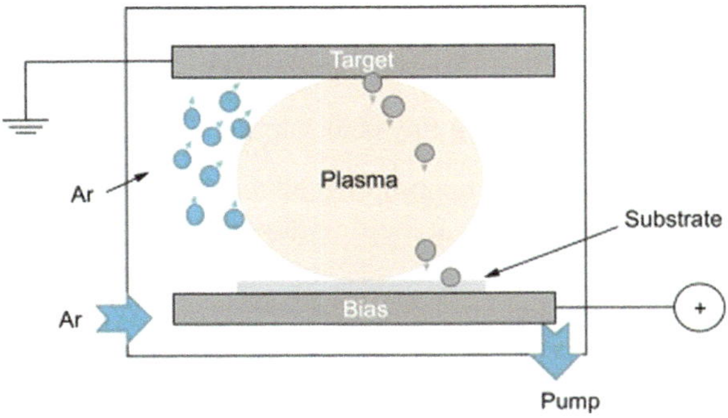

Fig. 16.3 Schematic diagram of standard sputtering deposition method (Zafar et al. 2019). Copyright 2019. Reproduced with permission from Elsevier

(EB-PVD) (Singh and Wolfe 2005), (c) thermal evaporation (Bae et al. 2022; Wang et al. 2018b).

(i) Sputtering

In the sputtering process, depicted schematically in Fig. 16.3, positive gas ions (commonly argon ions) generated in a glow discharge (gas pressure ranging from 10^{-6} to 10^{-10} torr) strike the target material (referred to as the cathode), causing groups of atoms to be dislodged, which then enter the vapor phase and deposit onto the substrate (Shohet 2016). Sputtering is an ineffective method for facilitating the transition of a solid into vapor. In the typical deposition method, a continuous flow of an inert gas like argon (Ar) is supplied to the vacuum chamber, where the gas molecules are transformed into positively charged ions that emit electrons when subjected to high pressure and voltage. The released electrons can further ionize additional gas atoms, maintaining the glow discharge and initiating a cascading effect until all gas molecules become ionized (Bunshah 1994). These ions collide with the perovskite material being deposited on the cathode, causing the ejection of molecules from its surface due to momentum transfer. This process is referred to as sputtering. For a 50-eV argon ion striking a metal surface, the typical yield (atoms ejected per incoming ion) is about one.

Consequently, the energy required for this phase change is 3–10 times greater than evaporations. The ion's speed and energy influence the interaction between inert gas ions and the target perovskite material. Electric and magnetic fields can manipulate the angle at which ions strike. The material molecules ejected from the target travel toward the substrate, with or without colliding with other ejected molecules and inert gas ions. This leads to the deposition of thin films on the substrate. Subsequent nucleation and film development occur on the substrate as the target atoms condense, altering the deposited layer's composition, microstructure, residual stress, and physical characteristics.

The properties of this deposited film can be affected by the bombardment of low-energy particles, energetic particles, and ions that are back-scattered from both the target and the inert gas (Kersten et al. 2001). This allows for control over the deposition properties. This method does not require thermal energy; therefore, it is also referred to as non-thermal vaporization deposition. An excellent and must-read review article reported by Thornton (Thornton 1973) explains the application of sputtering in deposition technology.

(ii) Electron Beam Physical Vapor Deposition (EB-PVD) Method

EB-PVD is an uncomplicated technique where a concentrated high-energy electron beam is aimed at melting the evaporant materials within a vacuum chamber. In this deposition process, an electron beam serves as the energy source, and the final film coating thickness can reach several hundred micrometers, substantially thicker than films produced by other PVD methods like sputtering. These substantial films result from the high deposition rates that EB-PVD can achieve compared to alternative PVD approaches. The equipment needed for this process is illustrated in Fig. 16.4. Electrons are induced by a high voltage toward the target, which is typically shaped as an ingot, due to the requirement of sufficient material to generate thicker film coating. A beam guidance system concentrates the beam onto the target, melting the material and creating a melt pool. When the energy is adequate, atoms and molecular fragments from the target can escape the melt pool and evaporate within the chamber, subsequently condensing on any cooler surfaces, including the substrate. Ingots are maintained in a water-cooled hearth and are fed through openings to manage the feed rate of the target material.

Multiple ingots can be positioned in the hearth for sequential or simultaneous evaporation to achieve intricate compositions, material mixtures, or layered film coatings. Nevertheless, generating complex compositions or evaporating unique new materials can pose challenges based on local vapor pressures, resulting in non-uniform coatings.

(iii) Thermal Evaporation Method

The thermal evaporation technique eliminates the need for solvents and facilitates the production of large-area devices with a smooth surface free of pinholes. Additionally, it allows for precise control over film thickness, which is challenging to achieve with the spin-coating method. It also prevents the formation of trap or defect sites that can occur due to solvents used in synthesizing precursors. In this method, a perovskite precursor is synthesized and then evaporated directly in a high-temperature tungsten boat, or perovskite powder is prepared separately in a tungsten boat and then evaporated. The formation of perovskite thin films is influenced by several factors, including the internal temperature of the evaporator, the source's pressure and evaporation rate, and the substrate temperature. For thermal evaporation, perovskite precursors can be utilized either as a single or multiple source. A single source consisting of a perovskite structure of ABX_3 can be prepared via methods such as single crystal growth, slow evaporation of solvent, temperature reduction techniques, or inverse temperature crystallization. In a multi-source thermal evaporator, two or more precursors can be loaded into a crucible and then deposited either sequentially or simultaneously to create a perovskite film. Thermal evaporation enables the growth of a uniform film, even on irregular or flexible substrates. A typical schematic of the thermal evaporation method is shown in Fig. 16.5.

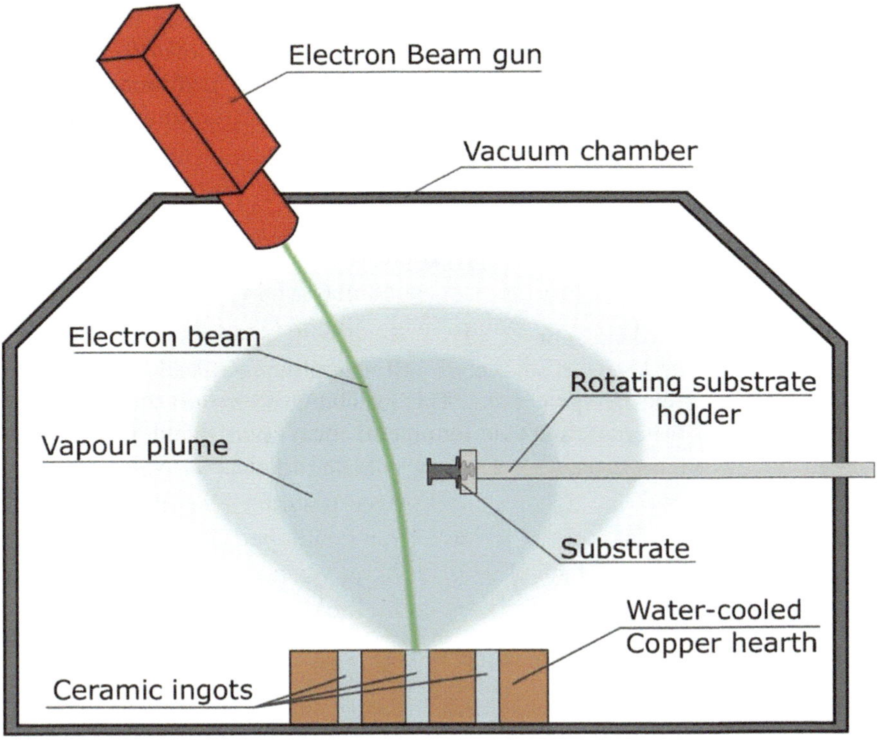

Fig. 16.4 Schematic representation of EB-PVD method (Chevallier et al. 2021). Copyright 2021. Reproduced with permission from Springer Nature.

Fig. 16.5 Schematic representation of the thermal evaporation method for creating thin films (Bashir et al. 2020). Copyright 2020. Reproduced with permission from Elsevier

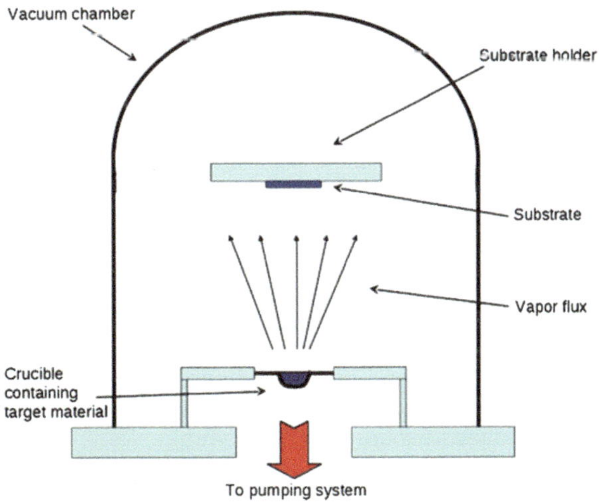

Moreover, because any compounds amenable to deposition can be applied, a range of perovskite films can be coated, making them appropriate for industrial use. However, controlling the source's evaporation rate and confirming the exact deposited amount poses challenges in the thermal evaporation method. By managing the evaporation rate, one can achieve a uniform film.

Chemical Vapor Deposition Method

The chemical vapor deposition (CVD) method is a prominent technique for depositing perovskite layers in solar cells, offering advantages such as scalability, uniformity, and solvent-free processing. This method involves the sequential deposition of precursor materials, allowing for precise control over the film's properties and structure, which is crucial for enhancing solar cell efficiency.

Mechanism of CVD for Perovskite Layers

(i) Precursor Selection and Preparation

Metal halides (like PbI_2 and SnI_2), organic halides (like methylammonium iodide or formamidinium iodide), or inorganic cations (like CsBr) are commonly used as precursors for perovskite production. Some of the selected perovskite precursors that satisfy the structural and stoichiometric requirements are $MAPbI_3$ and $CsSnBr_3$, and they may exist as liquids, solids, or vapors. They are continuously delivered in the vapor phase in CVD.

(ii) Precursor Vaporization

To evaporate the precursors, heating or other methods (such as sublimation or bubbling carrier gases through liquid precursors that float on the surface) are used. The carrier gases such as hydrogen (H_2), argon (Ar), or nitrogen (N_2) frequently transport the precursors in the vapor phase to the reaction chamber.

(iii) Transport to Reaction Chamber

Controlled flow rates are used to introduce the precursor vapors into the reaction chamber, guaranteeing an ideal mixing ratio. The chamber is often kept at a particular temperature and pressure to promote the reaction and deposition.

(iv) Surface Reaction and Nucleation

The perovskite deposition begins when the precursor vapors arrive at the heated substrate. The temperature of the substrate is crucial since it allows for:

(a) Decomposition of precursors into reactive intermediates.
(b) Variables diffusion of these intermediates onto the substrate surface.

(c) Nucleation and growth of the perovskite layer.

For example, during the deposition of $MAPbI_3$, PbI_2 reacts with CH_3NH_3I vapor to form $MAPbI_3$ crystals.

(v) Film Growth and Crystallization
(a) The continual surface reaction and precursor adsorption cause the film to grow.
(b) Variables like temperature, gas flow rates, and chamber pressure influence the morphology, homogeneity, and crystallinity of the film.
(c) By modifying these parameters, crystal size and orientation can be regulated, which is essential for improving the perovskite layer's optoelectronic properties.
(vi) Exhaust and Byproduct Removal
(a) The exhaust system removes reaction byproducts, such as unreacted precursors and gasses like HI or HBr.
(b) This provides clean deposition and avoids contamination.
(vii) Post-Deposition Treatments (if Needed)
(a) Post-deposition annealing can enhance the perovskite film's stability and crystallinity.
(b) Additional precursors or anti-solvent vapor treatments may be used to improve film quality.

Temperature and Environmental Control

The deposition temperature plays a crucial role; for instance, a temperature of 100 °C is optimal for achieving highly crystalline and uniform layers in two-dimensional perovskites (Magubane et al. 2023). Maintaining low pressure during the CVD process can improve the stability and phase control of the perovskite materials, which is essential for their performance in solar cells (Luo et al. 2017) (Fig. 16.6).

Advantages of CVD for Perovskite Layers

Using the CVD method for the fabrication of perovskite layers presents many advantages, which include enhancing their performance and stability. This method has precise control over the growth conditions, leading to high-quality, uniform films that are crucial parameters for photovoltaics and optoelectronics applications. The following points discuss the key benefits of using CVD for the deposition of perovskite layers.

(i) **Scalability of CVD Methods**
(a) ***Sequential CVD:*** It is a low-pressure method that ensures controlled growth of perovskite absorber layers to achieve uniformity and scalability. However, pressure conditions may vary depending on the specific material to be deposited. Under optimal conditions, compact layers with well-defined grains can be achieved, enhancing the overall stability of the material (Magubane et al. 2023).

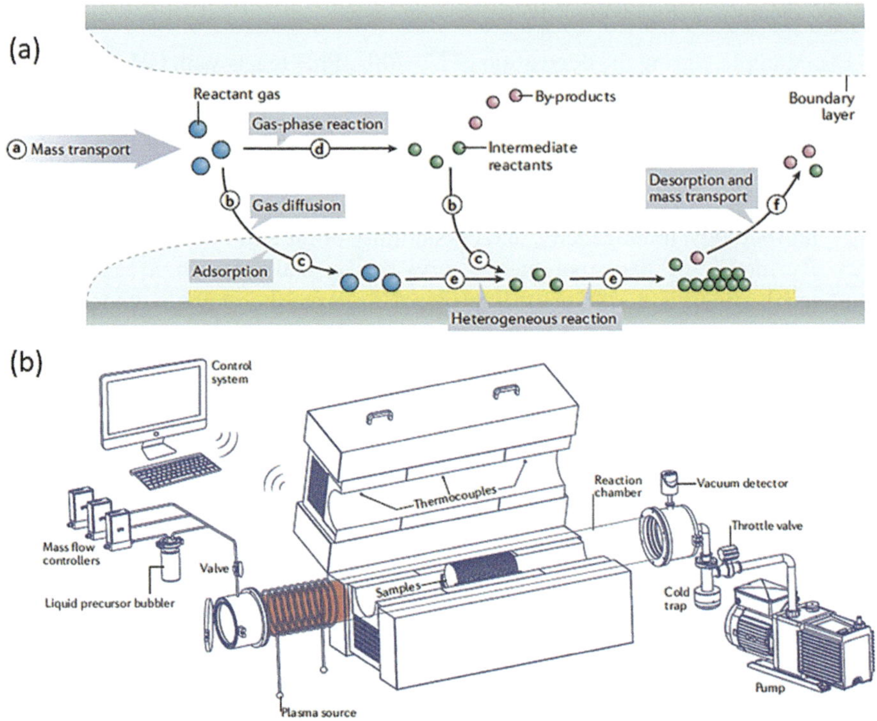

Fig. 16.6 **a** A schematic illustration outlining the basic elementary steps characteristic of a typical CVD method, and **b** schematic representation of a standard horizontal CVD setup featuring a gas feed unit, a quartz reaction chamber, a vacuum system, an energy input source, and an automated control system (Sun et al. 2021). Copyright 2021. Reproduced with permission from Springer Nature.

(b) ***Hybrid CVD:*** While it offers large-area uniformity, it faces challenges in power conversion efficiency due to nonradiative recombination losses and deep-level defects (Wu et al. 2023).

(c) ***Aerosol-Assisted (AA) CVD:*** It is a solution-based method that relies on precursor solubility instead of its volatility. In this method, the precursors are dissolved in a solvent to generate an aerosol using an ultrasonic humidifier. This novel approach reduces the complexity of equipment requirements and relies on environmentally friendly solvents, illustrating the potential for scaling up with high-quality film formation (Ratnasingham et al. 2021).

(ii) **Uniformity in Perovskite Layers**

(a) ***Crystallinity and Grain Size:*** Techniques like modified CVD enhance crystallinity and grain size, which are crucial for efficient charge transport (Qiu et al. 2021).

(b) ***Surface Quality:*** The uniformity of the perovskite layer is influenced by the underlying electron transport layers, which can improve the overall performance of solar modules (Liu et al. 2022).

(iii) **Environmental Stability**

The environmental stability of perovskite layers in the chemical vapor deposition (CVD) method is significantly enhanced through various strategies, including the use of protective interlayers and optimized deposition techniques. These advancements contribute to improved operational lifetimes and reduced degradation under environmental stressors.

(iv) CVD Techniques and Stability

(a) ***Ultra-Thin Polymer Interlayers:*** The application of a hydrophobic polymer, Parylene C, via CVD has been shown to enhance stability by reducing surface defects and acting as a barrier against moisture. This resulted in a degradation of only 11% after 500 h under ambient conditions (Zhou et al. 2023).

(b) ***Inorganic Layer Integration:*** Inorganic perovskite solar cells, such as $CsPbBr_3$, demonstrated remarkable stability when fabricated with an n-doped CdS interlayer. These cells maintained performance with minimal degradation after 600 h in the air (Sharikadze et al. 2022).

(v) Performance Metrics

(a) ***Efficiency Improvements:*** The integration of 3D/2D heterojunctions and optimized electron transport layers has led to power conversion efficiencies reaching up to 23.70% while maintaining stability (Tian et al. 2024).

(b) ***Long-Term Testing:*** Devices with CVD-fabricated layers exhibited stable resistive switching properties for extended periods, outperforming traditional spin-coated films (Yoo et al. 2019).

(vi) Enhanced Material Properties

CVD enables the growth of highly crystalline perovskite layers, as demonstrated by the formation of 2D Sn–Pb compound perovskites with well-defined grains up to 5 μm, which exhibit stable excitonic properties across a wide temperature range (Magubane et al. 2023).

While CVD offers significant advantages, challenges such as optimizing precursor materials and scaling up production processes to meet commercial demands remain. Addressing these issues will be crucial for the widespread adoption of CVD in perovskite technology.

Spray Coating Method

Spray coating is a sophisticated deposition technique that is widely known for its effectiveness and operation at lower temperatures. This method is often used to create perovskite absorbers for solar energy applications because of its capacity to produce high-quality thin films with remarkable uniformity. The method involves spraying finely atomized droplets of a precursor solution onto a pre-warmed substrate, which forms a wet film that is later annealed to obtain the desired crystalline structure. Pneumatic spraying is the most commonly used of the various spray methods due to its straightforwardness and efficient droplet generation via rapid gas flow. There are numerous technologies that can be classified as spray coating. One of the most common types is the ultrasonic spray nozzle. Ultrasonic spray nozzles are widely utilized for coating stents. The operational specifications for this use necessitate flow rates between 20 and 100 μL/min, spray diameters ranging from 0.5 to 2 mm, and tiny median drop sizes (Lamprou et al. 2018).

Key Steps of the Spray Coating Process

The spray coating process can generally be categorized into four key steps (Liu et al. 2016):

(i) **Atomization or Droplet Formation:** In the first phase, micron-sized droplets form at the nozzle. This process, referred to as atomization, influences the size and spread of the droplets, which are essential for consistent film creation. The nozzle and gas jet pressure selection are crucial for attaining ideal atomization.
(ii) **Droplet Transportation:** After atomization, the droplets are transported from the nozzle to the substrate via the gas flow (typically N_2). It is essential to precisely manage the distance between the nozzle and the substrate to avoid droplet dispersion or evaporation before their arrival on the surface.
(iii) **Droplets Coalescence:** When the droplets contact the substrate, they expand and merge to form a uniform wet layer. The characteristics of the substrate's surface, particularly its wettability, play a vital role in this process. The surface tension of the solvent and the contact angle of the substrate are key elements that affect the distribution of the solution over the surface.
(iv) **Drying, Annealing, and Film Formation:** Once coalescence occurs, the wet film is subjected to drying to eliminate any remaining solvents, followed by annealing. This heat treatment crystallizes the perovskite material, improving its structural and electronic characteristics.

Factors Affecting the Film Uniformity

The consistency and quality of the applied film rely on many important factors, such as:

(a) ***Design of Nozzle:*** Different types of nozzles affect the size and distribution of droplets, which directly impact the quality of the film.
(b) ***Gas Jet Pressure:*** The gas jet's pressure propels the droplets forward, influencing both their speed and path.
(c) ***Substrate-Nozzle Distance:*** Keeping an ideal distance guarantees that droplets reach the surface in the desired state, free from excessive evaporation or dispersion.

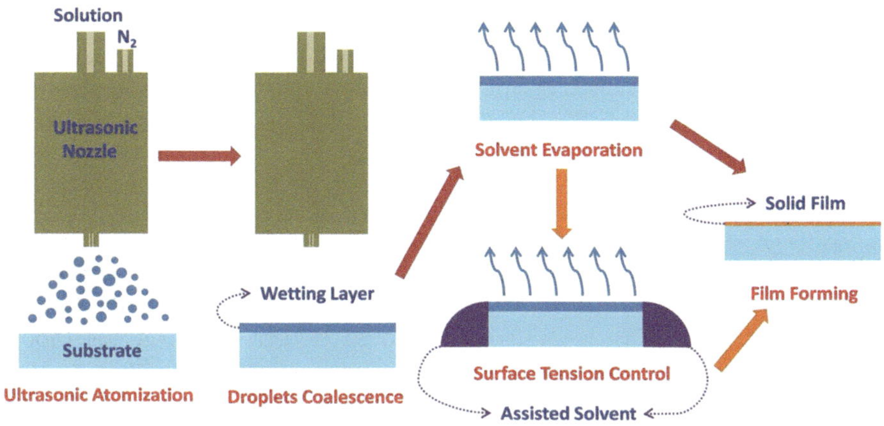

Fig. 16.7 A schematic illustration of the thin film deposition process through the spray coating method (Liu et al. 2016). Copyright 2016. Reproduced with permission from Springer Nature

(d) ***Substrate Temperature:*** Pre-heating the substrate prior to application improves the spreading of droplets and the evaporation of the solvent, resulting in enhanced film uniformity.

Finally, the film is created and ready for further testing. Further, a schematic representing the complete spray coating process is shown in Fig. 16.7, demonstrating all the steps involved.

Recent Developments

Over the recent span year (2016–2020), several research groups published papers on spray-cast devices fabricated using various methods, including scalable ultrasonic spray coating (Chou et al. 2018), mega sonic spray coating (Park et al. 2019), and electrospray deposition (Heo et al. 2019). However, none of these deposition methods outperformed the record efficiency reported in 2016 by (Heo et al. 2016). Later in the year 2019, another breakthrough was made by (Su et al. 2019), as they reported an excellent power conversion efficiency (PCE) of 18.5%. The range of deposition techniques presented reinforced the argument for adopting spray coating as a feasible deposition method for perovskite absorbers. This approach is beneficial for producing smooth, defect-free perovskite films with uniform thickness. The uniform droplet size attained through spray coating guarantees excellent film coverage and improved device performance.

Application of Spray Coating in Perovskite Absorbers

Spray coating has become increasingly popular in fabricating perovskite absorbers in perovskite solar cells (PSC) because of its scalability and compatibility with roll-to-roll manufacturing. This method facilitates the deposition of large-area films, rendering it suitable for large-scale industrial applications. Furthermore, the low-temperature process enables the employment of flexible substrates, opening opportunities for lightweight and bendable solar energy devices.

Electrodeposition Method

Electrodeposition is a highly adaptable and roll-to-roll-compatible method for fabricating perovskite solar cells (PSCs). Its efficiency, cost-effectiveness, and ability to deliver uniform films make it a promising approach for depositing perovskite layers (Chen et al. 2015). The primary attributes of this technique include:

(a) Applicability for diverse surface coatings.
(b) Low-temperature processing.
(c) High-speed deposition.
(d) Suitability for large-scale manufacturing.
(e) The cost-efficient operation.
(f) Precise control over film thickness.

Steps Involved in the Electrodeposition Method

(i) Electrolyte Preparation

The preparation of electrolytes typically includes metal halide precursors such as PbI_2 and $PbBr_2$, along with organic halides like methylammonium iodide and formamidinium iodide.

(ii) Film Deposition

A three-electrode system is required for the film deposition: a reference electrode, a working electrode (substrate onto which precursors are to be deposited, such as Si, FTO), and a counter electrode.

(iii) Post-Deposition Treatment

Following deposition, the films are subjected to annealing to transform the deposited materials into a high-quality perovskite phase, ensuring crystallinity and reducing defects.

Unlike the spin-coating technique, electrodeposition does not require substrate heating since heating may cause the development of uneven films. The application of heat can cause the film to deteriorate, leading to unpredictable island formations

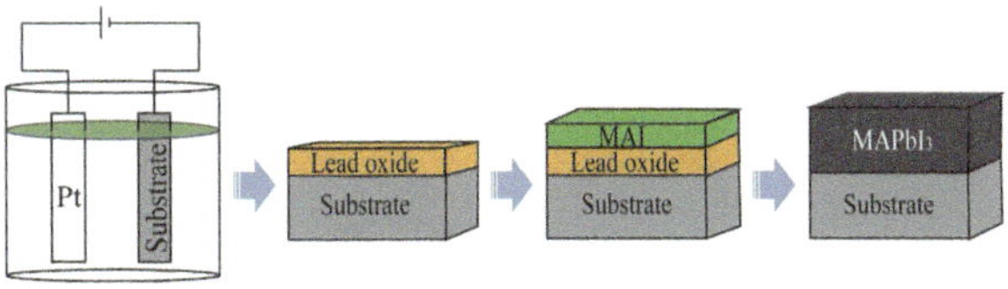

Fig. 16.8 A schematic illustration of the perovskite thin film deposition process from the electrodeposition method (Huang et al. 2015). Copyright 2015. Reproduced with permission from Springer Nature

on the surface of the substrate (Huang et al. 2015). A schematic illustration of PSC fabrication using the electrodeposition method is shown in Fig. 16.8.

This technique allows the fabrication of exceptionally uniform thin films that cover large areas and mitigate the effects of shear forces (Su et al. 2015; Koza et al. 2016). This method is especially beneficial for the deposition of perovskite layers onto substrates that have sophisticated shapes, which can be challenging to accomplish with the previously mentioned techniques. This adaptability positions electrodeposition as a highly effective option for large-scale production applications (Huang et al. 2015).

Challenges in Deposition Methods

Fabrication of high-quality perovskite absorber films comes with several crucial challenges that limit the overall performance of the perovskite solar cells. Table 16.1 summarizes the advantages and challenges associated with different deposition methods for perovskite absorbers:

Conclusion and Outlook

The fabrication methods of the absorber layer are crucial factors that influence the final effectiveness of a perovskite solar cell. Perovskite high-quality films that are free of defects result in a greater diffusion length for both electrons and holes, leading to an increased filling factor and reduced recombination (Khorasani et al. 2024). We have introduced several deposition methods to create high-quality perovskite absorber films by a spin coater, solution-based deposition (one-step and two-step), pulsed laser deposition, physical vapor deposition, chemical vapor deposition, spray coating, and electrodeposition methods. Each discussed deposition technique has its benefits and difficulties, and the selection varies based on several factors like stability, scalability, and cost. For instance, the PVD and CVD methods are more versatile and can create a high-quality and denser perovskite film compared to other deposition methods. PLD and spray coating methods are most suitable for stoichiometry and thickness-controlled deposition. Researchers and manufacturers are actively investigating and

Table 16.1 Summary of advantages and challenges with different deposition techniques

Deposition methods	Advantages	Challenges
Spin coating	Simple, affordable, used to make uniform films on small areas	Limited to small scale, wastage of material during deposition
Pulsed laser deposition	Control over stoichiometry and can be integrated with intricate device architectures	Expensive instrument, maintaining stoichiometric balance
Solution-based deposition	Versatile, partially scalable	Non-uniform films over large areas
Physical vapor deposition	Superior quality dense and uniform films, control over thickness	Expensive, partially scalable
Chemical vapor deposition	Uniform film, scalable, highly pure films	Slower deposition rate, temperature limitation
Spray coating	Fast and cheap, control over the thickness of the film	Wastage of material during deposition, meticulous optimization
Electrodeposition	Cost-effective, environment-friendly, scalable	Slow deposition rates, crystallinity issues, limited material compatibility

refining these techniques to enhance the efficiency and dependability of large-scale perovskite solar cell manufacturing.

References

Angmo D et al (2021) A lab-to-fab study toward roll-to-roll fabrication of reproducible perovskite solar cells under ambient room conditions. Cell Rep Phys Sci 2:1

Ava TT, Al Mamun A, Marsillac S, Namkoong G (2019) A review: thermal stability of methylammonium lead halide based perovskite solar cells. Appl Sci 9(1):188

Bae S-R, Heo DY, Kim SY (2022) Recent progress of perovskite devices fabricated using thermal evaporation method: perspective and outlook. Mater Today Adv 14:100232

Bashir A, Awan TI, Tehseen A, Tahir MB, Ijaz M (2020) Interfaces and surfaces. Chem Nanomater 51:87

Bing J, Huang S, Ho-Baillie AWY (2020) A review on halide perovskite film formation by sequential solution processing for solar cell applications. Energy Technol 8(4):1901114

Boix PP, Nonomura K, Mathews N, Mhaisalkar SG (2014) Current progress and future perspectives for organic/inorganic perovskite solar cells. Mater Today 17(1):16–23

Brenner TM, Egger DA, Kronik L, Hodes G, Cahen D (2016) Hybrid organic—inorganic perovskites: low-cost semiconductors with intriguing charge-transport properties. Nat Rev Mater 1(1):1–16

Bryant D et al (2016) Light and oxygen induced degradation limits the operational stability of methylammonium lead triiodide perovskite solar cells. Energy Environ Sci 9(5):1655–1660. https://doi.org/10.1039/c6ee00409a

Bunshah RF (1994) Handbook of deposition technologies for films and coatings: science, applications and technology. William Andrew
Burschka J et al (2013) Sequential deposition as a route to high-performance perovskite-sensitized solar cells. Nature 499(7458):316–319
Calisi N, Galvanetto E, Borgioli F, Martinuzzi SM, Bacci T, Caporali S (2022) Thin films deposition of fully inorganic metal halide perovskites: a review. Mater Sci Semicond Process 147:106721
Calisi N et al (2023) $CsPbBr_3$ perovskite thin films by magnetron sputtering: the role of the substrate on texture and morphology. Thin Solid Films 783:140057
Chen H, Wei Z, Zheng X, Yang S (2015) A scalable electrodeposition route to the low-cost, versatile and controllable fabrication of perovskite solar cells. Nano Energy 15:216–226
Chen B, Yang M, Priya S, Zhu K (2016) Origin of J–V hysteresis in perovskite solar cells. J Phys Chem Lett 7(5):905–917
Chevallier J, Isern L, Almandoz Forcen K, Chalk C, Nicholls JR (2021) Modelling evaporation in electron-beam physical vapour deposition of thermal barrier coatings. Emergent Mater 4(6):1499–1513
Chou L-H, Wang X-F, Osaka I, Wu C-G, Liu C-L (2018) Scalable ultrasonic spray-processing technique for manufacturing large-area $CH_3NH_3PbI_3$ perovskite solar cells. ACS Appl Mater Interfaces 10(44):38042–38050
da Silva Filho JMC, Ermakov VA, Marques FC (2018) Perovskite thin film synthesised from sputtered lead sulphide. Sci Rep 8(1):1563
Fu Q et al (2018) Recent progress on the long-term stability of perovskite solar cells. Adv Sci 5(5):1700387
Haider AJ, Alawsi T, Haider MJ, Taha BA, Marhoon HA (2022) A comprehensive review on pulsed laser deposition technique to effective nanostructure production: trends and challenges. Opt Quantum Electron 54(8):488
Hamakawa Y (1994) Recent advances in solar photovoltaic technology and its new role for environmental issue. Renew Energy 5(1–4):34–43
Han Y, Xie H, Lim EL, Bi D (2022) Review of two-step method for lead halide perovskite solar cells. Sol. Rrl 6(6):2101007
Haruyama J, Sodeyama K, Han L, Tateyama Y (2016) Surface properties of $CH_3NH_3PbI_3$ for perovskite solar cells. Acc Chem Res 49(3):554–561
Heo JH, Lee MH, Jang MH, Im SH (2016) Highly efficient $CH_3NH_3PbI_{3-x}Cl_x$ mixed halide perovskite solar cells prepared by re dissolution and crystal grain growth via spray coating. J Mater Chem A 4(45):17636–17642
Heo JH, Im K, Kim J, Im SH (2019) Efficient metal halide perovskite solar cells prepared by reproducible electrospray coating on vertically aligned TiO_2 nanorod electrodes. ACS Appl Mater Interfaces 12(1):886–892
Hong Z et al (2019) Completely solvent-free protocols to access phase-pure, metastable metal halide perovskites and functional photodetectors from the precursor salts. Iscience 16:312–325
Hu Z, Wang Z, Gao P (2024) Advancements in scaling up perovskite solar cells: from small-area devices to large-scale modules. ChemPhysChem 1:202400587
Huang J et al (2015) Direct conversion of $CH_3NH_3PbI_3$ from electrodeposited PbO for highly efficient planar perovskite solar cells. Sci Rep 5(1):15889
Im J-H, Kim H-S, Park N-G (2014) Morphology-photovoltaic property correlation in perovskite solar cells: one-step versus two-step deposition of $CH_3NH_3PbI_3$. Apl Mater 2(8):81510
Jamal MS et al (2018) Fabrication techniques and morphological analysis of perovskite absorber layer for high-efficiency perovskite solar cell: a review. Renew Sustain Energy Rev 98:469–488
Jung HS, Park N (2015) Perovskite solar cells: from materials to devices. Small 11(1):10–25
Kannan N, Vakeesan D (2016) Solar energy for future world: a review. Renew Sustain Energy Rev 62:1092–1105. https://doi.org/10.1016/j.rser.2016.05.022
Karavioti A, Chalkias DA, Katsagounos G, Mourtzikou A, Kalarakis AN, Stathatos E (2021) Toward a scalable fabrication of perovskite solar cells under fully ambient air atmosphere: from spin-coating to inkjet-printing of perovskite absorbent layer. Electronics 10(16):1904

Kersten H, Deutsch H, Steffen H, Kroesen GMW, Hippler R (2001) The energy balance at substrate surfaces during plasma processing. Vacuum 63(3):385–431

Khorasani A, Mohamadkhani F, Marandi M, Luo H, Abdi-Jalebi M (2024) Opportunities, challenges, and strategies for scalable deposition of metal halide perovskite solar cells and modules. Adv Energy Sustain Res 1:2300275

Kojima A, Teshima K, Shirai Y, Miyasaka T (2009) Organometal halide perovskites as visible-light sensitizers for photovoltaic cells. J Am Chem Soc 131(17):6050–6051. https://doi.org/10.1021/ja809598r

Koza JA, Hill JC, Demster AC, Switzer JA (2016) Epitaxial electrodeposition of methylammonium lead iodide perovskites. Chem Mater 28(1):399–405

Kumar NS, Naidu KCB (2021) A review on perovskite solar cells (PSCs), materials and applications. J Mater

Lal NN, Dkhissi Y, Li W, Hou Q, Cheng Y, Bach U (2017) Perovskite tandem solar cells. Adv Energy Mater 7(18):1602761

Lamprou D, Scoutaris N, Ross SA, Douroumis D (2018) Polymeric coatings and their fabrication for medical devices. In: Encyclopedia of biomedical engineering. Elsevier

Leupold N, Panzer F (2021) Recent advances and perspectives on powder-based halide perovskite film processing. Adv Funct Mater 31(14):2007350

Li Z et al (2018) Scalable fabrication of perovskite solar cells. Nat Rev Mater 3(4):1–20

Li F, Lin FR, Jen AK (2024) Current state and future perspectives of printable organic and perovskite solar cells. Adv Mater 36(17):2307161

Liu S, Zhang X, Zhang L, Xie W (2016) Ultrasonic spray coating polymer and small molecular organic film for organic light-emitting devices. Sci Rep 6(1):1–10

Liu X et al (2022) Area-scalable Zn_2SnO_4 electron transport layer for highly efficient and stable perovskite solar modules. ACS Appl Mater Interfaces 14(20):23297–23306

Luo P, Zhou S, Xia W, Cheng J, Xu C, Lu Y (2017) Chemical vapor deposition of perovskites for photovoltaic application. Adv Mater Interfaces 4(8):1600970

Magubane SS, Burns R, Ngqoloda S, Oliphant CJ, Miceli PF, Arendse CJ (2023) Sequential chemical vapor deposition of two-dimensional Sn–Pb compound perovskite thin films and its exciton transport. ACS Appl Electron Mater 5(10):5352–5361

Matteocci F et al (2014) Solid-state solar modules based on mesoscopic organometal halide perovskite: a route towards the up-scaling process. Phys Chem Chem Phys 16(9):3918–3923

Mbumba MT, Malouangou DM, Tsiba JM, Bai L, Yang Y, Guli M (2021) Degradation mechanism and addressing techniques of thermal instability in halide perovskite solar cells. Sol Energy 230:954–978

Ortega-San-Martin L (2020) Introduction to perovskites: a historical perspective. Revolut Perovskite Synth Prop Appl, 1–41

Park N (2015) Perovskite solar cells : an emerging photovoltaic technology. Biochem Pharmacol 18(2):65–72. https://doi.org/10.1016/j.mattod.2014.07.007

Park M et al (2019) Highly reproducible large-area perovskite solar cell fabrication via continuous megasonic spray coating of $CH_3NH_3PbI_3$. Small 15(1):1804005

Qiu L, He S, Ono LK, Qi Y (2020) Progress of surface science studies on ABX_3-based metal halide perovskite solar cells. Adv Energy Mater 10(13):1902726

Qiu L, He S, Jiang Y, Qi Y (2021) Metal halide perovskite solar cells by modified chemical vapor deposition. J Mater Chem A 9(40):22759–22780

Ranjan AK, Jha PK, Jha PA, Singh P (2024) "Catalyzing hydrogen production: exploring plasmonic effects in self-assembled CuO/Cu_2O thin films via pulsed laser deposition. J Appl Phys 135:18

Ratnasingham SR et al (2021) Novel scalable aerosol-assisted CVD route for perovskite solar cells. Mater Adv 2(5):1606–1612

Ren M, Qian X, Chen Y, Wang T, Zhao Y (2022) Potential lead toxicity and leakage issues on lead halide perovskite photovoltaics. J Hazard Mater 426:127848

Roy P, Sinha NK, Tiwari S, Khare A (2020) A review on perovskite solar cells: evolution of architecture, fabrication techniques, commercialization issues and status. Sol Energy 198:665–688

Schulze PSC et al (2020) Perovskite hybrid evaporation/spin coating method: from band gap tuning to thin film deposition on textures. Thin Solid Films 704:137970

Seo J et al (2014) Benefits of very thin PCBM and LiF layers for solution-processed p–i–n perovskite solar cells. Energy Environ Sci 7(8):2642–2646

Sharikadze S, Zhu J, Kottokkaran R, Akopian A, Dalal V (2022) Inorganic perovskite solar cells with very high voltage and excellent stability against thermal and environmental degradation. In: 2022 IEEE 49th photovoltaics specialists conference (PVSC). IEEE, pp 244–246

Shi Z, Jayatissa AH (2018) Perovskites-based solar cells: a review of recent progress, materials and processing methods. Mater Basel 11(5):729

Shohet JL (2016) Encyclopedia of plasma technology-two volume set. CRC press

Singh J, Wolfe DE (2005) Review Nano and macro-structured component fabrication by electron beam-physical vapor deposition (EB-PVD). J Mater Sci 40:1–26

Stuckelberger M et al (2016) Elemental distribution and charge collection at the nanoscale on perovskite solar cells. In: 2016 IEEE 43rd Photovoltaic specialists conference (PVSC). IEEE, pp 1191–1196

Su T-S, Hsieh T-Y, Hong C-Y, Wei T-C (2015) Electrodeposited ultrathin TiO_2 blocking layers for efficient perovskite solar cells. Sci Rep 5(1):16098

Su J et al (2019) Perovskite ink with an ultrawide processing window for efficient and scalable perovskite solar cells in ambient air. ACS Appl Mater Interfaces 12(3):3531–3538

Sun L et al (2021) Chemical vapour deposition. Nat Rev Methods Prim 1(1):5

Teixeira CO, Castro D, Andrade L, Mendes A (2022) Selection of the ultimate perovskite solar cell materials and fabrication processes towards its industrialization: a review. Energy Sci Eng 10(4):1478–1525

Thornton JA (1973) Sputter coating—its principles and potential. SAE Trans, pp 1787–1805

Tian S et al (2024) Enhancing the efficiency and stability of perovskite solar cells using chemical bath deposition of SnO_2 electron transport layers and 3D/2D heterojunctions. Small 20(47):2406929

Tsai H et al (2017) Effect of precursor solution aging on the crystallinity and photovoltaic performance of perovskite solar cells. Adv Energy Mater 7(11):1602159

Vidal R et al (2021) Assessing health and environmental impacts of solvents for producing perovskite solar cells. Nat Sustain 4(3):277–285

Wang W, Tadé MO, Shao Z (2015) Research progress of perovskite materials in photocatalysis- and photovoltaics-related energy conversion and environmental treatment. Chem Soc Rev 44(15):5371–5408

Wang M, Feng Y, Bian J, Liu H, Shi Y (2018a) A comparative study of one-step and two-step approaches for $MAPbI_3$ perovskite layer and its influence on the performance of mesoscopic perovskite solar cell. Chem Phys Lett 692:44–49. https://doi.org/10.1016/j.cplett.2017.12.012

Wang S, Li X, Wu J, Wen W, Qi Y (2018b) Fabrication of efficient metal halide perovskite solar cells by vacuum thermal evaporation: a progress review. Curr Opin Electrochem 11:130–140

Wu X et al (2023) Insight into the defect chemistry and ion migration in perovskite fabricated by hybrid chemical vapor deposition. ACS Appl Energy Mater. 6(24):12198–12205

Yang Y et al (2017) Enhanced crystalline phase purity of $CH_3NH_3PbI_{3-x}Cl_x$ film for high-efficiency hysteresis-free perovskite solar cells. ACS Appl Mater Interfaces 9(27):23141–23151

Yoo E et al (2019) Strategic extended air stability of organolead halide perovskite nonvolatile memory devices. J Alloys Compd 811:151999

Zafar MS, Farooq I, Awais M, Najeeb S, Khurshid Z, Zohaib S (2019) Bioactive surface coatings for enhancing osseointegration of dental implants. In: Biomedical, therapeutic and clinical applications of bioactive glasses, Elsevier, pp 313–329

Zhang K et al (2020) A prenucleation strategy for ambient fabrication of perovskite solar cells with high device performance uniformity. Nat Commun 11(1):1006

Zhou H et al (2014) Interface engineering of highly efficient perovskite solar cells. Science 345(6196):542–546

Zhou J et al (2023) An ultra-thin chemical vapor deposited polymer interlayer to achieve highly improved stability of perovskite solar cell. Chem Eng J 461:141914

Chapter 17
Perovskite Solar Cells: Summary and Future Directions

Abhishek Raj, Manish Kumar, and Avneesh Anshul

Abstract Perovskite solar cells (PSCs) have become a progressive photovoltaic technology because of their remarkable optoelectronic properties and superior power conversion efficiency (PCE). PSCs have made remarkable strides in scalability and PCE. However, this photovoltaic technology still faces several significant obstacles that need to be overcome before they can be widely used commercial applications. Enhancing compatibility with current industrial photovoltaic standards, addressing lead toxicity, creating scalable and economical manufacturing processes, and boosting long-term operational stability are some of these difficulties. This final chapter evaluate the development of PSC research, highlighting significant breakthroughs and pointing out outstanding issues. It offers a comprehensive viewpoint on the prospects and avenues for upcoming innovations, including investigating environmentally friendly and lead-free substitutes, incorporating perovskites with tandem architectures, and developing recycling and life-cycle assessment techniques. To support global renewable energy goals, this chapter aims to motivate researchers and stakeholders to close the gap between scientific discovery and the sustainable commercial deployment of PSCs by imagining these next steps.

Keywords Perovskite absorber · 2D materials · Multiferroic perovskite · Tandem perovskite · Stable perovskite

A. Raj · A. Anshul (✉)
CSIR-National Environmental Engineering Research Institute (CSIR-NEERI), Nagpur, India
e-mail: anshul@neeri.res.in

Academy of Scientific and Innovative Research (AcSIR), Ghaziabad, India

M. Kumar (✉)
Experimental Research Laboratory, Department of Physics, ARSD College, University of Delhi, New Delhi, India
e-mail: mkumar2@arsd.du.ac.in

M. Kumar and A. Raj (eds.), *Perovskite Solar Cells*, Springer Tracts in Electrical and Electronics Engineering, https://doi.org/10.1007/978-981-96-9463-1_17

Introduction

The advancement of perovskite based photovoltaic technologies opens the significant research rea in this field. Since their inception in 2009, PSCs have seen remarkable advancements, the device efficiency enhance from 3.8% to over 26%, positioning them as strong competitors to traditional silicon-based solar cells. This significant progress can be attributed to the distinctive characteristics of perovskite materials, such as their high absorption coefficients, adjustable bandgaps, extended carrier diffusion lengths, and suitability for solution processing (Olaleru et al. 2020; Park 2015; Stranks and Snaith 2015; Raj et al. 2021; Kumar et al. 2022).

The PSCs is a heterostructure device which consist of perovskite absorber, electron transport layer (ETL), hole transport layer (HTL) and electrode materials. In the perovskite photovoltaic technologies researchers are trying to develop a PSC without ETL or HTL. Studies are also going on to replace the highly expansive organic HTL to inorganic HTL materials (Dianyi et al. 2014; Liao et al. 2019; Zhu et al. 2016). The electrode material is another crucial factor which are responsible for overall cost of PSCs. The increasing interest in wearable and self-powered flexible electronics has led to a rising demand for flexible. The perovskite film is particularly beneficial for the development of flexible devices, as it can be processed at low temperatures and possesses inherent flexibility. Additionally, flexible PSCs are cost-effective due to their compatibility with roll-to-roll manufacturing processes, which not only reduce fabrication time but also conserve materials and facilitate large-scale production. Furthermore, the lightweight nature and ease of packaging of flexible PSCs contribute to lower transportation and installation costs. To design high-performance flexible PSCs (FPSCs), it is essential for the flexible substrate to possess significant flexibility to withstand repeated bending operations, extensive optical transparency to maximize light absorption, excellent thermal stability to endure the annealing process during the deposition of subsequent films, and effective hermeticity to prevent moisture intrusion. Commonly utilized flexible substrates in FPSCs include polymers such as polyethylene terephthalate (PET) and polyethylene naphthalate (PEN) (Xu et al. 2022; Aftab et al. 2024; Dong et al. 2024; Raj et al. 2023). Further, the development of low-band gap materials become popular as absorber material in PSCs (Liao et al. 2016; Hossain et al. 2021).

There has been a growing number of research regarding the integration of PSCs with another type of superior materials, specifically two-dimensional (2D) materials such as graphene, transition metal dichalcogenides (TMDs), and black phosphorus (BP). The remarkable physical and chemical characteristics of both perovskites and 2D materials have led to significant advancements in the development of PSCs through the incorporation of these 2D materials (You et al. 2019; Raj et al. 2024; Shi et al. 2013).

The significant advancements in the performance of silicon cells are primarily restricted by auger recombination and parasitic absorption. A promising strategy to significantly improve the PCE involves the integration of wide-bandgap metal halide perovskites on top of silicon heterojunction bottom cells in a tandem arrangement, effectively reducing carrier thermal losses. The perovskite/silicon tandem configuration has attracted considerable interest, as evidenced by recent demonstrations from various research groups achieving independently verified PCE surpassing 31%. The tandem solar cells primarily utilize an inverted *p–i–n* configuration, which influences the benefits of reduced parasitic absorption and a film structure featuring a gradient refractive index on the sun-facing front contacts. Nevertheless, *p–i–n*-type perovskite devices continue to experience significant interface recombination at the perovskite/ C_{60} interface during electron extraction (Li and Zhang 2020; He et al. 2023; Cheng and Ding 2021).

Nevertheless, despite these achievements, PSCs encounter several substantial obstacles that impede their widespread commercialization. The durability of perovskite materials in real-world environments, particularly when subjected to moisture, heat, and UV radiation, remains a major issue. Furthermore, the presence of lead-based compounds in many high-efficiency perovskites raises environmental and health concerns, highlighting the need for lead-free alternatives. Scalability also presents a significant challenge, as the transition from laboratory-scale devices to large-area modules necessitates overcoming issues related to uniformity, reproducibility, and production costs.

This chapter provide the progress made in PSC technology while pinpointing key challenges and opportunities for enhancement. It explores the potential of PSCs to transform renewable energy through innovative strategies such as tandem architectures, lead-free perovskites, and recycling initiatives. Additionally, the chapter underscores the necessity of interdisciplinary research, collaboration with industry, and supportive policies to effectively tackle these challenges. Perovskite solar cells have the potential to completely change the photovoltaic technology landscape as they develop.

Challenges Associated with PSCs

Although PSCs have achieved significant progress in terms of efficiency and manufacturing methods, there remain several technological obstacles that hinder their broad acceptance in the commercial solar energy market. These issues encompass a range of factors related to material science, device design, scalability, and ecological effects. The following is a comprehensive examination of these challenges:

The selection of appropriate perovskite absorber materials with suitable device configuration is a difficult task as the deviation from appropriate ETL, HTL with suitable perovskite absorber leads to enhance the charge carrier recombination and reduction in the device efficiency. The generation of defects, deviation from appropriate perovskite structure, bandgap alignment and interfacial defects are the real challenge associated with technology. Further, the market feasibility of PSCs relies on the development of electrode materials that are affordable, eco-friendly, and effective. The noble such as gold and silver, metals which as generally used in PSCs, are costly and hinder widespread production. The interface between electrodes and perovskite is liable to deterioration caused by chemical interactions, including the migration of halide ions, which accelerates material degradation and shortens the lifespan of the device. The front electrode material such as indium tin oxide (ITO) or fluorine-doped tin oxide (FTO) are crucial for capturing light but face challenges in achieving ideal conductivity while preserving transparency.

Transferring PSCs from laboratory-scale devices to larger modules necessitates deposition techniques for the perovskite layer that are reliable and free of defects. Current techniques, such as spin-coating, are inadequate for large-scale applications. Ensuring consistent device performance and efficiency across large production batches proves difficult due to variations in processing conditions. Moving to industrial-scale production requires the integration of low-cost, high-throughput methods like roll-to-roll processing, which is still evolving for PSCs. PSCs are extremely vulnerable to degradation when exposed to moisture, oxygen, UV radiation, and fluctuations in temperature. This instability restricts their performance in real-world conditions. The major problem associated with ion migration within the perovskite layer and phase segregation result in a decline in efficiency over time, impacting device reliability. In addition, PSCs frequently lack thermal resilience at elevated temperatures, which is essential for outdoor applications where devices can become significantly heated. Adopting effective and affordable encapsulation solutions that guarantee long-term stability while retaining transparency and conductivity presents a considerable obstacle. Figure 17.1 shows the major challenges related to PSCs are shown in.

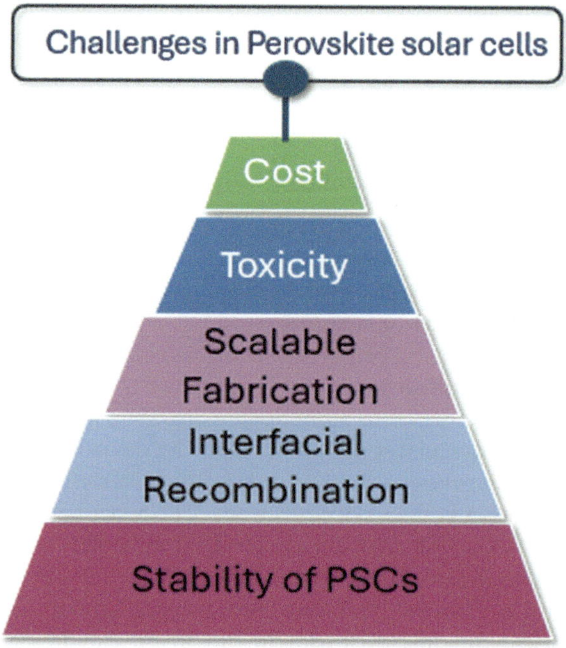

Fig. 17.1 Major challenges associated with PSCs

Recently, multiferroic based PSCs are also getting attention. However, producing high-quality, reproducible multiferroic perovskites that integrate ferroelectric, magnetic, and photovoltaic properties is a complex task. The structural instability of multiferroic materials is another challenge which negatively impacting device performance. The challenge of incorporating multiferroic perovskites into established solar cell designs represents a significant technological hurdle.

Perovskites absorber materials with narrow band gaps are susceptible to phase segregation when illuminated, resulting in a gradual loss of efficiency. The low band gap materials typically leads to increased defect densities, which subsidize to increase the non-radiative recombination and diminished device performance. The use of heavy metals like lead in low band gap perovskites raises critical environmental and health issues. The hold on precise control over the composition and processing of low band gap perovskites is a formidable task, especially for scalable production.

Addition of 2D material in PSCs is another technique to reduce the recombination losses and enhance the device efficiency. However, the effective charge movement across the interface of 2D perovskites is essential yet difficult to attain without considerable losses. Although 2D materials enhance the durability of perovskite layers, they frequently hinder the efficiency of charge extraction and the overall functionality of the device. Creating high-quality 2D materials and incorporating them intoPSCs at a commercial level continues to be expensive and technically challenging. Furthermore, the incorporation of 2D materials raises the complexity of fabrication and production expenses, restricting their use in markets where cost is a critical factor.

The tandem configuration solar cell has been surpassed the threshold presented by the Shockley–Queisser limit. However, the tandems device still struggling to hit the commercial market due to various issues. The optimal band gap alignment among the sub-cells is vital for maximizing efficiency, yet it remains a challenge to current time. The variations in thermal expansion coefficients of materials within tandem structures can cause mechanical stress and diminished durability. To produce tandem cells with minimal defects and high reproducibility necessitates advanced and costly manufacturing methods. The more intricate nature of tandem configurations renders them more vulnerable to rapid degradation in real-world operating environments.

The solar driven hydrogen generation are quite popular nowadays. But these technologies are facing many challenges. Perovskite materials tend to degrade swiftly in wet environments or under oxidative conditions, limiting their application in solar water splitting. The highly reactive surfaces of perovskites are susceptible to photo corrosion, which diminishes their photoelectrochemical efficiency. The band edges of perovskites must align precisely with the redox potentials involved in water splitting reactions, which is difficult to achieve without sacrificing stability.

The challenges outlined underline the technological obstacles that hinder the commercial success of perovskite photovoltaics. To overcome these issues, a collaborative approach involving materials science, engineering, and environmental research will be essential to fully comprehend the potential of this innovative technology.

Future Outlook

The advancement of PSCs is depending upon addressing significant challenges and practicing innovative strategies to realize their full potential across various applications. Research on ETL and HTL-free PSCs is concentrating on modernizing device architecture by removing electron and hole transport layers, which is anticipated to lower production costs and improve scalability. Progress in interfacial engineering and the development of self-passivating perovskite materials is expected to facilitate this innovation while ensuring high efficiency and stability. In the area of electrode materials, the focus is on creating affordable and sustainable substitutes for precious metals, including carbon-based electrodes and conductive polymers, as well as robust transparent conductors suitable for flexible and semi-transparent devices. Additionally, multiferroic perovskite materials are set to pioneer new possibilities by merging photovoltaic, ferroelectric, and magnetic characteristics into multifunctional devices, bolstered by advancements in material design and stability improvements. The incorporation of two-dimensional materials such as graphene, MoS_2, and 2D perovskites into PSCs is anticipated to enhance charge transport, stability, and flexibility of the devices, thereby facilitating their use in wearable and flexible solar panel applications. In the dominion of tandem PSCs, which include both perovskite-perovskite and perovskite-silicon configurations, research efforts are focused on achieving efficiencies exceeding 30%, while simultaneously addressing issues such as thermal mismatch and interlayer stability. These tandem systems are expected to be pivotal in realizing high energy outputs in large-scale photovoltaic applications. Furthermore, the exploration of solar-driven perovskite technologies for water splitting and CO_2 reduction is gaining traction. Safeguarding the stability of perovskites in aqueous and reactive conditions, along with the integration of effective catalysts, will facilitate their application in photoelectrochemical cells aimed at hydrogen production and solar fuels. Consequently, the future of PSCs is not solely centered on improving their efficiency and stability but also on broadening their range of applications, positioning them as a fundamental element of sustainable energy solutions.

Acknowledgements Author Abhishek Raj acknowledges the Council of Scientific and Industrial Research (CSIR), CSIR-HRDG, New Delhi, India, and its constituent laboratory CSIR-National Environmental Engineering Research Institute (NEERI), Nagpur, India, for supporting the research.

Authors Contributions **Abhishek Raj involved in** writing of original draft and formal analysis; **Manish Kumar participated in** writing of review and editing and formal analysis; **Avneesh Anshul involved in** conceptualization and writing of review and editing.

Declarations of Competing Interest The authors declare that they have no known competing financial interests or personal relationships that could have appeared to influence the work reported in this review article.

References

Aftab S, Hussain S, Kabir F, Aslam M, Rajpar AH, Al-Sehemi AG (2024) Advances in flexible perovskite solar cells: a comprehensive review. Nano Energy 120:109112. https://doi.org/10.1016/j.nanoen.2023.109112

Cheng Y, Ding L (2021) Perovskite/Si tandem solar cells: fundamentals, advances, challenges, and novel applications. SusMat 1:324–344. https://doi.org/10.1002/sus2.25

Dianyi L, Jinli Y, Timothy LK (2014) Compact layer free perovskite solar cells with 13.5% efficiency. J Am Chem Soc. https://doi.org/10.1021/ja508758k

Dong L, Qiu S, García Cerrillo J, Wagner M, Kasian O, Feroze S et al (2024) Fully printed flexible perovskite solar modules with improved energy alignment by tin oxide surface modification. Energy Environ Sci 17:7097–7106. https://doi.org/10.1039/D4EE01647E

He R, Wang W, Yi Z, Lang F, Chen C, Luo J et al (2023) Improving interface quality for 1 cm^2 all-perovskite tandem solar cells. Nature 618:80–86. https://doi.org/10.1038/s41586-023-05992-y

Hossain KM, Hasan MZ, Ali ML (2021) Narrowing bandgap and enhanced mechanical and opto-electronic properties of perovskite halides: effects of metal doping. AIP Adv. https://doi.org/10.1063/50039308

Kumar M, Prajapati B, Raj A, Anshul A, Chandra Sati P, Sahni M et al (2022) Progresses and challenges in the structural and magnetic properties of double perovskite La2NiMnO6 with their applications in solar energy. Mater Today Proc 49:3088–3092. https://doi.org/10.1016/j.matpr.2020.11.036

Li H, Zhang W (2020) Perovskite tandem solar cells: from fundamentals to commercial deployment. Chem Rev 120:9835–9950. https://doi.org/10.1021/acs.chemrev.9b00780

Liao W, Zhao D, Yu Y, Shrestha N, Ghimire K, Grice CR et al (2016) Fabrication of efficient low-bandgap perovskite solar cells by combining formamidinium tin iodide with methylammonium lead iodide. J Am Chem Soc 138:12360–12363. https://doi.org/10.1021/jacs.6b08337

Liao J-F, Wu W-Q, Jiang Y, Kuang D-B, Wang L (2019) Maze-like halide perovskite films for efficient electron transport layer-free perovskite solar cells. Sol RRL 3:1800268. https://doi.org/10.1002/solr.201800268

Olaleru SA, Kirui JK, Wamwangi D, Roro KT, Mwakikunga B (2020) Perovskite solar cells: the new epoch in photovoltaics. Sol Energy 196:295–309. https://doi.org/10.1016/j.solener.2019.12.025

Park N-G (2015) Perovskite solar cells: an emerging photovoltaic technology. Mater Today 18:65–72. https://doi.org/10.1016/j.mattod.2014.07.007

Raj A, Kumar M, Anshul A (2021) Recent advancement in inorganic-organic electron transport layers in perovskite solar cell: current status and future outlook. Mater Today Chem 22:100595. https://doi.org/10.1016/j.mtchem.2021.100595

Raj A, Kumar M, Anshul A (2023) Topical advances in fabrication technologies of perovskite solar cell heterostructures: performance and future perspective. Mater Lett 340:134171. https://doi.org/10.1016/j.matlet.2023.134171

Raj A, Kumar M, Singh DV, Anshul A (2024) A critical review on transition metal dichalcogenides (TMDs): an efficiency booster for perovskite solar cells. FlatChem 44:100629. https://doi.org/10.1016/j.flatc.2024.100629

Shi Y, Hua C, Li B, Fang X, Yao C, Zhang Y et al (2013) Highly ordered mesoporous crystalline $MoSe_2$ material with efficient visible-light-driven photocatalytic activity and enhanced lithium storage performance. Adv Funct Mater 23:1832–1838. https://doi.org/10.1002/adfm.201202144

Stranks SD, Snaith HJ (2015) Metal-halide perovskites for photovoltaic and light-emitting devices. Nat Nanotechnol 10:391–402. https://doi.org/10.1038/nnano.2015.90

Xu Y, Lin Z, Wei W, Hao Y, Liu S, Ouyang J et al (2022) Recent progress of electrode materials for flexible perovskite solar cells. Nano-Micro Lett 14:117. https://doi.org/10.1007/s40820-022-00859-9

You P, Tang G, Yan F (2019) Two-dimensional materials in perovskite solar cells. Mater Today Energy 11:128–158. https://doi.org/10.1016/j.mtener.2018.11.006

Zhu Q, Bao X, Yu J, Zhu D, Qiu M, Yang R et al (2016) Compact layer free perovskite solar cells with a high-mobility hole-transporting layer. ACS Appl Mater Interfaces 8:2652–2657. https://doi.org/10.1021/acsami.5b10555

GPSR Compliance

The European Union's (EU) General Product Safety Regulation (GPSR) is a set of rules that requires consumer products to be safe and our obligations to ensure this.

If you have any concerns about our products, you can contact us on ProductSafety@springernature.com

In case Publisher is established outside the EU, the EU authorized representative is:

Springer Nature Customer Service Center GmbH
Europaplatz 3
69115 Heidelberg, Germany

Batch number: 09193487

Printed by Printforce, the Netherlands